JONATHAN EISEN

AVERY PUBLISHING GROUP

Garden City Park • New York

Cover Designer: Eric Macaluso
In-House Editors: Peggy Hahn and Karen Hay
Typesetter: Elaine V. McCaw
Printer: Paragon Press, Honesdale, PA

Avery Publishing Group
120 Old Broadway
Garden City Park, NY 11040
1–800–548–5757
www.averypublishing.com

Cataloging-in-Publication Data

Eisen, Jonathan.
    Suppressed inventions and other discoveries :
revealing the world's greatest secrets of science and
medicine / Jonathan Eisen. — 1st ed.
        p. cm.
    Includes bibliographic references and index.
    ISBN 0-89529-809-0
    1. Discoveries in science.   2. Fraud in science.
    3. Cancer—Treatment.   4. Medical innovations.
    5. Renewable energy sources.   I. Title.

Q180.55.D57E57 1998   500
                                QBI98-1626

Printed in the United States of America

10   9   8   7   6   5   4   3   2   1

Contents

Appendices

I dedicate this book to Katherine Joyce Smith, with love and gratitude, and to Duncan Roads, my fellow editor, whose magazine, Nexus, has inspired thousands of people around the world.

Acknowledgments

I would like to thank Duncan Roads, the editor and publisher of *Nexus* magazine, for his tireless efforts to bring to light many of the stories of suppression and chicanery that have inhibited the progress of the human race and endangered the very survival of the planet. *Nexus*, in the company of other great magazines like *Exposure, Probe, Steamshovel Press*, and *Perceptions*, is essential reading for anyone concerned with exposing the Big Lies . . . and the little ones. In a world where common sense is considered radical, *Nexus* continues to publish information about the development of new and non-polluting technologies, and bravely champions independent thinking, provocative ideas, and feasible solutions. Many of the articles in this book were either first published or reprinted in *Nexus*.

I would also like to thank the good people at the Auckland Institute of Technology in New Zealand for granting me the resources to research, edit and publish an early "trial" version of this book for the New Zealand market. Without their help and encouragement, their financial support, and their willingness to entertain controversy in the interests of getting the truth out, this book would not have been published.

Much thanks to my publisher, Rudy Shur, for his patience, and his faith in this project. There are very few publishers—if any—in the world today with a list as consistently good and as consistently helpful as his, and it is an honor to be counted among his authors and his friends.

There are literally hundreds of people who helped to bring this book into existence, directly or indirectly. They know who they are and that I am eternally grateful for the work they've done. I would like to publicly remember my teachers, Charles Shulman, Richard Alpert, Sy Jacobi and Leonard Orr, as well as my friends, my first grade teacher Mrs. Poole, my father, whose 1948 discovery linking smoking and heart disease was ignored by the AMA, Stuart Troy, and Wilf Brinsbury. Wilf's encyclopedic knowledge of alternative energy is matched only by his enthusiasm, selfless sharing, and creativity. I also appreciate the help of Alan McLaughlin, publisher of the unique free energy catalogue, *Lost Tech Files*.

Most important, I would like to express my deepest and undying gratitude to my wife. Thank you, Katherine . . . for the lot.

Preface

We live in an age of marvels. Electronics has made us a global village; the Hubble can enable us to see to the beginning of time itself; we can pinpoint our location through satellite navigation systems and hold encyclopedias on a microchip.

We can do all these things, and yet something is radically wrong, terribly wrong. We keep polluting our magnificent home with the rancid waste from our chemical and petroleum industries, despoiling our planet and ourselves for the evanescent glory of the bottom line. We imbalance the most delicate of balances to conform with the logic of a system that is, to put it charitably, horribly out of whack.

Face it: Our entire global immune system is breaking down before our very eyes. Cancer, the defining disease of our time, inexorably increases in virulence, claiming millions of people every year; our climate is becoming more extreme with each passing season; and we seem to be losing the battle with the mighty virus as we breed it into our foodstuffs, our vaccines, and ourselves. Our antibiotics have helped to breed new super strains of bacteria that eat antibiotics for breakfast. Our vaunted educational systems produce graduates with great erudition in inconsequential matters, while illiteracy rises and the incompetent prevail.

And all the while nations become increasingly violent to each other as well as to themselves. The very worst role models are emulated, as some vestigial third brain reptile territoriality takes hold of our collective consciousness, selling itself as "free market economics" or some such nonsense.

In the immortal words of the Chinese curse, we have all been born into "interesting times."

Looking back over the past 100 years or so, when the industrial engine really began to get serious about eating the planet, it is tempting to ask whether or not the results really needed to have happened. Is there something fundamentally wrong with the human experiment, some genetic flaw, some cosmic misunderstanding that has made all this somehow inevitable? Or is it more mundanely political, that we have been taken over and overtaken by a materialist elite whose interests have overridden the common good so often as to be mistaken for the way it has to be? Perhaps there is just an unwritten conservatism that replenishes a reality construct over and over again until it becomes the paradigm which the culture as a whole accepts automatically, condemning alternatives to limbo, sublimely unconscious to the looming icebergs on the port bow.

People seem to know in their bones that that pouring good money after bad is not the way to save the earth. But through rewards and punishments from infancy on we are encouraged to kneel at the alter of denial, to negate our creativity except when that creativity is enlisted to support the system. We are told that this is all there is, that it was meant to be, and that we can't change anything fundamentally.

We seem to need the comfort of predictability, to be thankful when the next moment closely resembles the last, even if both bolster the common dysfunctionality. The "devil you know" kind of thing. And never mind that it has never really worked very well . . . and it is not likely to work now.

We all want acceptance, and to limit that cognitive dissonance between us and the people that matter and the system that really matters. But somewhere is that place where we know that all this is wrong, that it doesn't really have to be like this, doesn't really have to continue like this until we are all dead on a dead planet. Somewhere we know that within the human spirit is a place of creativity so powerful and so encompassing, that given half a chance we can change this course, change this moment and change history.

And that is what this book is about. It concerns itself with those inventions and ideas that have been developed over the last 100 years or so which, given enough encouragement, might well have led to a radically different culture and economics than the one in which we find ourselves today. Good—indeed, great—ideas have arisen and have been rejected by a society so mired in the dominant paradigm that it could not bestir itself to support its own survival.

Nevertheless, one perseveres in the often vain but necessary hope that success will eventuate. There really isn't any choice. The inventions and discoveries described and explored in this book may one day be developed for all to use and share. But in the meantime I believe that the first step may rest with the dissemination of the knowledge of what was, what might have been, and what may yet be.

Jonathan Eisen
October, 1998

Introduction

In 1979 a New Zealand inventor by the name of Archie Blue astounded the world—or at least that part of the world that was paying attention—with an invention that would allow any car with a gasoline engine to be fueled solely by water. He was awarded a patent for his work, and although he kept certain vital secrets out of his patent diagrams, he did demonstrate his device on numerous public and private occasions. Witnesses from England's Royal Automobile Club announced that the car did indeed run on water, and was in fact getting one hundred miles per gallon.

A group of English investors in the Channel Islands supported Blue. They brought him to the United Kingdom and tested the device, but then, mysteriously, progress was halted. Blue returned to New Zealand and stopped publicizing his invention. Immediately upon his death in 1991, his daughter and son-in-law cleaned out his laboratory, and brought what they described as "junk" to the garbage dump. Thus, Archie Blue's secrets died with him.

In 1996, I was invited to Australia to witness a demonstration of another mechanism that reputedly allowed a car to run on water. The inventor (who understandably wishes to remain anonymous) had received threats after having conversed with a magazine editor. He was told never to try to put his invention on the market or to write about it in a public media. However, no one stopped him from showing several of us how the device worked in a Ford Cortina.

Running on gasoline, the old Ford could barely manage 4,500 revolutions per minute before it screamed in obvious pain and blue smoke billowed out from the exhaust pipe. However, after the water device was connected, the engine went to 10,500 rpm with nothing but water vapor coming out of the pipe, and no smoke evident at all. Its acceleration was phenomenal. The engine still screamed, but it was obviously happier running on water. Unfortunately, the inventor's garage was later raided, and his equipment destroyed, making further development impossible.

Archie Blue and the aforementioned Australian inventor were not alone in developing their water-fueled automobiles. The first report of such an event was recorded in Dallas, Texas in 1934. Another version of the same idea turned up in 1936, witnessed by hundreds of spectators in England. In the 1950s, Guido Franch astounded automotive engineers with a chemical that allowed water to be burned in exactly the same manner as gasoline. The performance of a car running on this fuel was fantastic.

But despite the obvious successes of such prototypes, not one of these devices is on the market today. Countless inventors have been not-so-gently persuaded to abandon their projects through intimidation tactics such as sabotage and blackmail. Some have even been coerced into surrender by death threats. And should any inventor persist in making his work known, orthodox science can be counted on to intervene and effectively kill the project with rhetoric. Obtaining greater than 100-percent efficiency is, as any *sensible* scientist knows, impossible. Orthodox engineers would like to have you believe that these inventions somehow violate the "immutable" laws of physics by apparently producing more energy than they are consuming.

For the true innovator there is only theory, and so there are no laws that cannot be broken. Everything in Nature is a catalyst for wonder and discovery, and the authentic inventor welcomes the next moment as an opportunity for creation. Really significant advances have always grown out of the revelations of independent thinkers and tinkerers who were not learned enough to know that they were violating the laws or physics or any other branch of science. Or perhaps, in the pursuit to improve mankind's quality of life, they simply didn't care.

In our world of research and scientific advancement, it seems only logical to think that if an invention can further the cause of progress, it will eventually find its way into the mainstream of society. After all, the wonders of our post-industrial age are numerous and diverse, ranging from television to antibiotics. If a suppression syndrome has infiltrated our society, how could these modern-day marvels have come into existence?

On the surface, this would appear to be a valid argument. However, the point weakens under scrutiny. For example, television was suppressed for many years by companies with huge investments in the film industry, who believed that movies would become obsolete. Thanks to their pressure development was slowed, and more than thirty years passed after its discovery before television actually made it to the commerical mainstream—even though it was backed by large corporations like RCA.

Antibiotics were released for use on World War II battlefields only because the United States government made a deal with the pharmaceutical companies, granting them patent rights for something they had never even developed. This came after several years of negotiations, at the cost of thousands of lives.

For every once-revolutionary idea that is now commonplace in our daily lives, many more have been suppressed or witheld by those vested interests with a focus on profit or power. Pure self-interest results in strong opposition from multinational corporations, orthodox science, and even our own governments when innovation threatens the status quo. Wealthy and powerful individuals are not inclined to forfeit their fortunes or their authority, even though the human population as a whole would benefit greatly from new technologies.

The suppression of innovation and discovery is an overwhelming and frightening problem. I have put this book together in order to directly address this critical issue, which I believe deserves our utmost attention. These collected articles, some of which may surprise or even shock you, are highly varied, but each and every one is vital to our understanding of the nature of suppression—where it begins, who it affects, and how it is perpetuated. Because the suppression syndrome is so far-reaching, I have grouped the material into four sections, each detailing the struggle of specific ascendancies to maintain their funds and their jurisdiction.

Section I focuses on the suppression of alternative medicine. Powerful pharmaceutical companies and their agents, the orthodox medical societies, are not ready to lose millions of dollars by admitting that there are nontoxic, inexpensive treatments that are effective in the fight against diseases such as cancer and AIDS. Therefore, patients suffering from these and other degenerative illnesses are denied access to possible cures. Many remain unaware that these therapies even exist until there is little, if any, hope for recovery.

The efforts of organized science to suppress the independent researcher are detailed in Section II. Establishment science has yet to examine itself according to the stringent guidelines of its own Scientific Method, the doctrine by which all research and discovery is measured. It seems that if scientists assessed their work objectively, they would find that there is no monopoly on truth, a realization which could undermine their elevated status. What a sad commentary on a branch of knowledge whose constituents should humbly admit that they do not know all the answers—or even all the questions. How can any "radical" ideas find acceptance in a system whose aim is self-perpetuation, rather than the betterment of humanity?

In Section III you will discover that the public at large remains shockingly ignorant as to the extent of our government's involvement with UFOs and extraterrestrials. What if our highest powers are in fact subservient to higher powers? It is clear that the censorship of sensitive information regarding extraterrestrial life has been carefully orchestrated so as not to upset the power of our dominant social, religious, and political institutions.

Finally, Section IV will introduce you to some of the alternate energy resources that could potentially eliminate our dependence on fossil fuels, and curtail research into the deadly menace of nuclear power. We are not driving around in cars fueled by water, or tapping into the free energy in our atmosphere to light our homes, not because these things are impossible, but because power and petroleum monopolies would crumble if our world ran on the abundant, clean, and safe energy that some inventors were harnessing decades ago. It is therefore "in the best interest" of these monopolies to maintain a system that is destroying our environment and threatening our very lives.

The true nature of suppression is the willingness on the part of everyone with a stake in the system to uphold the power of that system. To ask if there is more out there than meets the eye is to question our very reality, and to ultimately upset the status quo. We don't really know our real power—the power of one ethical and courageous act, of speaking the truth. *Suppressed Inventions and Other Discoveries* is my attempt to empower concerned individuals, and to enlighten those who are unaware that there is need for concern.

Section I

The Suppression of Alternative Medical Therapies

Ralph Moss is, perhaps, the best medical journalist in the United States today. His book *The Cancer Industry* uncovered the corruption of the second most profitable business in the twentieth century—cancer. His latest book, *Questioning Chemotherapy*, is also a gem. Moss is very persuasive, although it may not take a genius to realize that if powerful drug magnates are sitting on the board of directors of Memorial Sloan-Kettering Cancer Center in New York City, and other major cancer hospitals and research centers, it will be difficult, if not impossible, to get any favorable results for non-pharmaceutical therapies.

In an interview with Gary Null (WBAI radio in New York City), Moss elaborated on this idea:

> What my research has shown is that many of the top directors (what they call "overseers") at Memorial Sloan-Kettering are also top directors at drug companies. For instance, Richard Furlow, who is the President of Bristol-Myers Squibb and Director of the Pharmaceutical Manufacturers' Association, is a top official of Memorial Sloan-Kettering. Richard Gelb, who is the Chairman of the Board of Bristol-Myers Squibb, is a Vice Chairman of Memorial Sloan-Kettering. James D. Robinson, a Director of Bristol-Myers Squibb, is another Vice President of Memorial Sloan-Kettering. The President of Memorial Sloan-Kettering Cancer Center, Paul Marx, is a Director of Pfizer [pharmaceuticals]. And others [officials of Memorial Sloan-Kettering] are Directors of Bio-Technology General, Life Technologies, Merck, and so forth.
>
> And so what happens, in effect, is that you have a . . . closed circle of people who are, on the one hand, directors of the world's largest cancer center; on the other hand, they are either officers or directors of the very companies that are producing the drugs which are used and advocated by these centers. . . . There are many, many ways that the drug industry influences the direction of cancer research, and of AIDS research. You have to look at it from an economic point of view . . .

Moss points out that nutritional therapies—impossible to patent—and therapies like ozone and vitamin C and many others, have been either suppressed or ignored by the cancer establishment, by people looking at the issue, as he says, "from an economic point of view." The only treatments chosen are the ones that, strangely enough, turn out to be toxic—because these are the only ones that can be patented.

And patenting new drugs is extremely expensive. So if you were the head of a pharmaceutical company spending millions of dollars on a new drug that could earn hundreds of millions of dollars in profits, would you want people relying on herbs and vitamins to treat their illnesses? Would you want a regulatory agency like the FDA to be an advocate for safe, nontoxic treatments, rather than chemical therapies?

Interesting alternatives like Hoxsey's herbal treatments, and Royal Raymond Rife's frequency machine, that reputedly scored a 100-percent cure rate on terminal cancer patients, are not available options for patients searching for cures less harmful than the disease itself. And you might be surprised to learn that people living with HIV may not have to inundate their systems with drug cocktails, but could instead benefit from the restorative powers of oxygen in its purest form. Yet because giant, far-reaching drug companies stand to lose so much, most patients are not made aware of the many alternative treatments that have been developed. To this day workable, testable alternatives to corporate medicine are not recognized by a system that is geared to maximize the profits of a pharmaceutical/medical establishment.

Censorship extends to information regarding the very nature of deadly viruses and crippling degenerative diseases. Mercury from dental fillings, for example, is actually toxic, and is thought to be the cause of some neurological disorders. And although we are confronted daily by the media with news about HIV, or even if we know someone who is living with AIDS, we probably don't know as much as we think about its origin. It's clear that what we don't know *can* hurt us. This is what happens when profits are prioritized over our health and our very lives.

This section contains descriptions of several revolutionary therapies for diseases that were formerly accepted as "untreatable" or "fatal." While the treatments described in this book have been shown to be effective for many people, they may not represent the best treatment for all cancers, immunodeficiency diseases, or other degenerative illnesses. Choosing a therapy for any life-threatening disease is a serious matter. You should read widely and discuss your options with a health professional in order to make an informed decision about which therapy or therapies you will use.

1 Does Medicine Have a Bad Attitude?

James P. Carter

*" . . . And besides, looking through those spectacles
gives me a headache."*

Professor Cesare Cremonini in 1610,
explaining why he would not look through Galileo's
telescope at the moons of Jupiter.

ARROGANT IGNORANCE

The sort of excuse above has delayed medical discoveries for decades, even half-centuries. Canadian nutritionist Dr. David Rowland describes this repression of medical innovation as a bad attitude which he termed "arrogant ignorance." This negative attitude toward many great discoveries represents a tremendous ego threat. Today such negativity is compounded with the industrialization of medicine, which has brought on that "greed is good (for me)" philosophy expressed in the recent movie *Wall Street*. Segments of the medical profession take what they want when they can get it.

Arrogant ignorance has followed science and medicine throughout history. Beginning with the learned colleagues of Galileo who refused to even look through the glass of his new invention, the telescope, because they believed they already knew all about the laws of physics, that not-invented-here attitude is alive and well at the dawn of the twenty-first century. Is it only a coincidence that "not invented here" shares initials with our government's National Institutes of Health?

Past suppressions—at least those safely back in past centuries—are readily admitted by contemporary medicine. French explorer Jacques Cartier, for example, in 1535 learned from the American Indians that pine-needle tea prevented and cured scurvy, a vitamin C deficiency disease. Upon his return to France, Cartier excitedly shared his discovery with French doctors, who turned a cold shoulder—such a primitive therapy was witchcraft. If we pass this off as Eurocentrism, we miss the similarities to present-day rejections of alternative healing methods that are getting the cold shoulder. The case of Dr. Charles Peres, M.D., of Ft. Meyers, Florida, provides an excellent example.

Dr. Peres was diagnosed with a stage D2 prostate cancer spread throughout his body. In lay terms, you can't have a gloomier prognosis. After he adopted a natural regimen based on a low-fat vegetarian macrobiotic diet, his cancer went into complete remission. Naturally overjoyed, upon his return to functional living he noticed that many of his medical colleagues actually appeared angry that he had survived. Would they rather he die than heal himself with this unorthodox treatment? This very same disdain has been noted by cancer patients who have sought out alternative cancer doctors and have gone into permanent remission, only to be told by their first doctor that they never had cancer to begin with (despite the complete diagnostic work-up that he had witnessed). Negative reactions range from obvious anger to feigned indifference. It must also be told that there are doctors who secretly recommend alternative treatments but warn their patients to never tell the wrong party lest the doctor get in trouble.

In 1747, James Lind, a surgeon's mate in the British Navy, conducted dietary experiments on board ship. He concluded that citrus fruits prevented and cured the killer disease scurvy which ravaged sailors. Captain Cook was one of the first ship commanders to supplement his sailors with rations of lime. The captain sailed throughout the world for over three years without a single death from scurvy—unprecedented for that time.

But it took forty-eight years before the British Admiralty made it official policy to distribute one ounce of lime juice daily for each sailor. This simple nutritional supplement of vitamin C was a factor in Britain's ascent to being the world's greatest sea power. It was as though they doubled their forces. Britain sailed farther than any other navy into uncharted territory, easily defeating weakened enemies who had lost many sailors to scurvy.

Now, neither the British nor the American Indian performed any double-blind, cross-over studies to arrive at their discovery. In their respective ways, they learned that it worked very well for their needs. James Lind had conducted empirical studies (based on observation) to determine that a citrus fruit could save naval forces from certain death. Ridiculed by their rivals for this use of lime juice, the British were derisively referred to as "limeys." Had they never conducted their simple experiment, or had another sea power done so, world history could have been altered.

Dr. Jenner, a British doctor, discovered in the early 1800s that milkmaids who had previously contracted cowpox were protected against smallpox. Jenner scientifically developed a vaccine from the crusty lesions of small-pox patients to inoculate others against smallpox. It took more than fifty years for the medical power structure to endorse his simple remedy for a killer disease.

In 1848, Dr. Semmelweis, a graduate of the prestigious University of Vienna Medical School, introduced a revolutionary idea while assisting in the Vienna Obstetrical Clinic: he required medical students to wash their hands in chlorine water before entering the clinic. There was an immediate and dramatic decrease in the high death rate from puerperal (childbirth) fever. The good doctor became an outspoken advocate, pleading with obstetricians to tend maternity patients only after proper hand washing. After a vicious attack on his personal and professional integrity, he was fired from the hospital where he had just eradicated a cause of death.

This courageous, principled doctor then spent ten years gathering evidence to prove that hand washing would prevent terrible misery and death from childbirth fever. He published his research in 1861 and distributed the medical text to the major medical societies throughout Europe. It was completely ignored. In one of those years, 40 percent of the maternity patients in Stockholm, Sweden, contracted the fever; 16 percent of those new mothers died.

The deadly fever continued to ravage women while the hand washing prevention/cure was "put on hold" by Organized Medicine. The poor doctor could no longer cope with the preventable death and misery of so many women. In 1865 he died after a mental breakdown; such tragedies still occur among gifted researchers whose great discoveries are ignored. So, from the safety of the next century, Dr. Semmelweis can be credited by the medical profession with his lifesaving discovery—hand washing.

In 1867, Dr. Joseph Lister introduced sanitation in surgery, but not without a big fight with the leading surgeons of nineteenth-century England. His paper, "On the Antiseptic Principle in the Practice of Surgery," was read before the British Medical Association in Dublin, Ireland. His noteworthy summary concluded:

> Since the antiseptic treatment has been brought into full operation, my wards though in other respects under precisely the same circumstance as before, have completely changed their character, so that during the last 9 months not a single instance of pyemia,* hospital gangrene or erysipelas** has occurred in them.

---

* Blood poisoning caused by pyogenic microorganisms (pus).
** A disease of the skin and underlying tissue caused by hemolytic streptococcus, a bacteria that destroys red blood cells.

Dr. Lister's contribution outraged the leading surgeons of the day. An 1869 conference of the BMA devoted the surgery address to a scathing attack on the antiseptic theory. What presumptuous London surgeon would believe a lowly provincial from Scotland who was telling them how to improve surgical protocol?

As evidence of similar incredible intolerance in the United States, U.S. Senator Paul Douglas related the following story, which was recorded in the Congressional Record in 1963:

> I spent a part of 1923 with Dr. W. W. Keen. In the Civil War he was a surgeon and had seen many men die from the suppuration of wounds after he had operated.
>
> He went to Scotland and studied under Lister. Dr. Keen came back from Scotland. He was referred to as a crazy Listerite. He was denied an opportunity to practice in every hospital in Philadelphia.
>
> Finally there was one open-minded surgeon in the great Pennsylvania General Hospital. He said, "Let us give this young fellow a chance!" So they let him operate.
>
> No one died from infection under Keen. Keen began to chronicle the results in statistical articles. He was threatened with expulsion from the Pennsylvania Medical Society.
>
> This was in the 1890s. Finally he was accepted as the greatest surgeon in the United States.

Next came Dr. Louis Pasteur, a chemist. His germ theory for infectious diseases provoked violent opposition from the medical community of the late 1800s. How could a mere chemist poach upon their scientific turf?

Dr. Harvey's monumental work on the theory of blood circulation was forbidden to be taught at the University of Paris Medical School twenty-one years after Harvey published his findings. And it doesn't end there.

Austrian botanist Gregory Mendel's theory of genetic composition was generally ignored for thirty-five years. His pioneering work was dismissed as that of an idle, rich dilettante by the leading scientists of his day.

Dr. Fleming's mid-twentieth-century discovery of the antibiotic penicillin was ridiculed and ignored for twelve years before this life-saver was admitted into the medical circle. Once scorned, Dr. Fleming was eventually knighted and received the Nobel Prize in Medicine for what had once been denounced.

As a final example, Dr. Joseph Goldberger unraveled the mystery of pellagra, a disease which ravaged especially the poor in the American South. Pellagra was at first thought to be an infectious disease causing the three Ds of dermatitis, diarrhea and dementia. Goldberger discovered that, like scurvy, pellagra was a vitamin-deficiency disease. The milling or refining process of corn removed important vitamins and minerals from the husk. Those people dependent on corn-based foods such as grits, corn bread, etc., became deficient in vitamins and minerals. Goldberger's recommendation—to re-fortify corn flour—is now a routine practice in refining most flours. But the foot-dragging over this minor business expense by the greedy flour barons of the time dragged on for fifty years.

These examples are just a handful of so very many courageous doctors and scientists who braved a battle with Organized Medicine on behalf of what could help patients. They have the honor and distinction of representing "The Enemy of the People" that was portrayed in Ibsen's drama of that name.

The role of deficiency in causing disease is carried a step further by Dr. Max Gerson in his text for doctors, *A Cancer Therapy—The Results of Fifty Cases*. He exposes the depletion of farm soil from chemical fertilization as early as the 1930s and concludes that the depletion subsequently affects nutritional levels in the plants growing in depleted soil.

IT'S THE SAME IN SCIENCE

Throughout the course of Western Civilization, there has been a strong resistance to new information in the other scientific fields. There is so much evidence of this bigotry that only a few brief examples are offered here.

Thomas Kuhn's book *The Structure of Scientific Revolution* (2nd Edition, University of Chicago Press, 1970) relates the typically bitter conflict between an independent science researcher who discovers something important and the current power structure which fights to maintain the status quo.

German biologist Hans Zimmer wrote, "Academies and learned societies are slow to react to new ideas, this is in the nature of things . . . The dignitaries who hold high honors for past accomplishments do not like to see the current of progress rush too rapidly out of their reach!"

In his 1966 book, DeGrazia recounted the mistreatment of scientist Immanuel Velikovsky for his theories in astronomy. Velikovsky had proposed that the catastrophic events recorded in the Old Testament and in Hindu Vedas and Roman and Greek mythology were due to the earth repeatedly passing through the tail of a comet during the fifteenth to seventeenth centuries, B.C.

DeGrazia wrote,

What must be called the scientific establishment rose in arms, not only against the new Velikovsky theories but against the man himself. Efforts were made to block dissemination of Dr. Velikovsky's ideas, and even to punish supporters of his investigations. Universities, scientific societies, publishing houses, and the popular press were approached and threatened; social pressures and professional sanctions were invoked to control public opinion.

The issues are clear: Who determines scientific truth? Who are its high priests, and what is their warrant? How do they establish their canons? What effect do they have on the freedom of inquiry and on public interest? In the end, some judgment must be passed upon the behavior of the scientific world, and if adverse, some remedies must be proposed.

Philosopher and professor of physical chemistry Michael Polanyi commented in 1969, referring to the persecution of Velikovsky, that new ideas in science are not accepted in a rational manner, based on factual evidence, but instead are determined by random chance, the ruling economic/political powers, or the ruling ideology.

A recent paper by sociologist Marcell Truzzi, "On the Reception of Unscientific Claims," delivered at the annual American Academy for the Advancement of Science, proposed that it is even harder today for new discoveries and ideas to break through, due to the escalating economics of research. Truzzi wrote, "Unconventional ideas in science are seldom positively greeted by those benefitting from conformity." Truzzi predicted that new forms of vested interest will emerge from today's programs that must compete for massive funding. He warned, "This has become a growing and recognized problem in some areas of modern science."

There is another reason for resistance to scientific discoveries. Many of the major advances have come either from a scientist in another scientific discipline or from researchers who just don't qualify for membership in the scientific elite (as in high school "in crowds"). No wonder advances so often come not from the "in crowd" who are blinded or corrupted by prevailing dogma.

ORGANIZED MED IGNORES SUCCESS

Currently there exists impressive statistical and clinical (case study) data on alternative approaches to reversing or controlling some cancers without the use of chemotherapy, radiation and surgery. But covert politicking and overly rigid systems of testing and approval suppress these biological approaches that Americans are increasingly accessing. Desperate cancer patients rarely learn about all their medical options; in fact, a full 80 percent of those who travel outside the U.S. for alternative cancer therapies are so terribly advanced in their diseases that it is too late, even for alternative approaches. This fact alone obscures the value of these therapies when they are promptly applied under competent medical supervision and not tried as a last resort, following, for example, chemotherapy. Chemotherapy alone can destroy a patient's immune system, and biological methods usually require a functioning immune system.

Sadly, in this supposed age of enlightenment, the ridicule of the medical orthodoxy and a rigid system of testing and approval (calcified by the same suspicions of alternative therapies that plagued discoveries for centuries) keeps these treatments from ill patients who might benefit from them, as De Felice, Director of the Foundation for Innovation in Medicine, in 1987, lamented:

One of the tragedies of our times is that over the past 20 years, a pervasive and aggressive regulating system has evolved that has effectively blocked the caring clinical innovator at nearly every step. Let there be no doubt that we have quietly, but effectively, eliminated the Louis Pasteurs of our great country.

DR. HORROBIN'S CALL FOR AN END TO THE SUPPRESSION OF INNOVATION

The *Journal of the American Medical Association (JAMA)* in March, 1990, published selections from the first International Congress on Peer Review in Biomedical Publications. Dr. David Horrobin presented "The Philosophical Basis of Peer Review and The Suppression of Innovation," a classic presentation. Dr. Horrobin stressed that the ultimate aim of peer review in biomedical science cannot differ from the ultimate aim of medicine—"to cure sometimes, to relieve often, to comfort always." (Believed to be a French folk saying of medieval origin, this beautifully simplistic description of medicine's intent is inscribed on the statue of Edward Trudeau at Saranac Lake, New York.)

Dr. Horrobin stated that the purpose of peer review should be nothing less than to facilitate the introduction of improvements in curing, relieving and comforting. Even in the fields of biomedical research that are remote from clinical practice, the peer reviewer should always ask whether the proposed innovation could realistically lead to improvements in the treatment of patients.

He notes the necessity for a creative tension between innovation on the one hand and quality control on the other. The innovators who generate the future are often impatient with the precision and systematic approach of the quality controllers. On the other hand, the quality controllers are often exasperated by the seeming lack of discipline and predictability of the innovators. If either side dominates, research progress falters.

The public is the ultimate source of money for medical research. They agree to this use of their money for the sole purpose of improving their medical care. When improvement does not progress satisfactorily, support for medical research (and medical journals) will dwindle. The public wants satisfactory progress; if such progress is not forthcoming, the present medical research enterprise will crumble. For satisfactory results, quality control must comprise only one side of the editorial equation. There must also be an encouragement of innovation. Presently, quality control is overwhelmingly dominant, and encouragement of innovation receives very little attention. Without appropriate balance, peer review fails its purpose.

Dr. Horrobin notes that, in the last six decades, the accuracy of medical articles has improved substantially but so has a failure to acknowledge innovation. Between 1930 and 1960, patient care improved dramatically. Many infectious diseases were controlled by drugs and immunization. Prototypes of drugs used today were discovered during that time. However, by 1960 (despite major developments, especially in the field of diagnosis), patients increasingly felt dissatisfied, and we must accept the fact that their dissatisfaction stems from our trading innovation for quality control.

Dr. Horrobin presents many situations in which, through peer review, Organized Medicine has tried to suppress an innovative concept but failed. He shows how the use of peer review influences journals, conference choices and grant awards.

Pathologist Charles Harris has written editorials about the "Cult of Medical Science" in which he says pseudo-science in medicine is currently a cult which inhibits innovation and considers participation in clinical drug trials (which have been designed by statisticians) as the work of scientists because these trials reject so-called anecdotal evidence based on clinical observations alone. But this narrow attitude is not real science which leads to discovery. It is merely indoctrination and a pledge of allegiance to the flag of pseudoscience.

Harris also asserts that diagnosis, which is supposed to be the determination of the nature of disease either by examination or by exclusion, is not being practiced as it should be. Diagnosis today too often does not consist of examination, exclusion, clinical or therapeutic trials; rather, it often consists only of a rushed referral to a medical specialist under the guise of a diagnosis. The specialist may accept and act on the initial diagnosis which was not valid in the first place. The initial diagnosis serves to justify referral and satisfy the CPT code in order for the doctor to get reimbursed.

The New York Times, on March 26, 1991, carried an article by Philip J. Hilts entitled "How Investigation of Lab Fraud Grew Into a Cause Celebre" recounting how scientists turned a tangled dispute into a defense of science. This article is about a draft report which had been recently released by the newly-established Office of Scientific Integrity at the National Institute of Health. This office had been investigating the case of

Dr. Thereza Imanishi-Kari and a paper she published with Dr. David Baltimore in the April, 1986, issue of the scientific journal *Cell* about the basis of an immune reaction. Questions about the paper arose when Dr. Margot O'Toole, a post-doctoral fellow in Dr. Imanishi-Kari's laboratory, went to Dr. Baltimore (who was then at the Massachusetts Institute of Technology) and told him her reasons for doubting the authenticity of the data in the article. She alleged that the paper made false statements, a conclusion she reached after seeing seventeen pages of data that supposedly, but did not, support claims in the paper. She persisted in her accusations, and, as a result, two scientific reviews of the paper were conducted in 1986—one at MIT, where the work was done, and the other at Tufts University, where Imanishi-Kari was seeking employment. Both of these reviews found problems with the work but found no reason to believe misconduct was involved. Dr. O'Toole, who was eventually fired from her job at MIT, had been told by Dr. Baltimore that she could publish her objections to the paper, but that if she did he would also publish his views of it.

The matter lay dormant for two years after the initial scientific reviews conducted at MIT and Tufts, until Representative John D. Dingell, who heads the House Subcommittee on Oversight and Investigations, asked the Secret Service to examine Dr. Imanishi-Kari's notebooks for their authenticity. This action raised the hackles of the scientific community. Supporters of Dr. Baltimore criticized Dingell for prying into the notebooks of science and, described his panel as the "science police." Dingell revealed that his committee was soon buried in letters from scientists concerned with the subcommittee's actions, but he also said that in perhaps 50 percent or more of the letters the scientists included disclaimers, saying that they did not know the facts of the case. What had begun as a small dispute within Dr. Imanishi-Kari's laboratory had become a national debate, pitting Dr. Baltimore and his many supporters in the scientific community against Dingell's House Subcommittee and generating bitter controversy over a period of five years.

The controversy was eventually addressed by the National Institute of Health's Office of Scientific Integrity and put to rest by its draft report. In that report, the OSI concluded that Dr. O'Toole's actions were heroic and that Dr. Baltimore's response was troubling because he, instead of ending the matter within weeks of its beginning, allowed it to mushroom into a national debate. Dr. O'Toole's allegations were vindicated, and most of Dr. Baltimore's supporters have withdrawn their objections to the Congressional action after confronting the evidence uncovered by the OSI and presented in their draft report.

This case of scientific fraud illustrates the need for an office such as the Office of Scientific Integrity. Dr. David Goodstein, Vice-Provost of the

California Institute of Technology, helped to write the rules for dealing with misconduct. He stated in regard to the Imanishi-Kari/Baltimore case, "The scientific community until recently was disposed to believe that fraud didn't exist. So, in the rare cases that it did come up, the community was not prepared for it."

Having established that fraud can exist in the scientific community and having acknowledged the need for an agency to investigate such fraud, we now need to address a disturbing question: What about fraud and deceit that is conducted by individuals who work for organizations such as the AMA and the FDA? Why doesn't the Office of Scientific Investigation inquire about what happened with the Koch reagents and how an injunction was issued by the FDA prohibiting interstate shipment and the making of any medical claims, without the FDA even investigating them? What about the recall by the FDA of all contaminated tryptophan products sold in health food stores while allowing the continued use of the contaminated product in infant formulas and in intravenous pharmaceutical preparations?

The statute of limitations has expired in the case of the Koch reagents, but it is arguable that there should be no statute of limitations in science, particularly regarding a therapeutically useful drug. In any case, the statute of limitations has not expired in the recent contaminated tryptophan case. Why is this case not investigated by the OSI? If their mandate is not to investigate cases like this, then what is it? Surely their mandate goes beyond an occasional nabbing of a cheating researcher. It appears that the Office of Scientific Integrity is prepared to investigate individual instances of fraud but not collusion and conspiracy within the ranks of the government itself.

The story of vitamin C and cancer was thoroughly researched by Dr. Evelleen Richards and published in "Social Studies of Science" in 1988. Her paper received much publicity. Dr. Richards documented in great detail the failure of two Mayo Clinic studies to test vitamin C in the correct manner proposed by Nobel winner Linus Pauling and his associate E. Cameron.

Richards noted the repeated refusal of the *New England Journal of Medicine* to publish letters and articles by Pauling and Cameron that demonstrated why the second Mayo trial was not a test of their hypotheses. Cameron showed that highly toxic treatments for cancer, including 5-fluorouracil for colon cancer, continued to be used despite their failure to demonstrate efficacy in placebo-controlled trials. Richards proposed a valid question: Why does the full weight of disapproval fall on vitamin C (which has low toxicity), when toxic drugs with no demonstrated efficacy are widely used?

Dr. Horrobin contends that the peer-review process harbors antagonism toward innovation. While this is not the norm, it certainly is not the exception. Editors must encourage innovation as much as they ensure quality control, and that will require a conscious effort of will. He points out that the hypercritical reviews and behaviors of many distinguished scientists are unwarranted and pathological. Such professionals are gate-keepers against innovation unless the new thought or discovery is their own.

Dr. Horrobin concludes with a call for editors to muster the courage to select reviewers of the highest caliber without vested interest, or at least to note when vested interest is present. Editors must stop rejecting innovative articles for minor details which never keep establishment-approved articles out of the journals. An editor must never lose sight of the ultimate aim of biomedical science—to improve the quality of patient care. Only after scrupulous study of both the article's contents and the peer review should the editor make an objective decision.

WE MUST STOP PERSECUTING ALTERNATIVE PRACTITIONERS

To refuse to learn from history is to repeat it. The medical profession continues to libel and slander innovative doctors. The term "quack" has no legal definition. It is often misused to libel a doctor who is bright, full of initiative and well-loved by patients, and who has made an original discovery or happened to acquire the non-toxic methods that in the U.S. are referred to as alternative. Real charlatans should certainly be stopped. But should there be these programs aimed at American doctors such as the chelation doctors or those who employ alternative methods for treating cancerous tumors or other chronic diseases such as arthritis, multiple sclerosis, etc.?

This moral injustice should be halted. The involved branches of business, government and medical profession will in the near future have to answer the well-documented evidence spanning the twentieth century, that, hidden from the trusting public, a horrible orchestration of doctor-bashing has occurred to destroy the competition.

And what does Organized Medicine say for itself? Why, they believe in an overly-rigid definition of what constitutes scientific proof. The Canadian agriculturalist who developed the double-blind study never intended for it to be used in such a rigid manner. It was intended to eliminate the subjective bias of scientific investigators and their research assistants, not to become the gold-standard bearer for scientific proof in clinical medicine. Most genuine scientists (the term excludes the majority of the medical profession) do realize this fact. Real scientists understand that all science starts with careful observation and the recording of events.

This point can be best illustrated with a story. In the time of Julius

Caesar, there was a legendary bandit by the name of Procrustes. Now, a Procrustean bed which bears his name is an adjustable hospital bed. Legend had it that Procrustes would kidnap people, bring them to his home and force them to lie down on his bed. If they were too tall to fit his bed, he just cut off their legs and they bled to death. Too short? He put them on a rack and stretched them until they died screaming. The highly regarded Dr. Edward Whitmont, homeopathic physician of New York City, likens the rigidity of the cruel Procrustes to the rigid adherence to a methodology that blinds one to an obvious truth in medicine—that an alternative treatment works. The obsession with a rigidly narrow definition of what constitutes scientific proof is more slavishly believed by physician scientists than by modern physicists. Recognizing only this rigid, narrow definition of proof, orthodox medicine holds a sword of Damocles* over their competition.

Do they really believe that they can keep alternatives out of medicine? Or do they know that the exclusion will end in the near future and so "make hay while the sun still shines"?

---

* A fourth century Greek courtier to Dionysius the Elder, who, according to legend, was forced to sit under a sword suspended by a single hair to demonstrate the precariousness of a king's fortunes.

2 The Great Fluoridation Hoax: Fact or Fiction?

Dr. Ronald S. Laura and John F. Ashton

DOES IT BENEFIT YOU? OR BIG BUSINESS!

The controversy surrounding fluoridation raises a number of important socio-ethical issues which cannot be overlooked. One of the most burning questions is whether the fluoridation programme represents a milestone in the advancement of community health or the opportunistic outcome of a powerful lobby concerned largely to advance its own vested interests at the expense of the interests of the public. The historical origins of fluoridation are revealing, though we shall for obvious reasons in what follows not interpret the revelation itself, but rather tease out a few of the truly remarkable coincidences which make those origins revelatory.

In a more direct approach to a related issue, however, we shall argue that the potential and actual health risks associated with fluoridation have not been sufficiently appreciated by those in favor of fluoridation. The intentional introduction of fluorides in drinking water has certainly not received the rigorous scrutiny and testing properly brought to bear on the wide array of available medical drugs, many of which can be bought without prescription. Finally, we urge that even if it were determined that the addition of a minimal amount of fluoride to our water supply was both safe and effective in the reduction of caries in the teeth of children, the relevant dosage of fluorides could not be satisfactorily restricted to ensure that the harmful effects of fluoride did not outweigh the alleged beneficial effects.

THE GENESIS OF FLUORIDATION

Many readers will be surprised to hear that fluorides have been in use for a long time, but not in the prevention of tooth decay. The fluorides we

---

Dr. Ronald S. Laura is a professor of education at the University of Newcastle and is a PERC Fellow in Health Education, Harvard University, and John F. Ashton teaches in the department of education at the University of Newcastle.

now, in the name of health, add to our drinking water were for nearly four decades used as *stomach poison,* insecticides and rodenticides. Fluorides are believed to exert their toxic action on pests by combining with and inhibiting many enzymes that contain elements such as iron, calcium and magnesium. For similar reasons fluorides are also highly toxic to plants, disrupting the delicate biochemical balance in respect of which photosynthesis takes place. Nor is there any reason to suspect that humans are immune from the effects of this potent poison. Even a quick perusal of the indexes of most reference manuals on industrial toxicology list a section on the hazards of handling fluoride compounds. In assessing the toxicity levels of fluorides Sax confirms that doses of 25 to 50 mg must be regarded as "highly toxic" and can cause severe vomiting, diarrhea and CNS manifestations.[1]

It is crucial to recognise from the outset that fluoride is a highly toxic substance. Appreciation of this simple point makes it easier to understand the natural reluctance on the part of some to accept without question the compulsory ingestion of a poison to obtain partial control of what would generally be regarded as a noncommunicable disease. The potent toxicity of fluoride and the narrow limits of human tolerance (between 1–5 ppm) make the question of optimum concentration of paramount importance.

FLUORINE WASTES—A MAJOR POLLUTANT

The fluoridation controversy becomes even more interesting when we realise that industrial fluorine wastes have since the early 1900s been one of the main pollutants of our lakes, streams and acquifers, causing untold losses to farmers in regard to the poisoning of stock and crops.

Fluorides such as hydrogen fluoride and silicon tetrafluoride are emitted by phosphate fertilizer manufacturing plants (phosphate rock can typically contain 3 percent fluoride). The industrial process of steel production, certain chemical processing and particularly aluminium production which involves the electrolysis of alumina in a bath of molten cryolite (sodium aluminium *hexafluoride*) all release considerable quantities of fluorides into the environment. The fluorides emitted are readily absorbed by vegetation and are known to cause substantial leaf injury. Even in concentrations as low as 0.1 ppb (parts per billion), fluorides significantly reduce both the growth and yield of crops. Livestock have also fallen victim to fluoride poisoning caused primarily by ingesting contaminated vegetation.[2] It is reported that the Aluminum Corporation of America (ALCOA) was confronted by annual claims for millions to compensate for the havoc wreaked by their fluorine wastes. It was in 1933 that the United States Public Health Service (PHS) became particularly concerned about the poisoning effect of fluoride on teeth, determining that dental fluorosis (teeth mottled with yellow, brown and even black stains) occurred amongst 25–30 percent of children when just over 1 ppm of fluoride was present in drinking water.[3] By 1942 the PHS, largely under the guidance of Dr. H. Trendley Dean, legislated that drinking water containing up to 1 ppm of fluoride was acceptable. The PHS was not at this stage *introducing* fluoridation—it was concerned mainly to define the maximum allowable limit beyond which fluoride concentrations should be regarded as contaminating public water supplies. Dean's research investigations also indicated that although 1 ppm fluoride concentration caused enamel fluorosis or mottling in a small percentage of children (up to 10 percent), it also served to provide partial protection against dental decay.[4]

HOW IT ALL STARTED

Dean was also well aware that fluoride concentrations of as little as 2 ppm could constitute a public health concern, causing severe dental fluorosis. Coincidentally, the U.S. PHS was at the time sponsored under the Department of the Treasury, the chief officer of which was Andrew Mellon, owner of ALCOA. In 1939 The Mellon Institute (established and controlled by the family of Andrew Mellon), employed a scientist, Dr. Gerald Cox, to find a viable market for the industrial fluoride wastes associated with the production of aluminium. Of this intriguing series of connections between the interests of ALCOA and the story of fluoridation Walker writes:

> In 1939, Gerald Cox, a biochemist employed by the University of Pittsburgh, was undertaking contract work for the Mellon Institute.
> At a meeting of water engineers at Johnstown, Pennsylvania, he first put forward his idea to add fluoride to public water supplies.
> By 1940, Cox had become a member of the Food and Nutrition Board of the National Research Council, and he prepared for this illustrious body a series of submissions strongly promoting the idea of artificial fluoridation.[5]

Dennis Stevenson also comments about this connection between Dr. Cox, ALCOA and fluoridation but somewhat more cynically. He writes:

> Dr. Cox then proposed artificial water fluoridation as a means of reducing tooth decay. What better way to solve the huge and costly problem of disposing of toxic waste from aluminum manufacturers than getting paid to put it in the drinking water? What an incredible coincidence— ALCOA and the original fluoridation proposal.[6]

Nor do the chain of seeming coincidences end here.
Caldwell refers to the very interesting testimony of Miss Florence Bir-

Facts About Fluoride

TOOTH DECAY IS NOT REDUCED BY WATER FLUORIDATION!

A computer analysis of the data from the largest dental survey ever done—of nearly 40,000 school children—by the National Institutes of Dental Research revealed no correlation between tooth decay and fluoridation. In fact, many of the non-fluoridated cities had better tooth decay rates than fluoridated cities. The city with the lowest rate of tooth decay was not fluoridated. Of the three with the highest rate of decay, two were partially fluoridated.

The Missouri State Bureau of Dental Health had conducted a survey of more than 6,500 lifelong resident second- and sixth-grade children in various parts of Missouri and found that overall . . . there were no significant differences between children drinking optimally fluoridated water and children drinking suboptimally fluoridated water.

—Albertt W. Burgstahler, Ph.D.
Professor of Chemistry, University of Kansas

. . . school districts reporting the highest caries-free rates, were totally unfluoridated. How does one explain this?

—A. S. Gray, D.D.S.
Journal of the Canadian Dental Association, 1987

mingham on May 25–27, 1954, before the Committee on Interstate and Foreign Commerce, which had organised a series of hearings on the fluoridation issue. As President of the Massachusetts Women's Political Club, Miss Birmingham was on the occasion representing some 50,000 women. She is recorded as saying:

In 1944 Oscar Ewing was put on the payroll of the Aluminum Company of America [ALCOA], as attorney; at an annual salary of $750,000. This fact was established at a Senate hearing and became part of the Congressional Record. Since the aluminum company had no big litigation pending at the time, the question might logically be asked, why such a large fee? A few months later Mr. Ewing was made Federal Security Administrator with the announcement that he was taking a big salary cut in order to serve his country. As head of the Federal Security Agency

. . . all surveys both here and in Western Europe show that the reduction in [dental] caries over the past 20 years is just as great in unfluoridated as in fluoridated communities.

—John R. Lee, M.D.

Even the *Journal of the American Dental Association* [states] that "the current reported decline in caries in the U.S. and other Western industrialised countries has been observed in both fluoridated and nonfluoridated communities, with percentage reductions in each community apparently about the same."

—*Chemical & Engineering News,* 1 August 1988

INFANT MORTALITY RATES ARE HIGHEST IN FLUORIDATED CITIES

Figures released by the National Centre for Health Statistics reveal that infant mortality is a big problem in the United States. The data shows that the ten cities with the worst rate of infant mortality have all been artificially fluoridated at least seventeen years or longer!

After the first full year of fluoridation Kansas City, Missouri's infant mortality increased 13 percent.

—*The Kansas City Star,* 21 November 1982

After the fifth year of fluoridation in Kansas City, infant mortality increased 36 percent.

—*The Kansas City Star,* 26 February 1987

(now the Department of Health, Education and Welfare), he immediately started the ball rolling to sell "rat poison" by the ton instead of in dime packages . . . sodium fluoride was dangerous waste product of the aluminum company. They were not permitted to dump it into rivers or fields where it would poison fish, cattle, etc. Apparently someone conceived the brilliant idea of taking advantage of the erroneous conclusions drawn from Deaf Smith County, Texas.* The Aluminum Company of America then began selling sodium fluoride to put in the drinking water.[7]

* In a footnote Caldwell comments on this point:

This refers to a widely circulated report published in a popular magazine in the early forties, in which Dr. George Heard, a dentist in Deaf Smith County, claimed he had no business because of the natural fluoride in the water. Later, when Dr. Heard found mottled teeth too brittle to fill and a rushing business after supermarkets moved in with processed foods, he tried in vain to set the record straight. He could find no publisher for his new information. His original article was entitled "The Town Without A Toothache."[8]

The series of events which thereafter led to the apparently inevitable implementation of fluoridation deserve also to be reviewed. In 1945 Grand Rapids, Michigan was selected as the site of the first major longitudinal study of the effects of fluoridation on the public at large. Comparisons were to be made with the city of Muskegon, Michigan which remained unfluoridated so that it could be used as a control.[9]

Although the experiment was supposed to be undertaken over the course of ten years to determine any cumulative side-effects which might result from the fluoridation of municipal water, Ewing intervened after only five years to declare the success of the study in showing fluoridation to be safe. As Walker puts it:

> In June, 1950, half-way through the experiment, the U.S. PHS under its Chief, Oscar Ewing, "endorsed" the safety and effectiveness of artificial fluoridation; and encouraged its immediate adoption through the States.[10]

One year later Ewing was able to convince the American Congress that fluoridation was a necessity, and a total of two million U.S. dollars (an enormous sum of money in those days) was immediately directed to promote the fluoridation program throughout the USA.[11]

While the circumstances surrounding Ewing's achievement were revealing, an even more intriguing set of interconnections was yet to be revealed. Miss Birmingham's testimony had included a statement that "Mr. Ewing's propaganda expert was Edward L. Bernays."[12] Her testimony continued:

> We quote from Dr. Paul Manning's article: "The Federal Engineering of Consent." Nephew of Sigmund Freud, the Vienna born Mr. Bernays is well documented in the Faxon book published in 1951 (Rumford Press, Concord, N.H.); *Public Relations: Edward L. Bernays and the American Scene.* The conscious and intelligent manipulation of the organized habits and opinions of the masses must be done by experts, the public relations counsels (Bernays invented the term); "they are the invisible rulers who control the destinies of millions . . . the most direct way to reach the herd is through the leaders. For, if the group they dominate will respond . . . all this must be planned . . . indoctrination must be subtle. It should be worked into the everyday life of the people—24 hours a day in hundreds of ways . . . A redefinition of ethics is necessary . . . the subject matter of the propaganda need not necessarily be true," says Bernays.

If the socio-ethical attitudes expressed in this testimony are associated with the fluoridation programme, it is clear that we have more than just health reasons to be concerned about fluoridation.

In 1979 *Chemical & Engineering News*[13] published a review of a well documented anti-fluoridation book by Waldbott.[14] The unashamedly pro-fluoridation review prompted a spate of letters criticising the tenor and content of the review, and re-asserted Waldbott's persuasive case against fluoridation. One letter complained that the reviewer was in fact explicitly urging readers *not* to take seriously the various reports of fluoride poisoning.[15] Another letter writer drew attention to another aspect of the review, saying:

> Waldbott does not base his objection to fluoridation merely on dental fluorosis but on the broader issue of individual clinical toxicity. Those of us in clinical practice (and our patients as well) have much to be grateful to Waldbott for in our attention to this aspect of fluoridation problems. The alert clinician who goes beyond the orthodox practice of making diagnoses keyed to organicity and providing symptomatic treatment will find in his practice *those individuals who are being made ill by fluoridation.* It is this insight that is Waldbott's greatest contribution . . .

A second major point bypassed in the book review is the fact of dramatically increased dietary fluoride exposure, as confirmed by the data of Rose and Marier (Canadian National Research Council), Herta Spencer, Wiatroski, and others, including my own food fluoride study . . . It boggles the mind to argue, as the U.S. Public Health Service does, that "optimal" water fluoridation levels should be the same in 1979 as they were in 1943 when food fluoride was essentially negligible.

It is ironic that if fluoridation were to be raised as new concept for the prevention of tooth decay today, the same government agencies that might employ reviewer Burt would reject the proposal without a second thought. It is only an accident of historical scientific naivete that fluoridation became an entrenched public policy. The fact that 100 million Americans (and a large percentage of them against their expressed desire) are subject to the unnecessary ecologic burden of water fluoridation does not make it right . . . [16]

Mandatory medication by fluoridation was not of course peculiar to the United States. Australians have for more than three decades been subjected to forced fluoridation of their drinking water. In 1953 the National Health & Medical Research Council of Australia lent its support to the mandatory mass-medication of Australians.[17] It is bizarre and disconcerting to find that the introduction of the fluoridation programme into our cities was also linked with political and industrial interplay. These connections have been deftly exposed by Walker and more recently by Wendy Varney in her book, *Fluoride in Australia—A Case to Answer.*[18]

Today, Australia has "distinguished" itself by promoting the fluoridation programme with such vigour that Australia now ranks as the most comprehensively fluoridated country in the world. More than 70 percent of Australians are obliged to drink water to which fluorides have been added. Brisbane is the only capital city which remains unfluoridated. Australia persists in its policy committment to artificial fluoridation, despite the fact that 98 percent of the world's population has either discontinued fluoridation programmes or never begun them.

Statistics show that less than 40 percent of the U.S. is currently fluoridated and less than 10 percent of England. Sweden, Scotland, Norway, Hungary, Holland, West Germany, Denmark, and Belgium have all discontinued fluoridation, to name only a few.[19]

CAN FLUORIDATION BE KEPT AT SAFE LEVELS?

Although 1 ppm is standardly defined as that level of fluoride concentration which provides maximal protection against dental decay, with minimal clinically observable dental fluorosis, controversy ranges widely as to adverse effects of prolonged fluoride exposure even at this level. As early as 1942, it was reported that in areas of endemic fluorosis with fluoride concentrations of 1 ppm or less, children with poor nutrition suffered skeletal defects, coupled with severe mottling of teeth.

Even if one grants that fluoride concentrations of 1 ppm are relatively safe, it has become increasingly clear that individual levels of *safe* fluoride ingestion cannot be adequately controlled. Drinking water dosages of fluoride, for example will depend partly upon variable factors such as thirst. Liquid intakes also vary according to age, work situation, climate and season and levels of exercise. Athletes, for instance, tend to consume more water than their non-athletic counterparts. Adjustments to municipal water supplies cannot accommodate satisfactorily the wide array of relevant individual differences of this kind.

In addition fluorides are ingested in varying quantities from many unsuspected sources. Fluoride tablets, seemingly innocuous mouthwashes, gels and even water-based tablets contribute to dangerous increases in fluoride levels well beyond the recommended 1 ppm contained in drinking water. Although the point has yet to be established definitively, it has been suggested that aluminum cooking utensils and non-stick cookware which are coated with Tetrafluoroethylene may exude fluoride into food, particularly if they have surface scratches or are overheated.[20] Even more surprising is the fact that tea leaves contain sufficient fluoride that by drinking three to eight cups daily, using fluoridated water, the total fluoride dosage is somewhere between four and six times the safe maximum recommended daily allowance.[21] In addition to endemic fluorides in the natural foods we eat, we are in many industrial cities forced to breathe fluorides derived from factory emissions.[22]

FLUORIDE CONTAMINATION
FROM BEVERAGE CONSUMPTION

By far the most common source of additional fluoride intake comes from beverage consumption. Beverages which contain fluoridated water include reconstituted juices, punches, popsicles, other water-based frozen desserts and carbonated beverages. Studies have shown that soft drink consumption in the U.S. has increased markedly over the last two decades, not only among teenage boys from 15–17 years of age, but among 1–2 year old children. Statistics show that in Canada soft drink consumption increased by 37 percent from 1972 to 1981.[23] The increase in soft drink consumption coincided with a decrease in the consumption of milk, thereby increasing the overall fluoride intake. A number of studies reveal that the dramatic increase in beverage consumption, coupled with fluoridation of municipal waters constitutes a potential health hazard.[24] Prolonged exposure to fluorides may actually increase rather than diminish the incidence of tooth decay. Enzymatic damage related to enamel mineralisation creates a parotic tooth far more susceptible to caries than would otherwise be the case.[25]

In a major study of adverse effects of fluoride, Yiamouyiannis and Burk reported in 1977 that at least 10,000 people in the U.S. die every year of fluoride-induced cancer. In the introduction to their work 17 research papers are cited which demonstrate the mutagenic effects associated with fluorides.[26] There is now side consensus within the scientific community that the mutagenic activity of a substance can be regarded as an important indication of its potential cancer-causing activity.

Since those provocative studies over a decade ago, a vast scientific literature has continued to accumulate which strongly indicates that the practice of fluoridating municipal water supplies is dangerous. In 1983 an Australian dental surgeon, G. Smith, reported a number of studies which suggest that there is now a serious risk to the public of fluoride overdose. He argues that "the crucial argument does not concern the fluoride level in a community water supply per se, but rather whether fluoridation increases the risk that certain people develop, even for a short time, levels of fluoride in the blood that can damage human cells and systems."[27]

In 1985 another Australian scientist, M. Diesendorf, drew attention to the discovery of a whole new dimension to the health hazards associated with the ingestion of fluorides. Sodium fluoride, for example, had been found to cause unscheduled DNA synthesis and chromosonal aberrations in certain human cells.[28] Other recent studies purport to reveal the actual mechanism by virtue of which fluoride can disrupt the DNA molecule and the active sites of the molecules of many human enzymes.[29]

When all is said, it is manifestly clear that the time has come for a serious and comprehensive review of the policy which mandates the compulsory fluoridation of our municipal water supplies. Such a review will no doubt require a multi-faceted approach in which reliable research investigations can be integrated with a philosophy of health education to assist their implementation. Through education it may be possible to appreciate that within nature itself are important patterns of design for an overall programme of health. In nature, for instance, fluorides are typically found in decidedly *insoluble* forms which are relatively safe. By deliberately intervening to make nature's *insoluble* forms of fluoride *soluble* we transform a relatively harmless natural substance into a concentrated and highly toxic substance which can then be indiscriminately dispersed throughout the environment as a poison. The subtle constellation of health clues which nature provides in respect of fluorides is further illustrated by the simple but elegant mechanisms of breastfeeding. Breastfed infants are actually protected from receiving more than extremely low concentrations of fluoride in breast milk by an inbuilt physiological plasma/milk barrier against fluoride.[30] There is much about health to learn from nature, but to do so we must be more concerned to join with nature in partnership than to stand back from nature to subdue and manipulate it.

Whether the fluoridation campaign must be indicted in the light of the evidence as one of the major public hoaxes perpetrated this century, is a judgement best reserved for the reader. Whatever the judgment, it is incontestable that the prevention of tooth decay is not the bottom-line of the fluoridation debate when the panacea has become the poison.

For more information on artificial fluoridation, we recommend to readers: *The Australian Fluoridation News*, GPO Box 935G, Melbourne, Vic, 3001. This is a bi-monthly publication, which costs $15 per annum.

REFERENCES

1. N.I. Sax, *Dangerous Properties of Industrial Materials*, 2nd ed. (New York: Reinhold Publishing Corp., 1963), p. 1187.

2. L. Hodges, *Environmental Pollution*, 2nd ed. (New York: Holt, Rinehart and Winston, 1977), p. 64.

3. G.S.R. Walker, *Fluoridation—Poison on Tap* (Melbourne: Glen Walker Publisher, 1982), p. 40.

4. H.T. Dean, "Studies on the Minimal Threshold of the Dental Sign of Chronic Endemic Fluorosis," *Public Health Rep*, 50:1719-1729, 1934.

5. Walker, op. cit. p. 115.

6. D. Stevenson, "Fluoridation, Panacea or Poison?," *Simply Living Magazine*, Vol. 3, #6 (1988), p. 102.

7. G. Caldwell and P.E. Zanfagna, *Fluoridation and Truth Decay* (California: Top-Ecol Press, 1974), p. 7.

8. Ibid.

9. W. Varney, *Fluoride in Australia* (Sydney: Hale & Iremonger, 1986), p. 14.

10. Walker, op. cit. p. 159.

11. Ibid.

12. Caldwell and Zanfagna, op. cit. p. 8.

13. B. Burt, *Chem & Eng News* (22 October 1979), p. 6.

14. G.L. Waldbott, *Fluoridation: the Great Dilemma* (Kansas: Coronado Press Inc., 1978).

15. D. Sherrell, *Chem & Eng News* (7 January 1980), p. 4.

16. J.R. Lee, *Chem & Eng News* (28 January 1980), pp. 4–5.

17. Walker, op. cit. p. 156.

18. Varney, op. cit.

19. Stevenson, op. cit. p. 103.

20. Ibid. p. 104.

21. Committee on Food Protection, Food and Nutritional Board National Research Council, *Toxicants Occurring Naturally in Foods* (Washington, DC: National Academy of Science, 1973), pp. 72–74.

22. Walker, op. cit. p. 308.

23. J. Clovis and J.A. Hargreaves, "Fluoride Intake from Beverage Consumption," *Community Dent Oral Epidemiol*, 16:14, 1988.

24. J. Mann, M. Tibi, and H.D. Sgan-Cohen, "Fluorosis and Caries Prevalence in a Community Drinking Above-Optional Fluoridated Water," *Community Dent Oral Epidemiol*, 15:293–294, 1987.

25. Ibid. p. 295.

26. J. Yiamoyiannis and D. Burk, "Fluoridation and Cancer. Age-Dependence of Cancer Mortality Related to Artificial Fluoridation," *Fluoride*, 10:102–123, 1977.

27. G. Smith, "Fluoridation—Are the Dangers Resolved?," *New Scientist* (5 May 1983), p. 286.

28. M. Diesendorf, "Fluoride: New Risk?," *Search*, 16, nos. 5–6:129, 1985.

29. Ibid.

30. Smith, op. cit. p. 287

3   Deadly Mercury: How It Became Your Dentist's Darling

Val Valerian

Exposure to mercury from "silver" dental fillings is slowly poisoning millions of Americans each year. In fact, chronic mercury toxicity from such fillings ranks among our most serious public health problems.

The modern dental amalgam, widely misnamed "silver" and used in fillings for more than 180 years, now accounts for 79–80 percent of all dental restorations.[1] In truth, however, it contains only about 35 percent silver by weight, compared to 50 percent mercury (with 13 percent tin, and small amounts of copper and zinc).[2]

Citing the silver-mercury ratio, Murray Vimy, professor in the Department of Medicine at the University of Calgary (Canada), notes that average amalgam fillings have a mercury mass of 750–1,000 milligrams (mg) and should more properly be called *mercury* fillings. They have a functional life of about 7–9 years, after which they are usually replaced with another one made of the same material.[3, 4]

Mercury is more toxic than lead or even arsenic. Considering the mountains of scientific information that have accumulated over the last 70 years, which clearly show the poisonous effects of mercury, using it today in dentistry is simply criminal. Yet each year worldwide, hundreds of tons of this toxic material are placed into patients' teeth, while some finds its way from dental offices into sewage and refuse systems, to poison the environment instead of the patients.

The American Dental Association (ADA) and government scientists *know* mercury's potential and actual harm, yet continue to promote its use. They thus make a direct, if covert assault on America's health while producing large profits for themselves and their special interest group. *Appropriately, such crimes are punishable by death* under the Crime Bill of 1994 and United Nations rules concerning genocide.[5]

Within the dental profession, the issue of mercury-filling safety has recurred periodically. Introduced in 1812 by British chemist Joseph Bell, the "silver paste"—a combination of old silver coins and mercury—became fashionable for tooth restoration. Since the coins were not pure silver, the material often expanded, fracturing teeth and/or giving patients a "high bite."

When it was first introduced in the United States (in 1833 in New York), dentists rebelled. They refused to use the "silver" because it caused *immediate symptoms of mercury poisoning*. Within the first 10 years most dentists denounced its poor filling qualities and toxic nature, forming the American Society of Dental Surgeons in 1840 to declare mercury usage malpractice. The society mandated that its members sign an oath not to use materials containing mercury.

Nevertheless, amalgam increased in popularity, particularly among poorly-educated practitioners; it was cheaper than gold—the standard until then—giving the renegades an economic edge over their colleagues who demanded higher quality. (In those days many nonprofessionals, including itinerant peddlers, were filling teeth for the pioneer population also.) Besides, amalgam fillings were user-friendly—for the dentist, not the patient—and durable in the mouth.

By 1856, the anti-amalgam society had lost so many members that it had to disband, while wealthy businessmen (not dentists) founded a new group to push the toxic material: the ADA. For a time the debate was dead. The poison had won; the patients had lost.[6, 7]

MERCURY AND THE BRAIN

In the 1920s another controversy erupted after Dr. Alfred Stock, a German chemistry professor, published articles and letters attacking mercury fillings for their possible toxic effects. By 1935, Stock's research proved that some of the mercury vapor coming from dental amalgams enters the nose, is absorbed by the mucosa and passes rapidly into the brain: It was found in the olfactory lobe and in the pineal gland. After a while, however, the furor surrounding Stock's findings also died down.

Now, nearly 150 years after its founding—and in the midst of its third "amalgam war"—the ADA is trying to kill the debate for all time: It has amended its code of ethics to condemn the removal of serviceable mercury fillings as *unethical*, if the reason is to eliminate a toxic material from the body and if the recommendation is made solely by the dentist.[8]

According to Professor Vimy, the ADA considers a dentist ethical if he places the poisonous substance and recommends its safety. However, if he suggests that mercury fillings are potentially harmful or that exposure to unnecessary mercury can be, he is acting unethically. Serviceable mercury fillings can be "ethically" removed if done for aesthetic reasons, at the request of the patient (without prompting) or a medical doctor.

In 1967, the inaugural issue of *Environmental Research* magazine featured "Mercury-Blood Interaction and Mercury Uptake by the Brain After Exposure," the research of L. Magos. His experiments with mice indicated that inhaled mercury reaches the brain.

Mercury amalgam continuously gives off vapor, and, inhaled, it has an incredibly high absorption rate—between 70 and 100 percent! Once in the brain, mercury oxidizes into its elemental form. Then the blood-brain barrier, designed to protect the brain from foreign substances in the blood, works against its own purpose, *preventing* the removal of the toxic metal from the brain.

Nonetheless, in 1976, the FDA pronounced its "acceptance" of mercury amalgam, since "it had been in use since 1840." It was added to the Generally Recognized As Safe (GRAS) list of pre-1930 drugs.

In 1979, perhaps as a reaction to the FDA ruling, research began appearing which documented that mercury vapor is *constantly* released from fillings, especially when they are stimulated by chewing, brushing or temperature shifts (such as drinking a hot liquid). Every time someone with amalgam fillings eats or brushes his teeth, the leaching of mercury into the body peaks. Afterwards, it takes almost 90 minutes for the vaporization rate to decline to its previous level.[9]

Also, the greater the number of fillings and the larger the chewing-surface area they occupy, the greater the toxic exposure.[10,11] Since hot liquids also increase vaporizing, individuals with amalgam fillings who frequently drink tea or coffee run a higher risk of mercury-related pathologies.

The average person, with eight biting-surface mercury fillings, is exposed to a daily-dose uptake of about 10 mcg (micrograms) of mercury from his fillings.[12] Some individuals may have daily doses 10 times higher (100 mcg) because of factors that exacerbate mercury vaporization, such as more frequent eating, chronic gum chewing, chronic tooth grinding (usually during sleep), chewing patterns, consumption of hot or acidic foods and drinks, and mouth acidity.[13]

Furthermore, Professor Vimy asserts the concentration of mercury in the brain correlates with the number of fillings present. In 1980, the World Health Organization (WHO) admitted, "The most hazardous forms of mercury to human health are elemental mercury vapor and the short-chain alkyl-mercurials."[14]

By 1984, autopsy studies were published that demonstrated that *the amount of mercury found in brain and kidney tissue directly relates to the number of mercury amalgam fillings in the teeth.* Research at the University of Calgary School of Medicine demonstrated that *mercury from dental fillings could be found in the blood and tissues of pregnant mothers and their babies within a few days of insertion.*

The Biological Effects of Mercury Amalgam: Scientific Facts and References

1. Mercury penetrates the blood-brain barrier which protects the brain, and as little as one part per million [ppm] can impair this barrier, permitting entry of substances in the blood that would otherwise be excluded.

—Chang and Hartman, 1972;
Chang and Burkholder, 1974

2. Mercury exposure from amalgams leads to interference with brain catecholamine reactivity levels, has a pronounced effect on the human endocrine system and accumulates in both the thyroid and pituitary glands reducing production of important hormones.

—Carmignani, Finelli and Boscolo, 1983;
Kosta et al., 1975; Trakhtenberg, 1974.

3. Mercury inhibits the synaptic uptake of neurotransmitters in the brain and can produce subsequent development of Parkinson's disease.

—Ohlson and Hogstedt,
"Parkinson's Disease and Occupational Exposure to Organic Solvents,
Agricultural Chemicals and Mercury," *Scandinavian Journal of Work
Environment Health*, 7, no. 4:252–256, 1981.

4. Mercury is nephrotoxic (toxic to the kidneys) and causes pathological damage.

—Nicholson et al., "Cadmium and
Mercury Nephrotoxicity," *Nature*, 304:633, 1983.

5. Mercury has an effect on the fetal nervous system, even at levels far below that considered to be toxic in adults. Background levels of mercury in mothers correlate with incidence of fetal birth defects and still births.

—Reuhl and Chang, 1979;
Clarkson et al., 1981; Marsh et al, 1980; Tejning 1968; W.D. Kuntz,
R.M. Pitkin, A.W. Bostrum, and M.S. Hughes, *The American Journal of
Obstetrics and Gynecology*, 143, no. 4:440–443, 1982.

6. Chronic exposure to mercury may cause an excess of serum proteins in the urine which may progress to nephrotic syndrome and peculiar susceptibility to infections that break into and modify the course of any pre-existing disease.

—Friberg et al. (1953),
"Kidney Injury After Chronic Exposure to Inorganic Mercury," *Archives of Environmental Health*, 15:64, 1967; Kazantis et al., "Albuminuria and the Nephrotic Syndrome Following Exposure to Mercury," *Quarterly Journal of Medicine*, 31:403–418, 1962; Joselow and Goldwater, "Absorption and Excretion of Mercury in Man and Mercury Content of 'Normal' Human Tissues," *Archives of Environmental Health*, 15:64, 1967.

7. Mercury fillings can contribute to a higher level of mercury in the blood and can affect the functioning of the heart, change the vascular response to norepinephrine and potassium chloride and block the entry of calcium ions into the cytoplasm.

—Abraham et al.,
"The Effect of Dental Amalgam Restorations on Blood Mercury Levels," *Journal of Dental Research*, 63, no. 1:71–73, 1984; Kuntz et al., "Maternal and Cord Blood Background Mercury Levels: A Longitudinal Surveillance,"*American Journal of Obstetrics and Gynecology*, 143, no. 4:440–443, 1982; Joselow et al., 1972; Mantyla and Wright, 1976: Trakhtenberg, 1968; Oka et al., 1979.

8. The effect of mercury on the nervous system selectively inhibits protein and amino-acid absorption into brain tissue.

—Yoshino et al., 1966; Steinwall, 1969; Steinwall and Snyder, 1969; Cavanagh and Chen, 1971.

9. Mercury induces the thyroid gland to absorb an increasing amount of nuclear radiation from the environment.

—Trakhtenberg, 1974.

10. Mercury has a distinct effect on the human immune system, especially the white blood cells. Mercury ions have been observed to cause chromosomal aberrations and alter the cellular genetic code. Mercury has the ability to induce chromosomal breakage, alter cellular mitosis, cause a drop in T-cell production and kill white blood cells.

—Vershaeve et al., 1976;
Popescu et al., 1979; Skerfving et al., 1970, 1974; Fiskesjo, 1970.

11. Mercury can impair the adrenal and testicular steroid hormone secretions, causing intolerance for stress and decreased sexual ability. In rats, it causes subnormal fertility and sperm production.

—Burton and Meikle, 1980;
Khera, 1973; Stoewsand et al., 1971; Lee and Dixon, 1975;
Thaxton and Parkhurst, 1973.

12. Mercury in the body can produce contact dermatitis and reduced function of the adrenal glands (Addison's disease), producing progressive anemia, low blood pressure, diarrhea and digestive disturbances.

—Alomar et al., 1983.

13. Mercury in the human body can contribute to intelligence disturbances, speech difficulties, limb deformity and hyperkinesia (hyperactivity resulting from brain damage). Abnormally small heads and retardation were present in 60 percent of cases.

—Amin-Zaki and Clarkson, et al., 1979.

SCIENTIFIC "TRUTH"—WHO'S RESPONSIBLE?

As the existence of the ADA has gradually become integrated into the public's awareness, the perception has grown up that its statements represent scientific truth. The association and its members know that a conditioned and ignorant public will rarely question any statement it makes. Thus, when it issues false statements that jeopardize public health, such as recommending that deadly toxins be placed in the body, it is party to a criminal act. Having third-party dentists do the dirty deeds does not excuse the association, especially since its members have spent years and much money studying in ADA-approved schools, and anyone who bucks the system risks his license, his income and his career. This constitutes extortion and racketeering by the ADA, punishable under the RICO Act.[15]

DISCOVERIES AND REVELATIONS IN THE 1990S

1990—On December 16, CBS's *60 Minutes* did a major story on the amalgam issue that generated the second-highest response for additional information in the history of the program. Most viewers thought it would produce a change in the system. Instead, the ADA launched a vast campaign to counter the knowledge *60 Minutes* had given the public.

First, the ADA and "responsible" government agencies invested huge sums of money: They sent special letters and news releases to every dentist in the United States discrediting the scientific information presented on the program and assuring the dentists that amalgam had been used for 150 years so it *must* be safe. They also sent press releases to all the major media and even created a special program for dentists entitled "What To Tell Your Patients When They Ask About Amalgam."

Also appearing in 1990 was the first controlled research on the effects of amalgam implants. It cited significant effects of mercury amalgam on various tissues and organs in experiments with monkeys and sheep.

1991—Sweden banned the use of mercury amalgam implants.

In March, the FDA recommended "further studies," and consumer groups and legislators began introducing *informed consent legislation* whereby dentists would be required to inform patients of the content of amalgam fillings (implants) and the potentially harmful effects of mercury.

In May, a medical research team at the University of Kentucky established "a probable relationship of mercury exposure from mercury-amalgam fillings to both Alzheimer's disease and cardiovascular disease." University of Georgia microbiologists determined that mercury from fillings inhibits the effectiveness of antibiotics.

The WHO reported that exposure to mercury from amalgams is higher than that of other environmental sources and that each amalgam filling releases from three to seventeen micrograms of mercury daily during eating alone. The WHO audited all available scientific data on the subject and concluded that mercury from dental fillings is the greatest source of human exposure, exceeding intake from fish by about 200 percent.

1993—The International Academy of Oral Medicine and Toxicology (IAOMT) developed a certification program for biocompatible dentistry [using materials compatible with each patient's body chemistry]. Dentists certified by the IAOMT will have "demonstrated proficiency in replacing amalgams safely and properly."

Swedish researchers discovered that gastrointestinal function improves after amalgam is removed. A citizens' petition requesting a ban on the use of amalgams was filed with the FDA.

The State of California passed a law requiring the Board of Dental Examiners to develop, distribute and update a fact sheet describing and comparing the risks and efficacy of the various types of dental restorative materials. Prior to this, a dentist risked losing his/her license for giving such detailed information to a patient. Presenting both sides of the dental mercury issue to a patient was considered unethical!

1994—In January, the government of Ontario demanded a probe of mercury dental fillings. In February, Sweden announced a total ban on the use of mercury amalgam fillings.

BBC'S "PANORAMA" PANS AMALGAM

On July 11, 1994, the British BBC-1 documentary program *Panorama* focused on the amalgam controversy. Dr. Lars Friberg, chief advisor to the WHO on metals poisoning, told *Panorama*, "The use of mercury in dental fillings is not safe and should be avoided." The program also reported on new scientific research which demonstrated clear links between the mercury released from dental fillings and serious illness, including Alzheimer's disease.

Dr. Friberg told *Panorama* that he "did not know why" the British Dental Association considers mercury levels in amalgam "safe" and that he thought, "They are wrong." He was concerned particularly about deposition of mercury from fillings into the brains of children.

According to Dr. Friberg,

> [W]e know that children are especially vulnerable to the amalgam. We know that it takes a few years after birth until the brain is developed, and we know the brains in children are much more sensitive than those of adults. I think that you should try to avoid the implantation of toxic metals in the mouth. There is no safe level of mercury, and no one has actually shown that there is a safe level. I say mercury is a very toxic substance. I would like to avoid it as far as possible.

Panorama investigated a number of new, independent studies, some unpublished, which point inexorably towards the health risks of amalgam fillings. The first is a new study by Dr. Boyd Haley, professor of biochemistry at the University of Kentucky. He discovered that small quantities of mercury from amalgams can produce changes in the brain identical to those caused by Alzheimer's disease. Mercury inhibits the efficiency of *tubulin*, a protein vital to brain cells.

According to Dr. Haley:

> To the best that we can determine with these experiments, mercury is a time bomb in the brain. We need to have an effect—if it's not bothering someone when they are young, especially when they age, it could turn into something quite disastrous. I still have one amalgam filling. When I had the others replaced, I had them replaced with nonamalgam fillings. I would not make the statement that mercury causes Alzheimer's disease, hut there is no doubt in my mind that low levels of mercury, present in the brain, could cause normal cell death that could lead to a dementia which would be similar to Alzheimer's disease.

Panorama also reported on the unique, ongoing study of a group of nuns by Dr. William Markesbery, professor of pathology and neurology and director of the Sanders Brown Center for Aging at the University of Kentucky. He is investigating the link between mercury and Alzheimer's disease and told *Panorama*:

> Mercury is a toxic substance. It is a neurotoxin—that is to say it causes nerve cells to degenerate if there is enough mercury present in the brain. The major problem in Alzheimer's disease is the degeneration of nerve cells. It is possible that mercury could add to the degeneration of nerve cells—to the death of nerve cells.

The reaction of the British Dental Association (BDA) to the 1994 *Panorama* program was predictable: It told *Panorama* it was unaware of the work of Haley and also unaware of the work of Professor Aposhian at the University of Arizona who discovered that 66 percent of the mercury deposits in the body come from fillings. In fact, John Hunt, chief executive of the BDA, told *Panorama*, "Amalgam is safe." The BDA fact sheet available to dentists states, "The scientific evidence *available to the BDA* does not justify banning the use of amalgam in young children."

Said Hunt, "I've treated my children with amalgam and have no doubt that when they have their own children, they will do the same." Asked about any link between Alzheimer's and mercury, he added:

> As far as I know, there is no association with mercury and Alzheimer's. We rely on expert advice. There is no evidence to suggest that merely because mercury is found in the kidneys of the fetus and young children, that it is a hazard to health. I don't see why we should necessarily worry the population at large if there are no proven arguments one way or another. The fact that it is there and it is detectable doesn't mean to say that it's potentially doing any damage. You can probably find a whole lot of substances in the brain that perhaps should not be there.

Special thanks to Professor Murray Vimy for several of the above references and to the Dental Amalgam Mercury Syndrome (DAMS) consumer advocacy group. For more information contact DAMS at (800) 311–6265.

REFERENCES

1. J.G. Baurer and H.A. First, *California Dental Association Journal*, 10:47–61, 1982.

2. E.W. Skinner and R.W Phillips, *The Science of Dental Materials*, 6th ed. (Philadelphia: W.B. Saundets Co., 1969), pp. 303, 332.

3. N. Paterson, *British Dental Journal*, 157:23–25, 1984.

4. R.W. Phillips, et al., *Journal of Prosthetic Dentistry*, 55:736–772, 1986.

5. The United States is exempt from the U.N. Genocide Treaty of March 1988, as it signed on the condition that "no nation shall sit in judgment of the United States." Technically, any form of genocidal activity is legal in the United States under U.N. mandates.

6. American Academy of Dental Science, *A History of Dental and Oral Science in America* (Philadelphia: Samuel White, 1876).

7. D.K. Bremmer, *The Story of Dentistry*, rev. 3rd ed. (Brooklyn: Dental Items of Interest Publishing, 1954).

8. "Representation of Care and Fees," *Principles of Ethics and Code of Professional Conduct* (American Dental Association, 211 E. Chicago Ave., Chicago, IL 60611).

9. J.J. Vimy and F.L. Lorscheider, *Journal of Dental Restorations*, 64:1072–1075, 1985.

10. Ibid. pp. 1069–1071.

11. Ibid. pp. 1072–1075.

12. J.J. Vimy and F.L. Lorscheider, *Journal of Trace Element Experimental Medicine*, 3:111–123, 1990.

13. Vimy and Lorscheider, *Dental Restorations*, pp. 1072–1075.

14. World Health Organization, "Recommended Health-based Limits in Occupational Exposure to Heavy Metals," 1980.

15. *Racketeer Influenced and Corrupt Organizations Act* (18 USC 1961 et seq.; 84 Stat 922).

4 The Alzheimer's Cover-Up

Tom Warren

*I know that most men, including those at ease
with problems of the greatest complexity,
can seldom accept even the simplest and
most obvious truth, if it be such as would
oblige them to admit the falsity of conclusions
which they have delighted in explaining to
colleagues, which they have proudly taught
to others, and which they have woven
thread by thread into the fabric of their lives.*

Leo Tolstoy

Suppose that you had dedicated your whole life to helping patients achieve better health. In time you were recognized as an eminent professional and became a board member of your medical association, one of the largest and, to the uninformed, most respected conglomerates in the world. Your associates produce a gross income of $50 million a day. Collectively, the board is a major economic-political power. If change were in the air, what would you be willing to do to protect your position?

Dominant members of society always resist change. The power structure looks at identical data disinterested parties see clearly, but filters out information that challenges its earning power.

A hundred and eighty years ago, when the problems we are talking about started, medical science was in its infancy. Technology to determine the efficacy of pharmaceuticals and treatment processes was, by today's standards, nonexistent.

During the Civil War, surgeons wiped blood off their knives onto their aprons to get ready for the next operation. Surgeons did not know that they needed to wash their hands before delivering a baby, so one in four mothers died from childbirth fever (Sepsis infection). Doctor Ignaz Semmelweis, a Hungarian obstetrician from Vienna, observed that mid-wives regularly washed their hands before delivery. It was common knowledge that midwives had a significantly lower Sepsis infection rate than surgeons. Doctor Semmelweis installed a sink at the entrance to the delivery room and surgeons were not admitted into the operating room until they carefully washed their hands. Sepsis infection disappeared. Doctor Semmelweis informed the medical community and was driven insane by their ridicule and harassment.

In the United States, not less than seven thousand researchers, scientists, physicians and dentists are aware of at least part of the following information. Approximately half of that number, those that practice mercury-free dentistry and the physicians who support them feel threatened by their medical societies and state medical and dental boards. Scott McAdoo, D.D.S., of Denver, Colorado, who is one of the most skilled dentists I had the pleasure of going to said, "To advertise that you are a mercury-free dentist is like standing in a foxhole." The Colorado State Dental Board is after his license.

The only scientists and physicians that does not feel themselves at risk are our toxicologists, who have been trying for years to tell the public that mercury in "silver" fillings is a deadly poison. The Toxicology Society has published dozens of research papers about the toxicity of mercury. Unfortunately, the American Dental Association (ADA) has enough economic-political power to shut down the very doctors the public should be listening to.

Most medical research is funded by the pharmaceutical industry and researchers are disinclined to shoot their wallet in the toe. For example, a ten day cure for peptic ulcers was discovered over eleven years ago. Most ulcers are caused by H-pylori bacterial infection. I wrote about H-pylori in *Beating Alzheimer's* in December of 1989. The October 1995 issue of *Readers Digest* reports less than half of United States doctors know how to treat their 25 million ulcers patients correctly. "Doctors may be unwilling to try the new methods, while drug companies may be (are) reluctant to abandon the lucrative antacid market." Zantac and Tagamet are two of the fastest selling medications for ulcers in the United States.

Doctor Joel Wallach, M.D., D.V.M., who performed over 27 thousand autopsies on animals as well as 10 thousand autopsies on humans told me whatever disease the cadaver exhibited, malnutrition was the underlying cause of death. He discontinued practice and sells Body and Mineral Toddy, liquid vitamin and mineral supplements. I was introduced to Doctor Wallach's products several years ago. I did not try Body Toddy on myself,

I gave it to our house plants. They grew so huge during the summer that I had to cut them down to dirt and start over. I use both products now.

Malnutrition is a sub-clinical symptom evolved in the Alzheimer's disease process. I have heard physicians say people waste their money on vitamins and minerals, that we just piss them down the drain. What do they think happens to pharmaceuticals? I prefer to piss fifty cents worth of vitamins and minerals every day than die from Alzheimer's, or to have wasted more than a hundred thousand dollars going to doctors who did not have the slightest idea of how to reverse Alzheimer's, or any other chronic disease. All the traditional doctors did was waste my time and take my money. None of them—not one—had the slightest idea of what to do. In addition, the traditional medical community harassed every doctor I went to who knew anything about reversing chronic disease. If a physician cures patients other doctors cannot, his peers shun him, and medical boards do their damnedest to ruin his reputation, his practice and take his license. The bottom line is arrogance, ignorance and money. Mostly, it's the money.

I have a friend who was perhaps, the most informed physician within his area of expertise, in America. He wrote more than eight hundred research papers that were published in peer review medical journals. He was personal physician to some of the most powerful men in the world. He is in *Who's Who* in the United States and *Who's Who* of the world. At the university where he taught, doctors have large practices. My friend was not bringing in as much money as the good old boys wanted so they started directing indigent patients his way. His patients had a higher recovery rate than his colleagues who were charging significantly higher fees. Their patients learned of this, and dropped their physicians to go to my friend. The university fired him. He sued and won $300 thousand. But he is out of the circle; he is not considered one of the good old boys anymore. The traditional medical community pretends he does not exist. He is shunned. The pressure caused his marriage to fall apart. He has taken to drugs and is very ill. His license has been revoked. All his tomorrows are yesterdays.

A PARALLAX VIEW

Suppose that you were a foreign intelligence officer looking for the perfect undetectable poison to destroy an enemy in a way that brought no repercussions to yourself. Suppose that your research scientists discovered a low-level, slow-acting poison, a liquid metal, that is the most lethal non-radioactive element in the Periodic Table. This toxin has an affinity for brain and neurological tissue, especially the sheath surrounding neurons that transmit thought process within the brain.

The onset of clinical symptoms are so insidious physicians are unable to detect the etiology of the patient's distress. One person may develop schizophrenia, another tremors, or Alzheimer's, because the poison always heads for the weakest organ and weakest gene. If you could induce your enemy's dentists to insert this diabolical toxin into their patient's oral cavity "two inches from the brain" in 85 percent of your enemy's population, you would wreak havoc on the whole fabric of their society. You could put hundreds of thousands of their citizenry in mental hospitals. You could destroy whole families. Sick people are inefficient, troublesome workers. They fill the welfare rolls. They fill the prisons. You could cause hundreds of billions of dollars of taxes to be spent supporting institutionalized mental patients and imprisoned no-accounts. You could cause your enemy to spend trillions of dollars on unnecessary medical bills before poisoned patients expire. In addition, sick under-employed workers cannot pay taxes. The best part is that the last person anyone would suspect of poisoning them is their dentist.

If the above parody were true, if it was discovered that any nation caused tens of millions[1] of our population the devastation I just described, we would declare war. This discovery would be considered "A Day in Infamy." Unfortunately, the only incongruity in the above scenario is that there is no foreign intelligence officer. I wish there were. It would make the following information more palatable.

This is what has been covered up:

1. The primary cause of Alzheimer's is iatrogenic disease, doctor-induced disease, from so-called "silver" dental fillings that poison the brain and nervous system. Mercury in silver dental fillings has an affinity for brain tissue and easily passes through the blood-brain barrier. Mercury vapor ions are too miniscule to be recognized by the autoimmune system until enough mercury accumulates upon the nerve sheath that surrounds the synapses. At that point our autoimmune system does not recognize the tissue as self and attacks the nerve sheath as if it were a foreign invader.

2. If an Alzheimer's sufferer can carry on a half-way reasonable conversation for five minutes, do not let uninformed physicians persuade you that Alzheimer's is not reversible. An anesthesiologist who read *Beating Alzheimer's* checked and rechecked my information with dentists and physicians for six months, then removed thirteen root canals. He recovered from Alzheimer's disease in two hours flat. Every once in a while someone tugs on my sleeve in a store, a restaurant or after a speaking engagement and tells me that I have saved their life.

It is impossible to estimate how many millions of people died from neurological diseases of unknown etiology. The correct name for Alzheimer's should be "Chronic Low-Level Mercurial Poisoning." Mercury poisoning from silver dental fillings is a causative factor in the majority of people diagnosed with Alzheimer's disease. And I charge that the ADA has covered up their culpability in the same way cigarette manufacturers and Dow Chemical, who produced silicone breast implants, denied their products' connection to disease.

Those of us who have had our lives turned upside down by mercury poisoning could live with the fact our dentists made an appalling mistake in not recognizing the toxicity of mercury sooner. What we will not tolerate is the fact that the ADA did everything within its power to stonewall the fact that silver fillings were poisoning their patients. They allowed over ten million silver amalgam fillings to be placed into our mouths, two inches from our brain, *after* there was evidence pointing to serious problems due to mercury poisoning from silver amalgam. The ADA tried to protect their profits and their reputations, not their patients. In the meantime thousands of dental patients have unnecessarily suffered catastrophic diseases. If it seems impossible to believe that our dentists would cover-up the fact that they have poisoned a minimum of 12 percent of our population, read the inset on page 47.

If physicians knew mercury was poisonous, dentists had to know mercury was poisonous. It's that simple. Since 1830, the hierarchy—the good old boys in the ADA and our dental professors, must have known that dentists were placing poison into their patients' mouths. They lied to their dental students. They lied to us. In polite company another name for a lie is a "misnomer." Dentists called mercury-amalgam, silver-amalgam. Silver filling is a misnomer. That is a cover-up by any definition of "cover-up" that I understand. And that's fraud.

The ADA admitted that five percent of dental patients were allergic to silver amalgam. There is no such thing as allergy here. The ADA is trying to argue that a poison is not a poison. Goodman and Gilman's 1990 edition of *The Pharmacological Basis of Therapeutics* says, "With very few exceptions, *mercury poisoning* is most often not diagnosed in patients because of the insidious onset of the affliction, vagueness of early clinical signs, and the medical profession's unfamiliarity with the disease."

It is impossible to estimate how many millions of people died from chronic diseases due to mercury poisoning. The ADA claimed only 5 percent[2] (13 million) of dental patients were allergic to silver amalgam—that is, until the Center for Disease Control pointed out that by definition: that's an epidemic. Then the ADA changed their tune to one percent. Even so, that figure is 2,600,000 dental patients they admitted poisoning. The last I heard the ADA was saying it is just a miniscule amount—less than a hundred.

Finally even mainstream media had enough. A CBS *60 Minutes* special entitled "Is There Poison in Your Mouth?" ran nationwide December 16, 1990 that included fantastic recoveries when mercury-amalgam fillings were replaced. "Is There Poison in Your Mouth?" was the most highly viewed television program *60 Minutes* ever produced. CBC's *Panorama* followed up with "The Poison In Your Mouth" that aired in London. These two programs destroyed the credibility of the dental industry. The ADA knew that if they could not put a spin on the revelations exposed in those two television news stories their days were numbered.

THE WITCH HUNTS

In two states, Colorado and Minnesota, the dental boards have used the Attorney General's Office and millions of dollars of public tax money to take away licenses of dentists and physicians who inform their patients that mercury-amalgam is the etiology of the disease they are suffering. The Colorado Attorney General's Office spent $4 million over a twenty-two year period investigating Doctor Hal Huggins and eventually revoked his license. Before Hal decided the fight was not worth the effort anymore he had spent over $700 thousand in legal fees. The AMA took Doctor Sandra Denton's Colorado medical license because she was associated with Doctor Huggins. Sandra Denton is one of the smartest, most dedicated physicians that I have ever known. She was set up. Sandy did not do anything wrong. The Colorado Medical Board took her license to undermine Doctor Huggins' support within the medical community.

I know many critically ill patients that Doctor Denton helped. One young woman was dying by the hour. Other physicians offered no explanation for, or understanding of her illness. Doctor Denton correctly diagnosed the woman's condition and canceled half of her appointments over a two week period to care for this one indigent patient. She had her nurses watch over the young woman constantly. Area surgeons refused to operate because the woman was so close to death. Doctor Denton found a corporate jet to fly her patient from Seattle to Colorado Springs. Two oral surgeons removed titanium implants within the lower jaw and all of her silver fillings. Both surgeons laughed at Doctor Denton. They told Sandra the patient would die on the table.

The young woman started to recover two hours after surgery. Six months later she had two titanium posts reimplanted into her lower jaw to anchor her lower dentures in place and rapidly became schizophrenic. Doctor Denton pleaded with her surgeons to remove the implants, and [the patient] quickly recovered. Several years have gone by and the woman leads a normal life.

Dentistry: Stepping Out of the 1830s

James E. Hardy, D.M.D.

Mercury's use in dentistry has been the subject of my studies for the past 14 years. I thought I knew the history of the mercury-amalgam controversy well, but yesterday I pulled a book down from my shelf that a patient of mine had given me. It was a bound volume of *Dental Students' Magazine*, dated 1942 and 1943. What I read gave me new insight into the history of mercury-amalgam and its intimate relationship with organized dentistry.

Dentistry in the U.S. did not develop as a separate profession until the period of 1780 to 1800. There were two types of "dentists" at this time. There were those who had medical training and practiced both professions. These were medical-dentists. Then there were those who were merely craftsmen and engaged in some other trade, such as barbering, carving of wood, ivory, and metals . . . among them were the itinerant tooth pullers. These were called craftsmen-dentists.

There were no American national dental organizations and no dental schools in existence before 1840. Dentists were either self taught by "trial and error" on patients or were apprenticed under a practicing medical-dentist.

The first dental school was chartered in the United States [on] February 1, 1840. It was called the Baltimore College of Dental Surgery. It is of interest to note that the Board of the College of Dental Surgery was composed of nine physicians and five Clergymen; no dentists. Originally, the school was to be a graduate program for medical schools as it was only one session long and did not include any chemistry or pathology. But as it turns out, this was the beginning of dental training separating itself from medical education.

Also founded in 1840 was the first national dental organization. It was called the American Society of Dental Surgeons. Its members were medical-dentists, not barbers, carpenters, or sculptors. The medical-dentists were vitally concerned with the medical, biological and mechanical aspects of dentistry. The craftsmen-dentists were concerned with the mechanical aspects and did not consider medical questions.

Mercury-amalgam was first brought into the U.S. from Europe

by the Crawcour brothers in 1833. They had a very strong, effective advertising campaign that promised to save decayed teeth by filling them without pain in minutes. The Crawcours were considered unethical charlatans by many medical-dentists. They removed gold fillings and replaced them with mercury-amalgams. They did not dry decayed teeth or even remove the decay before they packed the hole. There were even some reports of their packing amalgam between teeth when there were no cavities at all. This began the first amalgam war.

The amalgam war was the war between the craftsman idea of ease of manipulation and the medical intent to avoid the danger of systemic mercurial poisoning. Mercury was clearly known to be poisonous by the physicians in the 1830s.

In 1845, the American Society of Dental Surgeons banned the use of (mercury) silver-amalgams, making the members sign a pledge of non-use and promise to oppose its use under any circumstances or be expelled from the Society. Remember, this Society was restricted to medical-dentists.

A great number of craftsmen-dentists picked up the use of silver-amalgam. They were not concerned with the medical ramifications of mercury in the human system. Anatomy, chemistry, histology, pathology and physiology were considered irrelevant to the craftsmen-dentists. So many dentists began using amalgam (because it was easy to use and very profitable) that membership growth in the American Society of Dental Surgeons (ASDS) was curtailed. The ASDS decided to rescind their anti-amalgam resolution in 1850 in hopes of gaining more members. But the craftsmen-dentists had already decided to organize and associate with each other.

In 1859, a new dental organization arose composed largely of craftsmen-dentists. It was called the American Dental Association. The average dentist in the newly-formed ADA was no longer one who was in sympathy with medicine. As a result, physicians adopted an extremely adverse attitude toward the profession. During this time, dental schools dropped courses in physiology, pathology, and *materia medica* and did no anatomy except head and neck.

In *Dental Students' Magazine*, October 1942, an editorial states hopefully, "The dental profession eagerly awaits . . . their legitimate rights as members of the healing arts." To this day, these feelings linger.

For dentistry to step out of the 1830s and fully acknowledge its responsibility as a healing art, it must prohibit the current use of mercury. The use of mercury in dentistry has only recently been banned in Germany. (Austria, Sweden, Denmark and Canada have either banned, or curtailed the use of silver amalgam fillings. Switzerland has stopped dental schools from teaching about silver amalgam.) Some day it will be banned in the United States. We've known the truth all along. Mercury is poisonous.

> *All truth passes through three stages.*
> *First, it is ridiculed.*
> *Second, it is violently opposed.*
> *Third, it is accepted as being self evident.*
>
> —Schopenhauer

All quotations and background information for *Dentistry* was taken from *Dental Students' Magazine*, November 1942 and September 1942.

Doctor Denton moved to Anchorage, Alaska, where she was instrumental in getting the state law changed so it is more difficult for medical boards to revoke a non-traditional physician's license. Several years ago, I thought enough of her expertise to drive the Alcan highway from Seattle, Washington to Anchorage, Alaska in December to participate in her allergy desensitivity program. I have not needed to see any physician for allergies or short term memory loss since that time. *Cerebral allergic reactions to favorite foods and chemical sensitivities environmental physicians call brain-fag and brain-fog are involved in the Alzheimer's disease process.*[3]

There have been so many doctors in trouble with their medical boards for exposing chronic low level mercurial poisoning as the etiology of their patient's distress it would fill several volumes of books to narrate their stories. It would take more volumes of writing to tell the stories of poisoned dental patients who have recovered after removing their silver fillings. A past-president of the Toxicology Society told me, "We know mercury is a poison. We teach it out to students. The use of silver dental fillings is an economic-political decision. Don't blame the doctors."

ENDNOTES

1. To the uninformed, "tens of millions" probably seems ridiculously high. For the 18 decades that dentists have used mercury silver-amalgam dental fillings the above estimate is in fact, conservative. Seven percent of

our population is hospitalized in a mental institution at some time during their life. Seven percent of our present population would be approximately 18 million people. The number one sub-clinical sign of mercury poisoning is endogenous depression. Mercury-amalgam has very recently, irrefutably, been connected to many neurological diseases. If mercury-amalgam is the direct cause of only half of the mental patients being committed to mental hospitals that figure is 9 million dental patients whose lives have been devastated by their family dentist. In addition, there are 4 million Alzheimer's sufferers in the United States.

From past experience we expect that the ADA will do everything within its power to discredit the Adolph Coors Human Study Results, that will be released in 1997. This study, the first of its kind in the world, covers amalgam removal, replacement and removal again and silver-amalgam's effect on body blood chemistries. This research prove, beyond speculation, that chronic low level mercurial poisoning from so-called silver dental fillings is the root cause of many chronic diseases for which, until now, our physicians have had no explanation.

2. Doctor Alfred Zamm says a more accurate figure is 12 percent. Twelve percent is more than 30 million Americans whose autoimmune system is adversely affected by mercury poisoning. Alfred V. Zamm, MD, "Dental Mercury: A Factor that Aggravates and Induces Xenobiotic Intolerance," *Journal of Orthomolecular Medicine*, Second Quarter, 1991, Volume 6 Number 2.

3. Cerebral allergic reactions to favorite foods and chemical sensitivities causing brain-fag and brain-fog are part of the pathology of the Alzheimer's disease process. Brain-fag feels like a cranky sleep child that needs a nap. Brain-fog is best described as being spaced out.

REFERENCES

Hahn et. al. "Dental Silver Tooth Fillings: A Source of Mercury Exposure Revealed by Whole-body Image Scan and Tissue Analysis," University of Calgary, Alberta, Canada, August 3, 1990, 2641–2646.

Hahn et. al. "Whole-body Imaging of the Distribution of Mercury Released from Dental Fillings in Monkey Tissues," University of Calgary, Alberta Canada, August 3, 1990, 2641–2646.

Markesbery, W.R. "Trace Elements in Isolated Subcellular Fractions of Alzheimer's Disease Brains," *Brain Research* 533, 1990, 125–131.

Wenstrup et. al. "Trace Element Imbalances in Isolated Subcellular Fractions of Alzheimer's Disease Brains," *Brain Research* 533, August 28, 1989, 123–189.

Zamm, A. "Dental Mercury: A Factor that Aggravates and Induces Xenobiotic Intolerance," *Journal of Orthomolecular Medicine* Second Quarter, 1991, Vol. 6, No. 2.

5 Vaccinations: Adverse Reactions Cover-Up?

Washington, D.C., 2 March 1994—The National Vaccine Information Center (NVIC) operated by Dissatisfied Parents Together (DPT) says that a new Institute of Medicine (IOM) report on the association between DPT vaccine and permanent brain damage confirms that the vaccine can cause children to suffer acute brain inflammation which sometimes leads to death or permanent neurological damage. The parent-consumer activist group also charges that they have obtained evidence through the Freedom of Information Act that the Department of Health and Human Services (DHHS) is failing to properly monitor reports of death and injuries following vaccination and that doctors around the country are failing to report to DHHS deaths and injuries which occur after vaccination.

In a year-long investigation of the Vaccine Adverse Reaction Reporting System (VAERS) operated by the Food and Drug Administration, NVIC/DPT analyzed VAERS computer discs used by the FDA to store data on reports of deaths and injuries following DPT vaccination. A total of 54,072 reports of adverse events following vaccination were listed in a 39-month period from July 1990 to November 1993 with 12,504 reports being associated with DPT vaccine, including 471 deaths.

A wide variation in the numbers of reports associated with different lots of DPT vaccine were discovered, with some lots listing many more deaths and injuries than others. In one DPT vaccine lot, there were 129 adverse events and nine deaths reported between September 1992 and September 1993. Most adverse events occurred within a few days of vaccination and many reports also contained descriptions of classic pertussis [whooping cough] vaccine reaction symptoms. This particular lot met the FDA's criteria for triggering an "investigation" (i.e., report of one death or two serious injuries within a seven-day period) 11 times within a 12-month period.

"There are some lots of vaccine which are associated with many more

From the *Campaign Against Fradulent Medical Research Newsletter*, 1994.

deaths and injuries than other lots. These lots are often referred to as 'hot lots'. Even though the FDA's criteria for an investigation were triggered 11 times within a 12-month period on just one of the many lots we looked at, we know for a fact the lot was never recalled. The FDA has not recalled a suspicious lot of DPT vaccine because of high numbers of deaths and injuries associated with it for at least 15 years," said Kathi Williams, NVIC/DPT Cofounder and Acting Director. "That is because the position of those who operate VAERS is that the DPT vaccine does not cause death or injury. So the death and injury reports are ignored. It is a shocking example of how little we know about the true extent of vaccine-associated injuries and deaths."

The NVIC/DPT investigation was featured on the 2nd March 1994 NBC News "Now with Tom Brokaw and Katie Couric" show. At the end of February 1994, NVIC/DPT also conducted a survey of 159 doctors' offices in seven states, including Arkansas, California, Georgia, Illinois, Maryland, New York, and Texas. When asked the question, "In case of an adverse event after vaccination, does the doctor report it and, if yes, to whom?" only 28 out of 159, or 18 percent said they make a report to the FDA, CDC or state health department. In New York, only one out of 40 doctors' offices confirmed that they report a death or injury following vaccination.

"This shameful record of gross underreporting of adverse events following vaccination by doctors around the country coupled with the shameful cover-up of vaccine-associated deaths and injuries by the federal government is an example of why more and more parents are losing faith in the mass vaccination system. Many times our organization must help parents report their children's vaccine-associated death or injury because their doctor refuses to make a report," said Barbara Loe Fisher, NVIC/DPT Co-founder and President. "Parents are legally required to vaccinate their children. Doctors should be forced to live up to their legal duty to report, and DHHS should be forced to live up to its responsibility to seriously investigate every vaccine-associated death and injury and, especially, to identify and recall lots of DPT vaccine associated with high numbers of deaths and injuries."

In November 1991, while Fisher was a member of the National Vaccine Advisory Committee operated by the DHHS, she presented the Committee with a detailed summary of the stories of 90 families who had reported vaccine-associated deaths and injuries to NVIC/DPT. Most of the 90 families with children or grandchildren who had suffered deaths and injuries following DPT vaccination, said that their doctors refused to make a vaccine adverse event report to the DHHS. NVIC/DPT had to help the families make the report to DHHS.

Upon analysis of the VAERS computer discs, NVIC/DPT discovered that some of the deaths and injuries which NVIC/DPT had helped families report to DHHS were either (1) never recorded in the VAERS computer system, or (2) recorded but the information was inaccurate, or (3) not adequately followed up.

The IOM recently released a report that stated: "The committee concludes that the balance of evidence is consistent with a causal relation between DPT and the forms of chronic nervous system dysfunction described in the NCES in those children who experience a serious, acute neurologic illness within seven days after receiving DPT vaccine." NVIC/DPT has maintained for more than a decade that children can suffer permanent damage and die after suffering a neurological complication following DPT vaccination, and has always cited the validity of the data from the British National Childhood Encephalopathy Study (NCES) which was published in the early 1980s and upheld by IOM in their newest report.

The National Vaccine Information Center (NVIC) operated by Dissatisfied Parents Together (DPT) is a national, nonprofit organization located in Vienna, Virginia, USA. Founded in 1982, NVIC/DPT represents parents and health care professionals concerned about childhood diseases and vaccines and is dedicated to preventing vaccine deaths and injuries through education, and working to obtain the right of all citizens to make informed, independent vaccination decisions.

In the mid-1980s DPT worked with Congress, physician organizations, DHHS and vaccine manufacturers to create the National Childhood Vaccine Injury Act of 1986, which set up the nation's first vaccine injury compensation program and also mandated that doctors give parents information on vaccine benefits and risks, record vaccine lot numbers, and record and report to DHHS deaths and injuries following vaccinations.

6 AIDS and Ebola: Where Did They Really Come From?

Dr. Leonard G. Horowitz

During the 1960s and early 1970s the World Health Organization func-tioned as the omnipotent supplier and standardizing authority of the world's experimental pharmaceuticals. In the field of virology, the United States Public Health Service (USPHS) and National Institutes of Health (NIH) directed the National Cancer Institute (NCI) to become, along with the Centers for Disease Control and Prevention (CDC) the WHO's chief dis-tributor of viruses and antiviral vaccines. The WHO Chronicle noted by 1968—ten years into the WHO's viral research program—"WHO virus reference centres" had served as authorized technical advisors and suppli-ers of "prototype virus strains, diagnostic and reference reagents (e.g., antibodies), antigens, and cell cultures" for more than "120 laboratories in 35 different countries." Within a year of this announcement, this number increased to "592 virus laboratories . . . [and] only 137 were outside Europe and North America." Over these 12 months, the NCI and CDC helped the WHO distribute 2,514 strains of viruses, 1,888 ampoules of antisera mainly for reference purposes, 1,274 ampoules of antigens, and about 100 samples of cell cultures. More than 70,000 individual reports of virus isolations or related serological tests had been transmitted through the WHO-NCI network.[1,2]

At the NCI in Bethesda, Maryland, from the late 1960s to the present, the chief retrovirus research laboratory was associated with the Depart-ment of Cell Tumor Biology, and chaired by Dr. Robert Gallo—an esteemed member of the National Academy of Sciences (NAS) who was hailed by the U.S. Secretary of Health and Human Services Margaret Heckler in 1984 as the discoverer of the AIDS virus, HTLV-III. LAV, identical to HTLV-III had been isolated by Montagnier's French team and allegedly forwarded to Gallo in 1983.[3]

MILITARY ORDERS FOR AIDS-LIKE VIRUSES:
THE BIOLOGICAL WEAPONS CONTRACTORS

As early as 1970, the U.S. Department of Defense (DOD) appropriated at least $10 million to "initiate an adequate program through the National Academy of Sciences-National Research Council (NAS-NRC)" to make a new infective microorganism which could differ in certain important aspects from any known disease-causing organisms. Most important of these [aspects] is that it might be refractory to the immunological and therapeutic processes upon which we depend to maintain our relative freedom from infectious disease. Members of the NAS-NRC had instructed scientific leaders in the DOD this work might be accomplished "within 5 years."[4] This research was then carried out by American defense contractors despite the authorization and signing of the Geneva accord by Dr. Henry Kissinger and President Nixon outlawing the production and testing of such biological weapons.[5]

Also in 1970 Gallo and his co-workers presented research describing the experimental entry of bacterial ribonucleic acid (RNA) into human white blood cells (WBCs) before a special symposium sponsored by NATO.[6] The paper published in the Proceedings of the National Academy of Sciences discussed several possible mechanisms prompting the "entry of foreign nucleic acids" into lymphocytes—the cells principally attacked by HIV. Prior to this, Gallo et al. had published studies identifying:

1. mechanisms responsible for reduced amino acid and protein synthesis by T-lymphocytes required for immunosuppression;[7]

2. specific enzymes required to produce such effects along with a "base pair switch mutation" in the genes of WBCs to create immune system dysfunction;[8] and

3. methods by which WBC "DNA degradation" and immune system decay may be prompted by the "pooling" of purine bases and/or the addition of specific reagents.[9]

Subsequent studies published in 1970 by Gallo and co-workers identified "RNA dependent DNA polymerase" (i.e., the unique AIDS-linked enzyme, reverse transcriptase) responsible for "gene amplification . . . biochemical cytodifferentiation," (i.e., the development of unique WBC characteristics including cancer cell production) and "leukaemogenesis";[10] and identified L-Asparaginase synthetase—a key enzyme that, if repressed, will induce treatment resistant leukemias and other cancers.[11]

The year following the $10 million appropriation by the DOD for AIDS-like biological weapons research, the NCI acquired the lion's share of the facilities at America's premier biological weapons testing center, Fort Detrick in Frederick, Maryland.[12] Perhaps not coincidentally, the Cell Tumor Biology Laboratory's output increased in 1971 as measured by the publication of eight scientific articles by Gallo and co-workers compared to at most four in previous years. These reports included Gallo's discovery that by adding a synthetic RNA and feline (i.e., cat) leukemia virus (FELV) "template" to "human type C" viruses (associated with cancers of the lymph nodes), the rate of DNA production (and subsequent provirus synthesis) increased as much as thirty times. The NCI researchers reported that such a virus may cause many cancers besides leukemias and lymphomas including sarcomas.[13]

In this 1971 report Gallo et al. also reported modifying simian (i.e., monkey) viruses by infusing them with cat leukemia RNA to make them cause cancers as seen in AIDS patients.[13]

Furthermore, Fujioka and Gallo concluded from studies conducted in late 1969 or early 1970 that they would need to further "evaluate the functional significance of tRNA changes in tumor cells," by designing an experiment in which "specific tumor cell tRNAs" would be "added directly to normal cells." They explained that one way of doing this was to use viruses to deliver the foreign cancer producing tRNA to normal cells. The viruses which were then employed to do this, the researchers noted, were the simian virus (SV40) and the mouse parotid tumor (polyoma) virus.[14]

Such experiments clearly advanced immunodeficiency virus technology and even provided a model for the development of HIV, the AIDS virus—allegedly of simian virus descent—which similarly delivered unique enzymes and a foreign RNA to normal cells necessary to cause an acquired immunodeficiency syndrome in animals and humans.

DEVELOPING MORE AIDS-LIKE VIRUSES

In 1972, Gallo, his superiors and inferiors studied portions of simian viruses to determine differences in RNA activity between infected versus uninfected cancer cells, and whether the differences could be ascribed to the infection and related DNA alterations.[15] They stated that "by studying viral or cellular mutants or cell segregants . . . which have conditional variations in virus-specific cellular alterations, it should be possible to more precisely determine the biological significance of the RNA variation reported here."

Clearly, the group was working to determine the relevance of various viral genes on the development of human cancers and immune system collapse. They reported their desire to use this information to find a cure for cancer, but at this time their activity was more focused on creating various cancers as well as new carcinogenic viruses which could infect humans.

For example, Smith and Gallo published another National Academy of Sciences paper examining DNA polymerase (i.e., reverse transcriptase) activity in immature normal versus acute leukemic lymph cells. To do so, they evaluated the single stranded 70S RNA retrovirus found in chickens which causes prominent features of AIDS including WBC dysfunction, sarcomas, progressive wasting and death.[16] Borrow, Smith and Gallo injected this chicken virus RNA into human WBCs to determine if the cells were prompted to produce proteins (including new viruses) encoded by the viral RNA.[17] Robert, Smith and Gallo also evaluated the neoplastic effects of single stranded 70S RNA reverse transcriptase delivered by the cat leukemia virus (FELV) and the mason-Pfizer monkey virus on normal human lymphocytes (NHL).[18]

This work foreshadowed the observation made ten years later by the CDC's chief AIDS researcher, Dr. Donald Francis who noted the "laundry list" of feline leukemia-like diseases associated with AIDS.[3]

Other examples are detailed by Gallo and co-workers while discussing their adapting monkey, rat, and bird leukemia and tumor viruses for experimental use in a human (NC-37) cell line.[19] Wu, Ting and Gallo[20] discussed the synthesis of new RNA tumor viruses "induced by 5-iodo-2'-deoxyuridine, IdU (a constituent of RNA) in rodent cell cultures, and noted that chemotherapy might be used to halt the reverse transcriptase-linked viral reproduction.

However, had HIV been synthesized for military purposes from various species components, it would be difficult, if not impossible, to prove. As Gillespie, Gillespie and Gallo et al. noted in 1973 concerning the origin of the RD114—another cat/human bioengineered virus—"it can always be argued that" a virus which naturally jumped species (as HIV is alleged to have done from the monkey) would be expected to have antigens that differ "from the antigen found on the viruses of known" origin.[21]

LITTON BIONETICS: GALLO'S LINK TO THE DOD

Four years later, during a U.S. Senate investigation of illegal "biological testing involving human subjects by the Department of Defense," Senators learned that Bionetics, Bionetics Research Laboratories, and Litton Bionetics—an organization which, along with the NCI, administered and provided Dr. Gallo's research funding[10, 13, 15, 17-19, 22, 26] were not only acknowledged DOD biological weapons contractors, but their affiliated Litton Systems, Inc., was among the most frequently contracted institutions involved in biological weapons research and development between 1960 and 1970 (the end of the reported period).[23] Additional biological weapons contractors with whom Dr. Gallo and/or his co-workers associated during the late 1960s and early 1970s included the University of

Chicago,[24] Texas,[13] Virginia,[25] and Yale,[17] Merck and Co. Inc.,[20] and Hazelton Laboratory, site of the famous 1989 Marburg-Ebola-like (Reston) virus outbreak.[22]

NCI staff reports revealed that Litton Bionetics had been granted the service contract to supply all NCI researchers, worldwide, with virtually every primate cancer research material requested, including seed viruses, viral hybrids, cell lines, experimental reagents, and African colony born monkeys including M. mulatta and C. aethiops which were associated with the major monkey AIDS virus outbreaks in California's Davis Lab, and the 1967 Marburg virus outbreaks in three European vaccine production facilities.[15–17]

Litton Bionetics chief John Landon reported an experiment begun in 1965 when he inoculated 18 monkeys with rhaabdovirus simian—a rabies virus known to cause Ebola-like hemmorhagic fever in monkeys. "Nine [monkeys] died or were transferred," to allied laboratories or vaccine production facilities. This shipment was likely to have started the first hemmorhagic fever "Marburg virus" outbreak among European vaccine laboratory workers in 1967. As noted by the world's leading simian virus expert at the time, Dr. Seymour Kalter, the Marburg virus was apparently manmade.[23–25]

In fact, Litton Bionetics, the chief military and industrial supplier of primates for cancer virus experimentation during the 1960s and early 1970s, also maintained the colony of the specific genus of monkeys associated with all of the major monkey AIDS virus outbreaks in the United States.[23]

Through telephone interviews with Litton Bionetics and MedPath administrators, I learned that Bionetics Research Laboratories had been sold to MedPath Corporation—one of America's largest medical and blood testing laboratories—a division of Dow Corning. Dow Corning's parent, Dow Chemical Company, was also listed among the Army's chief biological weapons contractors during the 1960s and early 1970s.[26,27] Litton Bionetics, a subsidiary of Litton Industries, Inc. remains in business as a proprietor of the Frederick Cancer Research Center, a "privately owned, government contracted" facility. Bionetics, it was noted, currently acts as an agency under "contract to manage and operate the Frederick (Md.) [Fort Detrick affiliated] Cancer Research Center for the National Cancer Institute."[27] Besides administering research grants and government funds earmarked for the NCI, Litton Bionetics also developed a division which produced and marketed test kits for bloodborne, infectious diseases including mononucleosis, hepatitis B, and AIDS.[28] This division was sold to Organon Teknica in 1985.[26]

Apparently, the military-medical-industrial complex was well aware of Litton Industries' service as a DOD and NATO contractor. In 1978, the company was indicted for filing false claims for $37 million in cost-over-runs during the building of three nuclear submarines under one of several multi-billion dollar defense contracts.[29-34]

ZAIRE AND ANGOLA: A CIA MILITARY ARENA

Between 1970 and 1975, the period the NAS-NCR scientific advisors informed DOD decision-makers that AIDS-like viruses could be readied,[4] American cold war efforts focused on Zaire and Angola.[33-35] Following the withdrawal of American forces in Vietnam, Secretary of State Henry Kissinger ordered the CIA to begin a major covert military operation against MPLA (communist bloc backed) "rebels" in Angola.[36,37] Zaire, indebted by over $4.5 billion to the International Monetary Fund, and headed by President Mobutu—paradoxically regarded as one of the world's wealthiest men with "a personal fortune put at $2,939,200,000 [1984 estimate] banked in Switzerland—was wooed by NATO allies during the 1970s (principally the U.S.) to be a staging area for CIA backed, Portuguese, French, and mainly South African mercenaries.[38,39] "American corporate investment, notably in copper and aluminum, doubled to about $50 million following a 1970 visit by Mobutu to the United States. Major investors included Chase-Manhattan, Ford, General Motors, Gulf, Shell, Union Carbide, and several other large concerns."[43]

However, in 1975 Mobutu apparently turned against NATO allies and increased negotiations with China and Russia[40,41] He proclaimed his intention to nationalize foreign owned enterprises.[42,43] In June 1975, following the CIA's thwarted efforts to convince the U.S. Congress to appropriate more funds for Mobutu and the Angola programme (A total of $31.7 million had already been "drawn from the CIA's FY 75 contingency fund" which was "exhausted on 27 November 1975").[39] Mobutu expelled the American ambassador and arrested many of the CIA's Zairian agents, placing some under death sentences.[40,41]

The following year, in October 1976, the "Ebola Zaire virus" broke-out in "fifty five villages surrounding the [Yambuku] hospital" first killing "people who had received injections." Mobutu then ordered his army to "seal off the Bumba zone with roadblocks" and "shoot anyone trying to come out" so "no one knew what was happening, who was dying, [or] what the virus was doing."[44]

Shortly thereafter, Ebola victim specimens were sent to the CDC, Special (meaning "secret" within the American intelligence community) Pathogens Branch; to Porton, England's controversial chemical and biological weapons (CBW) laboratories;[45] and teams of WHO and CDC researchers were dispatched to the Ebola region in Mobutu's private, American supplied C-130 Buffalo troop-transport plane.[44]

By the end of 1976, the Zairian leader had reconciled his differences with the American intelligence and corporate communities believing that Zaire would continue to reap his non-communist allies' social and economic aid. On April 4, 1977, Mobutu suspended diplomatic relations with Cuba; on April 21, reduced ties with the Soviet Union; and on May 2, he cut ties with East Germany.[46] Meanwhile, according to John Stockwell, Former Chief of the CIA's Angola Task Force, "the United States was exposed, dishonored and discredited in the eyes of the world," as "15,000 Cubans were installed in Angola"[40] despite the CIA's best efforts and a continuing policy of lying "to State Department officials, Congressmen, American press, and world public opinion in varying degrees, depending on the need" about the CIA's covert military campaign in Angola and Zaire.[39]

Throughout 1976 and 1977, Mobutu, NATO, and the CIA, constrained by the Tunney, Cranston and Clark amendment which prevented the expenditure of American funds in Zaire/Angola, except to gather intelligence,[39] remained embroiled in the "Shaba rebellion" against allegedly Russian-backed "Katangan rebels."[46]

At the same time, perhaps not coincidentally, NATO ally West Germany was pouring financial aid into Zaire and white ruled South Africa.[47–49] The Northeast region of Zaire, believed to be the epicenter of the AIDS epidemic, was specifically targetted for West German economic aid and industrialization.[50]

The "long tradition of friendship between Germans and South Africans" London's African Development noted, "dates back from the first waves of white immigrants to South Africa and the feeling of solidarity between Germans and Afrikaners during the Boer War, continuing throughout the First World War and then the Second World War. Many of today's Nationalist leaders in Pretoria," the paper reported, "were Nazi sympathizers: many ex-Nazis settled in South Africa after 1945 . . ."[48]

THE WEST GERMAN COMPANY OTRAG

Not surprising then, on March 26, 1976 Mobutu signed what the British Broadcasting Company (BBC) reported was a "secret agreement" with the West German company OTRAG (Orbital Transport-und Raketen-Aktiengesellschaft) to lease 260,000 square kilometers—essentially the complete Kivu province—for military/industrial purposes for the sum of DM 800,000,000 (approximately $250 million at that time).[51–54] The contract "made OTRAG sovereign over territories once inhabited by 760,000 people"[54] in the Eastern and South-central portion of Zaire. The leased property positioned North of the Shaba militarized region, expanded Southeast along the Congo River—the countries main waterway, and south of the Kinshasa highway—better known as the "AIDS Highway"—

which runs through Zaire's Northeast corridor and across central Africa bypassing the "Isle of Plaques," Mount Elgon and Kitum Cave allegedly by the author of *The Hot Zone* to be the breeding ground for the Ebola, Marburg, and AIDS viruses.[44]

OTRAG was apparently authorized to conduct any "excavation and construction; including air fields, energy plants, communication systems, and manufacturing plants. All movement of people into and within the OTRAG territory was only with permission of OTRAG, [which was] absolved from any responsibility for damage caused by construction. Its people enjoy complete immunity from the laws of Zaire in the granted territory, until the year 2000."[56]

Believed to be of military intelligence gathering significance to NATO[46-48, 55, 56] OTRAG and its principals were traced back to the West German government and "to those Nazi scientists who worked on V1 and V2 rockets during World War II. For example, Dr. Kurt H. Debus, Chairman of the Board of OTRAG, once worked at the Peenemude V2 program and later, until 1975, worked as director of the Cape Canaveral (now Cape Kennedy) space program. Richard Gompertz, OTRAG's technical director and a U.S. citizen, once was a specialist on V2 engines and later presided over NASA's Chrysler space division. Lutz Thilo Kayer, OTRAG's founder and manager, when young was quite close to the Nazi rocket industry, often called 'Dadieu's young man,' a reference to Armin Dadieu, his mentor, who served as prominent SS officer and as Gorling's special representative for a research program on storing uranium. While working for OTRAG Kayser also acted as a contact for the West German government, a special advisor to the Minister of Research and Technology on matters concerning OTRAG. He was also on the ad hoc committee on . . . [America's] Apollo program."[56]

In 1979, under pressure from the Soviet Union and Zaire's neighboring countries, Mobutu announced OTRAG would "halt its rocket testing" program. It was clear, however, strong diplomatic, military, and economic ties between West Germany, the U.S. and Zaire continued.[49,57]

COLD WAR PROPAGANDA VERSUS THE HARD CORE FACTS

According to the latest United States Army (USA) report, the "outlandish claim" that the AIDS virus was developed as a biological weapon for the Pentagon was communist propaganda "disavowed in 1987 by then Soviet President Mikhail Gorbachev, who apologized to President Ronald Reagan for the accusation."[58] However, more recently, the high ranking Soviet press official Boris Belitskiy, offered an alternative account regarding the origin of such "propaganda."[59]

"Several U.S. Administration officials such as USIA [CIA] Director Charles Wick, have accused the Soviet Union of having invented this theory for propaganda purposes. But actually it is not Soviet scientists at all who first came up with this theory," Belitskiy reported. "It was first reported in Western journals by Western scientists, such as Dr. John Seale, a specialist on venereal diseases at two big London hospitals.

"Just recently a Soviet journalist in Algeria, Aleksandr Zhukov, managed to interview a European physician at the Moustapha Hospital there, who made some relevant disclosures on the subject. In the early seventies, this physician, an immunologist, was working for the West German OTRAG Corporation in Zaire. His laboratory had been given the assignment to cultivate viruses ordinarily affecting animals but constituting a potential danger to man. They were particularly interested in certain unknown viruses isolated from the African green monkey, and capable of such rapid replication that they could completely destabilize the immune system. These viruses, however, were quite harmless for human beings and the lab's assignment was to develop a mutant virus that would be a human killer."[7-11, 13-22]

"Did they succeed?" the announcer asked. "To a large extent, yes," Belitskiy replied. "The lab was ordered to wind up the project and turn the results over to certain U.S. researchers, who had been following this work with keen interest, to such an extent that some of the researchers believed they were in reality working not for the West German OTRAG Corporation but for the Pentagon."[59]

Two years after Belitskiy's announcement, in 1991, Dr. Jacobo Segal, professor emeritus of Berlin's Humboldt University told the international press that the Pentagon theory of AIDS made sense. He alleged the virus was likely developed "through gene technology" as a result of Pentagon sponsored animal research, "to permit the attack on human immune cells." Furthermore, he reported this theory is "supported by many European scientists and has not been refuted."[60]

In 1977, at the height of OTRAG's Zairian missile testing phase, Litton Industries units received contracts worth: $5 million for medical electronic equipment from its Hellige division, in Freiburg, West Germany;[61] $19.8 million worth of missile fire-control equipment from the Army;[62] $32.9 million for electronic reconnaissance sensor equipment from the Air Force;[63] and in 1978 $11.3 million worth of computerized communications systems for NATO.[64] Some of these military supplies may have been earmarked for OTRAG.[65]

Moreover, given the "cooperation between NATO and the World Health Organization with regard to the control and regulation of the international exchange of pharmaceutical products, [and] the possible necessity of facing the dangers created by the use of chemical or bacteriological weapons"[66] it appears noteworthy that the outbreaks of the world's most feared and deadly viruses Marburg, Ebola, Reston, and AIDS—share the dubious distinction of breaking out in or around zones of U.S. and West German, NATO-allied, military experimentation at the height of the cold war; increased political/military interest in Central Africa, and a burgeoning of WHO and NCI contracts for the supply of simians for "defensive" research.[44]

AIDS, GAYS, BLACKS AND THE CIA

It is also noteworthy that in 1975, five years following the signing of the Geneva accord by Nixon, congressional records revealed that the USPHS through the Special (i.e., covert) Operations Division of the Army continued to supply biological weapons including deadly neurotoxins and viruses to the CIA which illegally stored them in their Fort Detrick facility for future unsanctioned uses.[67] Though records of who initiated and directed this covert activity were destroyed by the CIA, along with the famous Watergate tapes,[68] Mr. Nathan Gordon, Former Chief of the CIA's Chemistry Branch, Technical Services Division confessed knowledge that some of the stored substances were to be used to "study immunization methods for diseases vis-a-vis—who knows, cancer."[69] Furthermore, following Nixon's resignation, President Ford and Henry Kissinger were made aware that the CIA maintained a residual supply of biological weapons, but neither ordered their destruction according to testimony provided by Former CIA Director Richard Helms.[70]

In subsequent congressional hearings before the Senate Subcommittee on Health and Scientific Research, it was revealed that George W. Merck, "of the prominent Merck pharmaceutical firm," directed America's biological weapons manufacturing industry for decades following World War II.[71] Merck & Company, Inc. was also listed among the DOD biological weapons contractors.[72] Merck, Sharp and Dohme provided major financial support for the earliest hepatitis B vaccination studies conducted simultaneously in Central Africa and New York City during the early 1970s. Several authorities have argued these vaccine trials might best explain the unique and varying epidemiological patterns of HIV/AIDS transmission between the U.S. and Africa. "The vaccine was prepared in the laboratories of the Department of Virus and Cell Biology Research, Merck Insti-tute for Therapeutic Research, West Point, Pennsylvania. The placebo, [was] also prepared in the Merck Laboratories."[72]

During the holocaust, Nazi scientists assayed non-Arian blood to determine race specific disease susceptibilities. Blacks and homosexuals, along with Jews, were persecuted by the Nazis. Over 10,000 gay men were murdered.[3]

Similarly, U.S. intelligence agencies have been targeting blacks and gays for assassinations, harassment, illegal wire taps, and counterintelligence campaigns from the McCarthy era in the 1950s through the Reagan era in the 1980s. American black and gay civil rights groups and their leaders were considered communist threats during the cold war years—particularly during the late 1960s and early 1970s when Nixon, Kissinger, and Hoover supported COINTELPRO funding for covert FBI and CIA activities aimed at neutralizing all such domestic and foreign black and homosexual threats.[74–78]

The use of Third World people and American blacks and prisoners for unconscionable pharmaceutical experimentation and covert economic, social, and environmental exploitation by the U.S. and other western countries has been repeatedly alleged by reputable sources.[79]

About the author: Leonard G. Horowitz, D.M.D., M.A., M.P.H. is a Harvard graduate and internationally known public health education authority. One of healthcare's most captivating motivational speakers, Dr. Horowitz is a prolific author with ten books and more than eighty published articles to his credit. This article is based on Dr. Horowitz's new and most controversial book *Emerging Viruses: AIDS and Ebola—Nature, Accident or Genocide?* (Tetrahedron, Inc., 1996; 592 pp.; Available through Adventures Unlimited Press, P.O. Box 74, Kempton, IL. 60946; $29.95). Please direct lecture requests to Dr. Horowitz, care of Tetrahedron, Inc., a nonprofit educational corporation, P.O. Box 402, Rockport, MA 01966; (800) 336-9266; Fax: (508) 546-9226; e-mail: tetra@tetrahedron.org.

For more information about his research link to URL #http://www.tetrahedron.org.

REFERENCES

1. World Health Organization Report. Five years of research on virus diseases. WHO Chronicle 1969;23;12:564–572; Communicable diseases in 1970: Some aspects of the WHO programme. *WHO Chronicle* 1971;25;6:249–255; see also: Mathews AG. Who's influence on the control of biologicals. *WHO Chronicle* 1968;23;1:3–15.

2. World Health Organization Report. Recent work on virus diseases. *WHO Chronicle* 1974 28:410–413.

3. Shilts R. *And the Band Played On: Politics, People and the AIDS Epidemic*. New York: Penguin Books, 1987, pp. 450–453.

4. Department of Defense Appropriations for 1970. Hearings Before a

Subcommittee on the Committee on Appropriations House of Representatives, Ninety-First Congress, Part 5 Research, Development, Test, and Evaluation, Dept. of the Army. Tuesday, July 1, 1969, p. 79. Washington: U.S. Government Printing Office, 1969, p. 79 of the public record and page 129 of the classified supplemental record obtained through the Freedom of Information Act.

5. Washington Correspondent. Gas and germ warfare renounced but lingers on. *Nature* 1970; 228;273:707–8.

6. Herrera F, Adamson RH and Gallo RC. Uptake of transfer ribonucleic acid by normal and leukemic cells. Proceedings of the National Academy of Sciences. 1970;67;4:1943–1950.

7. Gallo RC, Perry S and Breitman RT. The enzymatic mechanisms for deoxythymidine synthesis in human leukocytes. *Journal of Biological Chemistry* 1967;242;21:5059–5068.

8. Gallo RC and Perry S. Enzymatic abnormality in human leukemia. *Nature* 1968;218:465–466.

9. Gallo RC and Breitman TR. The enzymatic mechanisms for deoxythymidine synthesis in human leukocytes: Inhibition of deoxythymidine phosphorylase by purines. *Journal of Biological Chemistry* 1968;243;19:4943–4951.

10. Gallo RC, Yang SS and Ting RC. RNA dependent DNA Polymerase of human acute leukaemic cells. *Nature* 1970;227:1134–1136.

11. Gallo RC and Longmore JL. Asparaginyl-tRNA and resistance of murine leukaemias to L-asparaginase. *Nature* 1970;227:1134–1136.

12. Washington Correspondent. Biological warfare: Relief of Fort Detrick. *Nature* Nov. 28, 1970;228:803.

13. Gallo RC, Sarin PS, Allen PT, Newton WA, Priori ES, Bowen JM and Dmochoowski L. Reverse transcriptase in type C virus particles of human origin. *Nature New Biology* 1971;232:140–142.

14. Fujioka S and Gallo RC. Aminoacyl transfer RNA profiles in human myeloma cells. *Blood* 1971;38;2:246–252.

15. Gallaher RE, Ting RC and Gallo RC. A common change aspartyl-tRNA in polyoma and SV transformed cells. *Biochimica Et Biophysica Acta* 1972;272:568–582.

16. Smith RG and Gallo RC. DNA-dependent DNA polymerases I and II

from normal human-blood lymphocytes. Proceedings of the National Academy of Sciences 1972;69;10:2879–2884.

17. Bobrow SN, Smith RG, Reitz MS and Gallo RC. Stimulated normal human lymphocytes contain a ribonuclease-sensitive DNA polymerase distinct from viral RNA directed DNA polymerase. Proceedings National Academy of Sciences 1972;69;11:3228–3232.

18. Robert MS, Smith RG, Gallo RC, Sarin PS and Abrell JW. Viral and cellular DNA polymerase: Comparison of activities with synthetic and natural RNA templates. *Science* 1972;69;12:3820–3824.

19. Gallo RC, Abrell JW, Robert MS, Yang SS and Smith RG. Reverse transcriptase from Mason-Pfizer monkey tumor virus, avian myeloblastosis virus, and Rauscher leukemia virus and its response to rifamycin derivatives. *Journal of the National Cancer Institute* 1972;48;4:1185–1189.

20. Wu AM, Ting RC, Paran M and Gallo RC. Cordycepin inhibits induction of murine leukovirus production by 5-iodo-2'-deoxyuridine. Proceedings of the National Academy of Sciences 1972;69;12:3820–3824.

21. Gillespie D, Gillespie S, Gallo RC, East J and Dmochowski L. Genetic origin of RD114 and other RNA tumor viruses assayed by molecular hybridization. *Nature New Biology* 1973;224:52–54.

22. Wu AM, Ting RC, and Gallo RC. RNA-Directed DNA Polymerase and virus-induced leukemia in mice. Proceedings of the National Academy of Sciences 1973;70;5:1298–1302.

23. NCI staff. The Special Virus Cancer Program: Progress Report #8 [and #9]. Office of the Associate Scientific Director for Viral Oncology (OASDVO). J.B. Moloney, Ed., Washington, D.C.: U.S. Government Printing Office, 1971 [and 1972]. (The University of North Carolina, Chapel Hill, Government Documents Department Depository, Reference # HE20.3152:v81.) pp. 15–19, 20–26, 187–188; 273–289; [and in 1972 Progress Report #9, pp. 195–196, 326].

24. Fine DL and Arthur LO. Prevalence of natural immunity to Type-D and Type-C Retroviruses in primates. In: *Viruses in Naturally Occurring Cancers*: Book B. Myron Essex, George Todaro and Harold zun Hausen, eds., Cold Spring Harbor, NY: Cold Spring Harbor Laboratory, 1980, Vol. 7, pp. 793–813; see also Gallo RC, Wong-Staal F, Marhkam PD, Ruscetti R, Kalyanamaraman VS, Ceccherini-Nelli L,

Favera RD, Josephs S, Miller NR and Reitz, Jr. MS. Recent studies with infectious primate retroviruses: Hybridization to primate DNA and some biological effects on fresh human blood leukocytes by simian sarcoma virus and Gibbon ape leukemia virus. Ibid., 793–813.

25. Simmons ML. Biohazards and Zoonotic Problems of Primate Procurement, Quarantine and Research: Proceedings of a Cancer Research Safety Symposium. March 19, 1975, Conducted at the Frederick Cancer Research Center, Frederick, Maryland. DHEW Publication No. (NIH) 76–890, pp. 27;50–52.

26. Committee on Human Resources, United States Senate. Hearings before the Subcommittee on Health and Scientific Research, Biological Testing Involving Human Subjects by the Department of Defense, 1977: Examination of Serious Deficiencies in the Defense Departments Effort to Protect the Human Subjects of Drug Research. Washington, D.C.: U.S. Government Printing Office, May 8 and May 23, 1977, pp. 80–100; for the Army's list of biological weapons contractors for 1959, see: Department of Defense Appropriations For 1970: Hearings Before a Subcommittee of the Committee on Appropriations House of Representatives, Ninety-first Congress, First Session, H.B. 15090, Part 5, Research, Development, Test and Evaluation of Biological Weapons, Dept. of the Army. U.S. Government Printing Office, Washington, D.C., 1969, p. 689.

27. Personal conversation with administrator at Frederick Cancer Research Center/NCI who wished to remain anonymous. 301–846–1000.

28. Litton Industries, Inc., 360 North Crescent Drive, Beverly Hills, CA. 90210, 25th Annual Report Fiscal 1978. pg. 2; see also: Staff reporter. Litton, Saudis Agree on a system valued above $1.5 billion. *Wall Street Journal*, Monday Oct. 9, 1978, p. 16.

29. Staff reporter. Litton and Navy Settle dispute. *Wall Street Journal*, Wednesday, June 21, 1978, p. 4.

30. Staff reporter. Suit against Litton Industries involving Navy job dismissed. *Wall Street Journal*, Thursday, May 26, 1977, p. 15.

31. Staff reporter. Suit against Litton may be renewed says U.S. appeals court. *Wall Street Journal*, April 7, 1978, p. 12.

32. Staff reporter. Court lets stand an indictment of Litton unit. *Wall Street Journal*, Tuesday, Oct. 3, 1978, p. 4.

33. Agee P and Wolf L. *Dirty Work: The CIA in Western Europe.* Secaucus, NJ: Lyle Stuart, Inc. 1977.

34. Kumar S. *CIA and the Third World: A Study in Crypto-Diplomacy.* New Delhi: Vikas Publishing House PVT LTD., 1981.

35. Agee P. Dirty Work-2: *The CIA in Africa.* Secaucus, NJ: Lyle Stewart, Inc., 1979.

36. Molteno R. Hidden sources of subversion. In: *Dirty Work-2: The CIA in Africa.* Secaucus, NJ: Lyle Stewart, Inc., 1979. pp. 100–101.

37. Woodward, B. *VEIL: The Secret Wars of the CIA 1981–1987.* New York: Simon and Schuster, 1987, pg. 268. According to Woodward, "CIA ties with [Zaire's president] Mobutu dated back to 1960, the year the CIA had planned the assassination of the Congolese nationalist leader Patrice Lumumba."

38. Freemantle B. *CIA.* Briarcliff Manor, NY: Scarborough House/Stein and Day Publishers. 1983, pp. 184–185.

39. Kumar S. Op cit., p. 72;74–76;91–92.

40. Stockwell J. *In Search of Enemies: A CIA Story.* New York: W.W. Norton & Company, 1978, pp. 43–44;248.

41. Colby W. *Honorable Men: My Life in the CIA.* New York: Simon and Schuster, 1978, pp. 439–40.

42. Daily News, Dar es Salaam. Zaire: Mobutu assails government functionaries. In: *Africa Diary*, January 22–28, 1975, pg. 7287.

43. West Africa, London. Zaire: Mobutu and the Americans. In: *Africa Diary*, February 19–25, 1975, pg. 7322.

44. Preston R. *The Hot Zone: A Terrifying True Story.* New York: Random House, 1994, pp. 10;25–26;71;78;79;84.

45. Staff writer. Porton opened to the public. *Nature* 1968;220:426.

46. Times, London; West Africa, London; Le Monde, Paris; New York Times; Daily News, Dar es Salaam, Times of India, New Delhi. Zaire: Shaba rebellion Virtually Crushed. In: *Africa Diary*, June 18–24, 1977, pp. 8536–8538.

47. South African Digest, Pretoria. South Africa: Documents on NATO, S. African Co-operation. In: *Africa Diary*, June 25–July 1, 1975, pp. 7488–7489.

48. African Development, London. South Africa: W. Germany may become number one trading partner. In: *Africa Diary*, April 30-May 6, 1975, pp. 7418–7419.

49. Zaire Radio. Zaire: W. German financial aid. In: *Africa Diary*, July 23–29, 1978, pp. 9103–7419.

50. West Africa, London. Zaire: Accord on Bonn Aid. In: *Africa Diary*, July 9–15, 1974, pp. 7033–7034.

51. British Broadcasting Company, London. Zaire: Secret agreement with German rocket firm. In: *Africa Diary*, October 8–10, 1977, p. 8700.

52. West Africa, London. Zaire: New light on missile testing report. In: *Africa Diary*, June 4–10, 1978, p. 9033.

53. West Africa, London. Zaire: W. German rocket project to be halted. In: *Africa Diary*, July 9–15, 1979, p. 9592.

54. Staff writer. Germans go rocketing on the cheap. *New Scientist* 1977; 74:535.

55. Hussain F. Volksraketen for the Third World. *New Scientist* 1978; 77:802–803.

56. Informationsdienst Sudliches Africa. OTRAG: Missiles against liberation in Africa. In: *Dirty Work-2: The CIA in Africa*. Ray E, Schaap W, Van Meter K and Wolf L, eds. Secaucus, NJ: Lyle Stewart, Inc., 1979, pp. 215–219; Additional references cited: Gesellschaft fur Unternehmendberatung, Hamburg, 1976, Diagnosebericht OTRAG, p. 12; Der Speigel, August 4, 1978; *The Evening Standard*, February 13, 1978; Deutcher Bundestag, 8th Session, 98th Sitting, June 15, 19778; *Aviation Week and Space Technology*, September 12, 1975.

57. African Development, London. Africa General: U.S. sends more intelligence personnel to Africa. In: Africa Diary, February 5–11, 1977, p. 8335. Reference notes Mr. William H. Crosson, chief of two branches of counter-intelligence in Vietnam according to the Pentagon was appointed in 1977 to be the Director of US Peace Corps activity in Zaire.

58. Covert NM. Cutting Edge: A history of Fort Detrick, Maryland, 1943–1993. Fort Detrick, MD: Headquarters, U.S. Army Garrison, 1993, p. 54.

59. Moscow World Service in English. Belitskiy on How, Where AIDS Virus Originated. March 11, 1988. Published in *International Affairs*. Soviet FBIS-SOV-88-049, March 14, 1988, p. 24.

60. Havana International Service in Spanish. German claims AIDS virus created by Pentagon. January 25, 1991. Published in *International Affairs*: Caribbean FBIS-LAT-91-017, March 14, 1988.

61. Staff reporter. Litton Industries Unit Gets Job. *Wall Street Journal*. September 15, 1977, p. 4, Column 1.

62. Staff reporter. Litton Systems Inc. awarded $19.8 million Army contract for missile fire-control equipment. *Wall Street Journal*, December 19, 1977, p. 21, Column 1.

63. Staff reporter. Litton Systems Inc. was given a $32.9 million Air Force contract for electronic reconnaissance sensor equipment. *Wall Street Journal*, Friday, December 30, 1977, p. 6, Column 1.

64. Staff reporter. Litton Industries Gets Order. *Wall Street Journal*, Tuesday, February 14, 1978, p. 33, Column 2.

65. Brumter C. *The North Atlantic Assembly*. Cordrecht/Boston/ Lancaster: Marinus Nijhoff Publishers, 1986, p. 183.

66. Ibid., p. 139–140;195.

67. United States Senate. Intelligence Activities, Senate Resolution 21: Hearings before the Select Committee to Study Governmental Operations with Respect to Intelligence Activities of the U.S. Senate, Ninety-Fourth Congress, First Session. Volume 1, Unauthorized Storage of Toxic gents, September 16, 17, and 18, 1975. Washington, D.C.: U.S. Government Printing Office, 1975, pp. 2;5;6;20;35;40–41; 7_;81;119–120;254–255 (inventory list).

68. Ibid., pp. 22–23.

69. Ibid., p. 61

70. Ibid., p. 82;98.

71. Committee on Human Resources, United States Senate, Ob. cit., pp. 5;91.

72. Szmuness W, Stevens CE, Harley EJ, Zang EA, Oleszk__ WR, William DC, Sadovsky R, Morrison JM and Kellnes __. Hepatitis B vaccine: Demonstration of efficacy in a controlled clinical trial in a

high-risk population in the United States. *New England Journal of Medicine* 1980;303;15:833–841.

73. Figure on display at The U.S. Holocaust Museum, Washington, D.C.

74. Powers RG. *Secrecy and Power: The Life of J. Edgar Hoover.* New York: The Free Press, 1987.

75. Goldstein RJ. *Political Repression in Modern America.* Cambridge, MA: Schenckmann/Two Continents, 1978.

76. Von Hoffman N. Citizen Cohn: *The Life and Times of* _____ *Cohn.* New York: Doubleday, 1988.

77. D'Emilio J. *Sexual Politics, Sexual Communities: The Making of a Homosexual Minority in the United States,* 1940–1970. Chicago, University of Chicago Press, 1983.

78. West African Pilot, Lagos and West Africa, London. U.S. leader's death abruptly ends African-American relations meeting. In *Africa Diary,* April 16–22, 1971, pp. 5428–5429.

79. Falk R, Kim SS and Mendlovitz SH. The perversion of science and technology: An Indictment. In: *Studies on a Just World Order, Volume 1, Toward a just world order.* Boulder, CO: Westview Press, pp. 359–363.

7 Polio Vaccines and the Origin of AIDS: The Career of a Threatening Idea

Brian Martin, Ph.D.

When a virus from one species is able to survive in a different species, at first it is often quite virulent in the new species. For example, the myxoma virus causes little problem in the South American forest rabbit, its longstanding host, but it was devastating when introduced among European rabbits in Australia. As the virus rampages through the new species, susceptible individuals are killed, whereas the resistant ones survive and reproduce, and eventually virulence declines, as in the case of myxomatosis in Australia.[1]

Thus, when a new viral disease springs unannounced on humans, one possible suspect is animal viruses. In the case of AIDS, this soon became the most favored explanation among scientists. In 1983, Luc Montagnier and his colleagues reported isolation of a virus, later called human immunodeficiency virus or HIV, linked to AIDS. Two years later, a type of virus very similar to HIV was found in African monkeys. It was called simian immunodeficiency virus or SIV. Many SIVs cause no obvious disease in their host species, though they can be virulent if transmitted to a different, unaffected monkey species. The obvious explanation for AIDS was that SIV somehow was transmitted to humans, where it became or evolved into HIV.

The next question was how the SIV might have been transmitted from simians to humans. Before looking at the possible explanations, it is worth mentioning some other evidence. First, there are two major types of HIV, called HIV-1 and HIV-2. HIV-1 is the type found throughout most of the world; HIV-2 is found mostly in western Africa.

Brian Martin is a professor at the University of Wollongong, NSW, Australia.

There are also different SIVs, and in fact new ones continue to be discovered. There is one known SIV that is very similar to HIV-2, but none yet proven to be highly similar to HIV-1.

HIV, like any virus, has a genetic structure. Even within one type of HIV, such as HIV-1, there are many variations. In other words, the genetic structure is pretty much the same, but there are slight variations. The variations are due to mutations and selection as the virus spreads. By examining the spread of variants and working backwards, it is possible to estimate when HIV-1 first entered the human species. The usual estimate is just before 1960.

The other relevant information is evidence of AIDS in humans. One of the earliest known cases has been traced to Kinshasa in Africa in the late 1950s. The implication is that SIV entered humans in central Africa in or by the late 1950s and thereafter spread to other parts of the world.

But how did SIV enter humans? This is of more than intellectual interest. Knowing the process may help to prevent recurrences and to provide clues for developing a cure.

It is known that HIV does not survive easily outside the body and that the most effective means of transmission are via blood or mucosa. One explanation is that a hunter, in butchering a monkey, allowed monkey blood to enter a cut. Others are that a human ate some undercooked monkey meat, that monkey blood was injected into humans as part of certain sexual customs, and that a monkey bit a human.

An explanation along these lines is the standard view on the origin of AIDS. But there is one obvious question. Why did AIDS develop in the 1950s? A cut hunter or monkey bite could have occurred any time in the past thousands of years. The usual explanation is that urbanization and travel led to the wider spread of AIDS beginning in the 1950s.

There is, though, another theory available, that explains both the transmission and the timing: polio vaccination campaigns in central Africa in the late 1950s. This theory is simple and obvious. Polio vaccines are cultured on monkey kidneys. Many of the monkeys would have been carrying SIVs, and many of them would have shown no symptoms and thus not been rejected as ill. Thus it would not be too difficult for some batches of vaccine to be contaminated with SIVs. Since the SIVs were not discovered until 1985, there was no way to screen for them in the 1950s.

There is even a precedent for monkey-human viral transmission. In the early 1960s, some polio vaccines were found to be contaminated with a simian virus named SV40. This caused great concern at the time, since SV40 had been given to tens of millions of people in the United States and elsewhere. Henceforth, steps were taken to screen all vaccines for SV40 and other such viruses. (The health consequences of SV40 in humans is a separate issue that deserves study.)

So here was a theory waiting to be developed and tested. Polio vaccines were already known to have led to the spread of simian viruses to humans. Monkeys with SIVs were almost certainly used in polio vaccine preparation, and there was no screening for the SIVs. Finally, some of the earliest known cases of AIDS were near to the time and location of major polio vaccination campaigns in Africa in the late 1950s.

But this theory was not investigated by the medical research establishment. There is one obvious reason for this: the theory, if accepted as true, would be extremely damaging to the image of medical science. The theory might have been talked about but not seriously studied, as indicated by a report early in 1992 "A senior AIDS researcher said it has been an open secret to many AIDS researchers for at least four years that polio vaccines might have been contaminated by HIV or a related retrovirus," but no testing of vaccine stocks had occurred because, according to this researcher, "Everybody was afraid there would be a public panic or a scandal."[2]

PASCAL'S STUDIES

If the medical research establishment was reluctant to investigate the theory, others were not. One of them was Louis Pascal, an independent scholar in New York City. In 1987, he heard a radio talk show with guest Eva Lee Snead who proposed that polio vaccine contaminated with SV40 was responsible for AIDS. Pascal knew enough biology to realize that SV40 couldn't be the cause, but what about the SIVs? He decided to investigate.

By reading medical journals from the 1950s and 1960s and making comparisons with recent reports about the development of AIDS, Pascal soon had a powerful set of arguments suggesting that polio vaccination campaigns in Africa may have led to AIDS. He focused on a particular batch of vaccine used by Hilary Koprowski, a pioneer in polio eradication but less well known than Jonas Salk and Albert Sabin. Koprowski's CHAT Type 1 polio vaccine was given to some 325,000 men, women and children in central and west Africa from 1957 to 1960, plus a few thousand people elsewhere, such as Poland. Pascal found a remarkable geographical coincidence. The main use of CHAT was in central Africa, not far from the area of Africa with one of the highest incidences of AIDS in the world today. Significant doses of CHAT were also administered in the city of Leopoldville; today that city, now called Kinshasa, has an extremely high incidence of AIDS. Sabin later found this batch of vaccine to be contaminated by an unidentified virus.

Koprowski's vaccine was administered orally, by spraying a mist of vaccine into a person's mouth. This seems to raise an immediate objection: HIV, some later critics said, has not been shown to be transmitted orally, so it is unlikely that SIV could be transmitted to humans this way.

Pascal has two responses. First, HIV *can* be transmitted orally, most clearly from breast-feeding mothers to their children. All that is required is that the mucus [membranes] in a recipient's mouth have reduced immune response. Second, it is quite possible that some of the recipients of the vaccine had ulcers or cuts in their mouths, allowing SIV to enter the bloodstream.

Pascal's main interest was to track the origin of HIV-1. He attributes it to an undiscovered SIV that infected a small number of people in central Africa via Koprowski's CHAT vaccine, followed by the spread of HIV-1 elsewhere via person-to-person contact.

Pascal had one further argument. He notes that the immune system normally resists alien cells, or indeed any biological material with an unfamiliar genetic sequence. This of course is why it is necessary to suppress the immune system when transplanting organs. Pascal asks rhetorically, how better to spread a virus from one species to another than by giving it to large numbers of individuals, some of whom are likely to have impaired immune systems? He then points out that Koprowski's vaccine was given to large numbers of children, some of whom were less than 30 days old. Not only are young children's immune systems undeveloped; the youngest children were given 15 times the adult dosage of polio vaccine.

Pascal found much else in his search through the medical literature, enough to convince him that this theory was worth testing because of its serious implications. One immediate implication is that vaccines should not be cultured on monkey kidneys. There are a number of different SIVs and new ones continue to be discovered. Pascal speculated that a new SIV might be entering the human species every few years, potentially leading to a new type of HIV and causing the death of a million or more additional people. Because different HIVs have different rates of exponential spread, one or two types will usually dominate infection statistics. Nevertheless, the human consequences of a single further new HIV are considerable. Therefore, Pascal thought his theory deserved urgent consideration. After all, a delay of a few years might conceivably lead to the deaths of millions of people.

Another implication of Pascal's theory is the need for an urgent assessment of other possible methods for spreading disease from one species to another. One example is the recently carried-out transplantation of a baboon liver into a human. This provides an ideal opportunity for the spread of any virus in the baboon to the human, given the mixing of cells and blood and the use of drugs to suppress the recipient's immune system. Another example is some of the experiments with genetic engineering.

Pascal had a theory and had good reason to believe it deserved urgent consideration. If the theory could be proved wrong, then there was nothing to worry about; but if it proved correct (or possibly correct) then its implications should be dealt with immediately. He assumed that since the theory seemed so obvious, there would be others who would come up with it independently. But, just in case, he did what he could to make sure it received critical examination.

Pascal believed that if he wrote up his findings and sent them to scientists and to scientific journals, then—taking into account the important potential social implications of the theory—scientists would either refute his ideas or accept them. In other words, he expected his ideas to be considered objectively, irrespective of who he was or how he wrote up his material. Proceeding on this assumption, Pascal wrote an account of his theory, including plenty of references and logical argumentation so that others could check his facts and inferences. He sent his paper to a number of prominent scientists for their examination and also submitted it to a number of leading scientific journals.

From the prominent scientists, Pascal received only one cursory acknowledgment. From the scientific journals—*Nature*, *Lancet*, and *New Scientist*—he received the brushoff, either a rejection with little or no explanation, or year-long failures to answer.

Pascal thought that scientists and scientific journals would give his ideas a fair hearing. Unfortunately, the standard view that science is objective and open to new ideas—a view that is taught to science students in high school and university and to the general public through many popular treatments—is flawed. The reality is that being taken seriously by the scientific research establishment depends sensitively on who the writer is, what their institutional affiliation is, how they write their paper and, not least, what they have to say. To be taken seriously, it is a great advantage to be an eminent scientist, to write from a prestigious address, to write precisely in the standard journal style, and to say something that is just marginally original and not threatening to any powerful interest group. Pascal, by being an "independent scholar" with no institutional affiliation, by writing in a style that deviated somewhat from the standard passionless prose and not citing prominent scientists in quite the appropriate respectful way, and by presenting a highly threatening proposal, was never taken seriously.

Defenders of the system would say that Pascal should have couched his ideas in the standard format. If he wanted to be taken seriously, he had to play the game of scientific publication by the rules. From Pascal's point of view, this sort of attitude misses the point. It was he who was raising a serious issue for science and public health. He felt it was the responsibility of editors to deal with his concerns promptly and effectively. If he was wrong, nothing was lost; if he was right, many might suffer. Therefore the "scientific reception system," namely the system by which potential contributions to scientific knowledge are considered, certified, and published, was responsible for making sure his ideas received proper consideration, even if he didn't couch them precisely in orthodox form.

Cynically speaking, the system works reasonably well to serve the interests of career scientists, who have a strong incentive to play the game by the rules, since that is the way they obtain publication and thereby obtain jobs, grants, and promotions. But Pascal was not seeking a career in science, nor did he particularly care about having his name in print. He was primarily concerned about scientific ideas and the social implications of science. This lack of career motive and personal ambition can seem strange to professional scientists. Likewise the operation of the scientific reception system seems strange, indeed immoral, to someone like Pascal with different motivations and goals.

One of his correspondents, a philosopher, sent Pascal's paper to the *Journal of Medical Ethics*, whose editor then invited Pascal to submit a paper on the ethical issues associated with his case. After much labor, Pascal prepared a new paper, but it was rejected . . . for being too long.

In 1990 I began corresponding with Pascal and was quite impressed by his ideas, his grasp of the issues, and his thoroughness. After his paper was rejected by the *Journal of Medical Ethics*, I arranged for it to be published in a working paper series at my university.[3] As soon as it began to be circulated, it generated considerable interest among scientists and others. One of the responses was by the editor of the *Journal of Medical Ethics*, who wrote an editorial explaining why they had rejected it, making known its availability and commenting that Pascal's thesis "is an important and thoroughly argued one and ought to be taken seriously by workers in the AIDS field."[4]

OTHER INVESTIGATORS

Pascal had long said that he would not be surprised if others independently developed the same theory, since it was so obvious. As indicated by the quote from the AIDS researchers, it had indeed been considered, but apparently not investigated further because of reservations about the possible implications. Most of the scientific community remained ignorant of the theory, aided by unreceptive journals.

One exception was two South African scientists, Professors Gerasimos Lecatsas and Jennifer J. Alexander. Independently of Pascal, they wrote several letters and short pieces to scientific journals raising the possibility of AIDS arising from polio vaccines. Most of their early submissions were rejected, but not all.[5] However, this airing of the idea in a medical journal did not stimulate others to investigate more deeply. Instead, they were personally attacked in a reply to their letter in the *South African Medical Journal.*

Blaine Elswood, an AIDS activist and employee of the University of California at San Francisco, also developed the same theory independently of Pascal. Elswood worked with medical researcher Raphael Stricker and they prepared a carefully written scientific paper. It was rejected by the *British Medical Journal.* They next tried *Research in Virology.* After being given strong encouragement by Luc Montagnier, months passed. Then, in an apparent reversal, they were asked to shorten the paper, delete most of a section on SV40, and resubmit their material as a letter to the editor. Many more months passed before their letter was finally published.[6] It was followed by a rebuttal from the editorial board of the journal.

Clearly the mainstream scientific journals were not eager to give the theory much visibility. Elswood had anticipated this, and he had encouraged Tom Curtis, a free-lance journalist based in Houston, to investigate. Curtis was enthusiastic. Starting with materials obtained from Elswood, he delved further into the literature and also did interviews with many scientists, including Sabin, Salk, and Koprowski. He wrote a series of important stories in the *Houston Post* and a major piece published in *Rolling Stone.*[7]

Whereas the scientific journals had stalled on the story for years, Curtis' *Rolling Stone* story broke through the usual barriers. It became a news item not only in the press, radio and television, but also a story in the news columns of scientific journals.[8]

Koprowski wrote a response in the form of a letter to the editor of *Science.*[9] Curtis wrote a reply, but *Science* refused to publish it.

The Wistar Institute, headed by Koprowski until 1991, holds seed stocks of polio vaccines. Koprowski had earlier been asked by medical researcher Robert Bohannon to release its vaccines for testing. If vaccines from the 1950s African campaigns were found to be contaminated by SIVs, this would provide support for the polio vaccine-AIDS theory. But Koprowski failed at first to even answer Bohannon's letters. Bohannon also had little success with similar requests to the Food and Drug Administration.

Curtis' story in *Rolling Stone* made it harder for Wistar to refuse to cooperate. The Institute set up an independent advisory committee to advise it concerning the implications of the theory. The committee provided a brief 8-page typed report which concluded that the chance that AIDS had originated from polio vaccination campaigns was "extremely low."[10]

Unfortunately the committee never consulted Pascal, Elswood or Curtis in preparing its report. Even if, *a priori*, the chance of causing AIDS from polio vaccines was quite low, we know now that AIDS did develop somehow. Therefore, the key issue is not the absolute probability of AIDS developing from a particular sequence of events, but the relative probability, namely the probability compared to other ways that AIDS might have developed (cut hunters, monkey bites, and so forth). But the Wistar committee made no such comparisons.

The only bit of real evidence that the committee used to criticize the theory was the case of a Manchester seaman who died in 1959, in retrospect apparently having contracted AIDS. HIV was detected postmortem. Koprowski, in the letter to *Science*, also made a big issue of the Manchester seaman. Yet there are several possible explanations for this case which reduce its power as an objection to the theory.

First, the test for HIV in the seaman's remains may have been a false positive. In other words, the seaman may not have had AIDS at all, but instead the tests that showed HIV may have been contaminated. Aptly, the first four pages of Pascal's paper deal with how easy it is for cell lines to be contaminated, drawing on the famous case of HeLa.[11] (See the inset "The HeLa Affair" on page 81.) Pascal uses the example to show how easy it is for scientists to slip up and how eager they are to avoid acknowledging their mistakes.

Second, the seaman might have been infected by HIV during a trip to Africa or by contact with other seamen, and then have developed AIDS much more rapidly than usual, especially considering that he was given immune-suppressive drugs.

Third, the seaman might have contracted AIDS via some earlier vaccine experiments from the 1920s to the 1950s, at least one of which involved the injection of live monkey cells into thousands of people.[12] Pascal points out that there is evidence of experiments involving grafts of monkey or chimpanzee organs at least as early as 1916.[13] It is possible that monkey viruses could have been transmitted to humans on one or more of these earlier occasions, leading to anomalous cases of disease. This is compatible with polio vaccination campaigns in Africa being the cause of the AIDS *pandemic*.

It is now the conventional wisdom in the history and sociology of science that a single piece of evidence is not sufficient to reject a theory. Within any general picture, such as a scientific paradigm, there are always some anomalies. These anomalies are either explained away or ignored so long as there are compensating advantages or insights to be gained from the wider picture. This is not to say that anomalies should be dismissed as trivial. Quite the contrary. But they are not alone sufficient basis to reject a theory.

The importance placed on the Manchester seaman example by opponents of the polio vaccine-AIDS theory, and their lack of examination of alternative explanations, suggests the eagerness with which they have

The HeLa Affair

The HeLa affair is the story of the contamination of cell cultures around the world and the corresponding refusal on the part of mainstream science to face up to and deal with the problem that presented itself.

The HeLa affair begin in 1951, when the first human cells were grown in long term tissue culture. The HeLa cells were cervical cancer cells taken from a woman named Henrietta Lacks. Although Henrietta Lacks died of her disease, the cells from her tumor—given the shortened name HeLa—not only survived, but flourished.

Because her cells were so strong, they were sent to various other laboratories around the country and from there around the world for experimental purposes. Unfortunately, laboratory errors allowed HeLa cells to contaminate other tissue cultures, and the HeLa cells quickly overtook and replaced the other cells. However, much of the time, the colonization of other tissue cultures by HeLa cells went unnoticed, since the appearance of many tissue cultures is highly similar. Thus, scientists who believed that they were studying cells from human breast tumors or monkey heart cells, for example, were in many cases studying HeLa cells.

To add to the problem of the spread of HeLa cells, many researchers would share their particularly hardy line of "breast tumor" or "monkey heart" cells with their colleagues. It took only a few years for the problems of HeLa contamination of other cell lines to have reached crisis proportions. An investigation by geneticist Stanley Gartler found that of seventeen tissue cultures—obtained from a number of different laboratories—*all were HeLa cell cultures, contrary to their official designation as a variety of human cell lines.*

The problem of HeLa contamination of tissue cultures was finally tackled by Walter Nelson Rees, who was then the head of a cell bank at the University of California. Nelson Rees also held the position of vice president for the Tissue Culture Association—the profesional body to which scientists involved in tissue culture work belonged. When he confirmed Stanley Gartler's findings, Nelson Rees submitted long lists of contaminated cell lines to journals. However, instead of promptly publishing these important documents, many journals procrastinated while still others refused to publish Nelson Rees's lists at all.

The *Journal of the National Cancer Institute* published what independent researcher Louis Pascal described as a "cooked-up" case by workers previously discredited by Nelson Rees, using "illegitimate photographs of chromosomes" and "shoddy logic" to try and prove that the charges of contamination were not valid.

One major supplier of biolgical supplies, Microbiological Associates—later M.A. Bioproducts—reportedly continued to sell a HeLa contaminated culture for thirteen years after the company was first informed by Stanley Gartler that it was contaminated—and seven years after other scientists had confirmed the contamination.

Nelson Rees's campaign against the contamination of tissue cultures made him so many enemies that he was forced to retire in 1981, aged just 52. After his retirement, the National Cancer Institute ceased funding his laboratory, sounding the death-knell for the best run cell culture center in the United States.

sought ways to dismiss the theory. Curtis' interviews revealed the extreme hostility with which Koprowski, Salk, and Sabin responded to the theory. This is not surprising, considering the strong emotional investment that leading scientists have in their own ideas.[14]

There are of course many other arguments concerning the theory, ranging from the problems of gene sequences, the species of monkeys used in polio vaccine trials, the spread of AIDS in other countries, and much more. The aim here is not to address these complexities but to outline the theory and point out the failure of the mainstream scientific community to confront it adequately.

This failure has an intriguing self-righteous twist. Many scientists look down upon the mass media and consider that science is only proper when it takes place in professional forums. Koprowski, for example, said that "as a scientist, I did not intend to debate Tom Curtis when he presented his hypothesis about the origin of AIDS in *Rolling Stone*."[15] He did condescend to reply after a letter by Curtis appeared in *Science*. In another example, Luc Montagnier supported the decision of *Research in Virology* to request Elswood and Stricker to shorten their paper to a letter to the editor by referring to the "extensive publication" of their views in the lay press.[16]

This seems rather unfair, since the reason the story obtained attention in the "lay press" first is that scientists, knowing about the theory for some years, declined to investigate it and editors refused to publish submissions to scientific journals. In other words, the relevant scientific community failed to come to grips with a theory that deserved critical attention, even if only to refute it. Then, when individuals outside the scientific mainstream worked on the theory and obtained media coverage, their approach was denigrated.

Nevertheless, some inroads into mainstream practice may yet occur. The Wistar committee, in spite of its assessment of the polio vaccine-AIDS theory as highly unlikely, have recommended that polio vaccine no longer be cultured using monkey kidneys,* because "There may well be other monkey viruses that have not yet been discovered that could possibly contaminate vaccine lots."[17] This was exactly the thing that Pascal has been warning about for years. It took an article in *Rolling Stone* for scientists to take it seriously.

The implications are wider than just polio vaccines. All transfers of material from one species to another should be scrutinized. For example, it has recently been found that many cattle in the United States are infected by bovine immunodeficiency-like virus or BIV, which has a genetic structure similar to HIV. This is not a scientific curiosity, Pascal points out, because bovine hemoglobin is being used to manufacture substitutes for human hemoglobin. The danger of introducing new diseases to humans may be low, but at the very least it should be investigated.

Thus, even if the theory is wrong, it may be valuable in leading to discoveries or revised practices that will advance the understanding of AIDS, how to deal with it, or how to prevent similar diseases. That is the most that can be asked of any scientific theory.

I want to thank Tom Curtis, Blaine Elswood, and Louis Pascal for valuable comments on earlier drafts of this paper.

Editor's Note: In December 1992, Mr. Koprowski sued Tom Curtis and *Rolling Stone* magazine for defamation. This effectively quashed further investigation into the story by mainstream media. Brian Martin publicised the lawsuit through the Sci-Tech Studies electronic mailing list, and also wrote a letter to *Nature* about the dangers of allowing legal action to stymie scientific debate. This seemed to stimulate renewed interest in the theory within the scientific community. Finally, in November 1993, just before Mr. Koprowski was to undergo deposition, his lawyers settled out

---

* As of January, 1998, a spokesperson for the New Zealand Ministry of Health confirmed that at least some of the polio vaccine used in New Zealand is still grown on monkey kidney tissue. The alternative growth media is a human diploid cell line derived from aborted fetuses.

of court. *Rolling Stone*, facing legal costs of $500,000 paid Mr. Koprowski damages of $1.00.

CORRESPONDENCE:

Brian Martin
Department of Science and Technology Studies
University of Wollongong
NSW 2522, Australia
Phone +61-42-213763
Fax +61-42-213452

REFERENCES

1. Seale, U. Crossing the species barrier—viruses and the origins of AIDS in perspective. *Journal of the Royal Society of Medicine*, Vol. 82, Sept. 1989, pp. 519–523.

2. Curtis, Tom. Vaccines not tested for HIV? *Houston Post*, 18 March 1992, p. A-1.

3. Pascal, Louis. What happens when science goes bad. Science and Technology Analysis Research Programme Working Paper #9, Department of Science and Technology Studies, University of Wollongong, NSW 2522, Australia, December 1991.

4. Gillon, Raanon. A startling 19,000-word thesis on the origin of AIDS should the JME have published it? *Journal of Medical Ethics*, Vol. 18, 1992, pp. 3–4.

5. Lecatsas, G. and Alexander, J.J. Safety testing of poliovirus vaccine and the origin of HIV infection in man. *South African Medical Journal* Vol. 76, 21 October 1989, p. 451.

6. Elswood, B.F. and Stricker, R.B. Polio vaccines and the origin of AIDS. *Research in Virology* Vol. 144,1993, pp. 175–177.

7. Curtis, Tom. The origin of AIDS, *Rolling Stone,* issue 626, 19 March 1992, pp. 54–55, 61, 106, 108.

8. For example, Brown, Phyllida. US rethinks link between polio vaccine and HIV. *New Scientist*, 4 April 1992, p. 1; Cohen, Jon, Debate on AIDS origin *Rolling Stone* weighs in. *Science*, 20 March 1992, p. 1505.

9. Koprowski, Hilary. Letter. *Science*. Vol. 257, 21 August 1992, pp. 1024, 1026–1027.

10. Basilico, Claudio et al. Report from the AIDS/Poliovirus Advisory Committee, 18 September 1992.

11. Gold, Michael. *A Conspiracy of Cells: One Woman's Immortal Legacy and the Medical Scandal it Caused.* State University of New York Press, 1986.

12. Pascal, op. cit., pp. 34–35.

13. Voronoff, Serge. *Life*, E.P. Dutton, 1920, p. 106.

14. Mitroff, Ian L. *The Subjective Side of Science.* Elsevier, 1974.

15. Loprowski, op. cit., p. 1024.

16. Montagnier, Luc. Fax to Raphael B. Stricker, 10 September 1992.

17. Basilico et al., op. cit., p. 7.

8

Oxygen Therapies, The Virus Destroyers

Ed McCabe

There exists a little known yet very simple way to treat almost all diseases. It's so simple it befuddles the great minds. Unlike our healthy human cells that love oxygen, most of the primitive bacteria and viruses, including HIV and others found in cancer, AIDS, Ebola, flesh eating bacteria, chronic fatigue, tuberculosis, arthritis, and a long list of other diseases are like most lower life from viruses and bacteria that can't stand oxygen. Bacteria and viruses are almost all anaerobic. "Anaerobic" means these microbes cannot live in oxygen

What would happen to the primitive anaerobic viruses and bacteria that cause disease if they were to be completely surrounded with a very energetic form of pure oxygen for a long time? What if enough of this special form of oxygen (O_2), or its higher medical grade form called "ozone" (O_3), were to be slowly and harmlessly introduced directly into the body day after day, every day, over the course of several months, by high concentration methods that usually bypass the lungs, and yet saturate every body cell with oxygen? All the disease-related, or disease-causing, bugs and microbes that can't live in oxygen also can't live in oxygen saturated body tissues and fluids. See how simple it is? It's amazing you were never told this.

What would happen to any HIV in your body if it was caused to be continually surrounded with oxygen, or its higher form, ozone? See the October 11, 1991 issue of *The Journal of the American Society of Hematology*, "Inactivation of HIV type 1 by Ozone in vitro" (page 1881): "Ozone, a higher form of oxygen, inactivates the HIV virus 97–100 percent of the time, and is harmless to normal cells, when used correctly." What currently over-hyped and over-prescribed AIDS drug can make anywhere near these claims for effectiveness *and* safety?

I just got off the phone with another of the long list of AIDS sufferers who have unsuccessfully gone the route of toxic drugs, and has ended washed up at the door to ozone. Typically, his liver had been destroyed before he arrived. *Now* he is willing to look at alternatives (if they would only look sooner) but, unfortunately, denial, and letting someone else be responsible for your own personal well being, is strongly ingrained in our society. Ozone is unknown in this country by you or your doctor due to the influence the drug companies wield over our healthcare system and the media. They are not your friend. Ozone is. In the Greek ozone translates into "the Breath of God." Millions of dosages of ozone have been given in Europe by thousands of doctors over the past fifty years with complete safety. Why isn't its use taught in medical schools here in the United States?

All fifty or so oxygen therapies have the same method of action. The point in oxygen therapies is always to *slowly and increasingly flood the body* with Nature's single oxygen atoms. Singlet oxygen and its by-products are very energetic oxidizers—they "burn up" or oxidize disease-causing waste products, toxins and other pollution, and send them out of the body. Unless your body rids itself of toxins at 100 percent efficiency, some waste is left behind, still in the body. Over time it accumulates within you, setting the stage for diseases to occur as the inner pollution accumulates. This is why you get more and more diseases as you get older. This is the cause of virtually all disease. It is very important. Remember this as the viral plagues get worse, and the air becomes less and less able to sustain us. Your body no longer naturally has or gets enough clean energetic oxygen. Your body can no longer take out the trash that causes disease.

As an example of how dirty we have become internally, I have witnessed hundreds of AIDS, cancer, arthritis, and other patients getting oxygen/ozone therapy. When they start out their blood is filthy, diseased, and so empty of oxygen that it is almost BLACK. Put the medical oxygen/ozone in them daily, and after a few weeks of oxidizing/oxygenating detoxification their blood starts to turn a bright cherry RED, clean and full of LIFE.

Oxygen atoms also burn up germs (microbes) which can't protect themselves against its oxidizing powers. Unlike evolved and specialized normal human cells, germs, microbes, and all parasites are primitive lower life forms. Normal body cells protect themselves from the burning up or oxidizing effects of oxygen by naturally producing and using their own protective antioxidant coatings. Our normal cells surround themselves with this insulating and protective coating that primitive disease-related bugs, germs, and even cancer cells do not have. Seeking out anything that is not a normal healthy human cell to oxidize, oxygen is a natural hunter, the killer enemy of bacteria and viruses and also dead, dying, deformed, or diseased cells like those found in cancers.

Naturally, oxygen therapies are best used as supplemental maintenance dose-type preventatives. By keeping the oxygen level high in our body fluids, we keep the anaerobic diseases from ever being able to establish themselves. If they can't live in oxygen, they won't live in highly oxygenated clean tissues. If one uses these therapies correctly and long term, one's internal fluids would be oxygen rich. Most oxygenated people I know personally don't even catch colds anymore.

When oxygen therapies are employed as a method of putting the patient back on the road to health, they have completely proven themselves, according to the thousands of interviews I have conducted with the actual users of these therapies. In addition, the world's medical literature is replete with the proof of oxygen/ozone therapies' being both inexpensive and highly effective, and they have been proven very safe and natural when used as directed. For further documentation, including over 100 medical references, please contact *The Family News* in Miami, telephone (305) 759-9500.

The human body is 66 percent water. We are internally permeated with fluids, and our organs are mostly water and float in our internal sea. In medicine, oxygen/ozone gas is slowly micro-bubbled into your personal interior body fluids to purify them by a variety of methods.

The methods are simple. In the simplest physician applied method, doctors slowly inject specific quantities and concentrations of ozone into the blood, always while the patient lies prone. In "autohemotherapy" some blood is taken out, ozonated, and returned to the body. The Germans prefer this method, but I find it to be inefficient, costly, and slow when compared to the newer methods, which are also perfectly safe when used correctly. These newer recirculatory methods are my favorites, and according to all the interviews I have done, the most effective. Using them, the diseased patient blood is continuously circulated into ozone filled chambers and the leftover sludge of oxidized dead pathogens and toxins is filtered out. All this happens *outside* the body, and then the fresh cherry red blood is returned into the body in real time and continuously, as in the Medizone Inc., or Polyatomic Apheresis Ltd. proprietary systems. Proceeding this way, where most of the action happens outside the body is best. That way the body organs, like the liver, do not also have to shoulder the added burden of detoxifying the system, especially since the usual prescribed toxic drugs have already weakened or seriously damaged the body organs. Unfortunately, these new recirculatory methods are expensive, and only available outside the U.S. at the present time.

At home, many of the successful AIDS patients that I have interviewed report that their doctors direct them to combine injection, sauna bagging, and rectal insufflation. Most people buy correctly assembled and certified cold plasma type ozone generators (never using ultraviolet bulb genera-tors which have little strength) and hook them up to tanked pure medical oxygen as the input gas and produce their own 27 to 42 micrograms per cubic milliliter concentrations of pure oxygen/ozone gas locally. They buy the equivalent of European medical grade ozone generators in the U.S. "underground" because the FDA won't approve them, due again, to drug company suppression politics. These people put specific qualities and quantities of pure oxygen/ozone into their bodily openings, or soak most or parts of their body in a bag filled with the gas—never breathing the high concentrations of ozone, since our lungs detoxify too fast and get edema—the lung cells protectively swell up, if exposed to "too high" con-centrations of ozone, meaning anything above the level that's found in nature. Then they use this humidified gas to give themselves ozone via "rectal insufflation," where the ozone/oxygen gas is introduced at specif-ic concentrations, durations, pressures, and volumes into the empty colon which transfers the ozone/oxygen into the blood. Then there's the "ozone sauna bagging" method, which saturates the capillaries in the skin, or they have a competently trained person slowly inject them, again at specific concentrations and volumes. For safety, any such very slow injections must always be done while the patient is lying prone before, during, and after.

Some bathe in or drink very dilute solutions of food grade hydrogen peroxide (which breaks down into water and releases oxygen in the blood), or any of the new oxygenated liquids and powders. Each method has specific concentrations, volumes, durations and cautions, so work with a competently trained oxygen therapist to be safe. My books, tapes and videos explain all this in more detail.

Over the past eight years I have seen many, many people rid their bod-ies of infections and other problems from being on oxygen therapies or ozone therapies, but *they stuck to a full protocol*—getting it daily, usually IV, in the right dosage, and the right concentrations, and combining it with other significant oxygen based modalities, proper diet, antioxidants, col-loidal minerals, and enzymes. People who have only "dabbled" in it, or those who have never really tried it, end up being the very few who are negative about it. Even with such people around, by now the term I coined —"Oxygen Therapies"—also my first book title, is a household word among the well informed, and thousands of health advocates now use the therapies themselves and even promote them. There must be something to what I have been teaching.

If you decide to start along the happy oxygen trail, remember these facts: According to the sum of my thousands of interviews with success-ful oxygen therapies using people all over the world, all those who succeeded first studied and understood the philosophy and use of the therapies fully, and then they applied the therapies correctly, continually, gently, and long term.

Note: This is a condensed part of the larger works of Ed McCabe. His first work, the self-published best-seller *Oxygen Therapies* has sold over 150,000 copies by word of mouth. Books and audio/videotapes like his latest video "Ozone, and Disease" are available by mailorder from *The Family News* (305) 759-9500, M-F 9–5 EST.

9 Oxygen Therapy: The Empire Strikes Back

*Basil Wainwright has categorically invented
a process to purify whole donor blood
in the bag, and his invention of polyatomic
apheresis ozone technology has created
the most significant breakthrough in the treatment
of AIDS and degenerative diseases found
anywhere in the world to date.*

Richard Bernhard (Polyatomic Apheresis Inc.)

GN = Gary Null
SAT = Sue Ann Taylor
BW = Basil Wainwright

GN: This programme is *Natural Living*, and I'm Gary Null of WBAI, a public-supported radio station. Tonight I'll be talking to Sue Ann Taylor, an investigative journalist, and Basil Wainwright, a scientist and inventor of a particular ozone machine. Why is he in the Metropolitan Correction Center in Miami—the jail? Why hasn't he had a trial in three years? Why does the government not want his story to get out? More on that later.

Is HIV the cause of AIDS? HIV has never been found in any scientific studies anywhere in the world to be the sole cause of AIDS. No one can prove it. It is speculation. It is political and economic. The man who said in 1982 that HIV was the probable cause of AIDS (instantly it became dogma that it was)—did he also inform the public he was the primary beneficiary of a test for HIV, that he owns the patent and that millions of dollars have gone to him and his associates? No.

Did the press vigorously explore all the allegations of fraud and corruption? No. The alternative press did. We're the ones that brought you that information. They tell you don't challenge orthodoxy. We challenge you not to believe that but rather to believe the experience of those who are the ultimate authorities: the patients who are alive and well, having had the opportunity to intelligently review the best of both and see what works, and that's what we bring you.

You've heard previously from patients successfully treated using nontoxic therapies, you've heard from the physicians who've treated them. Now today, in this segment, Sue Ann Taylor, investigative journalist, welcome to our programme.

SAT: Hello!

GN: Sue Ann, you recently returned from the Philippines where you observed and recorded the effects of ozone treatment and a polyatomic apheresis therapy on a group of HIV-positive and AIDS patients. Would you give us the background of this and why it is so important that the people hear this story?

SAT: Well, I was researching for a documentary that I had been working on, called *Living Proof—People Walking Away From AIDS Healthy*, because I was finding more and more evidence that there were things that were in fact working for some AIDS cases and/or HIV-positive cases. In doing that research I came upon ozone therapy, and I also came upon all the controversy that surrounds it. So when I was offered the opportunity to actually watch a trial happen first hand, in the Philippines, I jumped at the chance.

I went to the Philippines and I was stunned with what I saw, because I was expecting the entire thing to take place in a sort of wing of a hospital, or something that looked a little bit more like what I expected medicine to look like. It was actually a clinic that was set up rather *ad hoc* to provide space to do justice to this trial, so I started out a little on the sceptical side, not knowing what I was getting into.

There were nineteen HIV-positive people there, five of whom had fullblown AIDS. Over the course of about three weeks I watched the patients, or participants as they preferred to be called—six of whom were in pretty bad shape—watched them go through some pretty remarkable transformations and I saw it happen before my very own eyes. There's no amount of journalists or medical people who can tell me that what I saw I didn't see. I saw people who were unable to walk, be able to walk again. I saw people who were very, very ill just get considerably better, and all of the treatment was cut short by a raid by the government.

The Philippine government came in and shut down the entire operation, after only about one-third of the prescribed amount of treatment had been accomplished. It was a trial, so remember there wasn't an absolute number on how much treatment they were going to need—that was part of what they were there to establish—but one-third of what they were expecting would be close to the magic number of hours on the machine, had been accomplished, and in that period of time remarkable reversals in these people's conditions were evident.

GN: Alright, describe the clinic.

SAT: The [Cebu] clinic itself was an upscale home in the Philippines. An upscale home in the Philippines looks kind of like an upscale home in America. It was a very large home, two storey, fairly large lot, and behind the home they had built grass huts, but it wasn't as crude as that makes it sound; it really had a vacation resort feel to it. It was not really unacceptable—and by Phillipines standards it was just fine. I had an opportunity to go to one of the Philippines hospitals, and the cleanliness within the clinic beat the cleanliness of the Philippines hospitals that I visited. All the Filipino staff were excellent—I would pit their training against any training of any nursing staff anywhere in the world. But some of the things we take for granted, like refrigeration and insect control, they just have really come to learn to live without those things. The clinic was, by our own standards, crude, but it was, you know, acceptable also. The materials were all new; it's just, again, it didn't meet my preliminary expectations.

GN: Who was working there?

SAT: There was a group from Australia—the clinic was actually owned by a couple named Bob and Rosanna Graham. The second group was PAI, the polyatomic apheresis unit group, and all they did was supply the equipment and people to train the Philippine staff to use the equipment; and the third group was the Philippine staff which consisted of two Philippine doctors and eleven nurses.

GN: And who were the patients?

SAT: The patients were twenty Australians, nineteen with HIV, one with multiple cancers.

GN: Is it illegal to enter the Philippines if you are an HIV-positive person?

SAT: My understanding is that it is illegal to go in HIV-positive, but Immigration does not question you; there is no testing and I don't know that the patients realized that it was illegal.

GN: Could you tell us of some the success stories of the patients?

SAT: The most dramatic success story was a man named Paul. Paul is 42 years old, he had been HIV-positive since 1984, has full-blown AIDS and Kaposi's sarcoma. The lesions, the Kaposi's sarcoma lesions on the bottom of his feet, were so great when he left for the Philippines that he couldn't walk. He was in slippers for over a year. He could not wear shoes. He gingerly walked on the outsides of his feet and it was very difficult for him to get around at all. After eleven hours of treatment on the machine, Paul's lesions went away. He was able to wear leather shoes and, most importantly to Paul, he was off morphine for the first time in four years. Prior to his going to the Philippines, the cancer hospital had told him that he had reached the maximum amount of radiation that he could receive safely, and he would have to simply continue to increase his morphine to deal with his increasing pain. Paul believed that he had experienced just miraculous treatment, that in eleven hours of that treatment the lesions on his feet went away and he could wear shoes and walk normally again.

GN: Describe what the treatment consisted of.

SAT: The polyatomic apheresis looks like the following: a patient sits in a chair that looks a little like a dentist's chair. It's a comfortable chair. There are intravenous needles inserted in both of their arms, the blood coming out of the left arm is pulled through a pump that is in synch with the heart rate, and a circuit of blood is created between the left arm coming out and the right arm coming in. The blood goes through a series of tubes, goes down through a cascade tube where it is met with ozone under pressure, and at that point that's where the viral kill happens. The blood continues down through an escape tube, through a filter, back into their right arm. What you see visually is the blood exiting the left arm is a very black colour; it is *black*. It goes down through this cascade tube, which is a wide bore cascade tube, about an inch in diameter, and it goes back into the arm, the right arm, a bright cherry-red colour. It comes out looking alarmingly different—this is with the HIV patients—alarmingly different from what you would expect.

Now, the first patient I saw on the machine was a person without HIV. She was a normal person who had an infected foot, and her blood came out looking like yours and mine would, and went back in only slightly differently than it came out; so what I witnessed was that the HIV patients' blood was considerably blacker than a normal person's and went back considerably lighter. That's, in a nutshell, what it is.

GN: Alright, now, what other parts of the therapy were included with this ozone treatment, and how does this ozone treatment differ from, let's say, one which would be done in New York where you pull out about, oh, a half pint of blood, ozonate it and put it back in the arm over a fifteen- to twenty-minute period?

SAT: I've never witnessed any of the other treatments that you're talking about. The only two ozone treatments that I've seen actually operate are the polyatomic apheresis and, using the same equipment, a process called rectal insufflation where the ozone gas is put in through a catheter into the rectum, which becomes an ozone enema, so to speak. Those two were used at the clinic and in conjunction with one another. Some of the participants in the study had experienced the treatment that you are talking about and had some success with it. What they believe from their own experience, what they told me, is that it was the difference between a Volkswagen and a Rolls Royce, from what they felt with the treatment, you're talking about getting in New York versus what they got in the Philippines.

GN: So, it was far more productive in the Philippines?

SAT: Correct.

GN: Now, what happened to these twenty patients? Where are they at now and have there been any additional protocols for these people to follow?

SAT: The turning point of everything was on March 19. The youngest participant was a 23-year-old woman named Jodi, and she had full-blown AIDS. It was a real tragedy because she really kind of represented all of our daughters, and her courage was phenomenal. She died in the clinic and that's when things started to tumble very quickly. She died from a series of complications. I'm not a medical expert but I believe she received two insufflations too close together and her body had trouble coping with the amount of ozone that she had taken in. She also received those against doctor's orders, so I guess it would have to be chalked up to human error rather than anything to do with the equipment. She received the ozone via the rectal insufflation.

GN: You mean the Philippine doctors had suggested she not take those?

SAT: Actually, it was the American doctor, the expert on the ozone, who had said this girl shouldn't have another until she recovers a little bit. She had remarkable success on the equipment, though. When I first arrived I was afraid Jodi was not going to make it until the equipment arrived. There were all kinds of customs hangups that prevented the equipment from getting into the country and getting set up on time. So the patients arrived ahead of the equipment, which was a real management error because it just added too much stress to the patients.

GN: By the way, who raided the clinic?

SAT: It was raided by the Department of Immigration.

GN: Was there any evidence the FDA had been involved in the raid?

SAT: There wasn't any evidence that the FDA had been involved; but what I was told was that the story really got underway when Australia's version of *A Current Affair* did a scathing story on the clinic and what the patients were about to experience, just as they were getting on the plane. I was told by another journalist in Australia whom I trust, that ACA is the one who went in to the Department of Immigration and tipped them off. I was also told that the producers were directed by their upper management to do a 'chuck job' on the ozone therapy. And no matter what they were told, no matter how much positive information they were given, it never aired; and I watched this happen time after time.

GN: So, in other words, there was a gross bias in the media, from your interpretation, to prevent positive stories about the success of ozone from getting back to the general population?

SAT: It's not even a question of interpretation. I watched it happen; I watched the participants give interviews; I gave interviews myself. We would turn on the TV and we would be shocked at what actually would show up. Paul, whom I was telling you about, would tell his entire story; he would show his feet, all of those things; and he made a comment in one of the television interviews where he said, "After I got going I could just feel it in my heart that this was working." That little snippet is the only thing that they would use, and then they would cut to the doctor saying, "Well, you know, there's a certain amount of mind over matter," and all that kind of stuff. So they were completely dismissing the science of it and trying to make it sound like their improvements were all in their own minds; but fifteen patients had improved T-cell counts, one as high as a 70-percent increase.

GN: I would like to shift gears, now, and bring in another individual to share a different perspective on this, and one that we haven't talked about in the past. Basil Wainwright, welcome to our programme.

BW: Thank you very much, Gary. I must congratulate you on running a super programme, and a very courageous one too.

GN: Basil, you are now incarcerated in Florida?

BW: That's right, so if any of your listeners hear any background effects, I must apologise for that. I am currently incarcerated down here in Miami.

GN: From what I understand, you are a scientist and you are the inventor of this polyatomic machine, this ozone machine, and that you have been incarcerated without trial for three years. Is that correct?

BW: Yes, I'm now well into my third year without trial and some seven violations of my basic human rights.

GN: What are those violations?

BW: Well, there's the 4th amendment and the 5th amendment, the 6th amendment has been violated, and the 8th, and 14th. So . . .

GN: What has happened to your attorney filing proper motions to get a fair and speedy trial? That's one of the constitutional provisions for people who are incarcerated. I haven't heard of people waiting three years except this particular political detainee who was here in New York, the IRA supporter who was held for some seven years.

BW: That is absolutely right. Well, it all started that—really, I suppose I should give you and your listeners a brief synopsis. I was working with Dr. Viebahn in Germany and I was brought into this project along with Medizone, and then got very much involved in the process. And I was somewhat intrigued to find that nobody had really done any specific testing i.e., looking at the cytotoxic levels or, that is, the concentration of ozone, looking at the specific atomic structures of that, and also the contacting time; so there were an awful lot of areas that particularly interested me. I worked with the University of Medicine and Dentistry and also the Mt. Sinai Hospital with Dr. Weinburg and with Dr. Michael Carpendale, and started to get very, very involved in the course.

It was very evident there were some phenomenal results being seen in the AIDS area and I started to look at it more in-depth. There were several controversies going on as to whether it was a function of free radical reaction or oxidation—but of course both of those functions occur extensively—and also this ionisation; and I wanted to determine the specific parameters of that, because when people refer to ozone you might just as well refer to a vehicle being involved in a collision because you're not really defining the atomic structure of ozone which can be multifold. There can be many aggregate combinations of molecules which can have very specifically different responses, and I wanted to determine this.

GN: Since 1985 you have been working with some German doctors including Dr. Viebahn that you talked about. Now, you had a way of determining that the ozone being used back then was not as effective as the way you could create a better ozone; they were using O_2 but you also saw O_3 and O_4.

BW: Yes.

GN: Now tell us about what you found with what you created concerning viral inactivation.

BW: Well, of course, I think it's very important for your listeners to know that the reason scientists refer to retroviruses' inactivation as opposed to being killed is because normal micro-organisms have metabolic mechanisms, whereas a retrovirus could almost be considered a piece of genetic material drifting around in the bloodstream. And so, it's rather difficult to kill a non-living thing, hence scientists refer to inactivation. We looked at these various techniques and procedures, including the study which we did with Biotest in Miami. We determined that the German process worked but wouldn't be dramatically effective because they were not treating high enough volumes of blood. We wanted to see that once someone had been taken back to HIV negative using polyatomic oxygen or ozone, they indeed remained negative. I think there is only one case that actually went back to positive so that was rather unique because all the doctors were saying, Okay, so what? You get somebody to negative, but in a couple of months' time they're going to go back to positive. Well, that was proven not to be the case, which I think even surprised the Germans. And it might well be that the immune system kicks back in, and when we say negative we're looking at nucleic acid response or PCR work to determine that; but certainly the patients were not going back to positive—that was very interesting.

So we thought, okay, if these patients are going to use autohemotherapy which you referred to earlier, Gary, where you take out half a pint of blood, treat it with ozone, and then reinfuse it back into the patient, that was taking typically eleven months, of course combined with a very rigid nutritional control as well. But using that process it was very evident that it's like chipping away at a mountain with an ice pick when you're looking at the view of this pandemic facing mankind; and it became very apparent in 1987 that the best way to go was with dialysis or a dialysis-type procedure. So I worked with dialysis equipment and in fact filed my first dialysis patents using ozone in 1988.

However, using ordinary dialysis equipment which is a hollow fibre membrane, we discovered there was too much homolysis occurring as a result of that; also, the thing that we refer to as mechanical shear. The very fact of pumping the blood round outside the body can cause all sorts of trauma to cells—there are thermal reactions, there are pressure zones, the pumping head itself can actually crush cells—so we had to look at a number of factors. And then, when we did more research, we found that O_4 in particular had some very unique responses. It has a phenomenal amount of electrons; as a matter of interest, in O_4 you have 40 electrons, and that makes it a very powerful negative ionising platform drifting around in your bloodstream. It was also far more stable than O_3 which again was completely the reverse of what everyone was projecting.

It was very evident that O_3 had a better oxidative effect, and that was very effective in eliminating infected cells, but O_4 had the ability because of its ionisation to break down, we believe, the RNA, and of course uracil, which is a very important sugar combination—the 5-carbon sugar in the virus RNA—was actually being broken down. Well, when we actually achieved this, we did our first study down at Biotest Laboratories here in Miami—hence my incarceration down here. We did this study and as far as I know, for the first time in history, using apheresis we successfully converted HIV-positive to negative, and we could do this time and time again using PCR. That's the reason we came here, actually, because Biotest Laboratories in conjunction with Miami University had this latest state-of-the-art equipment; and from that very moment, the FDA witch hunt started.

We tried to keep a relatively low profile but of course the word soon got around the system, and then one night I came home and the SWAT team descended, guns drawn, and eight of them sort of crashed in the front door. I was arrested and charged wih practicing medicine without a license, which of course is complete nonsense. But the SWAT team, instead of looking for anything that might indeed have been relevant to my practising medicine without a licence, all they did was dig out all my patent specifications, technical data and intellectual mechanisms. So they came with a very specific directive from the FDA, to seize all my intellectual property rights. From there I was thrown into prison. Eventually I had charges from the FDA which boil down to sending and selling ozone generators from interstate—interstate trading laws, etc. Unfortunately, a couple of months after I was in prison, it was discovered that I had a very severe heart condition. In fact, if this radio show had been yesterday I doubt very much if I could have done it. It's progressed to a point now where I'm collapsing and having blackouts and stuff, but still hanging in there. I've just recently done a technical paper.

Well, from that episode this series of things went on, and as you quite rightly say—and I certainly won't bore your listeners with the phenomenal list of violations against me—I'm now into my third year; come October I'll be commencing my fourth year without any trial. I've just recently been appointed some new attorney who is hopeful of trying to get me bond. In fact, Dr. Michael Carpendale and other doctors very courageously were flying into Florida for a major hearing in front of the judge. Everything was scheduled but at the very last moment the FDA stepped in again and the hearing was cancelled, and my research team had to frantically phone around and cancel everyone coming in. I did get bond, much to the amazement of the FDA, which was really an administrative error, and I was out for a few months. During that time we managed to get a number of apheresis systems put together and out into studies.

Most of the studies which were conducted in and around the United States of course have already had the FDA SWAT teams descend on them, close them down and seize equipment. And we've had things reported like seven P24-antigen negatives, a couple of PCR negatives, but at no time have we ever been able to get into the real completion of a study. In every case, I think the doctors would tell you they've seen absolutely dramatic results, and that's not from me because this information has been fed back to us. They are very concerned that they're prevented from pursuing this, since the process does really show some pretty dramatic potential. The only way we are ever going to get this out there is if the AIDS groups get up and demand polyatomic apheresis so that we can get these studies up and running. We've got a group working with two very, very prominent stars who hope to apply sufficient pressure to be able to get this achieved.

During our studies we managed to determine that protein aspects in the blood, in other words, high protein levels would have an inhibiting effect on the success of the procedure. The normal procedure that has been adopted by the Germans, i.e., introducing antioxidants—which is very popular over here too—was also negating the effects of ozone. Everyone in the United States can enjoy the wonderful efficacy of ozone; there is nothing against the law that you can't use it, and there are several ways of applying it. In our protocols, prior to treatment the patients will be receiving no antioxidants so that we get the maximum oxidative effect from the O_3 component which we use 2 percent by weight, and 6 percent by weight of O_4; and we have a pretty rigid nutritional programme too.

GN: So let me see if I can put this into perspective. Basil Wainwright is now in a jail in Florida for developing a special form of ozone machine that puts an O_4 into the body. There are a number of patients, estimated as high as 200, who have undergone this polyatomic apheresis treatment so far. These have included HIV, environmental and degenerative diseases, approximately thirty persons with AIDS. Of those thirty people, all show dramatic improvement, seven are P24-antigen negative, and two are PCR negative, meaning there is no HIV viral DNA found in their bodies, and the P24 means there is no active replication—all replication of the HIV is done. For the effort, you have been put in prison without trial. When the doctors did come to testify on your behalf, the FDA saw that the hearings were postponed. On a technical glitch you were allowed out, and then when they found out the technical glitch they put you back in; and you have been in violation of several due processes including a speedy trial. Why weren't the other doctors put on trial or arrested? Why were you the only person involved in this?

BW: Well, because I was the primary motivating force and the one that indeed held the patents in the United States office for polyatomic apheresis, which is quite unique. The only reason that I can think of is that I enjoyed the energy in working in the process. We have a wonderful team, they're all terribly dedicated to helping people, and we would like to think we are motivated in attempting to do God's work. Sue Ann and everyone else who have been involved have expressed love and compassion to all these patients, so it's been more than just a research project for me. I thoroughly enjoyed working with the patients. Of course, the pharmaceutical companies cannot file a patent on ozone, and you can only file patents on the intellectual property rights or the designs of the delivery mechanisms to the patient; and being as we have those, I suppose the best thing they could do and their only reaction was to throw me in prison, hoping that it would completely bring everything to a halt. It hasn't done that.

There's been a dedicated bunch of people out there; they definitely need more support. We would certainly provide equipment for AIDS

groups on the United States if they would only get up and demand polyatomic apheresis and demand studies which they could do. We would be
only too pleased to provide the equipment and, indeed, a number of very
top doctors are prepared to come along and offer their services and monitor and support these test studies. You undoubtedly know that Ed McCabe
has been doing some tremendous work in trying to open people's horizons
on these issues, and Ed of course has been very supportive and he's
become very supportive because he's been seeing the successes.
Unfortunately, a lot of the doctors that have been involved in the research
have had terrible pressure applied to them; in fact, their very jobs and
livelihoods have been threatened by the FDA, which is very, very sad. I
must admit when I first came to the States in 1987 on this particular project, the people told me this sort of thing existed in the United States and
I thought it was all James Bond stuff, but of course I soon learnt to the
contrary that indeed it was fact, and here I am. All I want to do in fact is
get out of here and research and work for the betterment of mankind and
just simply conduct God's work. In fact, I've just finished two scientific
papers while I've been incarcerated, and I've been working very, very hard.

A lot of good things: we've got a Middle East project which has been
confirmed which will be up and running very soon; the Canadian government with NATO of course, as you've probably read, indicated great interest. Well, they've actually approached us and we've had talks with them
about structuring a very special process which we've developed. It's from
the blood bag to the patient, so for the armed forces, if they get injured out
in the field and they're having delivery or transfusion of a unit of blood,
there's this process we've developed which goes in series or in line with
the IV to the patient, which actually purifies the blood with polyatomic
structures before it goes into the wounded soldier. So, despite my various
bouts of illnesses and I must admit it's been a bit touch and go at times,
I've certainly been keeping myself active, Gary, and as I've said I've certainly been following your programme with intent and your work with
intent, and I hope your listeners out there realise what a super person you
are and how you're projecting this work and making this awareness to the
people out there.

GN: Thank you Basil Wainwright, and let's hope for the best and that justice will be served by being fair and by seeing that your machine is tested. I want to thank you also for being on today, Sue Ann Taylor. Any closing thought for us?

SAT: Well, the closing thought that I have is, after the raid the mayor of
the city gave the Department of Health the opportunity that if they want-

ed the study to continue, he would make space available in a hospital and make the patients the guests of the city. For them to turn down that offer and shut it down without looking at the patients' records, of which the blood tests all showed improvements, or watching a demonstration— that's when I started to believe that there was some level of a conspiracy happening right before my eyes, because they had made up their minds in the face of an offer from the mayor and said let's finish it right here. The only other point that I wanted to make, that I found alarming, is that people who have the ability to make those decisions were that closed-minded about the patients' pleas that this could save our lives, that they shut the door in their faces.

GN: Sue Ann Taylor, you learned a good lesson, and that lesson unfortunately is a bitter one: not always do the patients count when there is a political or economic agenda ahead of their interest. Thank you very much. I am Gary Null, the programme is *Natural Living*.

10 The FDA

Hans Ruesch

*People think the FDA is protecting
them—it isn't. What the FDA is doing
and what people think it's doing
are as different as night and day.*

Herbert L. Ley, Jr., M.D.,
former Commissioner of the FDA

*The hearings have revealed police state
tactics . . . possibly perjured testimony
to gain a conviction . . . intimidation
and gross disregard for Constitutional Rights.*

Senator Edward Long,
U.S. Senate hearings on the FDA

The cancer conspiracy is led by the FDA-NCI-AMA-ACS hierarchy. The initials stand for the Food and Drug Administration (FDA), the National Cancer Institute (NCI), the American Medical Association (AMA), and the American Cancer Society (ACS). The cancer conspiracy also includes the large pharmaceutical companies and key research centers such as the Memorial Sloan-Kettering Cancer Center in New York City and selected university research labs. The key personnel move in and out of official positions within these organizations, sit on common boards or investigation committees, and have both formal and informal networks. When a researcher or alternative medicine advocate is identified as a threat to the power or even the official views put out by the ruling hierarchy, the maverick is placed on various published and unpublished blacklists. Funding is stopped, legal harassment often begins, public denunciation as a quack frequently follows, and if the outsider persists in offering or advocating a non-sanctioned treatment, then rougher, clandestine methods can be employed.

It would take thousands of pages to describe various individuals who have fought the cancer conspiracy and how their threat to the ruling powers was neutralized. These pages can only summarize some of the more famous cases and facts which reveal how the cancer conspiracy functions, but those who wish to know more can pursue the details on their own, using the names and references offered here as a starting point. The people and procedures described in these pages are by no means inclusive, only the most notable or most promising.

The FDA (Food and Drug Administration) is the government police force which approves experimental studies for those it favors and hinders approval for those it dislikes. It conducts semi-legal break-ins (constitutional procedures are often ignored), confiscates records so that critical documentation is often lost or at least unavailable for months or years, and at times has interfered with constitutional protections through conspiratorial relationships with private organizations who share the same suppressive goals. New medical breakthroughs that threaten the sanctioned and financially lucrative treatments are ignored or "studied" for years. The FDA thus frequently subverts its legislated purpose which is to promote and protect the public health. Having lived in Washington, D.C., I know that the FDA is regarded by many astute civil servants as the federal agency with the lowest morale. A dark cloud of oppressive inertia, corruption and bureaucratic sloth pervades its corridors.

Dr. J. Richard Crout, test director at the FDA Bureau of Drugs beginning in 1971, described the agency in Congressional testimony on April 19, 1976 as follows:

> There was open drunkenness by several employees which went on for months . . . crippled by what some people called the worst personnel in government. Here was intimidation internally by people . . . . People, I'm talking about division directors and their staff, would engage in a kind of behavior that invited insubordination—people tittering in corners, throwing spitballs, I am describing physicians, people who would . . . slouch down in a chair, not respond to questions, moan and groan with sweeping gestures, a kind of behavior I have not seen in any other institution as a grown man. . . . Prior to 1974, not one scientific officer in our place knew his work assignments, nor did any manager know the work assignment of the people under him.

In 1967, the FDA stopped the use of an experimental cancer vaccine which was producing significant results. It was developed by H. James

Rand, inventor of the heart defibrillator. J. Ernest Ayre, an internationally recognized cancer specialist (co-developer of the PAP test) and Dr. Norbert Czajkowski of Detroit, Michigan assisted Rand. Treating only terminal cancer patients, the Rand vaccine produced objective improvement in 35 percent of 600 patients while another 30 percent demonstrated subjective improvement. "One 65-year-old woman with spreading tumor" was "completely cured in four months." Another woman with extensive breast cancer was cured in six months. The FDA stopped the vaccine's use in a federal court hearing where neither the cancer patients *nor their doctors* were allowed to testify. U.S. Senator Stephen Young of Ohio protested—to no avail. Senator Young could get nowhere with FDA Commissioner James L. Goddard. Senator Young recalled:

> I could not move them. They would not even agree to a modification of the ruling (banning the Rand vaccine), which would at least allow the 100 (cancer) patients at Richmond Heights (Ohio) to complete their injections. The Justice Department was prepared to go along, but the FDA Commissioner, Dr. James Goddard, was adamant, even belligerent. It's wrong of the government to snatch away this hope when there is no evidence against its use offered in court. It's damnably wrong.

It is known that when FDA Commissioner Goddard's own wife had serious health problems and orthodox medicine could not help her, Goddard contacted alternative health practitioners who quietly healed his wife. But for the suffering victims of cancer who needed the Rand vaccine or some other nontraditional treatment, Goddard lowered the boom, using the federal courts to enforce his dictum. Such are the ways of the FDA.

Goddard's greatest disservice to the American people was his persecution of DMSO, a simple molecule which often brought miraculous pain relief and offered numerous possibilities for medical advancement in other areas, including cancer. One respected science writer suggested that Goddard crushed DMSO research in order to gain increased police powers from Congress. The FDA has never admitted its errors regarding DMSO although the positive studies from qualified scientists number over a thousand while the FDA's criticisms have been shown to be almost completely based on lies or unsubstantiated rumors. Yet by the late 1980s, twenty years later, the FDA continues to imprison DMSO advocates. The malignity of Goddard's arbitrary and conscienceless acts in 1966–1968 against reputable scientists, dedicated doctors and the public good is one of the darkest chapters of FDA history.

No one is sure of the real reasons why it happened and why it continues to be covered up twenty years later. It has been suggested that one or more drug companies sabotaged DMSO because it threatened so many of their profitable products. One drug company executive reportedly told the leading DMSO researcher:

> I don't care if it is the major drug of our century—and we all know it is—it isn't worth it to us.

Who had the power to keep such a miraculous drug off the shelves? Surely not just an FDA Commissioner flexing his muscle. Was it a combination of drug companies whose individual profits were threatened by the miracle drug's possibilities?

> [It is] not our [FDA] policy to jeopardize the financial interests of the pharmaceutical companies.
> —from testimony before Congress of
> Dr. Charles C. Edwards, at the time commissioner of the FDA

It has also been surmised that FDA Commissioner Goddard used DMSO in 1966 in an attempt to become the medical dictator of America. In the years that followed, FDA officials simply refused to expose the agency's "dirty laundry." Hence the on-going suppression of what many recognize as "the major drug of the century." In any case, Goddard instilled fear into honest researchers and physicians as no previous FDA Commissioner had done. He ruined careers. He introduced an intensified police force mentality into the FDA with his emphasis on hiring ex T-men and G-men. He consciously blacklisted scientists as punishment for opposing him. And members of his agency, either with his encouragement or his acquiescence, openly began ignoring the Constitution for the sake of promotions and power.

Pat McGrady, Sr.'s book, *The Persecuted Drug: The Story of DMSO*, detailed what Goddard's FDA did. McGrady described "the no-knock system, the photocopying of private papers, bugging, punitive investigations, slander and libel, character assassination, forgery, lying and blackmail." One scientist declared to McGrady:

> For the first time in my life I know fear. I'm afraid for my family and myself. I'm afraid for doctors and scientists. And I'm more afraid for our country. I can't believe these things are happening in the United States.

Another noted researcher maligned by Goddard's FDA observed:

> The academic community and industry are so completely intimidated that one cannot look for any leadership to counteract some of the punitive actions of the FDA. . . . I am very pessimistic concerning the future

status of medical research unless a mood arises to combat overzealous bureaucratic authority.

Dr. Walter Modell of Cornell University Medical College finally warned in a published article ("FDA Censorship" in *Clinical Pharmacology and Therapeutics*):

> When the nonexpert in-group of the FDA threatens to become the dictator of American medicine we believe it will lead medicine from its respected eminence to its ultimate decline.

A few years after the DMSO suppression, one of Goddard's top aides, Billy Goodrich, left the FDA with his pension and became president of a food association regulated by the FDA. A personal friend who had been president of the food association took over Billy Goodrich's position at the FDA. They simply switched jobs! Congressmen screamed in protest. It was such a blatant demonstration of the "musical shuffle" (which Congress had previously observed but ignored) that they had to make some noise this time in order to avert public wrath. Still after all the sound and fury, nothing happened.

Goddard himself became Chairman of the Board of Ormont Drug and Chemical Company a few years after leaving the FDA.

A study conducted by the U.S. Congress in 1969 revealed that 37 of 49 top officials of the FDA who left the agency moved into high corporate positions with the large companies they had regulated. A General Accounting Office (GAO) study of the FDA in 1975 revealed that 150 FDA officials owned stock in the companies they were supposed to regulate. The record of "conflict of interest" (or worse) within the FDA is deep and extensive.

In 1976, Dr. J. Richard Crout of the FDA admitted that "endless questions" was a favored technique within the agency to discourage any researcher who sought approval for an unorthodox cancer therapy. Bureaucratic obstruction is a weapon as deadly as a gun when the lives of innocent millions are at stake. It is a delusion to consider such institutionalized, orchestrated conduct, consciously chosen either because of orders from above or personal inclination, as anything other than white collar murder. In fact, it closely resembles the role carried out by the bureaucrats who pushed the paper in Nazi Germany. The policy-makers may not fully perceive the effect of their actions, but the horror has gone on for too many decades to allow a plea of ignorance to be totally convincing.

In 1972, Dean Burk, Ph.D., of the National Cancer Institute (head of their cytochemistry section and a veteran of thirty-two years at the agency) declared in a letter to a member of Congress that high officials of the FDA,

AMA, ACS and the U.S. Department of Health, Education and Welfare (now Health and Human Services or HHS) were deliberately falsifying information, literally lying, committing unconstitutional acts and in other ways thwarting potential cancer cures to which they were opposed.

Dr. Burk's famous May 30, 1972 letter to Congressman Louis Frey, Jr. dealt with the issue of why the FDA had revoked an Investigative New Drug (IND) application. The IND application, according to Dr. Burk, was superior to many routinely approved. But it involved testing laetrile, a controversial, nonpatentable product opposed by the California Medical Association (CMA). FDA approved the original application, then rescinded the license, apparently because of pressure from the surgeon general, a member of the CMA and a laetrile foe.

Dr. Burk was not an advocate of laetrile. He was, however, in favor of fair testing. He was totally opposed to what he bluntly called "misleading and indeed fraudulent" FDA reports. In his correspondence with Congress, he openly referred to "FDA corruption."

Corruption indeed. It takes several forms. Refusing to allow investigation of a non-toxic compound which might help cancer patients is one. Failing to assert itself when a drug tested on human beings was determined to cause cancer is another. Here are the facts of such a case:

In August 1969 it was learned that a drug called Cinanserin, produced by E. R. Squibb and Sons, Princeton, New Jersey, caused tumors in the livers of rats. Human testing of the drug was thus stopped. But Squibb's executives did not want to do follow-ups on the humans who had taken Cinanserim.

For three years, the FDA tried to *persuade* Squibb to do follow-up studies. (Compare this approach with what the FDA does to alternative cancer treatments which actually work—raids, confiscation of documents, jail, etc. With the large drug companies, FDA tries persuasion!)

Finally, in 1972, the FDA and the National Academy of Sciences set up a committee to examine procedures on follow-ups when a drug was found to be dangerous. Who was appointed to head the committee? The vice-president of Squibb whom the FDA had tried for three years to persuade to do follow-ups on those people who had been given the cancer-causing drug!

The FDA has a long history of ignoring dangerous drugs and chemical additives marketed by the big drug companies while using bureaucratic delays, legal harassment, unconstitutional procedures, and even falsified evidence to stop unorthodox cancer treatments. In 1964, the FDA initiated a multimillion dollar prosecution of Andrew Ivy, vice-president and professor of physiology at the University of Illinois. Ivy was former chairman of the National Cancer Institute's National Advisory Council on Cancer. He was an internationally recognized scholar and a prolific author of scientific papers.

His sin was that he supported a cancer-curing serum called Krebiozen. Over 20,000 cancer patients had supposedly benefited from Krebiozen. One United States Senate Committee lawyer personally assessed 530 cases and concluded that Krebiozen was effective.

Krebiozen has never been tested objectively. The FDA used illegal methods to stop it, methods which have been part of a conscious goal of the FDA to dictate what medicine a citizen is permitted to use and what he may not use. Combined with the questionable behavior of FDA officials, the stock links to the large drug companies, and the testimony of FDA employees that conscious cover-ups were common, the intention of the FDA to dictate individual medicine has to be recognized as one of the most dangerous threats to freedom that has ever existed.

Peter Temin, a professor at MIT, carefully studied FDA history and policy for his 1980 book, *Taking Your Medicine: Drug Regulation in the United States*. His conclusion, based on a very careful, close look at FDA is frightening:

> The most important facet of FDA regulation is the agency's expression of its conviction that individuals—both doctors and consumers—cannot make reasonable choices among drugs.
>   The agency tried with increasing success to deny drug prescribers and users the option of taking "innocuous" drugs, that is, to force them to use drugs the FDA regards as appropriate for their condition.

Despite evidence which extends for decades, revealing criminal behavior in the one agency that holds the power to permit tests of alternative cures for cancer, Congress has done nothing. One night in Washington, D.C., I found out why. I was introduced to the aide of one of the most powerful U.S. Representatives in Congress. His boss had been in Washington for many years. Yet, despite the Congressman's powerful committee position and ranking status in the majority party, he was unable to do anything with the health officials at FDA or NCI. After a number of drinks, this Congressman's aide told me that FDA and NCI were protected fiefdoms. They wrote their own legislation, permitting only minor changes by Congress. They ignored Congressional complaints. They were extensively tied to the big drug companies. "They know no one controls them. No one is able to take a sword and tell them where to go," the aide said. He leaned across the table and whispered, "Only national security procedures are as tightly controlled, without outside examination. Only national security. Does that tell you something?"

It told me that the monster was real and dangerous if some of the most powerful men in the U.S. Congress, with their massive egos and independent political bases, were afraid of it.

G. E. Griffin, author of *World Without Cancer*, made explicit the fundamental, systematic wrong which has emerged out of the various cross-currents that make up the FDA—underpaid civil servants playing it safe; drug companies and their Washington lawyers putting unending pressure on the bureaucrats; academic medicos controlling the approval process and restricting the individual doctor's choice; revolving door employment between FDA and universities/drug companies, and behind-the-scenes political deals. According to Griffin, the FDA did two things: (1) they "protected" the big drug companies and were subsequently rewarded; and (2) they attacked—using the government's police powers—those who threatened the big drug companies, be it a young company with a new product such as DMSO, or natural health store products such as food, vitamins, minerals or other self-healing (non-drug, non-doctor) methods.

Griffin wrote the following about the FDA:

First, it is providing a means whereby key individuals on its payroll are able to obtain both power and wealth through granting special favors to certain politically influential groups that are subject to its regulation. This activity is similar to the "protection racket" of organized crime: for a price, one can induce FDA administrators to provide "protection" from the FDA itself.

Secondly, as a result of this political favoritism, the FDA has become a primary factor in that formula whereby cartel-oriented companies in the food and drug industry are able to use the police powers of government to harass or destroy their free-market competitors.

And thirdly, the FDA occasionally does some genuine public good with whatever energies it has left over after serving the vested political and commercial interest of its first two activities.

There is only one solution. No reform will work. No changing of personnel will have any long term effect. No new laws dealing with regulations. Only one solution.

It was provided by a southern doctor now living in New York City who has observed the monster in action for many years. Raymond Keith Brown, M.D., outlined the solution in his book, *Cancer, AIDS and the Medical Establishment*. He described how the power which FDA has to approve drugs and technology has to be eliminated and replaced with the solitary role of testing for effectiveness and safety, the results being the basis for FDA labeling. The *individual physician and individual patient* would regain the responsibility to use or not use a given-drug or technology.

Dr. Brown recommended that

The FDA should follow a simple rating system for effectiveness and safety. Effectiveness would fall into one of three categories

"Effectiveness Unconditionally Proved," "Effectiveness Conditionally Demonstrated," and "Effectiveness Undetermined." Safety could also be categorized in the same manner and the appropriate designation then affixed to all products or containers. Judgment of the effectiveness of any medical product or device should not be vested in any governmental agency or institutions, but should be returned to the province of the individual physician. Freedom of choice for medical materials, therapy and methods must be put on the same footing as civil liberties and as vigorously protected.

One of the better scholars in this field—Robert G. Houston—says simply:

There should be curbs on the FDA—on its powers to intrude into the private practice of medicine . . . the FDA should not be dictating to doctors what they can and cannot do.

Richard Ericson, a dedicated husband of a cancer victim, eloquently concurred (*Cancer Treatment: Why So Many Failures?*):

A physician should be able to prescribe any type of cancer treatment that he considers best for the patient, with the patient's consent and knowledge, without stringent governmental regulations that are now in force. Congress should consider such problems when new guidelines are enacted.

Only when FDA concentrates on the blatant health menaces such as overtly misleading health product claims or drugs shown to cause death and injury; only when FDA ceases to be the bully boy for the big drug companies and other vested interests; and only when FDA again allows physicians, nonconventional healers and their patients their choice of therapeutic treatments . . . will it regain its legitimate government function. In its present form, it is like a malignant beast, harming society rather than serving it.

AMERICAN CLINICAL LABORATORY
CITES *TOWNSEND LETTER*

Excerpts (August 1992):

According to reports in the *Townsend Letter for Doctors* (#108, July 1992, p. 559) including the reprint of an editorial from the *Seattle Post-Intelligencer (P-I)*, U.S. Food and Drug Agency (FDA) agents wearing flak jackets and accompanied by a contingent of King County (Washington) police with guns drawn broke into the Tahoma Clinic (Kent,

Washington) at 9 A.M. on May 6th without knocking on the door or accepting an offer to have it unlocked. Commanded to "freeze!" and put their hands in the air, the employees of the clinic were escorted from the building and refused readmittance for 14 hours while an uncounted number of boxes of clinic records and equipment was taken to an unknown location.

The clinic is owned by Jonathan V. Wright, M.D. (Harvard University; University of Michigan Medical School), who has practiced medicine in Washington since 1970 and treated tens of thousands of patients, including 1,200 currently under his care. The warrant issued authorized the search and seizure of injectable materials, including vitamins; a vice; literature; and patient records. Also raided at the same time was the For Your Health pharmacy operated by Raymond Suen. As of mid-July, no charges had been filed against anyone.

Suen, together with Kent Littleton, Chief Chemist, Meridian Valley Clinical Laboratory (Kent, Washington), had collaborated with Wright to submit a well-documented article, "Testing for vitamin K: An osteoporosis risk factor." (*Am. Clin. Lab.* 8[2], 16 [1989]). Wright et al. traced the connection between vitamin K and bone formation in promoting the gamma carboxylation of glutamic acid in osteocalcin that binds calcium ions leading to bone calcification. (In its identical but much better known function in blood clotting, vitamin K promotes the binding of calcium toprothrombin in its transformation to thrombin.) While considerable vitamin K is presumably synthesized by normal intestinal flora, many clinicians are aware that normal microflora are much less frequent today than in the preantibiotic days. Care must be taken in sample preparation and in the frequently used HPLC analysis of vitamin K in serum, but Wright finds that appropriate dietary changes and supplementation almost always result in substantial improvement in serum levels in just a few weeks.

11 Harry Hoxsey: An Introduction to His Life and Work

Katherine Smith

Harry Hoxsey was born in 1902, into a family which had been successfully curing cancer for several generations using herbal medicine. According to Hoxsey, the healing power of the herbs used in the family's secret formula was discovered by his great grandfather in the nineteenth century.

In 1840 or thereabouts John Hoxsey had a horse which became sick with a cancerous sore. He fully expected the animal to die, but when it seemed to be miraculously gaining in strength, he observed the animal closely. He noticed that the horse would go and eat the wild herbs ("weeds") in a certain corner of its paddock. By the time the horse made a full recovery from the cancer which had threatened its life, John Hoxsey had become convinced it was the herbs which it had consumed which were responsible for its reprieve from death.

He refined the herbs over the years, treating other animals afflicted with cancer. In time he developed an herbal tonic to be taken internally as well as a powdered formula—which could be mixed into a paste—to be applied to external lesions and tumors.

When John Hoxsey died, he passed the knowledge of his cancer treatment to his son, who later passed the knowledge down to his son, Harry's father.

Harry Hoxsey was told the secret formula of the family's anti-cancer preparations by his father when the older man was on his death bed. (See inset "The Hoxsey Family Formulas" on page 116 for a listing of the herbs.) At that time, the eighteen-year-old Harry Hoxsey was already skilled in the application of the powdered formula to skin cancers. He was also familiar with the dosage for the herbal tonic, having assisted his father in the unofficial cancer clinic he had held in the evenings after his work.

With his father's death, however, Harry Hoxsey gained not only the responsibility of being the sole inheritor of an important cure for cancer, but also the responsibility for taking over his father's role of supporting the family.

Harry Hoxsey worked hard at a number of jobs, including coal miner, since he not only had to provide for his mother and younger siblings, but he also dreamed of going to medical school. Hoxsey believed that if he were to become a bona fide M.D., he would be able to carry on the family tradition of healing cancer without fear of censure, or being prosecuted for practising medicine without a license.

To this end, he resolved not to treat any people with cancer until he had qualified as a doctor. However, after he was approached by an old friend of his father who had cancer, Hoxsey concluded that he couldn't withhold a life-saving treatment from someone who needed his help. After this first unofficial patient was cured of his cancer, word spread and Hoxsey soon found his skills in demand.

As a consequence of successfully treating cancer patients as an unqualified healer, Hoxsey found that his dream of becoming a doctor was to remain unfulfilled. The word had been put out within the community of organised medicine to blacklist the untutored young man who was treating cancer patients more successfully than the vast majority of the medical profession of the time (with a few notable renegade exceptions such as Dr. Max Gerson). No medical school in the United States would accept Hoxsey as a student.

For the rest of his working life, Hoxsey by virtue of his lack of formal qualifications was officially employed in his own clinics as a technical assistant by other doctors who were brave enough to face AMA reprisals and work with him.

Hoxsey also spent the rest of his life being harassed by the AMA, whose actions periodically forced him to close the clinics where he was successfully treating cancer, and move on to another location. Hoxsey, charged only modest fees for his cancer treatment in accordance to a promise he had made to his dying father. He never charged anyone who could prove they were unable to pay, but he was rewarded for his humanitarianism by being hauled into court over one hundred times for "practicing medicine without a license." Representatives of organised medicine even manipulated Hoxsey's own family into bringing a civil suit against him, having duped them into believing that Hoxsey had stolen a valuable economic asset from the family when his father had relinquished the secret formulas to him.

Despite his continued persecution by the AMA, Hoxsey remained hopeful that his cancer curing formulas might eventually be accepted by the medical profession. He submitted documented case studies to various official bodies as proof of the efficacy of his treatment. He also put out a bold challenge to the medical profession: that if he were allowed to choose his own cancer cases so that he would not be faced with the impossible task of trying to treat people who were on death's doorstep, he would guarantee that 80 percent of those people he treated would be cured.

Organised medicine continued to persecute and prosecute Hoxsey, even though in court his enemies publicly acknowledged that his treatment had merit. Despite this admission, Hoxsey's treatment for cancer was nonetheless driven out of the United States in 1963. Like the unconventional nutritional theory devised by Dr. Max Gerson, those doctors—and patients—who wish to use Hoxsey's external powders and herbal formula have been forced to relocate their practice out of the United States, in Tijuana, Mexico—a long way away from the hundreds of thousands of people most in need of their assistance.

The Hoxsey Family Formulas

Originally developed in the nineteenth century by John Hoxsey, the Hoxsey Therapy for cancer has been passed down from generation to generation. This therapy is comprised of herbs and other substances believed to have antitumor effects. Some of these substances are combined to form a tonic that is taken internally, while others are used to make a salve designed to heal external lesions. The following shows the specific ingredients used in each of the Hoxsey therapies:

Internal tonic:
potassium iodide
red clover
buckthorn bark
burdock root
stillingia root
berberis root
pokeberries and root

licorice root
Cascara amarga
prickly ash bark

External paste:
zinc chloride
antimony trisulfide
bloodroot

For those who cannot travel to get these formulas, there are companies in the United States that sell similar combinations designed to have the same effect.

12 The AMA's Successful Attempt to Suppress My Cure for Cancer

Harry M. Hoxsey, N.D.

One of the landmarks of Taylorville was the three-story converted residence on Main Street occupied by the local Loyal Order of Moose. My friend and former patient Fred Auchenbach, an official of the Lodge, informed us that the board of trustees might be induced to sell the property if persuaded that it would be employed for a worthy purpose. There was plenty of room not only for our rapidly-expanding clinic, but for a twenty-five bed hospital. He said his bank would finance the entire transaction.

Accordingly one Sunday afternoon in March, accompanied by Dr. Miller, I appeared before a meeting of the membership in the auditorium of the Lodge and presented our proposal. To reinforce my arguments nearly a dozen of my cured patients were present including Larkin, McVicker, Hunter, Bulpitt, Mrs. Sleighbough, Mrs. Stroud and Fred Baugh, secretary of the Lodge. After hearing their testimony the members voted unanimously to sell us the building.

At this point a stranger in the audience demanded the floor. He identified himself as Lucius O. V. Everhard, an insurance broker and member of a Moose Lodge in Chicago. He said he'd recently written a large policy on the life of Dr. Malcolm L. Harris, chief surgeon at the Alexian Brothers and the Henrotin Memorial hospitals in Chicago, and a power in the American Medical Association (he later became its President).

"If half of what I've heard today is true," Everhard declared, "Taylorville is too small to hold this clinic. Cancer is a national calamity. If Hoxsey is willing I'll telephone Dr. Harris and try to get his support for a clinic in Chicago, where the Hoxsey treatment can reach a wider audience."

I was more than willing, I was excited and elated. Here was the answer to all my problems. With the backing of Dr. Harris, medical recognition of the Hoxsey treatment was a foregone conclusion. Moreover his recommendation would be an "Open, Sesame" to any medical school in the country.

Everhard immediately put through a long-distance call, reached the eminent AMA official at his home, poured out what he'd just seen and heard, and urged that Dr. Miller and I be permitted to demonstrate our treatment on patients at one of the Chicago hospitals. There was a pause, then came the reply:

"Have them meet me tomorrow morning at 8:30 at the south door of the Alexian Brothers Hospital!"

The distance from Taylorville to Chicago by road is more than 200 miles. Setting out in my car immediately after dinner that same night Everhard, Dr. Miller and I arrived at our destination just before midnight. We checked into the Sherman Hotel. Bright and early next morning we were waiting outside the hospital.

Promptly at 8:25 a shiny black Locomobile piloted by a chauffeur drew up at the door and Dr. Harris alighted. He was a thin, slightly-built gentleman (about 5 feet 6 inches tall) in his late fifties with steel-gray hair and a small, closely-cropped mustache. Well dressed and carefully groomed, he moved with the dignity and self-assurance of a man of distinction. As he gave me a limp hand and inspected me from head to foot I was painfully conscious of my rough, calloused miner's paws and ill-fitting store clothes, my Sunday best. Leading the way into the hospital, he said:

"I have a patient I want you to see. Frankly, he's a terminal case. We've done all we can for him, so there's no harm in experimenting. I don't expect you to cure him. But if your treatment produces no unfavorable reaction, we'll go ahead and try it on other cases not so far advanced.'"

That sounded fair enough. We took the elevator to the third floor where we were met by Dr. Daniel Murphy, the resident in charge. After a brief conference with Dr. Harris he took us to the room where the patient lay.

Thomas Mannix, 66-year-old former desk sergeant at the Sheffield Avenue Police Station, seemed more dead than alive. His cadaverous appearance was enhanced by a head completely bald except for a fringe of gray hair, sunken orbs, long sharp nose and grizzled mustache over a bony chin. The mottled skin hung loosely from his scrawny neck, his once-burly frame had shrunk to little more than 70 pounds. The chart at the foot of his bed showed that a prodigious amount of morphine was being administered at regular intervals to dull his pain. Unfastening the patient's gown, Dr. Murphy drew it away from the left shoulder and disclosed a hideous mass of diseased flesh about six inches in diameter. It was seared and baked by intensive X-ray treatment which had, however, failed to halt the progress of the disease.

After a minute examination of the patient, Dr. Miller drew me aside and told me that in his opinion there wasn't a ghost of a chance that Mannix would survive our treatment. I was less pessimistic. Bending over the bed, I said with deep conviction:

"Sarge, if you help us we'll pull you through. You can get well. It depends on how hard you fight. Do you understand me?"

He was too weak to reply verbally, but I was sure I detected a responsive flicker in his dull, faded eyes.

Dr. Harris and Dr. Murphy watched intently as I applied a thick coating of the yellow powder to the gaping lesion, and Dr. Miller put a dressing over it. We left a bottle of our internal medicine with directions that the patient receive a teaspoonful three times per day, and advised the two doctors that we would be back in a week to administer another treatment.

On our way back to Taylorville, Dr. Miller observed: "If we pull this one through, it'll be a miracle!"

I patted his arm. "Don't worry, Doc. Mrs. Stroud looked even worse when I first saw her, and she recovered. We've given Mannix hope. He'll fight for his life now. And that's half the battle."

My confidence was justified in full.

Within two weeks the surface of the pustulant sore turned black and started to dry, a sure sign that our medicine was working on the malignancy.

Within four weeks a hard crust had formed, the cancer was shrinking and pulling away at the edges from the normal tissue. Moreover the rapid improvement in the patient's general physical condition amazed all who saw him. He was able to sit up now, his eyes were bright and alert and the pain had vanished, he no longer needed morphine to sustain him, his appetite had returned, he was beginning to pick up weight.

Two weeks later he was walking around, taking care of his physical needs, champing at the bit and impatient to get back to work. When he saw us he chortled:

"I guess we fooled 'em, didn't we, Doc? Can't wait to get back to the station house and see the look on the faces of the boys. They was all set to give me an Inspector's funeral."

His daughter Kate Mannix, a registered nurse who had assisted in the care of her father, stopped us in the corridor to express the gratitude of the family.

"We feel just as if he's been raised from the dead," she declared. "We'd given up hope. Anyone who's seen a loved one dying of cancer will know what a nightmare we've been through. Now we're just about the happiest people on earth. We'll remember you in all our prayers. And if there's any other way we can show our appreciation, please let us know."

To me these simple, heartfelt words were the richest reward any man could ask.

That same day we informed Dr. Harris that necrosis of the cancerous mass in the policeman's shoulder was complete, it had separated from the normal tissue and could be lifted out within two days. He could scarcely believe his ears, insisted on examining the patient himself.

"This is something I want every doctor in the hospital to witness," he asserted. "Would you be willing to perform that operation in the amphitheatre before the entire staff?"

Dr. Miller and I welcomed the opportunity.

That Wednesday at 10 A.M. when we arrived at the amphitheatre of the Alexian Brothers Hospital we found it buzzing with excitement. The gallery of seats surrounding the operating pit was crowded with more than sixty interns, house physicians and visiting doctors. Scrubbed and gowned, we took up our positions in the pit beside Dr. Harris. He cleared his throat, and the gallery suddenly was silent.

He began with a concise review of the case, detailed the various treatments given the patient, described the latter's condition when he was turned over to us. Then he introduced us and explained our procedure. When he'd finished the patient was wheeled in and we took over.

Dr. Miller removed the bandages from Mannix's shoulder. Self-conscious and tense with awareness that scores of trained eyes were following every move under the bright operating light, I picked up the forceps, scraped and probed the black mass of necrosed tissue. It moved freely at the perimeter but was still anchored at the base. I worked it loose, lifted it out with the forceps, deposited it on the white enamel tray provided for that purpose. And that's all there was to the operation.

Dr. Harris inspected the cavity left by the tumor. There was no sign of blood, pus or abnormal tissue, clean scar tissue already had begun to form. "In time it will heal level with the surrounding flesh," I told him. "There will be no need for plastic surgery."

Shaking his head incredulously, he declared: "It's amazing, if I hadn't seen it I wouldn't believe it!"

Then, looking closer: "What about the necrosis in the clavicular bone?"

"That was caused by X-ray. It too will slough off."

Doctors and interns filled the pit and crowded around the operating table, inspecting the patient, examining the necrosed tissue, firing questions at Dr. Miller and me. The entire demonstration had taken less than half an hour but it was nearly noon before we could break away. Dr. Harris accompanied us to the door and asked where we were stopping.

I told him I was at the Sherman Hotel, and would remain there a couple of days before returning to Taylorville. He promised to get in touch with us before we left. On our way back to the hotel I was jubilant.

"We did it! Now they'll have to admit that we have a treatment that cures cancer!"

Dr. Miller smiled skeptically. "It's not that easy. Wait until Harris gets a chance to think it over and discusses it with other doctors. They'll come up with all kinds of reasons why our treatment wasn't responsible for the patient's recovery. There's more at stake here than you think—prestige, and money, millions invested in X-ray and radium equipment . . ."

At that time it sounded fantastic, and I quickly changed the subject.

Early next morning I was awakened by a telephone call from Dr. Harris. "I'd like to have a talk with you," he said. "I'm at my office in the Field Annex, about two blocks from your hotel. Can you come right over?"

Glancing at my watch, I discovered that it was just 7:15 A.M. I agreed to meet him in half an hour. Hastily I showered, shaved, threw on my clothes and—postponing breakfast—set off to keep the appointment.

Dr. Harris occupied an extensive suite on the seventh floor of the imposing office building. Bristling with early morning energy he met me at the door, ushered me into his private office, motioned me into a chair beside his desk. Surveying me appraisingly, he began:

"Hoxsey, the demonstration you put on yesterday has opened up an entirely new vista in the treatment of cancer. I spent most of last evening discussing the Mannix case with some of my colleagues, and they agree that his amazing recovery is convincing evidence that chemical compounds such as you use offer the best hope to eradicate this disease. It's not just the yellow powder you used; that I suppose is an escharotic.* It's the amazing improvement in his general condition as the result of the medicine you've been giving him."

This was it, the official recognition I'd been seeking so avidly. Giddy with triumph, I could scarcely control an impulse to jump up and gratefully shake his hand.

"Of course," he cautioned, "it's much too early to say that Mannix is cured of cancer. There may be a recurrence; we'll have to wait five years or more before we can reach a definite conclusion. Moreover we can't be sure that your treatment actually cures cancer until we've tried it out on hundreds of other patients and thus can evaluate its effectiveness."

I broke in eagerly: "We can show you hundreds of people who've already taken our treatment and been cured!"

He shook his head impatiently. "That's not scientific proof. We must set up a large-scale experiment under absolute medical controls. Our doctors must select the cases treated so that we're treating cancer; they must administer the treatment in order to determine the effective dosage, unfavorable reactions etc.; the patients must be kept under constant medical observation to ward off the possibility that some other factor may account for their recovery. It's a long-range project involving technical skill, hard work and considerable expense."

---

* A substance that causes a dry scab to be formed on skin.

He paused significantly.

"Dr. Harris," I assured him fervently, "I'll cooperate 100 percent in any experiments you care to make with the Hoxsey treatment. All I want is the opportunity to prove that it actually cures cancer, and is made available as widely as possible to relieve human suffering."

He nodded approvingly. Opening a drawer in his desk he produced a sheaf of papers fastened together with a clip and handed it to me. "I was sure you'd feel that way about it, so I had my lawyers draw up a contract. Read it, sign it and we'll get busy at once in setting up an organization to handle the experiments."

There were ten double-spaced, typewritten, legal sized sheets in the contract. I read slowly, struggling with the involved, unfamiliar legal terms. Dr. Harris arose and strolled over to the window, contemplating the vast expanse of Lake Michigan in the distance.

By the time I reached the bottom of the second page I discovered that I was to turn over all the formulas of the Hoxsey treatment to Dr. Harris and his associates, and relinquish all claims to them. They would become the personal property of the doctors named in the contract.

On the following page it specified that I was to mix and deliver 10 barrels of the internal medicine, 50 pounds of the powder and 100 pounds of the yellow ointment, and instruct a representative of the doctors in the method of mixing these compounds.

Farther along I agreed to close my cancer clinic and henceforth take no active part in the treatment of cancer.

My reward for all this was set forth on next to the last page. It appeared that during a ten-year experimental period I would receive no financial remuneration. After that I was to get 10 percent of the net profits. Dr. Harris and his associates would set the fees—and collect 90 percent of the proceeds.

Stunned and appalled by this incredible document, I turned back and reread the principal clauses to make sure my eyes weren't playing me tricks. There it was, all neatly typed in black and white. The eminent doctor turned away from the window, seated himself behind his desk and favoured me with a nonchalant smile.

"Well," he said heartily, "I trust it's all clear to you."

It was all too clear. He and his friends were trying to trick me out of the family formulas, abscond with the fame and prestige attached to the discovery of a real cure for cancer, and thereby enrich themselves fabulously at my expense and the expense of millions of helpless cancer victims. Disillusioned and angry, I could scarcely speak. Finally I found my voice.

"Before signing this," I said carefully, "I'd like to show it to a lawyer. Mr. Samuel Shaw Parks, who has offices in the Delaware Building on

Randolph Street, was my father's attorney. I'll consult him and be guided by his advice. Perhaps he'll suggest some changes. . . ."

Dr. Harris' smile turned frosty. "There won't be any changes," he snapped. "We've set forth the only conditions under which your treatment can be ethically established. Unless you accept them in their entirety, no reputable doctor will have anything to do with you or your treatment."

With considerable effort I kept a tight rein on my temper. He has a powerful organization behind him, I kept telling myself. You musn't antagonize him.

"In any case, I'll have to have some time to think over your proposition." I stood up.

His eyes, friendly as a cobra's, took my full measure. "Hoxsey," he said levelly, "until you sign that contract you can't see Sgt. Mannix again."

He picked up the telephone, called the hospital, asked for the superintendent, Brother Anthony. "This is Dr. Harris. Until further orders, neither Hoxsey nor Dr. Miller are to be admitted to your hospital, or to communicate in any way with the patient Thomas Mannix."

I waited until he hung up the receiver, then seized the telephone and called the Mannix home. Before I could be connected Dr. Harris reached over the desk and tried to take the telephone away from me. My left elbow flipped up, caught him squarely in the chest and sent him flying into his chair. It promptly toppled over, depositing him in a most undignified position on the floor.

Miss Mannix came on the wire and I explained the situation to her. "If you want your father to get well you'd better get him out of the hospital and take him home. I'll be over to see him this evening and change the dressings."

She assured me she'd get him home immediately.

Dr. Harris picked himself off the floor, his dignity considerably ruffled, his face as red as a boiled lobster.

"You'll never get away with this!" he shrilled. "If you as much as touch that patient I'll have you arrested for practicing medicine without a license. As long as you live you'll never treat cancer again. We'll close down your clinic, run you and that quack doctor of yours out of Illinois. Try and set up anywhere else in this country and you'll wind up in jail."

Without bothering to reply I walked out.

Returning to the hotel, I received a telephone call from Dr. Miller. He was in a booth across the street from the Alexian Brothers Hospital, where he'd gone to see our patient. They'd refused to let him in. I explained what had happened at Dr. Harris' office. When I finished, there was a long silence. Finally he sighed:

"Well, that does it. Harris won't rest now until he's put us out of business. You've made yourself a powerful enemy. It's not just a few local doctors you have to reckon with now, it's the whole Medical Association. They'll hound you and blacken your name, and that of everyone associated with you, from one end of the country to the other. You're young and brash, but how long do you think you can go on bucking the entire medical profession?"

I didn't hesitate a moment. "Until I prove to the world that I can cure cancer. As my Daddy once told me, there's one thing doctors can't do, and that's put back the cancers we remove. Don't worry about me, Doc. I can take anything they dish out. How about you?"

His voice came back strong over the wire: "I still say a doctor's first duty is to his patients. I'll string along with you, my boy."

He waited outside while Kate Mannix, over the strenuous objection of hospital authorities, signed her father out of the institution. When they finally emerged he helped them into a cab and escorted them home. There we continued to treat the policeman until he was fully recovered, three months later.

CARTER GLASS, VA., CHAIRMAN

United States Senate

COMMITTEE ON APPROPRIATIONS

June 2nd, 1947

Mr. Harry M. Hoxsey,
c/o 4507 Gaston Avenue,
Dallas, 4,
Texas.

Dear Mr. Hoxsey:

I have your favor of May 29th and note the further difficulties you are having in carrying on your work.

It seems that the medical fraternity is highly organized
and that they have decided to crush you and your institution, if at all possible. I have had a few "rounds" with the
heads of the medical organization as well as the Public Health
Service here in Washington and it seems that the public officials are afraid that if they make any move, or say anything
antagonistic to the wishes of the medical organization that
they will be pounced upon and destroyed. In other words, the
public officials seem to be afraid of their jobs and even of
their lives. This presents a most serious case and I am at
a loss to know how to proceed.

I am of the opinion that what the medical organization does
will be repeated by the several state organizations in the
event any Congressman or Senator started out to publicly oppose
their program. I have done what I could to have your remedy
or at least the record of your accomplishments considered and
passed upon but, to date, the authorities here have refused
to act.

If I find wherein I can do anything further to help out I
shall gladly avail myself of the opportunity. If you receive
a reply to your letter to the Pure Food and Drug Administration,
I shall be glad to have a copy of the same.

With all good wishes, I am

Sincerely,

Elmer Thomas

ET:f

Letter from U.S. Senator Elmer Thomas, explaining his failure to obtain official investigation of Hoxsey treatment.

13 | Royal Raymond Rife and the Cancer Cure That Worked!

Barry Lynes

In the summer of 1934 in California, under the auspices of the University of Southern California, a group of leading American bacteriologists and doctors conducted one of the first successful cancer clinics. The results showed that:

a) cancer was caused by a micro-organism;

b) the micro-organism could be painlessly destroyed in terminally ill cancer patients; and

c) the effects of the disease could be reversed.

The technical discovery leading to the cancer cure had been described in *Science* magazine in 1931. In the decade following the 1934 clinical success, the technology and the subsequent, successful treatment of cancer patients was discussed at medical conferences, published in a medical journal, cautiously but professionally reported in a major newspaper, and technically explained in an annual report of the Smithsonian Institution.

However, the cancer cure threatened a number of scientists, physicians, and financial interests. A cover-up was initiated. Physicians using the new technology were coerced into abandoning it. The author of the Smithsonian article was followed and then was shot at while driving his car. He never wrote about the subject again. All reports describing the cure were censored by the head of the AMA (American Medical Association) from the major medical journals. Objective scientific evaluation by government laboratories was prevented. And renowned researchers who supported the technology and its new scientific principles in bacteriology were scorned, ridiculed, and called liars to their faces. Eventually, a long, dark silence lasting decades fell over the cancer cure. In time, the cure was labeled a "myth"—it had never happened. However, documents now available prove that the cure did exist, was tested successfully in clinical trials, and in fact was used secretly for years afterwards—continuing to cure cancer as well as other diseases.

BACTERIA AND VIRUSES

In nineteenth-century France, two giants of science collided. One of them is now world-renowned—Louis Pasteur. The other, from whom Pasteur stole many of his best ideas, is now essentially forgotten—Pierre Béchamp.

One of the many areas in which Pasteur and Béchamp argued concerned what is today known as pleomorphism—the occurrence of more than one distinct form of an organism in a single life cycle. Béchamp contended that bacteria could change forms. A rod-shaped bacterium could become a spheroid, etc. Pasteur disagreed. In 1914, Madame Victor Henri of the Pasteur Institute confirmed that Béchamp was correct and Pasteur was wrong.

But Béchamp went much further in his argument for pleomorphism. He contended that bacteria could "devolve" into smaller, unseen forms—what he called microzyma. In other words, Béchamp developed—on the basis of a lifetime of research—a theory that micro-organisms could change their essential size as well as their shape, depending on the state of health of the organism in which the micro-organism lived. This directly contradicted what orthodox medical authorities have believed for most of the twentieth century. Laboratory research in recent years has provided confirmation for Béchamp's idea.

This seemingly esoteric scientific squabble had ramifications far beyond academic institutions. The denial of pleomorphism was one of the cornerstones of twentieth century medical research and cancer treatment. An early twentieth century acceptance of pleomorphism might have prevented millions of Americans from suffering and dying of cancer.

In a paper presented to the New York Academy of Sciences in 1969, Dr. Virginia Livingston and Dr. Eleanor Alexander-Jackson declared that a single cancer micro-organism exists. They said that the reason the army of cancer researchers couldn't find it was because it changed form. Livingston and Alexander-Jackson asserted:

> The organism has remained an unclassified mystery, due in part to its remarkable pleomorphism and its simulation of other micro-organisms. Its various phases may resemble viruses, micrococci, diptheroids, bacilli, and fungi.

THE AMERICAN MEDICAL ASSOCIATION

The American Medical Association was formed in 1846 but it wasn't until 1901 that a reorganisation enabled it to gain power over how medicine was practised throughout America. By becoming a confederation of state medical associations and forcing doctors who wanted to belong to their county medical society to join the state association, the AMA soon increased its membership to include a majority of physicians. Then, by accrediting medical schools, it began determining the standards and practises of doctors. Those who refused to conform lost their licence to practise medicine.

Morris Fishbein was the virtual dictator of the AMA from the mid-1920s until he was ousted on June 6, 1949 at the AMA convention in Atlantic City. But even after he was forced from his position of power because of a revolt from several state delegations of doctors, the policies he had set in motion continued on for many years. He died in the early 1970s.

A few years after the funding of his successful cancer clinic of 1934, Dr. R. T. Hamer, who did not participate in the clinic, began to use the procedure in Southern California. According to Benjamin Cullen, who observed the entire development of the cancer cure from idea to implementation, Fishbein found out and tried to "buy in." When he was turned down, Fishbein unleashed the AMA to destroy the cancer cure. Cullen recalled:

> Dr. Hamer ran an average of forty cases a day through his place. He had to hire two operators. He trained them and watched them very closely. The case histories were mounting very fast. Among them was this old man from Chicago. He had a malignancy all around his face and neck. It was a gory mass. Just terrible. Just a red gory mass. It had taken over all around his face. It had taken off one eyelid at the bottom of the eye. It had taken off the bottom of the lower lobe of the ear and had also gone into the cheek area, nose and chin. He was a sight to behold.
>
> But in six months all that was left was a little black spot on the side of his face and the condition of that was such that it was about to fall off. Now that man was 82 years of age. I never saw anything like it. The delight of having a lovely clean skin again, just like a baby's skin.
>
> Well, he went back to Chicago. Naturally he couldn't keep still and Fishbein heard about it. Fishbein called him in and the old man was kind of reticent about telling him. So Fishbein wined and dined him and finally learned about his cancer treatment by Dr. Hamer in the San Diego clinic.
>
> Soon a man from Los Angeles came down. He had several meetings with us. Finally he took us out to dinner and broached the subject about buying it. Well, we wouldn't do it. The renown was spreading and we weren't even advertising. But of course what did it was the case histories of Dr. Hamer. He said that this was the most marvelous development of the age. His case histories were absolutely wonderful.
>
> Fishbein bribed a partner in the company. With the result we were kicked into court for operating without a license. I was broke after a year.

In 1939, under pressure from the local medical society, Dr. R. T. Hamer abandoned the cure. He is not one of the heroes of this story.

Thus, within the few, short years from 1934 to 1939, the cure for cancer was clinically demonstrated and expanded into curing other diseases on a daily basis by other doctors, and then terminated when Morris Fishbein of the AMA was not allowed to "buy in." It was a practise he had developed into a cold art, but never again would such a single mercenary deed doom millions of Americans to premature, ugly deaths. It was the AMA's most shameful hour.

Another major institution which "staked its claim" in the virgin territory of cancer research in the 1930–1950 period was Memorial Sloan-Kettering Cancer Center in New York. Established in 1884 as the first cancer hospital in America, Memorial Sloan-Kettering from 1940 to the mid-1950s was the centre of drug testing for the largest pharmaceutical companies.

Cornelius P. Rhoads, who had spent the 1930s at the Rockefeller Institute, became the director at Memorial Sloan-Kettering in 1939. He remained in that position until his death in 1959. Rhoads was the head of the chemical warfare service from 1943–1945, and afterwards became the nation's premier advocate of chemotherapy.

It was Dr. Rhoads who prevented Dr. Irene Diller from announcing the discovery of the cancer micro-organism to the New York Academy of Sciences in 1950. It also was Dr. Rhoads who arranged for the funds for Dr. Caspe's New Jersey laboratory to be cancelled after she announced the same discovery in Rome in 1953. An IRS investigation, instigated by an unidentified, powerful New York cancer authority, added to her misery, and the laboratory was closed.

Thus the major players on the cancer field are the doctors, the private research institutions, the pharmaceutical companies, the American Cancer Society, and also the U.S. government through the National Cancer Institute (organizing research) and the Food and Drug Administration (the dreaded FDA which keeps the outsiders on the defensive through raids, legal harassment, and expensive testing procedures).

THE MAN WHO FOUND THE CURE FOR CANCER

In 1913, a man with a love for machines and a scientific curiosity, arrived in San Diego after driving across the country from New York. He had been born in Elkhorn, Nebraska, was 25 years old, and very happily married. He was about to start a new life and open the way to a science of health which will be honored far into the future. His name was Royal Raymond Rife. Close friends, who loved his gentleness and humility while being awed by his genius, called him Roy.

Royal R. Rife was fascinated by bacteriology, microscopes and electronics. For the next seven years (including a mysterious period in the Navy during World War I in which he travelled to Europe to investigate foreign laboratories for the U.S. government), he thought about and experimented in a variety of fields as well as mastered the mechanical skills necessary to build instruments such as the world had never imagined.

By the late 1920s, the first phase of his work was completed. He had built his first microscope, one that broke the existing principles, and he had constructed instruments which enabled him to electronically destroy specific pathological micro-organisms.

Rife believed that the minuteness of the viruses made it impossible to stain them with the existing acid or aniline dye stains. He'd have to find another way. Somewhere along the way, he made an intuitive leap often associated with the greatest scientific discoveries. He conceived first the idea and then the method of staining the virus with light. He began building a microscope which would enable a frequency of light to coordinate with the chemical constituents of the particle or micro-organism under observation.

Rife's second microscope was finished in 1929. In an article which appeared in *The Los Angeles Times Magazine* on December 27, 1931, the existence of the light-staining method was reported to the public:

> Bacilli may thus be studied by their light, exactly as astronomers study moons, suns, and stars by the light which comes from them through telescopes. The bacilli studied are living ones, not corpses killed by stains.

Throughout most of this period, Rife also had been seeking a way to identify and then destroy the micro-organism which caused cancer. His cancer research began in 1922. It would take him until 1932 to isolate the responsible micro-organism which he later named simply the "BX virus."

THE EARLY 1930s

In 1931, the two men who provided the greatest professional support to Royal R. Rife came into his life. Dr. Arthur I. Kendall, Director of Medical Research at Northwestern University Medical School in Illinois, and Dr. Milbank Johnson, a member of the board of directors at Pasadena Hospital in California and an influential power in Los Angeles medical circles.

Dr. Kendall had invented a protein culture medium (called "K Medium" after its inventor) which enabled the "filtrable virus' portions of a bacteria to be isolated and to continue reproducing. This claim directly contradicted the Rockefeller Institute's Dr. Thomas Rivers who in 1926 had authoritatively stated that a virus needed a living tissue for reproduction.

Rife, Kendall and others were to prove within a year that it was possible to cultivate viruses artificially. Rivers, in his ignorance and obstinacy, was responsible for suppressing one of the greatest advances ever made in medical knowledge.

Kendall arrived in California in mid-November 1931 and Johnson introduced him to Rife. Kendall brought his "K Medium" to Rife and Rife brought his microscope to Kendall.

A typhoid germ was put in the "K Medium," triple-filtered through the finest filter available, and the results examined under Rife's microscope. Tiny, distinct bodies stained in a turquoise-blue light were visible. The virus cultures grew in the "K Medium" and were visible. The viruses could be "light"-stained and then classified according to their own colours under Rife's unique microscope.

A later report which appeared in the Smithsonian's annual publication gives a hint of the totally original microscopic technology which enabled man to see a deadly virus-size micro-organism in its live state for the first time (the electron microscope of later years kills its specimens):

> Then they were examined under the Rife microscope where the filterable virus form of typhoid bacillus, emitting a blue spectrum color, caused the plane of polarization to be deviated 4.8 degrees plus. When the opposite angle of refraction was obtained by means of adjusting the polarizing prisms to minus 4.8 degrees and the cultures of viruses were illuminated by the monochromatic beams coordinated with the chemical constituents of the typhoid bacillus, small, oval, actively motile, bright turquoise-blue bodies were observed at 5,000 times magnification, in high contrast to the colorless and motionless debris of the medium. These tests were repeated 18 times to verify the results.

Following the success, Dr. Milbank Johnson quickly arranged a dinner in honour of the two men in order that the discovery could be announced and discussed. More than 30 of the most prominent medical doctors, pathologists, and bacteriologists in Los Angeles attended this historic event on November 20, 1931. Among those in attendance were Dr. Alvin G. Foord, who 20 years later would indicate he knew little about Rife's discoveries, and Dr. George Dock who would serve on the University of Southern California's Special Research Committee overseeing the clinical work until he, too, would "go over" to the opposition.

On November 22, 1931, *The Los Angeles Times* reported this important medical gathering and its scientific significance:

> Scientific discoveries of the greatest magnitude, including a discussion of the world's most powerful microscope recently perfected after 14 years' effort by Dr. Royal R. Rife of San Diego, were described Friday

evening to members of the medical profession, bacteriologists and pathologists at a dinner given by Dr. Milbank Johnson in honor of Dr. Rife and Dr. A. I. Kendall.

Before the gathering of distinguished men, Dr. Kendall told of his researches in cultivating the typhoid bacillus on his new "K Medium." The typhoid bacillus is nonfilterable and is large enough to be seen easily with microscopes in general use. Through the use of "Medium K," Dr. Kendall said, the organism is so altered that it cannot be seen with ordinary microscopes and it becomes small enough to be ultra-microscopic or filterable. It then can be changed back to the microscopic or nonfilterable form.

Through the use of Dr. Rife's powerful microscope, said to have a visual power of magnification to 17,000 times, compared with 2,000 times of which the ordinary microscope is capable, Dr. Kendall said he could see the typhoid bacilli in the filterable or formerly invisible stage. It is probably the first time the minute filterable (virus) organisms ever have been seen.

The strongest microscope now in use can magnify between 2,000 and 2,500 times. Dr. Rife, by an ingenious arrangement of lenses applying an entirely new optical principle and by introducing double quartz prisms and powerful illuminating lights, has devised a microscope with a lowest magnification of 5,000 times and a maximum working magnification of 17,000 times.

The new microscope, scientists predict, also will prove a development of the first magnitude. Frankly dubious about the perfection of a microscope which appears to transcend the limits set by optic science, Dr. Johnson's guests expressed themselves as delighted with the visual demonstration and heartily accorded both Dr. Rife and Dr. Kendall a foremost place in the world's rank of scientists.

Five days later, *The Los Angeles Times* published a photo of Rife and Kendall with the microscope. It was the first time a picture of the super microscope had appeared in public. The headline read, "The World's Most Powerful Microscope."

Meanwhile, Rife and Kendall had prepared an article for the December 1931 issue of California and Western Medicine. "Observations on Bacillus Typhosus in its Filtrable State" described what Rife and Kendall had done and seen. The journal was the official publication of the state medical associations of California, Nevada and Utah.

The prestigious *Science* magazine then carried an article which alerted the scientific community of the entire nation. The December 11, 1931 *Science News* supplement included a section titled, "Filterable Bodies Seen With The Rife Microscope." The article described Kendall's filterable medium culture, the turquoise-blue bodies which were the filtered form of the typhoid bacillus, and Rife's microscope. It included the following description:

> The light used with Dr. Rife's microscope is polarized, that is, it is passing through crystals that stop all rays except those vibrating in one particular plane. By means of a double reflecting prism built into the instrument, it is possible to turn this plane of vibration in any desired direction, controlling the illumination of the minute objects in the field very exactly.

On December 27, 1931, *The Los Angeles Times* reported that Rife had demonstrated the microscope at a meeting of 250 scientists. The article explained:

> This is a new kind of magnifier, and the laws governing microscopes may not apply to it . . . Dr. Rife has developed an instrument that may revolutionize laboratory methods and enable bacteriologists like Dr. Kendall, to identify the germs that produce about 50 diseases whose causes are unknown . . .

Soon Kendall was invited to speak before the Association of American Physicians. The presentation occurred May 3 and 4, 1932 at Johns Hopkins University in Baltimore. And there Dr. Thomas Rivers and Hans Zinsser, two highly influential medical men, stopped the scientific process. Their opposition meant that the development of Rife's discoveries would be slowed. Professional microbiologists would be cautious in even conceding the possibility that Rife and Kendall might have broken new ground. The depression was at its worst. The Rockefeller Institute was not only a source of funding but powerful in the corridors of professional recognition. A great crime resulted because of the uninformed, cruel and unscientific actions of Rivers and Zinsser.

The momentum was slowed at the moment when Rife's discoveries could have "broken out" and triggered a chain reaction of research, clinical treatment and the beginnings of an entirely new health system. By the end of 1932, Rife could destroy the typhus bacteria, the polio virus, the herpes virus, the cancer virus and other viruses in a culture and in experimental animals. Human treatment was only a step away.

The opposition of Rivers and Zinsser in 1932 had a devastating impact on the history of twentieth century medicine. (Zinsser's *Bacteriology*, in an updated version, is still a standard textbook.) Unfortunately, there were few esteemed bacteriologists who were not frightened or awed by Rivers.

But there were two exceptions to this generally unheroic crowd. Christopher Bird's article, "What Has Become Of The Rife Microscope?" which appeared in the March 1976 *New Age Journal*, reports:

> In the midst of the venom and acerbity the only colleague to come to

Kendall's aid was the grand old man of bacteriology, and first teacher of the subject in the United States, Dr. William H. "Popsy" Welch, who evidently looked upon Kendall's work with some regard.

Welch was the foremost pathologist in America at one time. The medical library at Johns Hopkins University is named after him. He rose and said, "Kendall's observation marks a distinct advance in medicine." It did little good. By then Rivers and Zinsser were the powers in the field.

Kendall's other supporter was Dr. Edward C. Rosenow of the Mayo Clinic's Division of Experimental Bacteriology. (The Mayo Clinic was considered then and is today one of the outstanding research and treatment clinics in the world. The Washington Post of January 6, 1987 wrote, "To many in the medical community, the Mayo Clinic is 'the standard' against which other medical centres are judged.") On July 5–7, 1932, just two months after Kendall's public humiliation, the Mayo Clinic's Rosenow met with Kendall and Rife at Kendall's Laboratory at Northwestern University Medical School in Chicago.

"The oval, motile, turquoise-blue virus were demonstrated and shown unmistakably," Rosenow declared in the "Proceedings of the Staff Meetings of the Mayo Clinic, July 13, 1932, Rochester, Minnesota." The virus for herpes was also seen. On August 26, 1932, *Science* magazine published Rosenow's report, "Observations with the Rife Microscope of Filter Passing Forms of Micro-organisms."

In the article, Rosenow stated:

There can be no question of the filtrable turquoise-blue bodies described by Kendall. They are not visible by the ordinary methods of illumination and magnification . . . Examination under the Rife microscope of specimens, containing objects visible with the ordinary microscope, leaves no doubt of the accurate visualization of objects or particulate matter by direct observation at the extremely high magnification (calculated to be 8,000 diameters) obtained with this instrument.

Three days after departing from Rife in Chicago, Rosenow wrote to Rife from the Mayo Clinic:

After seeing what your wonderful microscope will do, and after pondering over the significance of what you revealed with its use during those three strenuous and memorable days spent in Dr. Kendall's laboratory, I hope you will take the necessary time to describe how you obtain what physicists consider the impossible. . . . As I visualize the matter, your ingenious method of illumination with the intense monochromatic beam of light is of even greater importance than the enormously high magnification . . .

Rosenow was right. The unique "colour frequency" staining method was the great breakthrough. Years later, after the arrival of television, an associate of the then deceased Rife would explain, "The viruses were stained with the frequency of light just like colours are tuned in on television sets." It was the best nontechnical description ever conceived.

"BX"—THE VIRUS OF CANCER

Rife began using Kendall's "K Medium" in 1931 in his search for the cancer virus. In 1932, he obtained an unulcerated breast mass that was checked for malignancy from the Paradise Valley Sanitarium of National City, California. But the initial cancer cultures failed to produce the virus he was seeking.

Then a fortuitous accident occurred. The May 11, 1938 *Evening Tribune* of San Diego later described what happened:

> But neither the medium nor the microscope were sufficient alone to reveal the filter-passing organism Rife found in cancers, he recounted. It was an added treatment which he found virtually by chance that finally made this possible, he related. He happened to test a tube of cancer culture within the circle of a tubular ring filled with argon gas activated by an electrical current, which he had been using in experimenting with electronic bombardment of organisms of disease. His cancer culture happened to rest there about 24 hours (with the current on the argon gas-filled tube), and then he noticed (under the microscope) that its appearance seemed to have changed. He studied and tested this phenomenon repeatedly, and thus discovered (cancer virus) filter-passing, red-purple granules in the cultures.

The BX cancer virus was a distinct purplish-red color. Rife had succeeded in isolating the filtrable virus of carcinoma.

Rife's laboratory notes for November 20, 1932, contain the first written description of the cancer virus characteristics. Among them are two, unique to his method of classification using the Rife microscope: angle of refraction—12 3/10 degrees; colour by chemical refraction—purple-red.

The size of the cancer virus was indeed small. The length was 1/15 of a micron. The breadth was 1/20 of a micron. No ordinary light microscope, even in the 1980s, would be able to make the cancer virus visible.

Rife and his laboratory assistant E. S. Free proceeded to confirm his discovery. They repeated the method 104 consecutive times with identical results.

In time, Rife was able to prove that the cancer micro-organism had four forms:

1. BX (carcinoma);

2. BY (sarcoma—larger than BX);

3. Monococcoid form in the monocytes of the blood of over 90 percent of cancer patients. When properly stained, this form can be readily seen with a standard research microscope;

4. Crytomyces pleomorphia fungi—identical morphologically to that of the orchid and of the mushroom.

Rife wrote in his 1953 book: "Any of these forms can be changed back to 'BX' within a period of 36 hours and will produce in the experimental animal a typical tumor with all the pathology of true neoplastic tissue, from which we can again recover the 'BX' micro-organism. This complete process has been duplicated over 300 times with identical and positive results.

Rife had proved pleomorphism. He had shown how the cancer virus changes form, depending on its environment. He had confirmed the work of Béchamp, of Kendall, of Rosenow, of Welch, and an army of pleomorphist bacteriologists who would come after him and have to battle the erroneous orthodox laws of Rivers and his legions of followers.

Rife said, "In reality, it is not the bacteria themselves that produce the disease, but the chemical constituents of these micro-organisms enacting upon the unbalanced cell metabolism of the human body that in actuality produce the disease. We also believe if the metabolism of the human body is perfectly balanced or poised, it is susceptible to no disease."

But Rife did not have time to argue theory. He would leave that for others. After isolating the cancer virus, his next step was to destroy it. He did this with his frequency instruments—over and over again. And then he did it with experimental animals, inoculating them, watching the tumours grow, and then killing the virus in their bodies with the same frequency instruments tuned to the same "BX" frequency.

Rife declared in 1953:

These successful tests were conducted over 400 times with experimental animals before any attempt was made to use this frequency on human cases of carcinoma and sarcoma.

In the summer of 1934, sixteen terminally ill people with cancer and other diseases were brought to the Scripps "ranch." There, as Rife and the doctors worked on human beings for the first time, they learned much. In 1953 when Rife copyrighted his book, he made the real report of what happened in 1934. He wrote:

With the frequency instrument treatment, no tissue is destroyed, no pain is felt, no noise is audible, and no sensation is noticed. A tube lights up and 3 minutes later the treatment is completed. The virus or bacteria is destroyed and the body then recovers itself naturally from the toxic effect of the virus or bacteria. Several diseases may be treated simultaneously.

The first clinical work on cancer was completed under the supervision of Milbank Johnson, M.D., which was set up under a Special Medical Research Committee of the University of Southern California. 16 cases were treated at the clinic for many types of malignancy. After 3 months, 14 of these so called hopeless cases were signed off as clinically cured by the staff of five medical doctors and Dr. Alvin G. Foord, M.D., pathologist for the group. The treatments consisted of 3 minutes' duration using the frequency instrument which was set on the mortal oscillatory rate for "BX" or cancer (at 3 day intervals). It was found that the elapsed time between treatments attains better results than the cases treated daily. This gives the lymphatic system an opportunity to absorb and cast off the toxic condition which is produced by the devitalized dead particles of the "BX" virus. No rise of body temperature was perceptible in any of these cases above normal during or after the frequency instrument treatment. No special diets were used in any of this clinical work, but we sincerely believe that a proper diet compiled for the individual would be of benefit.

Date: December 1, 1953.

Other members of the clinic were Whalen Morrison, Chief Surgeon of the Santa Fe Railway; George C. Dock, M.D., internationally famous, George C. Fischer, M.D., Children's Hospital in New York; Arthur I. Kendall, Dr. Zite, M.D., Professor of Pathology at Chicago University, Rufus B. Von Klein Schmidt, President of the University of Southern California. Dr. Couche and Dr. Carl Meyer, Ph.D., head of the Department of Bacteriological Research at the Hooper Foundation in San Francisco, were also present. Dr. Kopps of the Metabolic Clinic in La Jolla signed all fourteen reports and knew of all the tests from his personal observation. In 1956, Dr. James Couche made the following declaration:

I would like to make this historical record of the amazing scientific wonders regarding the efficacy of the frequencies of the Royal R. Rife Frequency Instrument . . .

When I was told about Dr. Rife and his frequency instrument at the Ellen Scripps home near the Scripps Institute Annex some twenty-two years ago, I went out to see about it and became very interested in the cases which he had there. And the thing that brought me into it more quickly than anything was a man who had a cancer of the stomach. Rife was associated at that time with Dr. Milbank Johnson, M.D., who was then president of the Medical Association of Los Angeles, a very wealthy man and a very big man in the medical world—the biggest in Los Angeles and

A Brief Exposure

Bacilli Revealed By New Microscope

DR. RIFE'S APPARATUS, MAGNIFYING 17,000 TIMES, SHOWS GERMS NEVER BEFORE SEEN.

November 21, 1931, Special to *The New York Times*

LOS ANGELES, Nov. 21—A description of the world's most powerful microscope recently perfected after fourteen years' effort by Dr. Royal Raymond Rife of San Diego, was one of the features of a dinner given last night to members of the medical profession by Dr. Milbank Johnson in honor of Dr. Rife and Dr. Arthur I. Kendall, head of the department of research bacteriology of the Medical School of Northwestern University.

The Strongest microscopes in use magnify 2,000 to 2,500 times. Dr. Rife, by a rearrangement of lenses and by introducing double quartz prisms and illuminating lights, has devised apparatus with a maximum magnification of 17,000 diameters.

Dr. Kendall told of cultivating the typhoid bacillus on his new "medium K." This bacillus is ordinarily nonfilterable. By the use of Dr. Rife's microscope, Dr. Kendall said, the typhoid bacilli can be seen in the filterable or formerly invisible stage.

---

he had hired this annex for this demonstration over a summer of time.

In that period of time I saw many things and the one that impressed me the most was a man who staggered onto a table, just on the last end of cancer; he was a bag of bones. As he lay on the table, Dr. Rife and Dr. Johnson said, "Just feel that man's stomach." So I put my hand on the cavity where his stomach was underneath and it was just a cavity almost, because he was so thin; his backbone and his belly were just about touching each other.

I put my hand on his stomach which was just one solid mass, just about what I could cover with my hand, somewhat like the shape of a heart. It was absolutely solid! And I thought to myself, well, nothing can be done for that. However, they gave him a treatment with the Rife frequencies and in the course of time over a period of six weeks to two months, to my astonishment, he completely recovered. He got so well that he asked permission to go to El Centro as he had a farm there and he wanted to see about his stock. Dr. Rife said, "Now you haven't the strength to drive to El Centro."

"Oh, yes," said he. "I have, but I'll have a man to drive me there." As a matter of fact, the patient drove his own car there and when he got down to El Centro he had a sick cow and he stayed up all night with it. The next day he drove back without any rest whatsoever—so you can imagine how he had recovered.

I saw other cases that were very interesting. Then I wanted a copy of the frequency instrument. I finally bought one of these frequency instruments and established it in my office.

I saw some very remarkable things resulting from it in the course of over twenty years.

Note: Biophysicists have now shown that there exists a crucial natural interaction between living matter and photons. This process is measurable at the cellular (bacterium) level. Other research has demonstrated that living systems are extraordinarily sensitive to extremely low-energy electromagnetic waves. This is to say, each kind of cell or micro-organism has a specific frequency of interaction with the electromagnetic spectrum. By various means, Rife's system allowed adjusting the frequency of light impinging on the specimen. By some insight he learned that the light frequency could be "tuned" into the natural frequency of the micro-organism being examined to cause a resonance or feedback loop. In effect, under this condition, it can be said the micro-organism illuminated itself.

Rife extrapolated from his lighting technique, which we may be certain he understood, that specific electromagnetic frequencies would have a negative effect on specific bacterial forms. There can remain no doubt that Rife demonstrated the correctness of his hypothesis to himself and those few who had the courage to look and the perceptual acuity to see! The same new discoveries in biophysics not only explain Rife's principle of illumination; they also explain his process for selective destruction of bacteria. The latter phenomenon is similar to ultrasonic cleaning, differing in delicate selectivity of wave form and frequency. Recently, researchers whose findings have been suppressed, have caused and cured cancer in the same group of mice by subjecting them to certain electromagnetic fields. Rife's work was far more sophisticated. He selected specific microscopic targets, and actually saw the targets explode.

A body of recognised scientific evidence now overwhelmingly supports the original cancer theories articulated and demonstrated by Rife fifty years ago. This includes modern AIDS research.

In December of 1931, Dr. Royal Raymond Rife and his colleague, Dr. Isaac Kendal, published their findings in *California and Western Medicine.* The following article describes the outcome of their research, as presented to the scientific community.

Observations on Bacillus Typhosus in Its Filterable State

Arthur Isaac Kendall, Ph.D. and Royal Raymond Rife, Ph.D.

Presented at a meeting of the Bacteriological Section of the Los Angeles Clinical and Pathological Society, November 20, 1931.

It seems improbable that viable bacteria in the filterable state have ever been unequivocally seen. Nevertheless, the theoretical and practical importance of filterable forms of bacteria in theoretical and applied biology cannot be denied.

Recently, through the simultaneous availability of the Rife microscope, an instrument combining very high magnification with coordinated resolving power, and a simple procedure for inducing the filterable state in bacteria at will,[1] the possibility of actually demonstrating organisms in this hitherto illusive condition very obviously presented itself.

Two features of the Rife microscope, full details of which will be presented elsewhere, must be specifically mentioned here. First, the entire optical system, including not only the lenses but also the illuminating unit, is made of quartz. In addition, a double wedge quartz prism is mounted between the illuminating unit and the quartz Abbé condenser. The latter can be rotated, with vernier control, through 360 degrees, thereby affording readily controllable polarized light at any required angle. The import of this polarization unit will be discussed later. Inasmuch as this microscope magnifies from 5,000 to 17,000 diameters, it is obviously very necessary to have it mounted upon an immovable foundation.

The organism selected for these experiments was the well known Rawlings strain of *B. typhosus*. The immediate history of the culture used is as follows:

October 29, 1931. An agar slant was made of a thrice-plated culture of *B. typhosus*, Rawlings strain. (Editor's Note: This agar slant was made in the Laboratory of Research Bacteriology, Northwestern University Medical School, Chicago, Illinois.)

November 2, 4 P.M. Inoculated six cubic centimeters of K (protein) Medium[2] from the agar slant culture.

November 3, 10 A.M. Filtered this culture in K Medium of November 2, through a Berkefeld "N" filter. (The culture was diluted with four volumes of sterile physiological saline solution; the vacuum used was less than four inches of water; the total time of filtration was less than ten minutes.)

November 3. One drop of filtrate, representing one-fifth drop of the original culture, was introduced into six cubic centimeters of K Medium. Incubated at 37°C. The filtrate was also tested for purity as follows: (1) cultural reactions; (2) sugar fermentation reactions; (3) agglutination with specific typhoid serum. All were typical.

November 5. The forty-eight-hour culture of November 3 in K Medium was filtered, as above, through a Berkfeld "N" filter. One drop of the filtrate was added to six cubic centimeters of K Medium and incubated at 37°C. Growth was abundant November 7.

November 9. The culture was again transferred to K Medium.

November 12. Still another culture was made, in every instance using three loops of culture for the inoculum.

It is worthy of note that this thrice filtered culture of *B. typhosus* grew quite readily in K Medium as above outlined: after the second filtration it failed to grow in peptone broth. In other words, the organism, having become filterable and accustomed to the protein media (proteophilic), lost its ability to grow in ordinary peptone . . .

The cultures of November 9 and November 12 were examined under the microscope and there were no discernible bacilli, although the cultures were markedly turbid. Darkfield illumination revealed very small, actively motile granules, and direct observation of these with the oil emersion lens confirmed the presence of these motile granules, without, however, affording any indication of their structure. Therefore, these granules for obvious reasons could not be unequivocally diagnosed as the filterable form of the bacillus.

In this viable, filtered state the culture was taken to Pasadena, California, and, through the instrumentality of Dr. Milbank Johnson, the cooperation of Dr. Alvin G. Foord, and the courtesy of the Pasadena Hospital, the necessary space and equipment for mounting the microscope and continuing the cultures were made available. The subsequent developments, which are the immediate subject of this discussion, are as follows:

November 16. The cultures of November 12, made in Chicago, were transferred to fresh K Medium and incubated at 37°C overnight.

November 17. The Rife microscope was installed and the first cultures, those inoculated November 16, were examined. The preliminary observations of these cultures were made with a polarizing microscope with a spectroscopic attachment. It should be borne in mind that the entire optical system of this micropolarimeter was of quartz. A one-eighteenth-inch apochromatic oil immersion lens was used, with a 20x quartz ocular.

When a culture of *B. typhosus* in the filterable state, grown as above indicated in K Medium was examined with this micropolarimeter, it was observed that the plane of polarization of the light passing through the culture was deviated plus 4.8 degrees, with the simultaneous appearance of a definite blue spectrum. With this observation in mind, the culture was next studied with the Rife microscope at 5000 diameters.

The double wedge quartz prism referred to above was set by means of the vernier to minus 4.8 degrees. Examined in this polarized light this thrice filtered culture of *B. typhosus* cultivated in K (protein) Medium showed small, oval granules, many of them quite actively motile. These motile granules when in *true focus* appeared as bright turquoise-blue bodies, which contrast strikingly, both in color and in their active motion, with the noncolored, nonmotile debris of the medium.

These observations were repeated eight times, using in each instance growth of the filterable organisms in K Medium. The cultures examined were both twenty-four and forty-eight hours old. The qualitative results were always . . . the occurrence of small, oval, actively motile, turquoise-blue bodies in the cultures and the absence of these small, oval, actively motile, turquoise-blue bodies in the uninoculated control K Media.

From the two facts thus far arrived at, namely, that the small, oval, turquoise-blue bodies were actively motile and also that they were cultivable from K Medium to K Medium, it is surmised that these small, oval, motile, turquoise-blue bodies are indeed the filterable forms of the *B. typhosus*.

There is another even more direct procedure for establishing the identity of these small, oval, motile, turquoise-blue bodies. It has been shown in previous communications[3] that agar cultures, or better, broth cultures of *B. typhosus* inoculated into K Medium, become filterable within eighteen hours' growth at 37°C. It should follow, inasmuch as not all of the bacilli appear to become filterable under these conditions, that at least some of the bacilli should have similar turquoise-blue granules within their substance if they are indeed passing to the filterable state. Also the free swimming filterable forms, the small, oval, motile, urquoise-blue bodies described above, should be simultaneously present.

Darkfield examination of such a culture eighteen hours old revealed unchanged, actively motile bacilli, bacilli with granules within their substance, and free swimming, actively motile granules. This culture examined in the Rife microscope with the quartz prism set at minus 4.8 degrees and with 5,000 diameters magnification, showed very clearly the three types of organisms just described, namely:

First, unchanged bacilli: These were relatively long, actively motile, and almost devoid of color.

Second, long, actively motile bacilli, each with a rather prominent granule at one end. The granule in such an organism was turquoise blue, reminiscent in size, shape, and color of the small, oval, actively motile, turquoise-blue granules found in the protein medium (K Medium) where, it will be recalled, no formed (rod shaped) bacteria could be demonstrated. These bacilli having the turquoise-blue granules were colored only at the granule end, the remainder of the rod being nearly colorless, in this respect corresponding to the unchanged (nonfilterable) bacilli just mentioned.

Third, free swimming, small, oval, actively motile, turquoise-blue granules, precisely similar, apparently, in size, shape, and color to those seen in the granulated bacilli just described.

From the fact that these small, oval, turquoise-blue bodies could be seen both in the parent rod and free swimming in the medium, it is assumed that these small, oval, actively motile, turquoise-blue bodies are indeed the filterable form of *B. typhosus*.

REFERENCES

1. James A. Patten Lecture, *Northwestern University Bulletin*, Vol. 32, No. 5 (September 28), 1931

2. *Northwestern University Medical School Bulletin*, Vol. 32, No. 8 (October 19), 1931, for full details.

3. Op. cit.

The Persecution and Trial of Gaston Naessens

The True Story of the Efforts to Suppress an Alternative Treatment for Cancer, AIDS, and Other Immunologically Based Diseases

Christopher Bird

Most secrets of knowledge have been discovered by plain and neglected men than by men of popular fame. And this is so with good reason. For the men of popular fame are busy on popular matters.

Roger Bacon (c. 1220–1292),
English philosopher and scientist

This is about a man who, in one lifetime, has been both to heaven and to hell. In paradise, he was bestowed a gift granted to few, one that has allowed him to see far beyond our times and thus to make discoveries that may not properly be recognized until well into the next century.

If the "seer's" ability is usually attributed to extrasensory perception, Gaston Naessens's "sixth" sense is a microscope made of hardware that he invented while still in his twenties. Able to manipulate light in a way still not wholly accountable to physics and optics, this microscope has allowed Naessens a unique view into a "microbeyond" inaccessible to those using state-of-the-art instruments.

This lone explorer has thus made an exciting foray into a microscopic world one might believe to be penetrable only by a clairvoyant. In that world, Naessens has "clear-seeingly" descried microscopic forms far more minuscule than any previously revealed. Christened *somatids* (tiny bodies), they circulate, by the millions upon millions, in the blood of you, me, and every other man, woman, and child, as well in that of all animals, and even in the sap of plants upon which those animals and human beings depend for their existence. These ultramicroscopic, subcellular, living and reproducing forms seem to constitute the very basis for life itself, the origin of which has for long been one of the most puzzling conundrums in the annals natural philosophy, today more sterilely called "science."

Gaston Naessens's trip to hell was a direct consequence of his having dared to wander into scientific *terra incognita*. For it is a sad fact that, these days, in the precincts ruled by the "arbiters of knowledge," disclosure of "unknown" things, instead of being welcomed with excitement, is often castigated as illusory, or tabooed as "fantasy." Nowhere are these taboos more stringent than in the field of the biomedical sciences and the multibillion dollar pharmaceutical industry with which it interacts.

In 1985, Gaston Naessens was indicted on several counts, the most serious of which carried a potential sentence of life imprisonment. His trial, which ran from 10 November to 1 December 1989, is reported here.

When I learned about Gaston Naessens's imprisonment, I left California, where I was living and working, to come to Quebec to see what was happening. I owed a debt to the man who stood accused not so much for the crimes for which he was to be legally prosecuted as for what he had so brilliantly discovered during a research life covering forty years. To partially pay that debt, I wrote an article entitled "In Defense of Gaston Naessens," which appeared in the September-October issue of the *New Age Journal* (Boston, Massachusetts). That article has elicited dozens of telephone calls both to the magazine's editors and to Naessens himself.

Because the trial was to take place in a small French-speaking enclave in the vastness of the North American continent, I felt it important, as an American who had had the opportunity to master the French language, to cover the day-to-day proceedings of an event of great historical importance, which, because it took place in a linguistic islet, unfortunately did not make headlines in Canadian urban centres such as Halifax, Toronto, Calgary, or Vancouver, not to speak of American cities.

When the trial was over, Gaston Naessens asked me, over lunch, whether, instead of writing the long book on his fascinating life and work that I was planning, I could quickly write a shorter one on the trial based on the copious notes I had taken. He felt it was of great importance that the public be informed of what had happened at the trial.

I agreed to take on the task because I knew that a great deal was at stake, not the least of which are the fates of patients suffering from the incurable degenerative diseases that Naessens's treatments, developed as a result of his microscopic observations, have been able to cure.

The tribulations and the multiple trials undergone by Naessens will come to an end only when an enlightened populace exerts the pressure needed to make the rulers of its health-care organizations see the light.

DISCOVERY OF THE WORLD'S
SMALLEST LIVING ORGANISM

When the great innovation appears, it will
almost certainly be in a muddled, incomplete,
and confusing form . . . for any speculation
which does not at first glance look crazy, there is no hope.

Freeman Dyson, *Disturbing the Universe*

Early in the morning of 27 June 1989, a tall, bald French-born biologist of aristocratic mien walked into the Palais de Justice in Sherbrooke, Quebec, to attend a hearing that was to set a date for his trial. On the front steps of the building were massed over one hundred demonstrators, who gave him an ovation as he passed by.

The demonstrators were carrying a small forest of placards and banners. The most eye-catchingly prominent among these signs read: "Freedom of Speech, Freedom of Medical Choice, Freedom in Canada!"; "Long Live Real Medicine, Down With Medical Power!"; "Cancer and AIDS Research in Shackles While a True Discoverer is Jailed!"; "Thank you, Gaston, for having saved my life!"; and, simplest of all: "Justice for Naessens!"

Late one afternoon, almost a month earlier, as he arrived home at his house and basement laboratory just outside the tiny hamlet of Rock Forest, Quebec, Gaston Naessens had been disturbed to see a swarm of newsmen in his front yard. They had been alerted beforehand—possibly illegally—by officers of the *Sureté*, Quebec's provincial police force, who promptly arrived to fulfill their mission.

As television cameras whirred and cameras flashed, Naessens was hustled into a police car and driven to a Sherbrooke jail where, pending a preliminary court hearing, he was held for twenty-four hours in a tiny cell under conditions he would later describe as the "filthiest imaginable." Provided only with a cot begrimed with human excrement, the always elegantly dressed scientist told how his clothes were so foul smelling after his release on ten thousand dollars bail that, when he returned home, his wife, Francoise, burned them to ashes.

It was to that same house that I had first come in 1978, on the recommendation of Eva Reich, M.D., daughter of the controversial psychiatrist-turned-biophysicist Wilhelm Reich, M.D. A couple of years prior to my visit with Eva, I had researched the amazing case of Royal Raymond Rife, an autodidact and genius living in San Diego, California, who had developed a "Universal Microscope" in the 1920s with which he was able to see, at magnifications surpassing 30,000 fold, never-before-seen microorganisms in living blood and tissue.*

Eva Reich, who had heard Naessens give a fascinating lecture in Toronto, told me I had another "Rife" to investigate. So I drove up through Vermont to a region just north of the Canadian-American border that is known, in French, as "L'Estrie," and, in English, "The Eastern Townships." And, there, in the unlikeliest of outbacks, Gaston Naessens and his Quebec-born wife, Francoise (a hospital laboratory technician and, for more than twenty-five years, her husband's only assistant), began opening my eyes to a world of research that bids fair to revolutionize the fields of microscopy, microbiology, immunology, clinical diagnosis, and medical treatment.

Let us have a brief look at Naessens's discoveries in these usually separated fields to see, step by step, the research trail over which, for the last forty years—half of them in France, the other half in Canada—he has travelled to interconnect them. In the 1950s, while still in the land of his birth, Naessens, who had never heard of Rife, invented a microscope, one of a kind, and the first one since the Californian's, capable of viewing living entities far smaller than can be seen in existing light microscopes.

In a letter of 6 September 1989, Rolf Wieland, senior microscopy expert for the world-known German optics firm Carl Zeiss, wrote from his company's Toronto office: "What I have seen is a remarkable advancement in light microscopy . . . It seems to be an avenue that should be pursued for the betterment of science." And in another letter, dated 12 October 1989, Dr. Thomas G. Tornabene, director of the School for Applied Biology at the Georgia Institute of Technology (Georgia Tech), who made a special trip to Naessens's laboratory, where he inspected the microscope, wrote:

> Naessens's ability to directly view fresh biological samples was indeed impressive . . . Most exciting were the differences one could immediately observe between blood samples drawn from infected and noninfected patients, particularly AIDS patients. Naessens's microscope and expertise should be immensely valuable to many researchers.

It would seem that this feat alone should be worthy of an international prize in science to a man who can easily be called a twentieth-century "Galileo of the microscope."

With his exceptional instrument, Naessens next went on to discover in

---

* "What Has Become of the Rife Microscope?," *New Age Journal* (Boston, Massachusetts), 1976. This article has, ever since, been one of the *Journal's* most requested reprints. Developments in microscopic techniques have only recently begun to match those elaborated by Naessens more than forty years ago.

the blood of animals and humans—as well as in the saps of plants—a hitherto unknown, ultramicroscopic, subcellular, living and reproducing microscopic form, which he christened a *somatid* (tiny body). This new particle, he found, could be cultured, that is, grown, outside the bodies of its hosts (*in vitro*, "under glass," as the technical term has it). And, strangely enough, this particle was seen by Naessens to develop in a pleomorphic (form-changing) cycle, the first three stages of which—somatid, spore, and double spore—are perfectly normal in healthy organisms, in fact crucial to their existence. (See Figure 1.)

Even stranger, over the years the somatids were revealed to be virtually *indestructible*! They have resisted exposure to carbonization tempera-

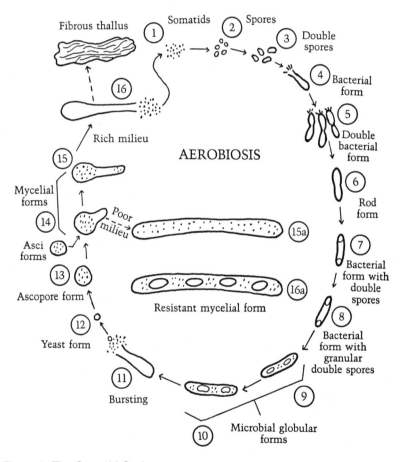

Figure 1. The Somatid Cycle
The figure above shows the complete somatid cycle first observed by Gaston Naessens. The somatid is a subcellular living, reproducing form which has been found to be virtually indestructible. This illustration shows the pleomorphic (form-changing) somatid going through sixteen separate forms.

tures of 200°C and more. They have survived exposure to 50,000 rems of nuclear radiation, far more than enough to kill any living thing. They have been totally unaffected by any acid. Taken from centrifuge residues, they have been found impossible to cut with a diamond knife, so unbelievably impervious to any such attempts is their hardness.

The eerie implication is that the new minuscule life forms revealed by Naessens's microscope are imperishable. At the death of their hosts, such as ourselves, they return to the earth, where they live on for thousands or millions, perhaps billions, of years!

This conclusion—mind-boggling on the face of it—is not one that sprang full-blown from Naessens mind alone. A few years ago, I came across a fascinating doctoral dissertation, published as a book, authored by a pharmacist living in France named Marie Nonclercq.

Several years in the writing, Nonclercq's thesis delved into a long-lost chapter in the history of science that has all but been forgotten for more than a century. This chapter concerned a violent controversy between, on the one side, the illustrious Louis Pasteur, whose name, inscribed on the lintels of research institutes all over the world, is known to all school-children, if only because of the pasteurized milk they drink.

On the other side was Pasteur's nineteenth-century contemporary and adversary, Antoine Béchamp, who first worked in Strasbourg as a professor of physics and toxicology at the Higher School of Pharmacy, later as professor of medical chemistry at the University of Montpellier, and, later still, as professor of biochemistry and dean of faculty of medicine at the University of Lille, all in France.

While laboring on problems of fermentation, the breakdown of complex molecules into organic compounds via "ferment"—one need only think of curdling milk by bacteria—Béchamp, at his microscope, far more primitive that Naessens's own instrument, seemed able to descry a host of tiny bodies in his fermenting solutions. Even before Béchamp's time, other researchers had observed, but passed off as unexplainable, what they called "scintillating corpuscles" or "molecular granulations." Béchamp, who was able to ascribe strong enzymatic (catalytic change-causing) reactions to them, was led to coin a new word to describe them: *microzymas* (tiny ferments).

Among these ferments' many peculiar characteristics was one showing that, whereas they did not exist in chemically pure calcium carbonate made in a laboratory under artificial conditions, they were abundantly present in natural calcium carbonate, commonly known as chalk. For this reason, the latter could, for instance, easily "invert" [ferment] cane sugar solutions, while the former could not.

With the collaboration of his son, Joseph, and Alfred Estor, a Montpellier physician and surgeon, Béchamp went on to study microzymas located in the bodies of animals and came to the startling conclusion that the tiny forms were far more basic to life than cells, long considered to be the basic building blocks of all living matter. Béchamp thought them to be fundamental elements responsible for the activity of cells, tissues, organs, and indeed whole living organisms, from bacteria to whales, and larks to human beings. He even found them present in life-engendering eggs, where they were responsible for the eggs' further development while themselves undergoing significant changes.

So, nearly a century before Gaston Naessens christened his *somatid*, his countryman, Béchamp, had come across organisms that, as Naessens immediately recognized, seem to be "cousins," however many times removed of his own "tiny bodies."

Most incredible to Béchamp was the fact that, when an event serious enough to affect the whole of an organism occurred, the microzymas within it began working to disintegrate it totally, while at the same time continuing to survive. As proof of such survival, Béchamp found these microzymas in soil, swamps, chimney soot, street dust, even air and water. These basic and apparently eternal elements of which we and all our animal relatives are composed survive the remnants of living cells in our bodies that disappear at our deaths.

So seemingly indestructible were the microzymas that Béchamp could even find them in limestone dating to the Tertiary, the first part of the Cenozoic Era, a period going back sixty million years, during which mammals began to make their appearance on earth.

And it could be that they are older still—far older. Professor Edouard Boureau, a French paleontologist, writes in his book *Terre: Mère de la Vie (Earth: Mother of Life)*, concerning problems of evolution, that he had studied thin sections of rock, over three billion years old, taken from the heart of the Sahara Desert. These sections contained tiny round coccoid forms, which Boureau placed at the base of the whole of the evolutionary chain, a chain that he considers might possibly have developed in one of three alternative ways. What these tiny coccoid forms could possibly be, Boureau does not actually know, but, from long study, he is sure about the fact they were around that long ago.

When I brought the book to Naessens's attention, he told me, ingenuously and forthrightly, "I'd sure like to have a few samples of moon rocks to section and examine at my microscope. Who knows, we might find somatid forms in them, the same traces of primitive life that exist on earth!"

Over years of careful microscopic observation and laboratory experimentation, Naessens went on to discover that if and when the immune system of an animal or human being becomes weakened or destabilized, the normal three-stage cycle of the somatid goes through thirteen of more successive growth stages to make up a total of *sixteen* separate forms, each evolving into the next. (See diagram on the somatid cycle.)

All of these forms have been revealed clearly and in detail by motion pictures, and by stop-frame still photography, at Naessens's microscope. Naessens attributes this weakening, as did Béchamp, to trauma brought on by a host of reasons, ranging from exposure to various forms of radiation or chemical pollution to accidents, shocks, depressed psychological states, and many more.

By studying the somatid cycle as revealed in the blood of human beings suffering from various degenerative diseases such as rheumatoid arthritis, multiple sclerosis, lupus, cancer, and, most recently, AIDS, Naessens has been able to associate the development of the forms in the sixteen-stage pathological cycle with all of these diseases. A videocassette showing these new microbiological phenomena is available. Among other things, it shows that when blood is washed to remove all somatids *external* to the blood's red cells, then heated, somatids latently present in a liquid state *within* the red blood cells themselves take concrete form and go on to develop into the sixteen-stage cycle. "This," says Naessens, "is what happens when there is immune system disequilibrium." It is not yet known exactly how or why or from what the somatids take shape. Of the some 140 proteins in red blood cells, many may play a role in the process. The appearance of somatids inside red blood cells is thus an enigma as puzzling as the origin of life itself. I once asked Naessens, "If there were no somatids, would there be no life?"

"That's what I believe," he replied.

Even more importantly, Naessens has been able to predict the eventual onset of such diseases long before any clinical signs of them have put in an appearance. In other words, he can "prediagnose" them. And he has come to demonstrate that such afflictions have a common functional principle, or basis, and therefore must not be considered as separate, unrelated phenomena as they have for so long been considered in orthodox medical circles.

Having established the somatid cycle in all its fullness, Naessens was able, in a parallel series of brilliant research steps, to develop a treatment for strengthening the immune system. The product he developed is derived from camphor, a natural substance produced by an East Asian tree of the same name. Unlike many medicinals, it is injected into the body, not intramuscularly or intravenously, but *intralymphatically*—into the lymph system, via a lymph node, or ganglion, in the groin.

In fact, one of the main reasons the medical fraternity holds the whole of Naessens's approach to be bogus is its assertion that intralymphatic injection is impossible! Yet the fact remains that such injection is not only possible, but simple, for most people to accomplish, once they are properly instructed in how to find the node. While most doctors are never taught this technique in medical school it is so easy that laypeople have been taught to inject, and even to self-inject, the camphor-derived product within a few hours.

The camphor-derived product is named "714-X"—the 7 and the 14 refer to the seventh letter "G" and the fourteenth letter "N" of the alphabet, the first letters of the inventor's first and last names, and the X refers to the twenty-fourth letter of the alphabet, which denotes the year of Naessens's birth, 1924. When skillfully injected, 714-X has, in over seventy-five percent of cases, restabilized, strengthened, or otherwise enhanced the powers of the immune system, which then goes about its normal business of ridding the body of disease.

Let us for a moment return to the work and revelations of Antoine Béchamp. As already noted, with the fairly primitive micorscopic technology available in Béchamp's day, it was almost incredible that he was seemingly able to make microbiological discoveries closely paralleling, if not completely matching, those of Naessens nearly a hundred years later. We have already alluded to the fact that the microzymas in traumatized animals did not remain passive, as before, but, on the contrary, became highly active and began to destroy the bodies of their hosts, converting themselves to bacteria and other microbes in order to carry out that function.

While the terminology is not exactly one that Gaston Naessens would use today, the principles of trauma and of destruction of the body are shared in common by the two researchers. Had Béchamp had access to Naessens's microscope, he, too, might have established the somatid cycle in all the detail worked out by Naessens.

So what happened to Béchamp and his twentieth-century discoveries made in the middle of the nineteenth century? The sad fact is that, because he was modest and retiring—just like Gaston Naessens—his work was overshadowed by that of his rival. All of Pasteur's biographies make clear that he was, above all, a master of the art of self-promotion. But, odd as it seems, the same biographies do not reveal any hint of his battle with Béchamp, many of whose findings Pasteur, in fact, plagiarized.

Even more significant is that while Béchamp, as we have seen, championed the idea that the cause of disease lay within the body, Pasteur, by denouncing his famous "germ theory," held that the cause came from without. In those days, little was known about the functioning of the immune system, but what else can explain, for instance, why some people survived the Black Plague of the Middle Ages, while countless others died like flies? And one may add that Royal Raymond Rife's microscope, like that of Naessens, allowed him to state unequivocally that "germs are not the cause but the result of disease!" Naessens independently adopted the view as a result of his biological detective work. The opposite view, which won the day in Pasteur's time, has dominated medical philosophy for over a century, and what amounted to the creation of a whole new worldview in the life sciences is still regarded as heretical!

Yet the plain fact is that, based on Naessens's medical philosophy as foreshadowed by Béchamp and Rife, up to the present time, Naessens's treatment has arrested and reversed the progress of disease in over one thousand cases of cancer (many of them considered terminal), as well as in several dozen cases of AIDS, a disease for which the world medical community states that there is no solution, as yet. Suffering patients of each sex, and of ages ranging from the teens to beyond the seventies, have been returned to an optimal feeling of well-being and health.

A layperson having no idea of the scope of Naessens's discoveries, or their full meaning and basic implications, might best be introduced to them through Naessens's explanation to a visiting journalist. "You see," began Naessens, "I've been able to establish a life cycle of forms in the blood that add up to no less than a brand new understanding for the very basis of life. What we're talking about is an entirely new biology, one out of which has fortunately sprung practical applications of benefit to sick people, even before all of its many theoretical aspects have been sorted out." At this point, Naessens threw in a statement that would startle any biologist, particularly a geneticist: "The somatids, one can say, are precursors of DNA. Which means that they somehow supply a 'missing link' to an understanding of that remarkable molecule that up to now has been considered as an all but irreducible building block in the life process."*

If somatids were a "missing link" between the living and the nonliving, then what, I wondered aloud in one of my meetings with Francoise

---

* Intriguing is a recent discovery by Norwegian microbiologists. On 10 August 1989, as Naessens was preparing for trial, the world's most prestigious scientific journal, *Nature* (United Kingdom), ran an article entitled "High Abundance of Viruses Found in Aquatic Environments." Authorised by Ovind Bergh and colleagues at the University of Bergen, it revealed that, for the first time, in natural unpolluted waters, hitherto considered to have extremely low concentrations of viruses, there exist up to 2.5 trillion strange viral particles for each litre of liquid. Measuring less than 0.2 microns, their size equates to the largest of Naessens's somatids. Much too small for any larger marine organism to ingest, the tiny organisms are upsetting existing theories on how pelagic life systems operate.

In light of Gaston Naessens's theory that his somatids are DNA precursors, it is fascinating that the Norwegian researchers believe that the hordes upon hordes of viruses might account for DNA's being inexplicably dissolved in seawater. Another amazing implication of the high viral abundance is that routine viral infection of aquatic bacteria could be explained by a significant exchange of genetic material. As Evelyn B. Sherr, of the University of Georgia's Marine Institute on Sapelo Island, writes in a sidebar article in the same issue of *Nature*: "Natural genetic engineering experiments may have been occurring in bacterial populations, perhaps for eons." What connection the aqua-viruses may have with Naessens's somatids is a question that may become answerable when Naessens has the opportunity to observe them at his microscope and compare them with the ones he has already found in vegetal saps and mammalian blood.

Naessens, would be the difference between them and viruses, a long debate about the animate or inanimate nature of which has been going on for years? There was something, was there not, about the somatid that related to its nonreliance and nondependence upon any surrounding milieu needed by the virus, if it were to thrive.

"Yes," agreed Francoise, "to continue its existence, the virus needs a supportive milieu, say, an artificially created test-tube culture, or something natural, like an egg. If the virus needs this kind of support for growth, either *in vivo* or *in vitro*—a 'helping hand,' as it were—the somatid is able to live autonomously, either in a 'living body,' or 'glass-enclosed.' This has something to do with the fact that, while the virus is a particle of DNA, a piece of it, the somatid, as we've already said, is a 'precursor' of DNA, something that leads to its creation."

To try to get to the bottom of this seemingly revolutionary pronouncement, I later asked Francoise to set down on paper some further exposition of it. She wrote:

> We have come to the conclusion that the somatid is no less than what could be termed a "concretization of energy." One could say that this particle, one that is "initially differentiated," or materialized in the life process, possesses genetic properties transmissible to living organisms, animal or vegetal. Underlying that conclusion is our finding that, in the absence of the normal three-stage cycle, *no cellular division can occur!* Why not? Because it is the normal cycle that produces a special growth hormone that permits such division. We believe that hormone to be closely related, if not identical, to the one discovered years ago by the French Nobel Laureate Alexis Carrel, who called it a *trephone*.

The best experimental proof backing up this astounding disclosure, Francoise went on, begins with a cube of fresh meat no different from those impaled on shish kebab skewers. After being injected with somatids taken from an *in vitro* culture, the meat cube is placed in a sealed vessel in which a vacuum is created. With the cube now protected from any contamination from the ambient atmosphere, and anything that atmosphere might contain that could act to putrefy the meat, the vessel is subsequently exposed during the day to natural light by setting it, for instance, next to a window.

Harbouring the living, indestructible somatids as it does, the meat cube in the vessel will, thenceforth, not rot, as it surely would have rotted had it not received the injection. Retaining its healthy-looking colour, it not only remains as fresh as when inserted into the vessel, but progressively increases in size, that is, it continues to grow, just as if it were part of a living organism.

Could a meat cube, animated by somatids, if somehow also electrically stimulated, keep on growing to revive the steer or hog from which it had been cut out? The thought flashed inanely through my mind. Maybe there was something electrical about the somatid? Before I could ask that question of her, Francoise seemed to have already anticipated it.

"The 'tiny bodies' discovered by Naessens," she went on, "are fundamentally electrical in nature. In a liquid milieu, such as blood plasma, one can observe their electrical charge and its effects. For the nuclei of these particles are positively charged, while the membranes, coating their exteriors, are negatively charged. Thus, when they come near one another, they are automatically mutually repulsed just as if they were the negative poles of two bar magnets that resist any manual attempt to hold them together."

"Well," I asked, "isn't that the same as for cells, whose nuclei and membranes are, respectively, considered to have plus, and minus, electrical charges?"

"Certainly," she replied, "with the difference that, in the case of the somatids, the energetic release is very much larger. Somatids are actually tiny living condensers of energy, the smallest ever found."

I was thunderstruck. What, I mused, would the great Hungarian scientist Albert Szent-Györgyi, winner of the Nobel prize for his discovery of ascorbic acid (vitamin C) and many other awards, have had to say had he, before his recent death, been aware of Naessens's discoveries? For it was Szent-Györgyi who, abandoning early attempts to get at the "secret of life" at the level of the molecule, had predicted, prior to World War II, when still living and working in Hungary, that such a secret would eventually be discovered at the level of the electron, or other electrically related atomic particles!*

Probing further into the world of the somatid and its link to life's basis and hereditary characteristics, I asked Francoise if Naessens had done any experiments to show how somatids might produce genetic effects on living organisms.

"I'll tell you, now, about one experiment we have repeated many times," she answered, "whose results are hard for any orthodox biologist to swallow. Before describing it, let me add that it is our belief—as it was also Antoine Béchamp's—that each of our bodily organs possesses somatids of varying, as yet indescribable, natures that are specific to it alone. But the whole ensemble, the 'family' of these varying forms, collectively circulates, either in the circulatory or the lymph system. On the basis of this experiment, we hold that, as a group, they contain the hereditary characteristics of each and every individual being."

---

* For more recent discoveries relating to the electrical basis for life, readers are also referred to two fascinating books by Dr. Robert O. Secker, *The Body Electric* (New York: Quill, William Morrow, 1985) and *Cross Currents* (Los Angeles: J.P. Tarcher, 1990).

As described by Francoise, the experiment begins by extracting soma-tids from the blood of a rabbit with white fur. A solution containing them is then injected, at a dose of one cubic centimetre per day, into the blood-stream of a rabbit with black fur, for a period of two weeks running. Within approximately one month, the fur of the black rabbit begins to turn a grayish color, half of the hairs of which it is composed having turned white. In a reverse process, the fur of a white rabbit, injected with soma-tids from a black one, also begins to turn gray.

Astonishing as this result, with its "genetic engineering" implications, might be, the effect of such "somatid transfer" from one organism to another also, said Francoise, produces another result offering great insight into the role played by the somatid in the immunological system. "When a patch of skin," she continued, "is cut from the white rabbit and grafted onto the empty space left after cutting a patch of similar size from the black rabbit, the graft shows none of the signs of rejection that normally take place in the absence of somatid transfer." What this might bode for the whole technique of organ transplant, attempts at which have been bedeviled by the "rejection syndrome," we shall let readers—especially medically trained readers—ponder.

GASTON NAESSENS: LIFE AND WORK

*Is it not living in a continual mistake to look upon diseases,
as we do now, as separate entities, which must exist, like cats
and dogs, instead of looking at them as conditions, like a dirty
and a clean condition, and just as much under our control;
or rather as the reactions of a kindly nature, against
the conditions in which we have placed ourselves?*

Florence Nightingale, 1860
(seventeen years before Pasteur announced his germ theory),
cited in *Pasteur: The Germ Theory Exploded* by R. B. Pearson

Even a single discovery as striking as those made by Naessens in five inter-linked areas could, by itself, justifiably be held remarkable. That Naessens was able to make all five discoveries, each in what can be termed its own dis-cipline, might seem to be a feat taken from the annals of science fiction.

And that is exactly the point of view adopted by the medical authori-ties of the province of Quebec. Worse still, those same authorities have branded Naessens an out-and-out charlatan, calling his camphor-derived 714-X product fraudulent and the whole of his theory about the origin of degenerative disease and the practice of its treatment, not to add the rest of his "New Biology," no more than "quackery."

Spearheading the attack was Augustin Roy, a doctor of medicine, but one who—like Morris Fishbein, M.D., for many years "Tsar" of the American Medical Association—actually practiced medicine for only a brief period of his life.

How did a researcher such as Gaston Naessens, endowed with genius, come to land in so dire a situation? Let us briefly review some of the story of his life and work, about which, during repeated trips to Quebec from the United States, I came to learn more and more.

Gaston Naessens was born 16 March 1924, in Roubaix, in northern France, near the provincial capital of Lille, the youngest child of a banker who died when his son was only eleven years old. In very early childhood, Gaston was already showing precocity as an inventor. At the age of five, he built a little moving automobile-type vehicle out of a "Mechano" set and powered it with a spring from an old alarm clock.

Continuing to exhibit unusual manual dexterity, a few years later Gaston constructed his own home-built motorcycle, then went on to fashion a mini airplane large enough to carry him aloft. It never flew, for his mother, worried he would come to grief, secretly burned it on the eve of its destined takeoff.

After graduation from the College Universitaire de Marcen Baroeul, a leading prep school, Gaston began an intensive course in physics, chemistry, and biology at the University of Lille. When France was attacked and occupied by Nazi forces during World War II, young Gaston, together with other fellow students, was evacuated to southern France. In exile near Nice, he had the highly unusual opportunity to receive the equivalent of a full university education at the hands of professors also displaced from Lille.

By the war's end, Naessens had been awarded a rare diploma from the Union Nationale Scientifique Francaise, the quasi-official institution under whose roof the displaced students pursued their intensive curriculum. Unfortunately, in an oversight that has cost him dearly over the years, Naessens did not bother to seek an "equivalence" from the new republican government set up by General Charles de Gaulle. He thus, ever since, has been accused of never having received an academic diploma of any kind.

Inspired by his teachers, and of singular innovative bent, Gaston, eschewing further formal education—"bagage universitaire" [academic baggage] as he calls it—set forth on his own to develop his microscope and begin his research into the nature of disease. In this determination, he was blessed by having what in French is called a *jeunesse dorée*, or gilded childhood—"born with a silver spoon in his mouth," as the English equivalent has it. His mother afforded him all that was needed to equip his own postwar laboratory at the parental home.

His disillusion in working in an ordinary laboratory for blood analysis spurred Gaston into deciding to go freelance as a researcher. Even his mother was worried about Gaston's unorthodox leanings. She clearly understood that her son was unhappy with all he had read and been taught. As he was to put it: "She told me what any mother would tell her son: 'It's not you who will make any earth-shaking discoveries, for there have been many, many researchers working along the same lines for decades.' But she never discouraged me, never prevented me from following my own course, and she helped me generously, financially speaking."

Gaston Naessens knew that there was something in the blood that eluded definition. It had been described in the literature as crasse sanguine (dross [waste products] in the blood) and Naessens had been able to descry it, if only in a blurry way, in the microscopic instruments up to then available to him. What was needed was a brand new microscope, one that could see "farther." He thought he knew how to build one and, at twenty-one, he determined to set about doing so.

In the design of the instrument that would open a vista onto a new biological world, Naessens was able to enjoin the technical assistance of German artisans in the village of Wetzlar, in Germany, where the well-known German optical company Leitz had been located before the war. The artisans were particularly helpful in checking Naessens's original ideas on the arrangement of lenses and mirrors. The electronic manipulation of the light source itself, however, was entirely of Gaston's own private devising. When all aspects of the problem seemed to have been solved, Naessens was able to get the body of his new instrument constructed by Barbier-Bernard et Turenne, technical specialists and military contractors near Paris.

Readers may fairly ask why Naessens's "Twenty-first-century" instrument, which has been called a "somatoscope" due to its ability to reveal the somatid, has never been patented and manufactured for wide use. To understand the difficulty, we should "fast forward" to 1964, the year Naessens arrived in Canada. Hardly having found his footing on Canadian soil, he received a handwritten letter, dated 3 May, from one of the province's most distinguished physicists, Antoine Aumont, who worked in the Division for Industrial Hygiene of the Quebec Ministry of Health.

Aumont, who had read about Naessens's special microscope in the press, had taken the initiative of visiting Naessens in his small apartment in Duvernay, near Montreal, to see, and see through, the instrument with his own eyes. Aumont wrote:

> Many thanks for having accorded me an interview that impressed me far more than I can possibly describe.
> I have explained to you why my personal opinions must not be con-

sidered as official declarations. But, after thinking over all that you showed, and told me, during my recent visit, I have come to unequivocal conclusions on the physical value of the instrumentation you are using to pursue your research.

As I told you, if my knowledge of physics and mathematics can be of service to you, I would be very glad to put them at your disposition.

It can be deduced that Aumont's enthusiasm for what he had seen caused a stir in the Quebec Ministry of Health, for on 17 July, Naessens received an official letter from that office stating that the minister was eager to have his microscope "officially examined" if its inventor would "furnish in writing details concerning this apparatus, including all its optical, and other, particularities, as well as its powers of magnification, so that experts to be named by the minister can evaluate its unique properties."

In reply to this letter, Naessens's lawyer sent a list of details as requested and stated: "You will, of course, understand that it is impossible for Monsieur Naessens to furnish you, in correspondence, with the complete description of a highly novel microscope which is, moreover, unprotected by any patent." Then, to explain why no patent had yet been granted, he added a key phrase: "since its mathematical constants have, up to the present, not been elucidated in spite of a great deal of tiresome work performed in that regard." In other words, it seemed that Aumont and his colleagues had been unable to explain the superiority of the microscope in terms of all the known laws of optics and it still seems that, so far, no one else has been able to do so.

There have been interesting recent reports on new microscopes being developed that apparently rival the magnification powers of Naessens's somatoscope. It would seem, however, that the 150 angstroms of resolution achieved by Naessens's instrument has not yet been matched.

The Los Angeles-based World Research Foundation's flyer, presenting its autumn (1990) conference "New Directions for Medicine . . . Focusing on Solutions," announced the development of an Ergonom-400x microscope, used by a German *Heilpraktiker*, or healer, Bernhard Muschlien, who paid a visit to Naessens's laboratory in 1985. While his microscope is apparently capable of achieving 25,000-fold magnification, its stated resolution is 100 nanometres (1000 angstroms), or several orders of magnitude less than the 150 angstroms developed with the somatoscope.*

In the July 1990 issue of *Popular Science*, an article "Super Scopes" refers to an extraordinary new technology in microscopy engineered at Cornell University under the direction of Professor Michael Isaacson, and also in Israel. The technology uses not lenses but apertures smaller than

---

* One nanometre is one-billionth of a metre; one angstrom is ten-billionths of a metre, or one-tenth of a nanometre.

the wave lengths of visible light to achieve high magnification. Isaacson is quoted as saying: "Right now, we can get about 40 nanometres (400 angstroms) of resolution," though he hopes to heighten that "power" to 100 angstroms "down the road." The 150 angstroms capacity built into Naessens's microscope over forty years ago still seems to lead the field.

Returning to the biography of Naessens, during the 1940s, the precocious young biologist began to develop novel anticancer products that had exciting new positive effects. The first was a confection he named "GN-24" for the initial letters of his first and last names, and for 1924, the year of his birth. Because official medicine had long considered cancerous cells to be basically "fermentative" in nature, reproducing by a process that, while crucial to making good wine from grape juice, produces no such salutary effect in the human body, Naessens's new product incorporated an "anti-fermentative" property. The train of his thinking, biologically or biochemically speaking, will not be here elaborated lest this account become too much of a "scientific treatise." What can be mentioned is that the new product, GN-24, sold in Swiss pharmacies, had excellent results when administered by doctors to patients with various forms of cancer.

As but one example of these results, Naessens cited to me the case of his own brother-in-law, on the executive staff of the famed Paris subway system, the Métropolitan. In 1949, this relative, the husband of a now ex-wife's sister, was suffering through the terminal phase of stomach cancer and had been forced into early retirement. After complete recuperation from his affliction, due to GN-24, he resumed work. Only recently, Naessens, who had lost contact with him for years, was informed that he was alive and well.

Another 1949 case was that of Germaine Laruelle, who was stricken with breast cancer plus metastases to her liver. A ghastly lesion that had gouged out the whole of the left section of her chest had caused her to go into coma when her family beseeched Naessens to begin treatment. After recovering her health, fifteen years later, she voluntarily came to testify on behalf of Naessens, who, as we shall presently see, had been put under investigation by the French Ordre des Médecins (Medical Association). She also allowed press photographers to take pictures of the scars on the left side of her breast-denuded chest. In 1969, twenty years after her initial treatment, she died of a heart attack.

Seeking a more imposing weapon against cancer, Naessens next began developing a serum. This he achieved by hyperimmunizing a large draft horse by means of injecting the animal with cancer-cell cultures, thus forcing it to produce antibodies in almost industrial quantities. Blood withdrawn from the horse's veins containing these antibodies, when purified, was capable of fighting the ravages of cancer. It proved to have therapeutic action far more extensive than that obtained by GN-24, and led to a restraint or reversal of the cancerous process, not only in cases of tumours but also with various forms of leukemia. Many patients clandestinely treated by their doctors with the new serum, called Anablast (*Ana*, "without," and *blast*, "cancerous cells"), were returned to good health.

One patient, successfully so treated, was to play a key role in Naessens's life. This was Suzanne Montjoint, then just past forty years of age, who, in 1960, developed a lump the size of a pigeon's egg in her left breast. Over the next year, the lump grew as large as a grapefruit. After the breast itself was surgically removed, Montjoint underwent a fifty-four-day course of radiation that caused horrible third-degree burns all over her chest. Within six months, she began to experience severe pain in her lower back.

Chemical examination revealed that the original cancer had spread to her fifth lumbar vertebra. More radiation not only could not alleviate the now excruciating pain, but caused a blockage in the functioning of her kidneys and bladder. When doctors told her husband she had only a week or so to live, Suzanne said to him, "I still have strength left to kill myself . . . but, tomorrow, I may not have it anymore."

Summoned by the husband, one of whose friends had told him about the biologist, Naessens began treating Madame Montjoint; who, by then, had lapsed into a semicoma. Within four days, all her pains disappeared and she had regained clarity of mind. By April 1962, after an examination of her blood at his microscope, Naessens declared that the somatid cycle in Suzanne Montjoint's blood had returned to normal. As she later told press reporters, "My recovery was no less than a resurrection!"

When these successful treatments, plus many others, came to the attention of French medical authorities, Naessens was twice brought before the bar of justice, first for the "illegal practice of medicine," next for the "illegal practice of pharmacy." On both occasions, he was heavily fined, his laboratory sealed, and most of its equipment confiscated, though, happily, he was able to preserve his precious microscope.

With all the harassment he was enduring (while at the same time saving the lives of patients whose doctors could afford them little, or no, hope for recovery), Naessens was almost ready to emigrate from his mother country and find a more congenial atmosphere in which to pursue his work, with the privacy and anonymity that he had always cherished and still longs for. An opportunity came when he was invited by doctors in the Mediterranean island of Corsica, whose inhabitants speak a dialect more akin to Italian than to French. With a long history of occupation by various invaders before it actually became part of the French Republic, its population has ever since been possessed of a revolutionary streak that, on occasion, fuels an urge toward secession from the "motherland."

In Corsica, Naessens established a small research laboratory in the village of Prunette, on the southwest tip of the island. What happened next, in all its full fury, cannot be told here. Reported in two consecutive issues of the leading Parisian illustrated weekly *Paris-Match*, the story would require, for any adequate telling, two or more chapters in a much longer book.

Suffice it to say that, having developed a cure for various forms of degenerative disease, Naessens saw his ivory tower invaded by desperate patients from all over the world who had learned of his treatment when a Scots Freemason, after hearing about it during a Corsican meeting with international members of his order, leaked the news to the press in Edinburgh. Within a week, hundreds of potential patients were flying into Ajaccio, the island's capital, some of them from as far away as Czechoslovakia and Argentina.

The deluge immediately unleashed upon Naessens the wrath of the French medical authorities, who began a long investigation in the form of what is known in France as an *Instruction*—called in Quebec an *Enquete préliminaire*—a kind of "investigative trial" before a more formal one.

All the "ins and outs" of this long jurisprudential process, thousands of pages of transcripts about which still repose in official Parisian archives, must, however regretfully, be left out of this narrative. Its denouement was that Gaston Naessens, together with key components of his microscope preserved on his person, left his native land in 1964 to fly to Canada, a country whose medical authorities he believed to be far more open to new medical approaches and horizons than those in France. His abrupt departure from the land of his birth was facilitated by a high-ranking member of France's top police organ, the *Sureté Nationale*, whose wife, Suzanne Montjoint, Naessens had successfully treated.

Hardly had Naessens set foot on Canadian soil than he was faced with difficulties, in fact a "scandal," almost as, if not just as, serious as the one he had just left behind.

During the French *Instruction* proceedings in 1964, one René Guynemer, a Canadian "war hero" of uncertain origin and profession, had accosted Naessens in his Paris home to beg him to come to Canada in order to treat his little three-year-old son, René Junior, who was dying of leukemia.

Though puzzled about a certain lack of "straightforwardness" in the supplicant, Naessens, ever willing to help anyone in distress, and with the approbation and assistance of the Canadian ambassador to France, immediately flew to Montreal, where he hoped, as agreed by Guynemer *père*, to be able to treat *fils* in complete discretion. Upon his arrival at Montreal's Dorval Airport, however, Naessens was aghast to see a horde of representatives of both the printed and visual media, creating, in anticipation of his arrival, what amounted to a virtual mob scene.

The Quebec "Medical College" had, at the time, agreed, for "humanitarian" reasons, to allow the treatment of the Guynemer child, in spite of the fact that Anablast had not been licensed for use in Canada. Various tests, lasting for several weeks, were made on the product at Montreal's well-known microbiological Institut Armand Frappier to confirm the presence of gamma globulin in it, the presence of which purportedly thorough French examinations had failed to detect.

Virtually at death's door, the Guynemer child was said to have been given nine injections of Anablast. Naessens himself was never given official confirmation that the injections had actually been administered. Nor was he permitted to make any examination of the little patient's blood at his microscope, or even to meet him face to face. After the little boy succumbed, the Quebec press exploded with stories that, in their luridness matched the ones that had been appearing all over France after the Corsican "debacle."

Some of the mysteries of the "Guynemer connection" will likely never come to light. Only later did it become clear that the true name of the leukemic child's father was actually Lamer, a man who had claimed that, in past years, he had been an officer in the Royal Canadian Air Force and a "secret agent" attached to the French "underground" during World War II. To the Naessenses, the question has always remained: If he was an "agent," then for whom, or for what?

In the spring of 1965, Naessens journeyed to France for his trial. When he returned to Quebec in the autumn of that year, he retired from the public scene to live incognito in Oka, a Montreal suburb, with a newfound friend, Hubert Lamontagne, owner of a business selling up-to-date electronic devices, whom he had met while looking for electrical components for his microscope in 1964. As a person skilled in electronics, Naessens was able to be of great assistance to his host, who also operated a large "repair shop" throughout the winter and the following summer, when, on tour with a troupe of comedians, he was put in charge of solving all the acoustical problems in the many provincial cabarets and theatres hosting the troupe's performances. Deprived, for several years, of any support to pursue his life goals, Naessens was constrained to utilize his skills as "Mr. Fixit," able to repair almost anything, from automobile engines to rectifiers.

After five years of working in electronics, Naessens had a stroke of luck, perhaps the most important of his career, when, in 1971, through a friend, he was introduced to, and came under the protective wing of, an "angel" who saw in Naessens the kind of genius he had for a long time been waiting to back.

That "angel" was the late David Stewart, head of Montreal's prestigious MacDonald-Stewart Foundation, which for many years had funded, as it still continues to fund, orthodox cancer research. Despondent about the recent death from cancer of a close friend, and in serious doubt that any of the cancer research he had so long supported would ever produce any solution, Stewart's guiding precept and motto was "In the search for a remedy for cancer, we shall leave no stone unturned." The philanthropist therefore decided personally to back Naessens's research. But after setting up a laboratory for the biologist on the Ontario Street premises of the well known MacDonald Tobacco Company, which Stewart's father had inherited from its founder, tobacco magnate Sir William MacDonald, David Stewart came under such violent criticism by leaders of orthodox cancerology that he advised Naessens to move his research to a low-profile provincial retreat.

Having, by that time, established a "liaison" with his bride-to-be, Francoise Bonin, whose parents lived in Sherbrooke, Naessens was, by 1972, able to take over the elder Bonin's summer house on the banks of the Magog River in Rock Forest, "winterize" it, and establish a well equipped laboratory in its basement. And there, the Naessenses, who were married in 1976, have ever since been located. Of his wife, Naessens has said to me:

> She was persuaded from the very start about the intrinsic value of my research and at once saw the truth of it. Just as then, so now, years later, she continues her loyal assistance to get this truth out. Some ask if it's moral support. Yes, it could be called that. We have the same kind of attitudes about things. Both of us, for instance, believe that if something new produces good results, it's got to be pursued to the bitter end. This is not ambition, but moral honesty. When one gets to know her, one realizes that she doesn't just repeat the things I think and say, but is convinced about them because of what she has seen and experienced.

Because legal restrictions applying to foundations and their grants prevented David Stewart from transmitting monies directly to Naessens, the foundation director arranged for them to be funneled via the Hotel Dieu— a leading hospital affiliated with the Université de Montreal that specializes in orthodox cancer treatment and research. Accused by Augustin Roy as a "quack," Naessens has consequently had his work modestly funded by checks made out by a hospital at the heart of one of Canada's cancer establishment's most prestigious fund-granting institutions. No more anomalous a situation exists anywhere in the worldwide multibillion-dollar cancer industry.

Given the importance of the foundation's assistance, it is all the more curious that Augustin Roy had not made the slightest mention of the foundation's loyal support of the biologist over the years. Instead, at a press conference held after Naessens's arrest to present traditional medicine's case against Naessens, Roy, perhaps unknowingly, demonstrated the "Catch-22" that any alternative medical, research, or frontier scientist faces. Roy stated that if Naessens were a "true" scientist he would have long since submitted his results to proper authorities for check, but when asked by journalists whether the Quebec medical community had thoroughly investigated the biologist's claims, Roy inscrutably replied, "That's not our job." In answer to another reporter's query about the assertions of many cancer patients that the Naessens treatment had completely cured their affliction, Roy added, "I just can't understand the *naiveté and imbecility* of some people."

To get a more complete idea of the full impact of Roy's attitude with respect to a brand new treatment and patients benefiting from it, we here excerpt some of his additional statements made during an interview on McGill University's Radio Station in the summer of 1989.

When, to open the interview, Roy was asked his opinion about what the interviewer termed a "remarkable new anticancer product, 714-X," the medical administrator replied, "I have been aware of Monsieur Naessens for twenty-five years. In 1964, he arrived from France with a so-called cancer treatment, Anablast, the very same medicinal he's now using under another name—714-X."

That anyone in a position as elevated as Roy's could publicly propagate so obvious an error is surprising. For Anablast, which, as we have seen, is a serum, has nothing to do with 714-X, a biochemical product. Yet here was the head of the Quebec medical establishment falsely stating that 714-X, developed over thirteen years in Canada, was nothing but the older French product bearing a new name, a statement tirelessly, and erroneously, repeated by journalists in the press.

As for Naessens himself, Roy told his radio audience: "That man's professional knowledge is equal to zero! You should know that he has, behind him, in France, an imposing, even 'heavy,' past involving serious judicial procedures and condemnations." It seems truly amazing that a doctor who, over a quarter of a century, had never met Naessens, or once visited his laboratory, or taken the trouble to investigate why hundreds of cancer patients had survived because of his new treatment, could so peremptorily reduce the biologist's knowledge to nil.

Was Roy really being impartial when he said, "I've got to be a bit careful because Naessens is currently under legal prosecution . . . But the fact remains that he was in serious trouble with the French legal authorities. Let's just say he's a 'slick talker,' one who knows how to address an audience. But, I ask you, why is it that he's been working in secret for so long?" In asking this question, Roy was obviously not in the least ashamed to be adding a second error to the one he had already propagated. For the truth was, and is, that Naessens, far from having worked "in secret", has at all times—as I have repeatedly witnessed over the years—kept his laboratory open to "all comers" and has stood ready to discuss his research with any of them. "It's so obvious," Roy disparagingly continued, "that all this man's affirmations and allegations just don't have a leg to stand on . . ."

"But," ingenuously interrupted his young interviewer, "haven't there been several people who have testified in writing, or on TV, that they've been cured by 714-X?"

Roy's unhesitating answer was breathtakingly categoric: *"No one's personal testimony has any value whatsoever!* All such testimonies are purely suggestive and anecdotal. Let's show a little common sense, after all! Common sense indicates that if Naessens had a real treatment for a malady such as cancer, it would have been criminal not to put it at the disposition of the whole world! I don't understand what he's up to, and I have even less understanding of those who go about publicizing his reputed treatment, which is pure quackery." Given the hyperbole on Roy's part, one could well wonder what hope there might be for any kind of new discovery in the health field ever to become authorized, or even known. For years, Naessens had been assiduously, but unsuccessfully, trying to "put his discovery at the world's disposition."

Unabashed by the weight of her interviewee's authority, the interviewer was not loath to press in on Roy again: "There have, however, been certain doctors who have been most surprised at how terminal patients have been brought back to good physical shape with 714-X. Would that not make anyone eager to verify the facts with respect to those recovered patients?"

"Not at all!" Roy's rejoinder was a virtual explosion. "It's not my job, or that of the Medical Corporation, to check on pseudocures of that kind! So what, if two, three, four, or half a dozen doctors, in their isolation, have something good to say in support of it? No matter where they come from, their statements are worthless!"

To get a countervailing idea of what Naessens might have said in rebuttal in Roy's presence, we shall next excerpt part of an interview with the biologist by the same interviewer on the same radio station a few days later.

"Gaston Naessens," she began, "is your 714-X really effective?"

Naessens: Absolutely! It builds up the immune system so that all the body's natural defenses can regain the upper hand. I don't make the claim in a void, because there are a lot of people around who were gravely ill with cancer who can now state they have gotten well due to my treatment.

Interviewer: If your product really works, why hasn't Dr. Roy been interested in doing an in-depth study of it? Does he know you at all?

Naessens: Many people have asked me both those questions. If you ask him the latter question, he will pull out a thick file on me and he'll tap it, and say, "Sure, I've known him since 1964." But the fact is he has never met me in person, never visited my lab, and never investigated my work! So, he is absolutely incapable of making any judgment whatsoever on whether that work has a solid foundation, or not!

In his lengthy reply, uninterrupted by the fascinated interviewer, Naessens, after a brief pause, began to reveal the essence of the difficult situation in which he had been placed over the years:

Naessens: Let's get to the heart of this matter: The medical community, on the one hand, and I, on the other, speak *completely different languages*. That anomaly connects to the important fact that all approved anticancer therapies are focused only on cancer tumours and cancerous cells. The reigning philosophy, medically speaking, is that a cytolytic (cell-killing) method must be used to destroy all cancer cells in a body stricken with that disease.

But I, on the contrary, have developed a therapy based on what has been called the body's whole terrain! To understand that, you have to realize that, every day, our bodies produce cancerous cells in no great amount. It's our healthy immune system that gets rid of them. My 714-X allows a weakened, or hampered, immune system to come back to full strength, so that it can do its proper job!

If medical "experts" pronounce my product worthless, it might even be admitted that, in terms of their own scientific philosophy, they are making some sense. This is largely because, when they examine my product for any *cytotoxic* effect it might have, they find none!

Interviewer: Is the Medical Corporation interested in sitting down and talking with you, or running tests to verify your product?

Naessens: No! Because they firmly believe that any success it might have is due to some kind of "psychological" effect, and they say that the product itself contains nothing that could possibly be of benefit.

Interviewer: Where did they get that idea?

Naessens: It seems that, with officialdom, it's always a case of misinformation, or of bad faith. If this whole affair were limited to patients I've successfully treated, patients who might have remained silent, I would still have small hope that my research will one day be recognized. But, now, a crucial turning point has been reached. I'm back in the international limelight. My arrest, incarceration, and indictment are important if only because, immediately following them, people "in the know" have begun to take action on my behalf. That being so, the medical community's negative reaction is no longer the only, or the dominant, one! It may be too bad that all this has to be thrashed out not in a scientific forum, but in a court of law. But that's the way it is. In my upcoming trial, many of my patients' cases will be examined, one by one, and exposed in full detail, in the courtroom! So the medical "authorities" will no longer be the sole judges.

After continuing on with this theme for several minutes longer, Naessens came to a firm conclusion: "I wouldn't want you to think that I'm even trying to boast when I say that my work represents a brand new horizon in biology! I have found a successful way of adjusting a delicate biological mechanism. I have no pretensions beyond that! If I can be of service to anyone, my laboratory is always open."

Gaston Naessens was brought to trial in Quebec, where he was acquitted and completely exonerated.

15 Dr. Max Gerson's Nutritional Therapy for Cancer and Other Diseases

Katherine Smith

The story of Doctor Max Gerson and the nutritional therapy he developed for cancer and other diseases is another sad chronicle of the suppression of a therapeutic programme which has the power to help—if not cure—many people who would otherwise suffer continuing illness and death.

Born in 1881 and raised in Germany, Dr. Gerson began the development of his nutritional therapy in an effort to find relief from the crippling migraine headaches from which he suffered as a young man. Working on a hunch that a chemical imbalance in his body might be responsible for the painful headaches which plagued him, Gerson decided to alter his diet and see if his condition improved. After trying a milk-based diet, using the rationale that milk was the primary food of mammals, he tried treating himself with a diet comprised mainly of raw foods, and found that his migraine headaches disappeared. Dr. Gerson then tried out the therapy on those of his patients who suffered from migraines, and found that they too found relief, and this painful condition disappeared.

One of the people Dr. Gerson treated for migraine headaches also suffered from lupus vulgaris—a so-called "incurable" disease. To Gerson's surprise, not only did this patient's migraine attacks disappear after beginning the nutritional therapy, but his lupus was also healed.

Dr. Gerson successfully treated other people suffering from lupus with his diet therapy. Then, since lupus vulgaris is also known as tuberculosis of the skin, Gerson had the inspiration to begin treating people suffering from other forms of tuberculosis. In 1933, he published his book, *Dietary Therapy of Lung Tuberculosis*. Unfortunately, the rise of Hitler to power in Germany meant that he was unable to publicly demonstrate his discoveries to the Berlin Medical Association. Faced with a deteriorating political situation in his homeland Dr. Gerson went to work in Vienna and

France, as well as giving lectures throughout Europe. Finally, as the clouds of war gathered ever more ominously over Europe, Gerson left Europe in 1936 to begin a new life in America.

Unfortunately for Dr. Gerson—not to mention the thousands upon thousands of people who could have been helped by his therapy—the U.S., while a haven from Hitler, was far from being the land of the free. Gerson found that publishing his work—which was a relatively easy proposition in Europe—was an almost impossible task in the United States.

Perhaps part of the reason why Gerson's work was not enthusiastically supported by his medical peers in the United States may have been that he was German, and therefore to be treated with suspicion, as a member of an enemy nation, even though he had qualified to practise medicine in the United States in 1938. However, a more important reason was that his treatments for cancer challenged the orthodox methods. In the 1930s and 1940s, according to the orthodox mind-set, cancer was to be treated in two basic ways: surgically to remove the offending tumour (when it was operable) and then with radiation to kill the cancerous cells.

Dr. Gerson's conception of cancer went far beyond merely viewing the cancer as a spontaneous eruption within a healthy body. Rather he saw cancer as the end result of generalised degradation of the bodily systems, especially the liver. Such concepts were quite foreign to the vast majority of the medical profession at that time, when doctors could not adequately account for the cause of cancer, nor inform people how to avoid this life threatening disease.

According to Gerson, the way to prevent cancer was by ". . . preventing damage to the liver. The basic measure of prevention is not to eat the damaged, dead, poisoned food which we bring into our bodies. Every day, day by day, we poison our bodies." Gerson's nutritional therapy worked on the principle that in order to cure a serious disorder such as cancer, treatment of the symptoms of the disease was not sufficient to restore the patient to health. He wrote in his book *A Cancer Therapy: Results of Fifty Cases* in 1958 that the "whole body" or "whole metabolism" had to be treated to "correct all the vital processes" in order to effect a cure.

The basis of Dr. Gerson's nutritional programme to strengthen the body to allow healing to take place is a diet comprised mostly of raw foods, especially freshly-made fruit and vegetable juices, green salad, and a soup cooked at a very low heat. Some cooked fruit and vegetables are also permitted in the first six weeks of his dietary plan. However, no canned, salted, pickled, bleached, sulphured, frozen or smoked foods—in short no denatured foods of any kind—are permitted at any time during the Gerson regime.

The therapy Dr. Gerson devised was also designed to be high in potassium and low in sodium. The soup mentioned above is especially high in potassium, which helped to correct a too-high ratio of sodium to potassium suffered by many people with cancer, especially those with moderate or advanced cancer. Dr. Gerson discovered that restoring a favourable potassium/sodium balance could reverse some of the cell damage caused by an excess of sodium.

The function of the freshly-made fruit and vegetable juices in the programme is to detoxify the body and provide oxidising enzymes to assist in the rehabilitation of the liver. Other techniques to support the liver and detoxify the body are also used in the programme, including coffee enemas to stimulate the flow of bile and safely dispose of toxins, the juice of raw calves' liver, and injections of crude liver extract. (Liver juices and extracts are no longer used by people following a Gerson programme in the 1990s, due to the contamination of the liver with pesticides and bacteria. Spirulina and carrot juice may be taken instead to provide nutritional iron and pro-vitamin A. Desiccated liver tablets may be used instead, since these are thought to contain fewer toxins.)

Gerson also supplemented the diet of people in his care with additional potassium salts, as well as organic and inorganic iodine. Fluoride-contaminated water or other products are forbidden because of fluoride's toxicity to valuable enzymes. Animal fats are excluded. Dr. Gerson's programme was originally completely free of fats and oils (excluding the small amount of fat present in the calves' liver), but after experimentation, Gerson modified his programme to include a small amount of flax seed oil to supply essential fatty acids.

After six weeks of detoxification using the diet and supplements outlined above, patients in Gerson's care graduated to a diet which included small amounts of the protein foods such as yogurt, cottage cheese, and natural buttermilk. (Foods containing protein had been previously restricted to allow the body adequate time to detoxify and begin to break down tumour tissue.)

These, then, were the basic theories and therapy which Dr. Gerson had developed by the time that he came to the United States in 1936. In January 1948—almost twenty years after he had first successfully treated cancer in Germany in 1928—he went to work at New York's Gotham Hospital. However, Gerson's efforts to publish his discoveries consistently met with a negative response from the publishers of medical journals.

His article "Cancer, A Deficiency Disease" was rejected by the *New York State Journal of Medicine* in 1943. The next year another paper, "Dietic Treatment of Malignant Tumours," was also rejected by every medical journal to which it had been sent. In 1945, he finally succeeded in publishing "Dietary Considerations in Malignant Neoplastic Disease," which appeared in the November-December edition of *Review of Gastroenterology*. His work might have been destined to obscurity forever, but for an investigative reporter who discovered the good doctor working quietly to cure cancer with his most unorthodox therapies, and determined to bring Dr. Gerson's life-saving discoveries to public attention.

Raymond Swing, an ABC radio journalist, proposed that Dr. Gerson be called to testify before the Senate which was debating a bill to allocate funds for cancer research. Raymond Swing's efforts on Dr. Gerson's behalf were successful and the doctor, together with five of his patients, went before a sub-committee of the Senate in 1946 and told their stories. All of the five patients had had a positive response to Dr. Gerson's therapy, and had been told by their former doctors that there was no longer any hope for them. They included: a woman with breast cancer who had undergone mastectomy and radiation treatments to no avail. Her cancer had disappeared after nine months of the Gerson therapy; a fifteen year old girl had been paralysed by a tumour in her spinal cord. Her tumour had vanished after 8 months of Gerson therapy; a soldier with an inoperable tumour which had grown from his neck into his skull, making radiation treatment impossible because of the risk of brain damage—a year after commencing the Gerson therapy he was completely free of cancer; and a woman who had suffered from a malignant sarcoma. Prior to beginning the Gerson therapy, she had large tumours in her groin, neck and abdomen. After a year on the Gerson therapy she was completely free of cancer.

Unfortunately, Dr. Gerson's successful treatment of these and other patients who otherwise had been doomed to die did not earn him the respect and recognition from the medical community that he deserved. Quite to the contrary. The public display of Dr. Gerson's successful but unorthodox treatment of cancer victims further alienated him from mainstream medicine. An abusive editorial in the *Journal of the American Medical Association* in November 1946 followed Dr. Gerson's appearance in front of the Senate sub-committee. The editorial celebrated the unfortunate fact that, despite the amazing and incredibly newsworthy results of Dr. Gerson's therapy, his presentation before the Senate sub-committee had received "little, if any newspaper publicity"—as if the lack of mainstream publicity itself was an indictment of the treatment! The editor further denigrated his work by splitting hairs as to what precisely constituted a "cure." Dr. Gerson, he wrote, "admits lack of any actual cure, claiming only that patients seemed improved in health and that some tumours were delayed in growth or became smaller." In one final coup against Dr. Gerson, who for years had been submitting work to the journal for publication without success, he wrote that "the journal has on several occasions requested Dr. Gerson supply details of the method of treatment but has thus far received no satisfactory reply."

This editorial was just the beginning of a concentrated campaign of harassment against Dr. Gerson and the people who were working with him in his Research Foundation. Between 1946 and 1954, Dr. Gerson was investigated five times by the Medical Society of the County of New York. After each investigation, the Research Foundation requested that the investigators give a statement, and in each case the request was denied.

However, in 1948, Dr. Gerson and his Research Foundation were left in no doubt as to what the medical establishment thought of them when a review of their work was published in the *Journal of the AMA*. The review was entitled "Frauds and Fables," in which the journal suggested that Gerson was a fraud. The Research Foundation threatened to sue the AMA and were able to stop reprints of the damaging article. However, as a consequence of the damaging publicity, Gotham Hospital refused to allow Dr. Gerson to work on its premises after 1950.

Moreover, Dr. Gerson was not able to restore his good name within the medical profession in America. It became impossible for him to publish a single piece of research in any medical journal from the end of 1949 until the end of his life, despite (or perhaps because of) his thousands of success stories. In addition, Gerson was prevented from presenting patients at a hearing of the House of Commerce Committee in 1953, which was investigating therapies for cancer and other diseases. Despite requests from his patients that he be allowed to present his findings, as well as a letter from Dr. Gerson himself, the chairman of the committee failed to offer Dr. Gerson the chance to demonstrate his findings.

With his work increasingly under fire in the United States, Dr. Gerson went to Europe in order to publish his discoveries. A German journal, *Medizinische Klinik*, published two of the reports which U.S. journals had refused to print: "Cancer: A Problem of Metabolism" and "No Cancer in Normal Metabolism." He was also invited to the University of Zurich in 1952, after attending the International Cancer Congress in Berchtesgaden.

When Dr. Gerson returned to the United States from Europe, however, he faced still more hurdles. In 1957, he was investigated by the Licensing Board of New York State. Even more damaging to his work, his malpractice insurance was terminated. In 1958 Gerson was suspended from the Medical Society of the State of New York, and the laboratories which Dr. Gerson's Research Foundation used for X-rays, blood, and urine analyses were warned that should they continue to do work for Dr. Gerson and his patients, they would be put out of business. Dr. Gerson died in 1959.

The harassment of Dr. Gerson by the medical establishment, while both unethical and immoral, is understandable within a commercial context. If

Gerson's methods of curing cancer had replaced the "conventional" cancer treatments, the profession's investment in expensive equipment such as surgical facilities and radiation treatment apparatus would have been lost, to say nothing of prestige.

However, it was not just Dr. Gerson and his colleagues who had to endure harassment. Patients who sought out the Gerson treatment in preference to orthodox medicine were also harassed by members of the orthodox medical community. In many cases, doctors harangued patients so persistently that they abandoned Gerson's therapy—even when it appeared to be helping them—and accepted conventional medical treatment for their cancer. These tactics compelled Gerson to write to a close friend in 1957 that:

> The most difficult and inhuman part of the measures taken against me is that the physicians approach the best and almost completely cured patients and try to have them returned to their hospitals. Here they manage with orthodox treatments to kill them. I lose in this manner somewhere between 25 and 30 percent of my best cases.

This sort of harassment of people using the Gerson therapy occurred even after Dr. Gerson's death, with doctors going as far as phoning patients in residence at the La Gloria Hospital in Mexico, which was set up in 1977 to provide Dr. Gerson's therapy on an in-patient basis.

In 1998, the climate of mainstream medical hostility towards Dr. Gerson's unorthodox therapy programme has not changed. "Anti-quackery" laws forbid the practise of the Gerson therapy in California and other states of the U.S. by medical doctors. People with cancer who reject the options of surgery, radiation and chemotherapy offered to them by the major cancer hospitals must either struggle to pursue a Gerson-type programme in their own homes without adequate medical support, or find the money necessary to travel to Mexico and pay for treatment at La Gloria Hospital, near Tijuana.

The parents of a growing number of children with cancer are even less able to choose what they believe is the most suitable treatment option for their son and daughter, *since by law all children in the United States who have cancer must be given chemotherapy*, or their parents may be imprisoned. Even a demonstrable improvement in the child's condition using non-toxic methods of cancer treatment will not forestall the application of this barbarous law.

Gerson's therapy remains on the "Unproven Methods List" of the American Cancer Society, despite ample evidence in Dr. Gerson's book *A Cancer Therapy: The Results of Fifty Cases*, as well as testimony from former cancer patients whose cancerous conditions have been healed by

Gerson's techniques. Dr. Gerson's therapy has now been practised for over 60 years, and patients of La Gloria Hospital experience an average improvement rate of 80 percent in early to moderate cancer and even more amazingly a 40–50 percent rate of improvement in people with so-called "terminal" cancer. Other benefits of the Gerson therapy for cancer sufferers included a marked reduction in pain, and also control of the acute infections which often led to the death of cancer patients. (The clinic does however, have a general rule of not accepting people who have previously undergone chemotherapy, since due to the damage that chemotherapeutic drugs inflict upon the liver and other organs of the body, sustained improvement in the condition of these cancer patients is much less likely. The medical director has also stated that people with tumours which have spread into the brain and begun to damage the delicate regulatory mechanisms within it are also less likely than most patients to respond to the Gerson treatment.)

In 1995, the Gerson Research Association and the Cancer Prevention and Control Program of the University of California published the results of a fifteen-year retrospective study which evaluated the success of Gerson therapy in treating malignant melanoma.

The results of this study showed that people with melanoma who used Gerson therapy survived longer than people using conventional therapy.

One most encouraging finding was that 100 percent of 14 people with stage I and II (localized melanoma) survived for five years, compared to 79 percent of 15,798 people who did not follow the Gerson program.

For people with melanoma classified as IIIA and IIIB (regionally metastasized), 70 percent of 33 Gerson patients lived for five years, compared to 41 percent of 134 melanoma patients under the care of the Fachklinik Hornheide.

Of those patients with melanoma which had metastasized to distant lymph nodes, skin areas or subcutaneous tissue—39 percent of 18 Gerson patients were alive after five years. By comparison, just 6 percent of 194 patients under the care of the Eastern Cooperative Oncology Group survived five years.

Despite these impressive results of Dr. Gerson's therapies, mainstream medicine is even less receptive to his ideas and treatment plans now than when he began publishing the results of his work in the United States in the 1940s.

In January 1945, the then manager-director of the American Cancer Society Mr. C. C. Little wrote (to a doctor):

It seems to me that since Dr. Gerson has frankly stated in detail what his diet is and in addition has given the theory on which he personally believes its claimed efficacy is based, that his material should receive publication and proper attention and criticism by the medical profession. I sincerely hope it will be possible to arrange this.

When both the American Cancer Society and the National Cancer Institute were approached about the Gerson therapy in the 1980s however, both organisations denied having even seen a copy of *A Cancer Therapy: The Results of Fifty Cases*, despite the fact that due to the heroic efforts of Dr. Gerson's daughter, Charlotte Gerson Strauss, the landmark book has remained in print for 40 years. In 1984, the American Cancer Society, along with the House of Representatives Select Committee on Aging declared that the "Gerson method of treating of cancer is of no value."

Although the "Unproven Methods List" is updated every six months, Gerson's therapy is not likely to be deleted from the list in the near future. The Unproven Methods Committee, according to the director of the Unproven Methods Office, G. Congdon Wood, supposedly makes its decisions on the medical literature. More recent information which supports his therapies, such as that published in the 1978 *Journal of Physiological Chemistry and Physics* seems to have been ignored—a spokesperson for the ACS explaining that they "had not seen" the article.

Government agencies such as the FDA. are also consulted in the review process. Unfortunately for people's health, the FDA. is notorious for its prejudice against vitamins and other natural therapies. When Charlotte Gerson Strauss was attempting to find a publisher for her father's book, some of the publishing houses considering the book received threats from the FDA, which had recently (May 1992) made a raid on the Tahoma (natural health) Clinic in Washington State, and seized vitamins and patient records, among other things. It is obviously not the sort of agency you would expect to endorse Gerson's therapy any time in the near future.

The National Cancer Institute is another agency that gives the Committee information about therapies on the Unproven Method's List. This agency long ago rejected Dr. Gerson and his work. Would you reasonably expect a prestigious national institute to sully its good name as the Castle of the Valiant Knights in White Coats battling the twentieth century scourge of cancer by associating with a "quack" who was expelled from his own State Medical Society? Hardly.

Section II

The Suppression of Unorthodox Science

The history of science is hardly the history of free inquiry. Rarely does science engage in self-examination, whether scientifically or simply reflectively. Occasionally we may benefit from the perspectives of those observers and historians from outside this branch of knowledge, who seek to bring to the world some solid wisdom. More frequently they fail to awaken interest within a rigid system that believes, as Organized Science does, that all mistakes were committed in the past. ("We might have got it wrong with Galileo and Semmelweis, but that was then.") Seldom is truth met with unconditional acceptance in professions that are not renowned for their engaging humility and willingness to embrace information that conflicts with their cherished and well-defended beliefs.

And so Freud was leaned on to radically alter his seduction theory. Under duress from colleagues, he lost concern for the welfare of sexually abused and beaten children. Rather, from 1894 onward he helped to found a system that blames the victim, turning his original thinking on its head—*it's the children who try to seduce the parents!* Only then could he find acceptance in the Viennese community of psychologists who then launched him on his stellar career.

Freud's case was not isolated. Wilhelm Reich's books were publicly burned by the FBI in a New York City incinerator in 1957; Immanuel Velikovsky's work was trashed by the U.S. scientific establishment, his publisher leaned on to offload his contract—in the middle of a bestseller; and Julius Hensel's pioneering work on "rock dust" fertiliser was suppressed by the NPK people who had something big to sell the world. So what if everyone is now lacking in essential trace minerals as a result?

Pat Flanagan's Neurophone patent was confiscated by the U.S. government and held for fourteen years—for "national security reasons"—while this most brilliant of brilliant American scientists was starved out. And what threat did his invention pose? It enabled deaf people to hear sounds through the nerves in their skin.

Are these examples mere abberations in an otherwise inclusive organization, or is there is a system-wide suppression syndrome? And if suppression is the norm in our supposedly objective scientific establishment, what exactly have we lost? I believe that we will probably never know what we have lost, or at least the extent of the loss. That's because who we become is a reflection of the attenuation of our available options by a system in which greed is valued above the human creative potential, and even the life force itself. This system's natural response is to suppress that which threatens its stake in the status quo.

Science is funded by giant corporations that do not have a vested interest in, say, organic agriculture, water as a fuel, or good nutrition and sanitation as ways of improving health rather than vaccinations and antibiotics. Science is not pure, nor has it ever been. The "Scientific Method" exists only for the purpose of censoring the innovations of independent thinkers.

The unconventional scientist, the person who comes up with something that threatens a billion dollar industry, will find him or herself either very rich or very dead. Or possibly both. Still, some courageous souls do try, despite the risks, to make their knowledge public. These truly great researchers and inventors are the pure scientists—the ones with a better idea, a new periodic table, a fresh perspective in looking at the universe, a cure for cancer. They represent thousands of other free thinkers who remain anonymous because their ideas and inventions have been bought up, suppressed, forgotten.

Common sense dictates that the quality of life of the human population would be greatly improved if only good ideas would triumph in a free marketplace of ideas.

But there is no existing free marketplace of ideas, and so good ideas do not triumph in the end. Thus it seems that, despite the vigorous protests of skeptics and others who profit from existing conditions, the evidence would indicate that suppression is the norm.

The current reality of a world in which creativity and independent thinking are stifled portends a dismal future. Is there any hope with a view like this one? Perhaps not. But then, perhaps it is up to us to change our outlook for the years to come.

16 Science as Credo

Roy Lisker

It seems to me that there are too many people in today's intellectual agar-agar who discovered at some early stage that they could feather their nest egg by the interminable cranking of a handful of dependable algorithms in obsessive-compulsive fashion in the same way as the Hindu peasant chews his betel-nuts, the cracker-barrel philosopher his wad of chaw, or the elderly Jewish housewife in Miami Beach her bag of sunflower seeds—and thereby concluded that any real effort towards a higher spiritual or cultural life is a waste of time.

For a great deal of science is nothing more than such forms of compulsive cud-chewing. Truly original ideas are few; many famous scientists have built their entire careers on one or two ideas.

In mathematics (the science with which I have the greatest familiarity), those who developed two original and entirely unconnected trains of thought are given special mention in the bibliographies and histories of the science: Bernhard Riemann, for work in both complex variables and differential geometry; or Gauss for work in number theory, probability and physics.

Really independent ideas are difficult to come by in any field—and by "idea" I mean something like "evolution" or "the square root of minus one," or "the atom." Consider Thomas Hardy, capping a successful career as a novelist with a second career as a poet. Serving us as the exception which proves the rule, his poetry, though much of it is of a high quality, is monotone in its affect of dreary gloom. He is fond, for example, of grieving the miseries of children who aren't even born yet!

Most scientific work, to return to the point, is mechanical, methodical, repetitive and dull. A person may turn out several hundred papers in his lifetime of work without the grace of a single idea worthy of the name. It must be stressed that this in no way negates his competence, dedication or "credibility." He can indeed be quite a good scientist.

Yet one retains the impression, buttressed by numerous historic encounters with every sort of bully in scientist's clothing, that a lifetime of this sort will reinforce an impoverishment of the soul, stinginess of the heart and narrowness of mental vision that is hardly any different from that of the medieval monk, scribe, soldier or peasant . . .

A few months ago, I attended a poetry reading given by a Czech poet/neurophysiologist Miroslaw Holub, at the Lamont Library of Harvard University. I liked his poetry quite a bit; I am sure he is a good neuro (etc.), and know him also as a prominent activist in the years between Dubcek and Havel. Commenting on the differences between literary theory and scientific work, Holub related this conversation between Paul Valery and Albert Einstein.

Valery asked Einstein: "Albert, answer me this: When you get a new idea, do you run to your notebooks to write it down as fast as you can before it's forgotten?" To which Einstein replied: "In our profession, Paul, a new idea arises so very rarely, that one is not likely to forget it, even years later."

To support my thesis that the scientists of the modern world are in no sense the torchbearers of true civilization, but are little different (in the majority) than the brain-dead scholastics of the Middle Ages, I have identified a Credo of thirteen articles resembling the dogmatic catechisms of various cults and creeds, such as the words of the Mass, the laws of Leviticus, the Nicene Creed, the Benedictine Rule, the Confessions of Faith, the Book of Common Prayer, and the like:

THE SCIENTIST'S CREDO

I. *That research be its own justification, whether its purpose be noble, silly or malevolent.*

We see this in particular in research on animal subjects, however there are many examples to be taken from all the sciences. The truism that many discoveries which were useless at the time they were made turned out to be of some use, even a century or two later, has, in our day, been elevated into the above principle, which asserts that "All research must be valuable because it may be useful." Such an argument would, in the older religious credos, be equivalent to an exhortation to monks to commit murder because they might find something which, thirty years later, will give them some good reasons to instruct novices in the evils of murder.

II. *That there are hidden laws of Nature which guarantee that the fruits of all research must ultimately be of benefit to mankind.*

This is a stronger version of Article I, however, the emphasis here is on the "hidden laws," which posit a kind of ultimate "Moral Essence," or

"Unconditioned Virtue" in research. There has been no attempt, as far as I know, made by anyone to discover these laws or to derive them from raw data. I may myself approach the NSF [Nation Science Foundation] to underwrite a few decades of Research to validate or invalidate the belief that Ultimate Goodness lies at the bottom of All Research.

III. *That the unbelievable amounts of suffering inflicted on living creatures, including human beings, through research in biology, medicine, psychology, and related Sciences, have been as necessary to our Salvation as torturings were necessary to the Salvation of the victims of the Inquisition.*

The definition of salvation changes from one era to the next, but the facts of power and sadism undergo little alteration. As long as there exist so many highly qualified professionals in respected fields who enjoy causing suffering to the helpless, it matters little that they toil in this service of some given creed or another one. Ten minutes of rational judgement could easily cancel 50 percent of all the experiments in which living creatures are subjected to such horrible tortures. (It is my belief that this figure can be raised to 100 percent, but that constitutes another essay.)

Still, there is no arguing with Salvation.

IV. *That there exists a well-defined methodology known as the "Scientific Method," and that every intelligent person not only knows what it is, but has exactly the same idea of what it is.*

We are here confronted with yet another classical barge before the tugboat dilemma: the standard definition of intelligence as that mental factor which understands and uses the "Scientific Method." The vulgar definition of this method, that which is adhered to by most members of the scientific community, is some dreary mix of Positivism and Empiricism. Positivism claims that Universals can be proven by the accumulation of Particulars, while Empiricism claims that facts, and facts alone, are self-evident.

In point of fact, this author knows quite a large number of intelligent people who don't buy either of these viewpoints, but they are also not among the legions who recite the Credo every morning upon rising.

V. *That science is not responsible for its creations.*

We all know that Szilard, Fermi, Ulam, Oppenheimer, etc., didn't make the A-bomb: God made the A-bomb. One is reminded of the famous remark of Pope Clement II in the fifteenth century, when he was asked how he and his friends might, in good faith, throw all the gold plates used during their daily feast through the windows of the Vatican and into the Tiber River, while at the same time most of Europe was starving:

"God made the papacy; it's our business to enjoy it."

VI. *That science has absolute control over its creations.*

Most of us go to sleep secure in the knowledge that genetic engineers are following all those guidelines (that they, in their superior wisdom also established), and that therefore Godzilla will not spring out of a test tube, at least not while we're alive.

It might appear to the discerning that Principles V and VI cannot both be true: yet that is the nature of true religion, which cannot be imagined without paradox and contradiction! Read, for example, Rudolph Otto's "The Idea of the Holy."

VII. *That the lifelong gratification of idle curiosity must produce all the raptures experienced by the mystics of the Middle Ages.*

What indeed is this much jubilated "Scientific Method," if not the promise of some delectation of infinite and perpetual bliss in the discovery, for example, that (Catalan's Conjecture) the Diophantine Equation, $xy - uv = 1$, has only one non-trivial solution in integers, namely $x = 3$, $y = 2$, $u = 3$; or that the uncovering of counter-examples, if there are any, would require more computer capacity than that presently available over the entire planet!

Alas, that Plotinus, Meister Eckhart, Heinrich Suso, Thomas a Kempis, St. John of the Cross, and so many others were not born in our glorious age of scientific faith, so that they might achieve union with the Ultimate Reality through computing 20 million roots of the Reinmann Zeta on the line $s = 1 + iy$, or through bashing in the brains of a thousand monkeys to learn about head injuries, or through counseling the world for more than half a century that it must find some way of copulating with its mothers to achieve psychological health, or through using the inhabitants of Bikini Atoll as guinea pigs for the study of radiation sickenss, or through elaborating very complex and involuted theories with no experimental basis, no predictive power, and hardly any theoretical purpose, such as string theory in particle physics.

Twenty years of wasted effort in the elaborate gymnastics of string theory must be worth, in the free market, at least a dozen visions of the Virgin Mary in tenth century gold crowns.

VIII. *That Science is value-free.*

Most of these abominations are justified, sooner or later, by arguments to the effect that Science is unable to determine values. There is, in other words, a limit even to the great powers of the Scientific Method. A book of matches is also value-free; this hardly give us the right to use it for the purposes of burning down someone's house.

The "ultimate benefit" argument, and the "value-free" argument are frequently employed by the same official personages, usually in the same paragraph.

IX. *That Science is the highest value.*

The metaprinciple that there is no contradiction in contradictory principles, is invoked with a high frequency in all organized religions; and, as a religion, Science is nothing if not organized, perhaps the most highly organized in the history of organized religion. One can well imagine, for example, that the author of this essay, sick unto dying from the gangrene of functional employment, would derive quite a lot of satisfaction and a good income by joining the ranks of Walter Sullivan, James Gleick, Gina Kolata, Isaac Asimov and so on, by writing a science column for some magazine or daily newspaper.

This is indeed true, the trouble being that he is unable to pay homage to the drivel demanded by the Religion of Science, a spiritually emaciated cult worship of such universal acceptance that "science writing," "science proselytizing," and "science worship," are inseparable in the public consciousness.

The Article of Faith which requires us to believe that "Science," as a metaphysic and mass opiate, is the highest and most enduring value, has prevailed over the past two centuries so that it has turned almost all of our schools and colleges, and certainly all of our big universities, into either technical schools or research institutes. Things have changed very little since twelfth-century Sorbonne, when Theology was lord of all, and all other intellectual endeavors had to go begging. It is only the name of the game which is different.

In today's schools, Philosophy has been reduced to an inane obsession with sententious doubt. Letters apologizes for its very existence. There's no money in an English degree, and the teaching of Languages for any profession outside the diplomatic corps has fallen to such a low level that even the pampered scientists of our day are in danger of losing their grasp on the scientific treasures of the past five hundred years, almost all of which were written in Latin and Greek—indeed, scientists in today's America can't even speak a good French, German or Russian, something unimaginable seventy years ago.

Culture is ridiculed with a sorry yawn; mathematicians, physicists, biologists, or even chemists who imagine themselves on the slashing edge of knowledge will make comments about modern art, music or poetry that a poor lonesome cowboy, far from the centers of learning and art, would be ashamed to utter.

Such is the power of *faith.*

X. *That non-scientific thought is ignorant, superstitious or crazy and merits ridicule and even persecution.*

Read Stephen Hawking's *Brief History of Time.* His account of the history of Science is factually threadbare—yet quite valuable in presenting the "Standard Model" of European Science: *every advance was halted by obscurantist monks and popes who burned Giordano Bruno, silenced Galileo, taught the unlettered that the Earth was flat, and so on.*

While not disputing the validity of these charges, it it very clear that the things which Hawking, or *Star Trek*, or *Nova*, or *The Shape of the World*, or Asimov, or Sagan (Carl, not Francoise), or Hofstader, or hosts of others really don't like about the Medieval Church, is the presence of a strong and well-organized competition. This myopic view of history also fails to understand that the kind of world that Science has created for us, and the kind of spiritual desert it wishes all of us to live in, is driving hundreds of thousands, millions of the "ignorant" into the arms of these simplistic, foolish, backward yet in so many ways more spiritually enriching faiths, such as Creationism, which people like Sagan and Stephen Jay Gould waste their time in hating and fearing.

As long as there is a well-entrenched, powerful intellectual Establishment trying to teach all of us that the pointless and sterile accumulations of silly facts has more spiritual merit than the compassion of a Mother Teresa or the courage of a Mahatma Ghandi, the legions of the "ignorant" and, presumably, the "damned," are going to swell.

XI. *That anything but the latest theory ("the paradigm") is ignorant, superstitious or crazy and should be ridiculed or even persecuted.*

(I am indebted for this example to Dr. Andreas Ehrenfeucht, at the University of Colorado.) We know that the father of the theory of Drifting Continents, Hans Wegner, was ridiculed and ignored throughout most of his scientific career for his belief in this theory.

Imagine today, however, that there is a geologist who for lots of good reasons believes that this theory is false.

He would probably be given much the same treatment that Galileo received, less brutal in its methods, perhaps, but with exactly the same results: a black-listing and a silencing.

XII. *That social involvement interferes with pure thought.*

Why should the priesthood, the social elite who are carried on the backs of the society like Hindu Brahmans of old in the hoodhahs of elephants, worry themselves about the cow dung that the elephants have to step in? Go to half a dozen science conferences and you will see that the academic scientific world lives in a kind of permanent merry-go-round from lectures to banquets to receptions to luxury hotels to jetliners to grants to awards to citations to publishing contracts to . . .

XIII. *That Science is pure thought.*

Few words in our vocabulary are quite so impure as the word "pure." The Burmese Buddhist tradition maintains that any person who is so advanced as to have no more than one sexually unclean thought each month is already a very high holy man and should be accorded deep veneration.

How much less can we expect of our modern day Western scientist? How often, even in a single day, does he (most of them being men, but this applies also to women), think of the path of the electron, or the structure of DNA, or the classification of all finite groups, or the hibernation of grizzly bears, without at least one reflection on how much money it can make him, or how many conferences he can travel to with it, or how much flattery his colleagues will give him, or how big his pension is going to be, or how handsome he will look in that photograph in the *Encyclopedia Brittanica* of the year 2024, or how much closer he is to the Nobel Prize, or how much better his theory is than that of the x, y, z group over in Illinois, or how his children will look up to him, or how bored his wife will be when he explains it to her, or how, even though it has little about it that appears useful in any way, *somebody might just, in two hundred years, discover a practical application that will eventually earn him posthumous praise as a benefactor of Mankind.*

Of such does the purity of Science consist. It has about the same rating as the purity of the monks in the medieval monasteries, of which we have read so many accounts. We see indeed that the "Credo of Science" is nothing but a long list of delusions on a par with the parting of the Red Sea, the immortal snakes of the Polynesian islands, the bodily ascension of Elijah, the material Ascension of the Virgin, the rebirth of Quetzacoatl, the immortality of the Pharoahs [sic], and the like. It is therefore hardly surprising that the scientific community (apart from the many individual exceptions), has contributed nothing to the advance of civilization beyond its barbarian precursors.

17 Sigmund Freud and the Cover-Up of "The Aetiology of Hysteria"

Jonathan Eisen

In 1896, the young psychiatrist Sigmund Freud presented the first major paper he had ever written to his colleagues at Vienna's Society for Psychiatry and Neurology. Freud considered that his paper, entitled "The Aetiology of Hysteria," was of the utmost importance, since it proposed what he believed to be an irrefutable cause for the neuroses suffered by many of his patients. Quite simply, when listening sympathetically to his women patients, Freud had heard that as children they had suffered sexual assaults, and he believed that it was these acts of violence which had led to the victims' mental illness later in life.

The point of the paper was that sexually abused children, many of whom had come from "respectable" middle class homes, displayed significant "hysterias" later on in life—an observation that today would pass as obvious to the point of banality, but something that in 1896 provoked a backlash among Freud's older colleagues.

All the strange conditions under which the incongruous pair continue their love relations—on the one hand the adult, who cannot escape his share in the mutual dependence necessarily entailed by a sexual relationship, and who is at the same time armed with complete authority and the right to punish, and can exchange the one role for the other to the uninhibited satisfaction of his whims, and on the other hand the child, who in his helplessness is at the mercy of this arbitrary use of power, who is prematurely aroused to every kind of sensibility and exposed to every sort of disappointment, and whose exercise of the sexual performances assigned to him is often interrupted by his imperfect control of his natural needs—all these grotesque and yet tragic disparities distinctly mark the later development of the individual and of his neurosis, with countless permanent effects which deserve to be traced in the greatest detail.

In fact, as author and former Freud Archives Director Jeffrey Masson discussed at some length in his controversial bestseller *The Assault On Truth: Freud's Suppression Of The Seduction Theory* (Farrar Straus and Giroux, 1984), the pressure that was brought to bear on Freud was strong enough to make him change his mind completely about the validity of the sexual assault theory. In a dramatic about-face, he formulated his "seduction theory," in which children themselves became the seducers rather than the victims.

Freud's inaugural paper, "The Aetiology of Hysteria" was singled out from all the other papers presented in Vienna in 1896 as the one paper that was not published in the "Wiener Klinische Wochenschrift," the peer journal for the newly forming school of psychoanalysis. Unlike all the other papers delivered, there was *no summary and no discussion* of Freud's work.

According to Masson, Freud wrote a letter to his close friend Wilhelm Fliess that "A lecture on the aetiology of hysteria at the Psychiatric Society met with an icy reception . . . and from Krafft-Ebing the strange comment: It sounds like a scientific fairy tale. And this after one has demonstrated to them a solution to a more than thousand year old problem, a 'source of the Nile.'"

According to Masson, "The prospect of being ostracized by medical society was negligible in the face of his knowledge that he had discovered an important truth." At this point Freud believed what his patients were telling them, namely that they had been sexually assaulted, usually by their fathers, but sometimes by their mothers, and were living in shame and pain and self-loathing.

> . . . The behaviour of patients while they are reproducing these infantile experiences is in every respect incompatible with the assumption that the scenes are anything less than a reality which is being felt with distress and reproduced with the greatest reluctance.

Masson states that Freud went to some pains to assert his own objectivity and admitted that "he too had to overcome resistances before accepting the unpalatable truth," and was therefore somewhat prepared for his colleagues' negative reaction to his paper.

When the reaction did come it was swift and severe, and conveyed the impression that unless Freud recanted, his future as a psychotherapist would be in jeopardy.

"I am as isolated as you could wish me to be: the word has been given out to abandon me, and a void is forming around me," Freud wrote to Fleiss. And slowly began the transformation that would result in his repudiation of the earlier theory of sexual trauma, to be replaced by the convoluted theory of the infants' fantasy of sexually seducing the parent.

Freud's recantation reads like something out of Stalin's trials of the 1930s when Freud writes of his patients . . .

> I believed (their) stories, and consequently supposed that I had discovered the roots of the subsequent neuroses in these experiences of sexual seduction in childhood . . . If the reader feels inclined to shake his head at my credulity, I cannot altogether blame him.

In fact Freud went so far as to say that ". . . I was at last obliged to recognize that these scenes of seduction had never taken place, and that they were only fantasies which my patients had made up." In other words, his patients had lied to him, and he had been naive to believe them. Rather than having been victims of sexual advances from their parents, they had made up stories "to cover up the recollection of infantile sexual activity . . ." He continues: "The grain of truth contained in this fantasy lies in the fact that the father, by way of his innocent caresses in earliest childhood, has actually awakened the little girl's sexuality (the same thing applies to the little boy and his mother)."

According to Masson, "giving up his 'erroneous' view allowed Freud to participate again in a medical society that had earlier ostracized him. In 1905 Freud publicly retracted the seduction theory. By 1908, respected physicians had joined Freud: Paul Federn, Isidor Sadger, Sandor Ferenczi, Max Eitingon, Karl Jung. . . . The psychoanalytic movement had been born but an important truth had been left behind."

When Masson went on to publish his beliefs about why Freud had abandoned the seduction theory, the psychoanalytic community did not at all take kindly to his indictment of the foundations of Freudian psychoanalysis. *The Assault On Truth* became itself the object of derision and pressure from the psychoanalytic community which refused to believe the evidence that Masson was publishing.

The first indication of trouble ahead came from Freud's daughter, Anna Freud, who voiced her displeasure when Masson began pressing her for the reasons why the letters quoted above had never been published. But the full fury of the psychoanalytic establishment was to come after the publication of preliminary papers divulging the author's discoveries, particularly those surrounding Freud's studies at the Paris Morgue in the 1880s. There he was likely to have witnessed autopsies performed on children who had been sexually mutilated and murdered by adults.

What Masson was doing in his research for *The Assault On Truth* was nothing less than uncovering evidence so damning that it called into question the whole foundation of psychoanalysis itself. Anna Freud virtually admitted that she had deleted her father's crucial letters dealing with the seduction theory and childhood rape. Masson wrote:

I began to notice what appeared to be a pattern in the omissions made by Anna Freud in the original, unabridged edition. In the letters written after September 1897 (when Freud was supposed to have "given up" his "seduction" theory), all the case histories dealing with sexual seduction of children were excised.

When Masson's book was finally published, he was already cast out, and the reason is obvious: He was accusing Freud, the founder of psychoanalysis, of having sold out. Moreover,

I believe that Freud is largely responsible for . . . having given intellectual sophistication to a wrong view (that women invent rape) [and] for the perpetuation of a view that is comforting to male society.

He was also saying that the doctrines of modern psychoanalysis rest on a very shaky foundation indeed:

The psychoanalytic movement that grew out of Freud's accommodation to the views of his peers holds to the present position that Freud's earlier position was simply an aberration.

Masson was attacked, as he says, with more vitriol and personal "ad hominem" arguments than he was with anything substantive, and he wound up taking to court one reviewer, Jill Malcolm and her magazine, *The New Yorker*, in a famous libel suit—which he won. Robert Goldman, writing in *The California Monthly*, would probably have agreed with the decision when he wrote:

. . . Malcolm's account of Jeffrey Masson is a tendentious, dishonest, and malicious piece of character assassination, all the more pernicious because of its studied tone of mildly amused detachment. Had her articles (and now book) never appeared, the arguments of Masson's book surely would have been given a fairer and more dispassionate hearing than is now seemingly possible.

With Masson's study of Freud we find a very clear indication that the so-called intellectual community is as much a part of the suppression syndrome as any other, despite pretensions to considered rationality or intellectual stewardship. The roughing up that people like Masson receive only serves to indicate how fundamentally insecure is our existential human condition. Our hold on honesty is tenuous; we seem ready to sell out when push comes to shove.

This goes far to explain why we have come so little way from the witch burnings of Salem. Masson is a classic whistle-blower; the child who brings our attention to the nakedness of the emperor; the fire stealer who has his liver pecked out every day while chained to the proverbial rock.

Masson's arguments and evidence are certainly convincing, coming as they do from primary sources either suppressed or ignored. If he is right, we begin to see psychoanalysis itself as politically determined and fundamentally flawed. If he is wrong, a lot of people have spent a lot of time trying to defame him. To this day there have been no refutations that we have been able to locate.

The academics must open their minds and accept the truths presented. Histories of countless individuals have gone unheard because classic Freudian psychoanalysis has turned a deaf ear to them. Perhaps it is time to turn the tables, and disempower psychoanalysis.

18 The Burial of Living Technology

Jeane Manning

Threatening to hang the fifty-eight-year-old man and to harm his family if he did not cooperate, Adolf Hitler forced an Austrian inventor to build a flying craft which levitated without burning any fuel. The inventor had previously produced electrical power from a unique suction turbine by the same implosion principles, using air or water in creating the force. The Third Reich wanted these inventions developed quickly. But the inventor took his time; understandably he did not want to give Hitler a technological advantage.

The Austrian, Viktor Schauberger, was known in his time as the Water Wizard. The courageous inventor built prototype examples of beneficial technology, in his effort to turn humanity away from deathdealing technologies. He defended Earth's water, air and soil, but at the end he was out-manoeuvred by people with lesser motives.

Schauberger was a big full-bearded man and could be ferociously gruff; he had no patience with greed-motivated fools. But he was untiringly patient when learning from his teacher—the natural world. In Alpine forests, along rivers and in the fields of wise old traditional farmers, the forester/scientist learned about a life-enhancing energy which enters a substance such as water or air through inward-spiralling movements of the substance. During his lifetime of persevering study he copied nature's motions in his own engineering.

"Prevailing technology uses the wrong forms of motion. It is based on entropy—on motions which nature uses to break down and scatter materials. Nature uses a different type of motion for creating order and new growth," he admonished in a voice stern with conviction.

The prevailing explosion based technology—fuel-burning and atom-splitting—fills the world with expanding, heat-generating centrifugal motion, he warned. On the other hand, energy production and other technologies could instead use inward-moving, cold-generating centripetal motion, which nature employs to build and enliven substances.

Even hydroelectric power plants use destructive motion, he said; they pressure water and chop it through turbines. The result is dead water. His suction turbine, on the other hand, invigorated water. The result, he said, was clean healthy water.

His stubborn certainty angered academics who assumed superiority over a largely self-educated man. It is not surprising that he was sometimes abrasive; the Schauberger heritage included defiant courage. His ancestors were privileged Bavarian aristocracy with a manor named Schauburg, and in the thirteenth century this ancient family lost its royal privileges by publicly defying a powerful Bishop.

IN TUNE WITH NATURE

A few centuries later, about 1650 A.D., a family member moved to Austria and began a branch of the Schaubergers which specialized in caring for forest and wildlife. Breathing the scent of sun-warmed pines, generations of Schaubergers then lived their family motto of *fidus in silvis silentibus*— faithful to the silent forests. Viktor's father was master woodsman in Holzschlag at Lake Plockenstein, and Viktor absorbed accumulated wisdom of generations of forest wardens. His mother also taught him to tune in to nature—to listen to its singing in a mountain stream as well as its whispering through the treetops, and to learn its cycles and rhythms.

The family's closeness to their environment was not only on a spiritual or poetic level; it was based on practical observations. For example, Viktor's elder relatives respected a certain vigour which they found in cool unpolluted water. So, instead of irrigating meadows in warm sunlight when the water was sluggish, they spent moonlit nights lifting gates on their irrigation canals so that the liveliest [most life-giving] water would flow onto their land. It grew noticeably more grain and grasses than did the neighbouring lands.

From childhood Viktor aspired to be a forest warden like his father, grandfather and a line of great-grandfathers. As a boy he explored nearby woods and then roamed farther. He came to know the rumbling rivers and the musical streams which feed them, just as other young people know streets and hallways and sounds of their childhood. However, he noticed that natural waterways rarely flowed in straight corridors. Instead, a river undulates through the landscape, swerving to one side and then to the other. Within the larger meandering caused by Earth's turning, water coils around a twisting central axis as it sweeps downstream. Keeping in mind this inward-spiralling motion, Schauberger later developed the basis for a technology in tune with nature.

When Viktor reached university age, his father wanted him to train as an arboriculturist. The young man resisted the pressure to limit his outlook to the academic viewpoint. He quit university, but later did graduate from forest school with state certification as a forest warden, and then apprenticed under an older warden. Throughout his life he continued to learn, from books and wise observers as well as directly from nature.

ROYAL GAMEKEEPER

Schauberger had the opportunity—rare in this century—of living for years in a vast unspoiled forest. After the First World War ended, Prince Adolf von Schaumburg-Lippe hired him to guard 21,000 hectares [51,870 acres] of mostly virgin forest in a remote district. As he patiently observed rhythms of life in this huge watershed, Schauberger saw phenomena which may be impossible to find today. One terrifying example, which in the end impressed him with the self-regulation of nature, was a landlocked lake which rejuvenated itself before his eyes. One warm day he was about to strip and swim in the isolated lake, when it roared with sudden movement. Whorls appeared on the surface and half-submerged logs started to move. The debris circled, faster and faster while a massive whirlpool formed in the middle of the lake. Then the huge logs sucked into the centre upended and disappeared into the whirlpool. After the waters stilled momentarily, a gigantic waterspout startled Schauberger even more. Turning as it rose, the spout reached as high as a house then settled back, and the waters began to rise on the shore. The young gamekeeper ran; he had seen enough. But the incident added to the mystery of this substance which fascinated him—water. Schauberger was well-placed for developing his unique understanding of water; his workplace was big enough for interconnected life processes to mesh without hindrance there. Life forms interacted in balance; it was still an unbroken web of life.

Six foot tall Viktor at that time of his life was said to be a picture of contentment—muscular good health from hiking the high country, and alert intelligence described in his facial features—farseeing eyes, the slight curve of his nose reminiscent of an eagle's beak, and the determined but good-humoured set to his mouth. He wrote that this was a happy time, while he watched the larger animals migrate with the seasons and observed salmon and trout in cold mountain streams. Countless hours of studying the fish in motion gave him insights which later led to one of his inventions, called the trout turbine. Picture him at rest on a summer afternoon, his long frame stretched on a grassy riverbank. Sunlight filters through a canopy of leafy branches overhanging the river. Deep in this pristine mountain setting, the combination of his sharply observant eyes and his intuition was synthesizing new knowledge.

LEARNING FROM THE SOURCE

He learned that water swirling over rocks in a tree-shaded natural setting carries a vitality which is real as an electric current carried by wires. And minerals carried along on that vitalized inward-curling water enrich the trees whose rootlets seek the mud. Trees and water, water and trees. Each needs to have the other growing in a natural state.

The young forest warden once hiked up a mountain with some hunters, old men who were familiar with the area. High on the mountain they found a heap of rocks which had been part of a stone hut which had arched over a mountain spring for as long as anyone could remember. Hikers traditionally would duck into the cool interior of the hut and ladle a drink of refreshing water. Now, however, someone had dismantled the hut and exposed the spring to sunlight. To the surprise of the old hunters who came there seasonally, the now exposed water shrank back into the earth; the spring dried up for the first time, and it stayed dry. After months and much head-scratching, they decided to rebuild the stone hut. Eventually the spring returned and continued to flow, season after season.

Incidents such as this taught Schauberger that water needs to be cool—about 4°C [celsius]—even as it bubbles out of the ground. Without a shaded exit, he found, water will not "grow" to a great height underground and emerge as the mountaintop spring. As well as temperature, time spent maturing in underground rocks provides minerals which help make water sparkle with energy.

Schauberger noticed beautiful vegetation growing around natural springs —an indication of "mature" mineralized energetically-charged water. These concepts, of water having qualities such as strength and maturity, were not found in any textbooks or lecture notes. The brash forester later told hydrologists to abandon their microscopes and testing laboratories, and instead study water holistically in its environment. He found natural watercourses to be alive with inherent intelligence, and not to be mere movements of a chemical substance.

Another mystery which fascinated him was the sight of large trout and salmon lying nearly motionless in a stream while facing into a swift current. When the forester moved and startled the fish, they darted upstream headlong into the rushing current. Why didn't they go with the obvious flow and escape downstream? Was there some invisible channel of energy running opposite to the current?

He decided to experiment on a sizable stream with rapids where a large trout often lay. Schauberger sent his woodsmen 500 metres upstream to build a bonfire. He instructed them to heat about a hundred litres of water and pour it in the stream on signal. This infusion of warm water made no noticeable difference in the overall temperature of the stream. But the position of the large trout downstream immediately weakened, and despite thrashing its tail and fins, it was swept downstream. Schauberger was then sure of the connection between water temperature and some unknown flow of energy in the water.

This reinforced his belief that the sheltering tangle of willow branches overhanging a river is crucial; without cooling shade, excess warming would cause the water to lose an electrical-type potency.

One moonlit night brought both danger and a magical sight. He was sitting beside a waterfall waiting to catch a notorious fish poacher. To pass the time he watched trout swim in the crystal-clear pond below. Suddenly a much larger trout arrived and dominated the scene with a twisting underwater dance. It headed under the main fall of water, and soon reappeared for an instant, spinning vertically under a glittering cone-shaped stream of water. To Viktor's amazement, the lone fish then stopped spinning and instead floated upward to a higher ledge of the waterfall. There it fell into the rush water and disappeared again with a swish of its tail.

The dangerous poacher was forgotten, after the spectacle of a silvey fish floating up the moonlit waterfall. Schauberger filled his pipe and slowly, thoughtfully, walked home. Again, it seemed the wild stream must generate some type of energy. Years later, Schauberger would devise an experiment which clearly demonstrated an electric charge present in moving water.

COULDN'T BELIEVE HIS EYES

Another clear night, in late winter, he again rubbed his sharply observant eyes in disbelief. Exploring a rushing stream in bright moonlight, he stood on the bank looking down into a deep pool. The water was so clear that he could see the bottom, several metres below the surface. Large stones on the bottom were jostling about. Even more amazing, an egg-shaped stone about the size of a human head started circling in the same way as a trout does before jumping a waterfall. Suddenly the rock broke the surface of the pond, and slowly a circle of ice formed around the floating stone. Was this a cold-generating instead of a heat-generating process? Then one by one nearly all the egg-shaped stones circled up and appeared on the surface. Stones of other shapes remained unmoving on the bottom.

What metals did the dancing stones contain? Why the egg shape? What force develops in this pristine water? What is motion, anyway?

Schauberger had a lot of solitude for mulling these questions, and eventually he developed a theory about different types of motion. He saw that water needed freedom to move in a vortexian motion (three dimensional spiralling).

He saw the spiralling shape in the growth of vines, ferns, snail shells, whirlpools, galaxies and countless other formations. The hyperbolic spiral was everywhere, as if acting out some underlying universal motion. In uncaged rivers, the spiral was seen in the horizontal tightening twists of the layered current. He became certain that the contracting vortex created a very real energy in the water as it flowed.

Schauberger learned how colder, denser, stronger water in streams carried heavy natural debris without silting, and how undisturbed rivers managed seasonal torrents without seriously eroding their banks.

Schauberger proved to be a skilled engineer who turned his insights into practical devices. But even his first invention was controversial.

PRINCE NEEDED CASH

While Schauberger was studying nature's habits, outside the forest others were more entranced by worldly ways. The aging prince who owned the wilderness had a young wife who liked to gamble, so he needed quick cash to pay his wife's debts. The prince eyed his remote forests and saw lumber which could be sold. The prince's predicament placed a challenge before his forester—could Schauberger make a miles-long wooden waterslide which would carry logs from the high mountain slopes down to the valley?

Experts said it was impossible—heavy logs would scrape to a halt on the wooden slide. Or if they somehow gathered speed, they would smash the sides of a flume. However, from his father and from observing wild rivers, Schauberger knew how to bolster the strength of water just as nature does, so that even heavy beechwood would ride high on the shallow stream. He hired men to build a strange structure which curved and twisted down the steep mountain. At points along the route, his design included valves for inlets and outlets which poured in cold water from other streams and released sun-warmed water from the chute.

The day before the deadline, a log started down the new chute for a test run, then it stalled and stuck in place. The workmen snickered, they had no faith in this zigzagging construction.

Schauberger sent them home so that he could think. While sitting on a rock looking down at his log-sorting dams, he felt a snake under his leather trousers. After he jumped up and threw it away, it landed in the dam. Observing it through binoculars, he wondered how a snake can swim so quickly without fins. As if in answer to his problem of transporting logs, the snake twisted in both vertical and horizontal curves.

"Understand Nature, then copy Nature," was Schauberger's motto. From the sawmill he ordered lengths of wood, and his workers hammered all night, nailing short timbers within the curves of the flume to add the up-down snakelike motion to the water.

When the Prince and Princess and other dignitaries arrived for the demonstration the next day, there had been no time for a test run. None of the men believed the flimsy-appearing structure could carry even one of the massive logs without disaster. But it did work. The cold water floated heavy logs and the shape of the chute spiralled the water, which swept the logs always toward the centre of the current and away from the sides of the wooden flume. The serpentine movement was a success.

PROFESSIONALS JEALOUS

In gratitude the Prince appointed Schauberger as head warden of all his hunting and forest districts. Then Schauberger was awarded a further honour—the position of State Consultant for Timber Flotation Installations. Not everyone was pleased, however. Experts with academics degrees resented the fact that a non-academic had landed such a high-salaried position, and the fact that they had to consult with him. Finally the pay-scale furore reached high levels, and the federal minister who hired Schauberger had to cut his salary in half. Schauberger was welcome to stay on the job, though, and the minister offered to make up the missing half of his wages out of the minister's "black funds." Schauberger would have nothing to do with such sleazy practices, however, and he immediately resigned.

He was then hired by a private building contractor to construct log flumes in various European countries until 1934, when Schauberger again criticized an employer's manipulations.

Why would a natural philosopher like Schauberger get involved in log transport, anyway? The answer is complex. Earlier as a forester, it was his job to plan how to move wind-felled timber from high slopes down to valleys where people could use it for firewood and building. Schauberger opposed what he saw as exploitation of horses; he objected to the practice of forcing draft animals to burst their sinews pulling heavy logs down mountainsides. Also, his biographer Olaf Alexandersson writes, Schauberger naively tried to restrict tree-cutting by reducing transport costs—the companies would not need to cut as many trees in order to make the same amount of profit.

At the same time as he was flume-building, he gave speeches and wrote articles about the result of clearing a forest area totally—loss of healthy water downstream and, eventually, drought.

"Every economic death of a people is always preceded by the death of its forests," he warned.

Forests were not as checkered with clearcuts at that time, and local sawmills were not all bought up by large companies which were to become voracious in their appetite for timber. However, Schauberger was alarmed at what he saw forthcoming—"Reckless deforestation results in the drying out of mountain sources, dying of whole forests, uncontrollable mountain streams, silting of water and the sinking of subterranean water stores near where human interference took place."

"Water follows the same laws as the blood in our bodies and the sap in plants; it has analogically the right of being treated as the blood of earth."

He sharply criticized hydrologists—the experts on water—and said that they had only their own careers in mind and had failed radically to understand what was happening in watercourses. "They did nothing, except reinforce . . . quite haphazardly, some banks of rivers and brooks, but managed to forget everything about the water itself as if it had no concern."

OFFICIAL EXPERTS JEER

Hydrologists scorned Schauberger's non-academic warnings. He had learned that river water is made up of layers of different densities and the lamination has a purpose in generating a charge in healthy water. Water is not merely a chemical compound, he insisted; it should not be recklessly chopped up in hydro-electric turbines, much less injected with chlorine or unnecessarily exposed to heating.

The experts hooted when he pointed out that in a person, a temperature change of only a tenth of a degree celcius could mean sickness or health. Was he comparing a planet with a person? Did he think Earth was a living organism with biologically-active bloodstream? They ignored the heretical concepts.

Schauberger offered to organize a job creation project to rebuild watercourses. If artificially-channelled rivers were to be uncaged and restored to their meanders and oxbows sheltered by vegetation, would the rivers again keep their own channels clean and stop their own wild flooding? Schauberger was never given the chance to find out. He was realistic enough to look for a more feasible way of rebuilding, and in 1929 he patented a system of braking barriers to be inserted along a troublesome watercourse. The barriers would redirect the axis of flow toward the middle of a stream, reducing the amount of soil carried away from the banks. Another complex Schauberger patent offered to both control the action of outlet water from holding dams and to strengthen the dams by including factors of temperature and motion.

Was anyone from academia listening? One renowned hydrologist eventually was; he started out by denigrating Schauberger and ended up following him around in the woods and even into a chilly river. Professor Forcheimer literally waded into Schauberger's teachings about the laws governing water's behaviour, and the professor decided that the self-educated man actually based theories on facts. Unlike colleagues who were in the middle of academic careers, Forcheimer would not lose financially by championing a heretic; the professor was in his seventies and, as it turned out, near the end of his life.

Regardless of his bitter battles with the scientific community, Schauberger believed in the scientific method. He experimented on liquids and gases in a small laboratory he set up. His aim however, was to develop a science which actually worked [on principles opposite to the orthodox viewpoint]. "Humanity has committed a great crime by ignoring the use of cycloidal motion of water," he said. For example, the current water-pumping devices were not only uneconomical, he said, "they cause water to degenerate by depriving it of its biological values."

Attempts to explain connections between cycloidal motion and levitation to a scientist are useless, Schauberger said bitterly. Nor are world leaders any help "because they lean on the ignorance of the masses, including the scientists, as well as . . . current physical laws, to safeguard their vested interests and positions."

Conventional energy conversion—burning of fossil fuels or atom-splitting—turns order into chaos. Schauberger proposed processes which would add order and energy to substances such as water, instead of destroying it, while generating useful electric power.

POWER FROM THE UNKNOWN

Schauberger believed that an invisible field structure permeated everything and was necessary for life, but he observed that technologies could propel the unknown field structure into either motions harmful to biosystems or helpful to biosystems. In other words, he held technical planners responsible for the life or death of biological systems.

How did he prove his ideas?

Not one to stay at the vapourware [designed but not yet produced] level of ideas, Schauberger picked up his tools and built hardware. From watercourses to agricultural implements, his constructions attracted praise from users. Then he turned to extracting electrical energy directly from the flow of water and air. "They contain all the power we need."

Hitler had heard of the Living Water Man through an industrialist. After Germany took over Austria in 1938, word came to Schauberger that he would be hired to plan log flotation structures in Bavaria, Bohemia and North Austria, and that furthermore he could use a professor's laboratory in Nuremburg for his research.

Viktor Schauberger sent for his son Walter (born July 26, 1914). Walter had studied physics in university and found that some of his father's concepts were foreign to the way he had been taught to think. However,

Walter's scepticism crumbled during the experiments they conducted. Walter contributed useful techniques himself, and the duo were soon extracting 50,000 volts from fine jets of water at low pressures. A physicist from a nearby technical college came; his first action was to search for hidden wires. When he could find none, he lost his temper and asked Walter where he had hidden the electrical leads. Eventually he had to admit that there was no trick involved; the experiment was valid. However, he could not explain such a high charge from water.

The Second World War interrupted their experiments, and Walter [was] drafted. Viktor was ordered to undergo a physical examination supposedly related to his forthcoming pension. However, says biographer Alexandersson, "it looked like an engineering and architectural association was behind this demand for a check-up."

Viktor Schauberger unsuspectingly showed up, but was whisked away to another clinic. He was told it was for a special exam, but to his horror he found himself being questioned in a psychiatric clinic. He forced himself to answer the questions in a peaceful non-abrasive way; if he displayed anger he might be locked up. Two doctors tested him and found him perfectly sane as well as highly intelligent. They never found out who had arranged to get him into the mental hospital.

"BUILD MACHINES, OR DIE"

He himself was drafted in 1943, despite his age. After a brief stint as commander of a parachute group in Italy, he was ordered by Himmler [Hitler's chief lieutenant] to the Mauthausen concentration camp. Himmler's greeting, passed on by the camp's military leader, gave him a choice—death by hanging, or develop machines which used the energy he had discovered. He was told to lead a scientific team of the best engineers and stress-analysts from among the prisoners.

The work was based on Schauberger's discovery of how to develop a low-pressure zone at the atomic level. This had happened in seconds when his laboratory device whirled air or water "radially and axially" at a falling temperature. He referred to the resulting force as diamagnetic levitation power. He emphasized that nature uses indirect—what Schauberger called reactionary—suction force.

He insisted that the technical team from the concentration camp be treated as free men would. After their research headquarters was bombed, they were transferred to Leonstein and started a flying disc project to be powered with his trout-inspired turbine which rotated air into a twisting type of oscillation resulting in a buildup of immense power causing levitation. A small model which crashed against the ceiling glowed blue-green at first as it rose, then trailed a silvery glow.

According to researcher Norbert Harthun, his devices were no more than laboratory models by the end of the War. However, the American military officers who showed up a few days after the model hit the ceiling seemed to know what he was doing. They seized everything. He was interrogated by a high-ranking officer, and put in "protective custody" for six months. The officers also heavily questioned his helpers. Russian members of the team later returned to the Soviet Union.

Alexandersson's book quotes a letter from Schauberger saying he was confined by the occupying forces for nearly a year because of his knowledge of atomic energy (even though his research was directed toward implosion—which was labelled fusion—rather than toward the destructive fission approach to the atom).

A few tantalizing bits of lore about Hitler's "flying saucers" rose into public awareness years later. The July 27, 1956 Munich publication *Da Neue Zeitalter* said that ". . . Viktor Schauberger was the inventor and discoverer of this new motive power—implosion, which, with the use of only air and water, generated light, heat and motion." The first unmanned flying disc was tested February 19, 1945 near Prague, the German periodical claimed; the disc could hover motionless in the air and could fly as fast backwards as forwards. "This 'flying disc' had a diameter of 50 metres."

Viktor wrote to a friend in 1958 that the craft test-flown near Prague was built according to the model he made at the concentration camp, and it rose to 15,000 metres in three minutes. It then flew horizontally at 2,200 kilometres per hour. "It was only after the war that I came to hear, through one of the workers under my direction, a Czech, that further intensive development was in progress; however, there was no answer to my enquiry."

There is no doubt Viktor Schauberger knew how to build an implosion device which levitated. His problem was how to brake it. Test models generated so much energy that an entire engine lifted itself off the floor, levitated in the high-ceilinged test hall, and crashed against the ceiling.

At the end of the Second World War, American and Russian military confiscated his models, diagrams and even the materials he used. Reportedly the Russians even burned his apartment in case they had missed any technological secrets hidden there. Did anyone carry on the levitation-craft work after Schauberger's wartime research team was split up? The answer may be buried in some country's classified defense files.

After the Far East Treaty was signed, Schauberger took up his research again. He had lost his financial assets in the war, but he stubbornly persisted from his home at Linz, and took out patents. Despite having no money, he thought he could help the world by turning his inventive genius and his insights toward agriculture.

Bitter about the effects of both the chemical industry and deforestation upon agriculture, he stated, "The farmers work hand-in-hand with our foresters. The blood of the earth continuously weakens, and the productivity of the soil decreases."

When forests can no longer nurture water sources which supply vitality, then farmland downstream cannot build up a voltage in the ground which is necessary for keeping parasitic bacteria in balance, he observed. Noticing that soil dried out after being ploughed with iron ploughs, he built copper-plated ploughs. The ploughs successfully increased crops, but the greed of special-interest groups stopped the venture.

Schauberger continued to come up with innovations to help grow healthy crops, until all his work was halted in 1958. Walter and Viktor were in the United States from June 26 through September 20, 1958, living together day and night, and Walter emerged from the experience with a new appreciation of Viktor's knowledge. But their joint attempt to get his implosion generator funded and developed was derailed.

PROMISES FROM THE USA

Little is known publicly about their trip to America except a few key aspects. In the winter of 1958 two men, which European researchers refer to as "American agents," visited Viktor and convinced him to go to America for what they promised would be only three months. He was led to believe that the purpose would be to finally convert his knowledge into the manufacturing of beneficial devices.

It turned out to be an ordeal which the father and son had not expected. They were flown to a sweltering hot climate—Texas in summer—which stressed Viktor's health. He was now nearly 73 years old. Over the months Viktor became increasingly angry because the men and their associates now were in no hurry to set up a facility and develop implosion motors to generate clean power. "Now we have plenty of time," was their reply.

At first trusting the sincerity of his hosts, Schauberger had brought all his documents and devices to Texas, and was then asked to write down everything he knew. He co-operated and the material was sent to an atomic technology expert who met with the Schaubergers for three days in September. According to Olaf Alexandersson, the expert from New York said ". . . The path which Mr. Schauberger in his treatise and with his models has followed is the biotechnical path of the future. What Schauberger proposes and asserts is correct. In four years, all this will be confirmed."

The two Schaubergers expected to go home now; three months had passed. But the Texas group apparently demanded that the father and son remain in the United States of America and live in the Arizona desert. The Schaubergers refused. After much argument, the Americans relented and said Viktor could travel home, but first he had to sign a contract and agree to take a course in English. Unfortunately the contract was in English and Viktor did not know the language. His biographers say he was pressured to sign quickly; their flight would leave shortly and there was no time to quibble.

Viktor at that point only wanted to get out of the hellish heat and away from these deceptive people. He signed. Walter refused to sign. He would be on dangerous ground with immigrant authorities if he signed such a contact, for one thing.

After Viktor gave in and signed, suddenly there was ample time before they needed to go to the airport. Champagne corks popped and their hosts celebrated.

One can only imagine the conversation between father and son on the flight home. At last we can go home; get away from those thieves. But what have we done?

Walter probably had the heartbreaking task of spelling it out to his father. "Yes, it is as I told you when they were pressuring you to sign; the contract says that now you can't write about or even talk about your past-and-future discoveries, and you are bound to give everything you know to that boss of the Texas consortium. Their contract says they now have all the rights to the 'Schauberger business' as they put it."

Was Schauberger's implosion process considered by the American officials to be "cold fusion"? The Austrian observer of nature apparently did arrive at results related to modern sub-atomic research. In the late 1980s, an independent researcher tried to get information on the Texas incident. Erwin Krieger's attempt to get information through the Freedom of Information Act failed; he was told by a form letter that the material may be related to national security.

"I DON'T EVEN OWN MYSELF"

Viktor Schauberger was at the end a despairing man. In the last few days of his life he reportedly cried over and over, "They took everything from me, everything. I don't even own myself!" Stripped of hope, he died five days after they returned home.

His passion for learning nature's ways and then applying that knowledge to life-enhancing technology, and his efforts to interest those who could fund its development, had let him a long way from the peaceful forest. The more recent loss was the legal right to work on his implosion technology. But how did that compare to what seemed like the loss of his lifetime of hard-won insights? The world had ignored warnings—from him and others—about what would happen if natural forests disappeared en masse, and his planet's weather, water, soil and air deteriorated as a result. Nature was thrown out of balance. Too much of the life-destructive motions and not enough of the life-creative motions? In Schauberger's despairing view, humanity was headed towards a mental and spiritual sluggishness, easily controlled by dictators who step in at a time of food shortages.

More than thirty-five years after Viktor Schauberger's death, there is a surge of concern for the planet's health. The health of its inhabitants—in the sea and on land—is in turn deteriorating. Will humanity turn toward Viktor Schauberger's insights? There are signs: maverick scientists are developing theories such as how a subtle energy (unknown field structure) may be drawn into use by shapes and vortexian movements. In Europe, new books and magazines bring out Schauberger's teachings; nonconventional scientists teach that the opposite poles in nature (light and dark, warm and cold, pressure and suction, male and female and so on) are necessary to create movement. Further, these books say, without movement there is no life, and the force created in healthy moving water is the life force.

Cambridge-educated John Davidson of England looks at "a possible similarity between magnetic alignment of atoms in iron, and alignment of molecules of water moved in Schauberger-advocated hyperbolic spirals . . . we create effects which were not apparent beforehand."

Across the Atlantic, nuclear physicist Dan Davidson suggested mathematical research into natural river meanders, naturally occurring spirals and other geometric patterns in nature, to find equations for tapping the diamagnetic forces which Viktor Schauberger used.

Meanwhile in Europe, Walter Schauberger snubbed Americans who tried to communicate with him; so deep was his anger at the way his father was treated. But Walter is reportedly doing all he can to carry on his father's work, at his secluded private institute. Among other teams doing scientifically-rigorous related research are the Scandinavian Institutes of Ecological Technique.

In New Mexico, William Baumgartner dedicated years to experimenting on building implosion hardware such as a version of Schauberger's "trout motor" and a water-energizing device, and he expects to have a reliable suction turbine built by the time this is in print. Baumgartner also lectures on Schauberger's innovations for agriculture and water treatment, as does Callum Coates in Australia and others in Europe and Canada.

Life-oriented technology may yet arrive in time.

REFERENCES

Alexandersson, Olaf, *Living Water: Victor Schauberger and the Secrets of Natural Energy*, Turnstone Press Ltd., Wellington, Northamptonshire, 1982.

Baumgartiner, Williams, Energy Extraction from the Vortex, *Proceedings of the International Symposium on New Energy*, Denver 1993.

Baumgartiner, Williams, *Energy Unlimited Magazine* and *Causes Newsletter*, numerous articles on vortexian mechanics and Schauberger technology, based on Baumgartiner's hands-on experience, 1970s and 1980s, Albuquerque, New Mexico.

Brown, Tom, Editor, *More Implosion than Explosion*, Borderland Sciences, Garberville CA, 1986.

Coats, Callum, "The Magic & Majesty of Water: The Natural Eco-Technological Theories of Viktor Shaubauger," *Nexus Magazine*, Australia, June-July 1993.

Davidson, Dan A., *Energy: Breakthroughs to New Free Energy Devices*. Rivas Publishing, 1990.

Davidson, John, *Secret of the Creative Vacuum*.

Frokjaer-Jensen, Borge, "Advances with Viktor Schauberger's Implosion System," *New Energy Technology*, The Planetary Association for Clean Energy, Ottawa, 1988.

Frokjaer-Jensen, Borge, The Scandinavian Research Organization On Non-Conventional Energy and The Implosion Theory of Viktor Schauberger, *Proceedings of the 1st International Symposium on Non-Conventional Energy Technology,* Toronto, 1981.

Harthun, Norbert, Systems in Nature: Models for Technical Conversion of Energy—Statements by Viktor and Walter Schauberger, *Proceedings of The Second International Symposium on Non-Conventional Energy Technology*, Cadake Industries, Atlanta, 1983.

Kelly, D.A., *The Manual of Free Energy Devices and Systems*, Vol. 11., Cadake Industries, 1986.

Lindemann, Peter A., *A History of Free Energy Discoveries*, Borderland Sciences, Garberville CA 1986.

Manning, Jeane, "Vortex Mechanic," *Explore More Magazine* No. 6, Mt. Vernon WA, 1990.

New Energy Technology, The Planetary Association for Clean Energy Inc., Ottawa, 1990.

Resines, Jorge, *Secret of the Schauberger Saucers: A Theoretical Analysis of Available Information on this Rare and Suppressed Technology,* Borderland Sciences, California, 1988.

Schauberger, Viktor, (translated by Dagmar Sarkar), "'Unfathomable Water,'" *Energy Unlimited Magazine*, Issue 24, Alburquerque, New Mexico.

Schauberger, Viktor, (articles translated by W.P. Baumgartner and Albert Zock) *Causes Newsletter* 1988–91, Albuquerque, New Mexico.

19 Egyptian History and Cosmic Catastrophe: The Ideas of Dr. Immanuel Velikovsky

"Gerard"

Dr. Immanuel Velikovsky was one of the twentieth century's great scholars. He sought to solve a mystery and in the process generated enormous controversy in the fields of archaeology, astronomy and cosmology. The attack by many members of the scientific community on his work, their attempts to intimidate his publisher and suppress his evidence have made Velikovsky the Galileo of our time.

The story starts in 1939. Then Dr. Velikovsky, a practising psychoanalyst who had studied with Freud, went to the U.S. to research a book on three dominant figures of the ancient Mediterranean—Moses, Oedipus and Akhnaton. When he was nearly complete a question arose about the time of the Exodus of the Hebrews from Egypt. Although recorded in detail by the Hebrews there was no equivalent record in Egyptian history. Why?

Under the conventional chronology of Egyptian history, the time period usually considered for the Exodus causes problems. For many other events in Hebrew history there appear to be no Egyptian counterparts either—but these two nations existed beside each other for centuries, according to the Hebrews. Velikovsky's answer to this lack of correlation was to suggest that the accepted chronology of ancient Egypt was off by five hundred years. He noted that our dating of Egyptian periods came from the dynastic records handed down by Aegyptus and Agrippa, and that the reigns of the Pharaohs used to date Egyptian history had been strung together one after another. The key to the "missing" half millenium was that there were many "co-regnal" periods where the reigns of monarchs overlapped or were indistinct. In his revised chronology, many events in Hebrew history were found to have their counterpart in the Egyptian record.

The new chronology was built from the archaeological evidence found in the ruins of ancient Egypt. From remaining papyrus and pottery, tomb paintings and monuments a new story emerged. Because the histories of Greece, Assyria, Babylon and Judea were all dated from the dynastic records of Egypt, Velikovsky's work became highly controversial. In 1945 this was limited to a specialist field, but his next publication in 1950, *Worlds in Collision*, aroused widespread controversy, and was Velikovsky's explanation of the cause of this amnesia in our collective memory.

THE COMET VENUS

The ancient civilisation of Egypt was nearly destroyed in a cosmic catastrophe that endangered the entire planet, according to Velikovsky. Everywhere, huge resources were devoted to study of the skies. It's widely known that ancient civilisations in Asia, the Americas, Europe and the Middle East were highly advanced in astronomy. While we accept this as a common feature of our past, why were so many people interested in the study of the movements of the planets? Why is the alignment of astronomical instruments found in Babylon 2.5 degrees out from the present alignment of the Earth? Why did calendars constructed between the middle of the second millenium BCE* and 800 BCE have 360 days and months of thirty days? Why do even earlier calendars have days, months and years of different lengths again?

Velikovsky's answer was that the Earth and Mars had been involved in repeated near collisions with a gigantic comet since our recorded history began. The events described in the Exodus and in Egyptian papyri are a vivid description of an age in chaos—plagues, turmoil and darkness, and the flight of the Hebrews from Egypt toward a "column of fire" in Sinai.

The Earth was momentarily slowed down and its axis slightly altered as the comet passed by. Electrostatic forces caused discharges to arc between the Earth and the comet turning the skies to fire and the forests to flame. The crust was rent, volcanoes erupted, earthquakes rocked and darkness enveloped the world—the time of the Exodus. Seven hundred years later Isiah, Joel and Amos described another series of upheavals; the Sun appeared to stand still in the sky. Although slightly dislodged from its axis and orbit again, the Earth fared better this second time. These were, in fact, the last two acts of a cosmic drama; the earliest act of which we have records is called The Deluge.

> All cosmological theories assumed that the planets have evolved in their places for billions of years . . . Venus was formerly a comet and joined the family of planets within the memory of mankind . . . We claim that the Earth's orbit changed more than once, and with it the length of the

---

* Before the Common Era

year; that the geographic position of the terrestrial axis and its astronomical direction changed repeatedly and that at a recent date the polar star was in the constellation of the Great Bear.

—*Worlds in Collision,* p. 361.

Velikovsky believed that the origin of the comet that was responsible for changes in the Earth's orbit was in the proto-star we know as Jupiter. This idea outraged the scientific community. But his theories about the natures of Jupiter and Venus have not yet been proven wrong. He said that because Venus was younger than the other planets, its surface temperature would be much hotter and its atmosphere denser than astronomers believed; these predictions were proven correct.

He predicted Venus would be found to have orbital anomalies in relation to the other planets; Venus has since been found to rotate on its axis in reverse direction to the other planets, and its day is longer than its year. We now know that parts of the atmosphere of Venus rotate in 4 days (with winds of up to 400 km/h) while the planet itself rotates in 243 days. Both these rotations are retrograde. One of Velikovsky's hypotheses for the slowing of the Earth's rotation which made the Sun appear to stand still was that the planet was engulfed in the extended atmosphere of the comet Venus. Some of the diurnal rotation of the Earth was imparted to this dust-cloud according to Velikovsky, which fits the eccentric characteristics of the Venusian atmosphere.

The comet spiralled past the Earth in an ever-decreasing path around the Sun before taking up its present orbit as the planet Venus. He further cites evidence to show that the Earth interacted with Mars on a number of occasions when writing was better developed than during the Venusian encounters, after Venus flipped Mars out of its orbit. Disturbances caused by the passages of Mars consisted of earthquakes and electrical discharges. Most of the "Mars events" took place within a ninety-year period. This may sound far-fetched, but Velikovsky's evidence and the predictions he made from it have stood the test of nearly four decades of investigation. As in his previous work, Velikovsky amassed an impressive range of evidence to support his case.

MYTHIC KEYS

To support his interpretation of the Hebrew and Egyptian histories, Velikovsky searched the records of the civilisations of the eighth and fifteenth centuries BCE. In his last book he described his many years of research as sitting at the feet of sages "to listen to those who lived close to the events of the past . . . I realised very soon that the ancient sages lived in a frightened state of mind." What became quickly apparent was the similarity of the events these peoples had experienced, and the fear that global upheavals associated with planetary encounters had inspired.

The legends of the past are folklore, but the similarity of motifs from five continents and Pacific Ocean islands is striking; witches on brooms, the dragon and the scorpion, an animal with many heads and winged body, a woman whose veils stream behind her—these images are universal cosmic myths recording the characteristic shapes possessed by comets.

Velikovsky tracks the motif of the sun being trapped in its movement through the tales of the Polynesians, Hawaiians and North American Indians. Like the Middle-East civilisations they have the story of the sun being snared and freed by a mouse. In the Hawaiian version Mauii caught and beat the sun, which begged for mercy and promised to go more slowly ever after. At the same time new islands appeared. The Ute Indians tell of a piece of sun setting fire to the world, which was broken off by a rabbit after the sun rose, went down and rose again.

The legend of the cosmic battle of the planetary gods is familiar to us all. In the Homeric epics the Greeks choose Athene/Venus for their protector, the Trojans Ares/Mars. A similar situation existed in ancient Mexico. The Toltecs worshipped Quetzal-cohuatl/Venus, but the later Aztecs revered Huitzilopachtil/Mars. The identity, conflict and features of the planetary gods are consistent across the ancient world.

Chinese chronicles record two suns doing battle in the sky and the disturbance of the other planets this caused. Mars was pursued by Venus, the Earth shook, glowing mountains collapsed, "the customs of the age are thrown into disorder . . . all living beings harass one another." An old textbook of Hindu astronomy has a chapter on planetary conjunctions. A planet can be struck down or utterly vanquished, and the victor in these encounters is usually the planet Venus. A juncture of the planets is called a *yuga* in Hindu astronomy; the ages of the world are also called *yugas*.

An association for the planet Mars with the wolf is also common. In Babylon one of the seven names of Mars was wolf. An Egyptian god with the head of a wolf prowled the land. The Romans used the wolf as the animal symbol for Mars. Slavic mythology has a god in the shape of a wolf, Vukadlak, who devoured the Sun and Moon. In the Icelandic epic *The Edda*, the god that darkens the Sun is the wolf Fenris, who battled the serpent Midgard in the heavens above. A Chinese astronomical chart quotes ancient sources in saying "once Venus ran into the Wolf-Star."

NATURE'S EVIDENCE

In *Earth in Upheaval*, Velikovsky excluded all references to ancient literature, traditions and folklore:

This I have done with intent, so that careless critics cannot decry the
entire work as tales and legends. Stones and bones are the only witnesses.

All over the coast of Alaska there are great heaps of smashed bones of
extinct animals mingled with uprooted trees and the occasional flint spearhead. Four layers of volcanic ash can be found in these remains of splin-
tered trees and dismembered bodies. In the polar regions of Siberia and on
the Arctic islands there are hills of broken wood piled hundreds of feet
high, and beyond them hills of mammoth bones cemented together by
frozen sand. On one island the bones of these animals were found with
fossilised trees, leaves and cones. When the mammoth lived in Siberia
there was abundant vegetation.

Spitsbergen is nearly 79 degrees north; yet fossil flowers and corals and
beds of coal thirty feet thick have been found. Antarctica is known to have
seams of coal at a latitude of 85 degrees. For this coal to have formed, the
polar regions must have had great forests in the past. How can relatively
recent and sudden changes in the Earth's climate and simultaneous widespread destruction of plant and animal species be explained?

The violence of this destruction can be seen across Western Europe
where every major rock fissure is filled with the bones of animals, splin-
tered and smashed into fragments. One 1,400 foot hill in France is capped
by the remains of mammoths, reindeer, horses and other animals. America
has beds of fossil bones containing 100 bones per square foot, deposited
in sand. Some of these are over 200 feet high. The hills of the Himalayas
and Burma contain similar beds of bones. In China, among these fractured
bones, the skeletons of seven humans were found. European, Melanesian
and Eskimo types were lying together. Extinct and extant species of animals have been found mixed together in English deposits.

RECORDS

The conventional theory of slow and uniform geological processes cannot
explain these deposits—instead, they are evidence of major catastrophes
which have struck the planet. Velikovsky suggests a giant tidal wave
which engulfed the world that picked up and carried plants and animals
over a great distance and smashed them intermingled into common
graves. This and the transformation of the Earth's climate are explained as
consequences of the rapid change of the Earth's axis brought about by a
near-collision with another planet.

The geological record tells a similar compelling story to that which
paleontologists have unearthed. At 1,400 feet (400 metres) altitude in the

Andes there are high water surf marks lined with undecayed seashells. There are many ruins surrounded by terraces for cultivation on the dry west side of the Andes. On the east side, terraces continue far past the permanent snowline. Before the last lava sheet spread over Columbia there were human settlements there, the remains of which have been found. That the Andes mountains were raised in fairly recent times by unimaginable forces is one conclusion.

The ocean floor around the globe also bears witness to flows of lava and volcanic ash which covered a violently shifting bedrock while tidal waves battered the continents. There were once dry land and beaches in many places where the Atlantic Ocean now lies. The bottom of the seas show that the Earth has been showered with meteorites on a very large scale, leaving clay deposits rich in nickel, radium and iron.

When the Earth's axis was shifted by the interaction of the Earth's and the [rogue] planet's magnetic fields, magnetic eddy currents formed in the atmosphere. These generated great heat and melted rocks on the surface. As this rock cooled, it reformed with a different magnetic polarity *to surrounding strata*. All over the world, similar local rock formations are found with their magnetic polarisation reversed. For this to be the case the Earth's magnetic field must have been reversed when these rocks were formed. Also, rocks with this inverted polarity are far more strongly magnetised than the Earth's magnetic field alone can account for.

Why was volcanic activity so common in the recent past? How was the sea floor raised and lowered around the world? As the Earth's axis shifted in earlier times, the inertia of air and water caused hurricanes and tidal waves; the stress on the planet caused volcanism and an outpouring of magma, sending up clouds of volcanic ash that threw a cloak of darkness over a sunless world.

The heat generated by these forces evaporated seas. In some places torrential downpours formed great streams running through recently opened fissures in the Earth's crust which suddenly eroded the landscape. Elsewhere, snow fell and covered the land with continental ice sheets. At the poles, a permanent snowcover grew as the land cooled.

Climatic changes, ice cover, mountain building and the reverse magnetic orientation of rocks are explained by Velikovsky's theory of cosmic catastrophe. However, the accepted view of the Earth's geological history is known as *uniformitarianism*; where the gradual workings of natural forces has produced the world as we know it. Needless to say, Velikovsky aroused as great a controversy in geology as he had previously in archaeology and astronomy. The defenders of uniformitarianism disliked Velikovsky's ideas at least as much as their fellow scientists had.

VELIKOVSKY ATTACKED

Publication of *Worlds in Collision* caused a violent reaction; astronomers everywhere denounced and decried the book. They threatened to boycott Macmillan, the publisher, who was forced to withdraw the book from circulation. Under pressure, Macmillan transferred publication rights to Doubleday, who did not have a textbook department and burned their unsold copies.

In reviews in reputable journals and public statements, academics and scientists even criticised some of Velikovsky's works before anyone had read the manuscript. Conferences were held to show Velikovsky's theories were wrong.

Velikovsky died on November 17, 1979 at the age of 84. As more is learned about our solar system, some scientists have realized that his theories might conflict with accepted ideas but not actually conflict with the facts.

RISING FROM AMNESIA

Velikovsky theorised that humanity suffered a collective amnesia on the subject of catastrophes. As a reaction to the repeated near-destruction of human civilisation, a deep scar has been left on the human psyche. Although the solar system has been settled for 2,700 years, he notes a 700 year cycle in the human collective consciousness. Christianity in the first century A.D. and Islam in the seventh [century] were both founded on apocalyptic visions of the transformation of the world by fire. The four-teenth century was the time of the Black Death and the Hundred Years War which reduced the population of Western Europe by two-thirds.

Velikovsky's fear was that in the twenty-first century this trauma would be re-enacted by humanity, who is now in possession of the means of its own destruction.

An examination of the facts may help the recall of our memory, the suppression of which could be the cause of great violence in our history.

THE BOOKS OF VELIKOVSKY:

Worlds in Collision, 1950

Ages in Chaos, 1952

Earth in Upheaval, 1955

Oedipus and Akhnaton, 1960

Peoples of the Sea, 1977

Ramses II and His Time, 1978

Mankind in Amnesia, 1982

The design pictured at the top is from Assyria and is several thousand years old. That at the bottom is from the Dogon tribe and is contemporary. The Dogon say their fishtailed figure is from Sirius, and astronomer Temple claims that the Assyrian design shows the same extraterrestrial with a fishtail.

20 Archaeological Cover-Ups?

David Hatcher Childress

Who controls the past, controls the future.
Who controls the present, controls the past.
George Orwell, *1984*

Most of us are familiar with the last scene in the popular Indiana Jones archaeological adventure film *Raiders of the Lost Ark*, in which an important historical artifact, the Ark of the Covenant from the Temple in Jerusalem, is locked in a crate and put in a giant warehouse, never to be seen again, thus ensuring that no history books will have to be rewritten and no history professor will have to revise the lecture that he has been giving for the last forty years.

While the film was fiction, the scene in which an important ancient relic is buried in a warehouse is uncomfortably close to reality for many researchers. To those who investigate allegations of archaeological cover-ups, there are disturbing indications that the most important archaeological institute in the United States, the Smithsonian Institution, an independent federal agency, has been actively suppressing some of the most interesting and important archaeological discoveries made in the Americas.

The Vatican has been long accused of keeping artifacts and ancient books in their vast cellars, without allowing the outside world access to them. These secret treasures, often of a controversial historical or religious nature, are allegedly suppressed by the Catholic Church because they might damage the church's credibility, or perhaps cast their official texts in doubt. Sadly, there is overwhelming evidence that something very similar is happening with the Smithsonian Institution.

The Smithsonian Institution was started in 1829 when an eccentric British millionaire, by the name of James Smithson, died and left $515,169 to create an institution "for the increase and diffusion of knowledge among men." Unfortunately, there is evidence the Smithsonian has been more active in the suppression of knowledge . . . than the diffusion of it for the last hundred years.

The cover-up and alleged suppression of archaeological evidence began in late 1881 when John Wesley Powell, the geologist famous for exploring the Grand Canyon, appointed Cyrus Thomas as the director of the Eastern Mound Division of the Smithsonian Institution's Bureau of Ethnology.

When Thomas came to the Bureau of Ethnology he was a "pronounced believer in the existence of a race of Mound Builders, distinct from the American Indians." However, John Wesley Powell, the director of the Bureau of Ethnology, a very sympathetic man toward the American Indians, had lived with the peaceful Winnebago Indians of Wisconsin for many years as a youth and felt that American Indians were unfairly thought of as primitive and savage.

The Smithsonian began to promote the idea that Native Americans, at that time being exterminated in the Indian wars, were descended from advanced civilizations and were worthy of respect and protection. They also began a program of suppressing any archaeological evidence that lent credence to the school of thought known as *Diffusionism*, a school which believes that throughout history there has been widespread dispersion of culture and civilization via contact by ship and major trade routes.

The Smithsonian opted for the opposite school, known as *Isolationism*. Isolationism holds that most civilizations are isolated from each other and that there has been very little contact between them, especially those that are separated by bodies of water. In this intellectual war that started in the 1880s, it was held that even contact between the civilizations of the Ohio and Mississippi Valleys was rare, and certainly these civilizations did not have any contact with such advanced cultures as the Mayas, Toltecs, or Aztecs in Mexico and Central America. By Old World standards this is an extreme, and even ridiculous idea, considering that the river system reached to the Gulf of Mexico and these civilizations were as close as the opposite shore of the gulf. It was like saying that cultures in the Black Sea area could not have had contact with the Mediterranean.

When the contents of many ancient mounds and pyramids of the Midwest were examined it was shown that the history of the Mississippi River Valleys was that of an ancient and sophisticated culture that had been in contact with Europe and other areas. Not only that, the contents of many mounds revealed burials of huge men, sometimes seven or eight feet tall, in full armour with swords and sometimes huge treasures.

For instance, when Spiro Mound in Oklahoma was excavated in the 1930s, a tall man in full armour was discovered along with a pot of thousands of pearls and other artifacts, the largest such treasure so far documented. The whereabouts of the man in armour is unknown and it is quite likely that it eventually was taken to the Smithsonian Institution.

In a private conversation with a well-known historical researcher (who shall remain nameless), I was told that a former employee of the Smithsonian, who was dismissed for defending the view of Diffusionism in the Americas (i.e., the heresy that other ancient civilizations may have visited the shores of North and South America during the many millennia before Columbus), alleged that the Smithsonian at one time had actually taken a barge full of unusual artifacts out into the Atlantic and dumped them in the ocean.

Though the idea of the Smithsonian's covering up a valuable archaeological find is difficult to accept for some, there is, sadly, a great deal of evidence to suggest that the Smithsonian Institution has knowingly covered up and "lost" important archaeological relics. The *Stonewatch Newsletter* of the Gungywamp Society in Connecticut, which researches megalithic sites in New England, had a curious story in their Winter 1992 issue about stone coffins discovered in 1892 in Alabama which were sent to the Smithsonian Institution and then "lost." According to the newsletter, researcher Frederick J. Pohl wrote an intriguing letter in 1950 to the late Dr. T. C. Lethbridge, a British archaeologist.

The letter from Pohl stated:

A professor of geology sent me a reprint (of the) Smithsonian Institution, *The Crumf Burial Cave* by Frank Burns, U.S. Geological Survey, from the report of the U.S. National Museum for 1892, pp. 451–454, 1984. In the Crumf Cave, southern branch of the Warrior River, in Murphy's Valley, Blount County Alabama, accessible from Mobile Bay by river, were coffins of wood hollowed out by fire, aided by stone or copper chissels. Eight of these coffins were taken to the Smithsonian. They were about 7.5' long, 14" to 18" wide, 6" to 7" deep. Lids open.

I wrote recently to the Smithsonian, and received reply March 11th from F. M. Setzler, Head Curator of Department of Anthropology. (He said) We have not been able to find the specimens in our collections, though records show that they were received.

David Barron, President of the Gungywamp Society was eventually told by the Smithsonian in 1992 that the coffins were actually wooden troughs and that they could not be viewed anyway because they were housed in an asbestos-contaminated warehouse. This warehouse was to be closed for the next ten years and no one was allowed in except Smithsonian personnel!

Ivan T. Sanderson, a well-known zoologist and frequent guest on Johnny Carson's *Tonight Show* in the 1960s (usually with an exotic animal like a pangolin or a lemur), once related a curious story about a letter he received regarding an engineer who was stationed on the Aleutian island of Shemya during World War II. While building an airstrip, his crew bulldozed a group of hills and discovered under several sedimentary layers what appeared to be human remains. The Alaskan mound was in fact a graveyard of gigantic human remains, consisting of crania and long leg bones.

The crania measured from 22 to 24 inches from base to crown. Since an adult skull normally measures about eight inches from back to front such a large crania would imply an immense size for a normally proportioned human. Furthermore, every skull was said to have been neatly trepanned (a process of cutting a hole in the upper portion of the skull).

In fact, the habit of flattening the skull of an infant and forcing it to grow in an elongated shape was a practice used by ancient Peruvians, the Mayas, and the Flathead Indians of Montana. Sanderson tried to gather further proof, eventually receiving a letter from another member of the unit who continued the report. The letters both indicated that the Smithsonian Institution had collected the remains, yet nothing else was heard. Sanderson seemed convinced that the Smithsonian Institution had received the bizarre relics, but wondered why they would not release the data. He asks, ". . . is it that these people cannot face rewriting all the text books?"

In 1944 an accidental discovery of an even more controversial nature was made by Waldemar Julsrud at Acámbaro, Mexico. Acámbaro is in the state of Guanajuato, 175 miles northwest of Mexico City. The strange archaeological site there yielded over 33,500 objects of ceramic [and] stone, including jade, and knives of obsidian (sharper than steel and still used today in heart surgery). Jalsrud, a prominent local German merchant, also found statues ranging from less than an inch to six feet in length depicting great reptiles, some of them in active association with humans—generally eating them, but in some bizarre statuettes an erotic association was indicated. To observers many of these creatures resembled dinosaurs.

Jalsrud crammed this collection into twelve rooms of his expanded house. There, startling representations of Negroes, Orientals, and bearded Caucasians were included as were motifs of Egyptian, Sumerian and other ancient non-hemispheric civilisations, as well as portrayals of Bigfoot and aquatic monsterlike creatures, weird human-animal mixtures, and a host of other inexplicable creations. Teeth from an extinct Ice Age horse, the skeleton of a mammoth, and a number of human skulls were found at the same site as the ceramic artifacts.

Radiocarbon dating in the laboratories of the University of Pennsylvania and additional tests using the thermoluminescence method of dating pottery were performed to determine the age of the objects. Results indicated the objects were made about 6,500 years ago, around 4,500 B.C. A team of experts at another university, shown Jalsrud's half-dozen samples but unaware of their origin, ruled out the possibility that they could have been modern reproductions. However, they fell silent when told of their controversial source.

In 1952, in an effort to debunk this weird collection which was gaining a certain amount of fame, American archaeologist Charles C. DiPeso claimed to have minutely examined the then 32,000 pieces within not more than four hours spent at the home of Julsrud. In a forthcoming book long delayed by continuing development in his investigation, archaeological investigator John H. Tierney, who has lectured on the case for decades, points out that to have done that, DiPeso would have had to have inspected 133 pieces per minute steadily for four hours, whereas in actuality, it would have required weeks merely to have separated the massive jumble of exhibits and arranged them properly for a valid evaluation.

Tierney, who collaborated with the late Professor Hapgood, the late William N. Russell, and others in the investigation, charges that the Smithsonian Institution and other archaeological authorities conducted a campaign of disinformation against the discoveries. The Smithsonian had, early in the controversy, dismissed the entire Acámbaro collection as an elaborate hoax. Also, utilizing the Freedom of Information Act, Tierney discovered that practically the entirety of the Smithsonian's Julsrud case files are missing.

After two expeditions to the site in 1955 and 1968, Professor Charles Hapgood, a professor of history and anthropology at the University of New Hampshire, recorded the results of his eighteen-year investigation of Acámbaro, in a privately printed book entitled *Mystery In Acámbaro*. Hapgood was initially an open-minded skeptic concerning the collection but became a believer after his first visit in 1955, at which time he witnessed some of the figures being excavated, and even dictated to the diggers where he wanted them to dig.

Adding to the mind-boggling aspects of this controversy is the fact that the Instituto Nacional de Antropologia e Historia, through the late Director of Prehispanic Monuments, Dr. Eduardo Noguera, (who, as head of an official investigating team at the site, issued a report which Tierney will be publishing), admitted "the apparent scientific legality with which these objects were found." Despite evidence of their own eyes, however, officials declared that because of the objects "fantastic" nature, they had to have been a hoax played on Julsrud!

A disappointed but ever-hopeful Julsrud died. His house was sold and the collection put in storage. The collection is not currently open to the public.

Perhaps the most amazing suppression of all is the excavation of an Egyptian tomb by the Smithsonian itself in Arizona. A lengthy front page story of the Phoenix Gazette on 5 April 1909 (see page inset "Explorations in Grand Canyon" on page 222), gave a highly detailed report of the discovery and excavation of a rock-cut vault by an expedition led by Professor S. A. Jordan of the Smithsonian. The Smithsonian, however, claims to have absolutely no knowledge of the discovery or its discoverers.

The World Explorers Club decided to check on this story by calling the Smithsonian in Washington, D.C., though we felt there was little chance of getting any real information. After speaking briefly to an operator, we were transferred to a Smithsonian staff archaeologist, and a woman's voice came on the phone and identified herself.

I told her that I was investigating a story from a 1909 Phoenix newspaper article about the Smithsonian Institution's having excavated rock-cut vaults in the Grand Canyon where Egyptian artifacts had been discovered, and whether the Smithsonian Institution could give me any more information on the subject.

"Well, the first thing I can tell you, before we go any further," she said, "is that no Egyptian artifacts of any kind have ever been found in North or South America. Therefore, I can tell you that the Smithsonian Institution has never been involved in any such excavations." She was quite helpful and polite but in the end, knew nothing. Neither she nor anyone else with whom I spoke could find any record of the discovery or either G. E. Kinkaid and Professor S. A. Jordan.

While it cannot be discounted that the entire story is an elaborate newspaper hoax, the fact that it was on the front page, named the prestigious Smithsonian Institution and gave a highly detailed story that went on for several pages, lends a great deal to its credibility. It is hard to believe such a story could have come out of thin air.

Is the Smithsonian Institution covering up an archaeological discovery of immense importance? If this story is true it would radically change the current view that there was no transoceanic contact in pre-Columbian times, and that all American Indians, on both continents, are descended from Ice Age explorers who came across the Bering Strait. (Any information on G. E. Kinkaid and Professor Jordan, or their alleged discoveries, that readers may have would be greatly appreciated.)

Is the idea that ancient Egyptians came to the Arizona area in the ancient past so objectionable and preposterous that it must be covered up? Perhaps the Smithsonian Institution is more interested in maintaining the status quo than rocking the boat with astonishing new discoveries that overturn previously accepted academic teachings.

Historian and linguist Carl Hart, editor of *World Explorer*, then obtained a hiker's map of the Grand Canyon from a bookstore in Chicago. Poring over the map, we were amazed to see that much of the area on the north side of the canyon has Egyptian names. The area around Ninety-four Mile Creek and Trinity Creek had areas (rock formations, apparently) with names like Tower of Set, Tower of Ra, Horus Temple, Osiris Temple, and Isis Temple. In the Haunted Canyon area were such names as the Cheops Pyramid, the Buddha Cloister, Buddha Temple, Manu Temple and Shiva Temple. Was there any relationship between these places and the alleged Egyptian discoveries in the Grand Canyon?

We called a state archaeologist at the Grand Canyon, and were told that the early explorers had just liked Egyptian and Hindu names, but that it was true that this area was off limits to hikers or other visitors, "because of dangerous caves."

Indeed, this entire area with the Egyptian and Hindu place names in the Grand Canyon is a forbidden zone—no one is allowed into this large area.

We could only conclude that this was the area where the vaults were located. Yet today, this area is curiously off-limits to all hikers and even, in large part, park personnel.

I believe that the discerning reader will see that if only a small part of the "Smithsoniangate" evidence is true then our most hallowed archaeological institution has been actively involved in suppressing evidence for advanced American cultures, evidence for ancient voyages of various cultures to North America, evidence for anomalistic giants and other oddball artifacts, and evidence that tends to disprove the official dogma that is now the history of North America.

The Smithsonian's Board of Regents still refuses to open its meetings to the news media or the public. If Americans were ever allowed inside the "nation's attic," as the Smithsonian has been called, what skeletons might they find?

Front page of the
*Ph*oenix *Gazette*, 5 April 1909

EXPLORATIONS IN GRAND CANYON

Mysteries of Immense Rich Cavern Being Brought to Light

JORDAN IS ENTHUSED

Remarkable Finds Indicate Ancient People Migrated From Orient

The latest news of the progress of the explorations of what is now regarded by scientists as not only the oldest archeological discovery in the United States, but one of the most valuable in the world, which was mentioned some time ago in the *Gazette*, was brought to the city yesterday by G. E. Kinkaid, the explorer who found the great underground citadel of the Grand Canyon during a trip from Green river, Wyoming down the Colorado, in a wooden boat, to Yuma, several months ago. According to the story related to the *Gazette* by Mr. Kinkaid, the archaeologists of the Smithsonian Institute, which is financing the expeditions, have made discoveries which almost conclusively prove that the race which inhabited this mysterious cavern, hewn in solid rock by human hands, was of oriental origin, possibly from Egypt, tracing back to Ramses. If their theories are borne out by the translation of the tablets engraved with hieroglyphics, the mystery of the prehistoric peoples of North America, their ancient arts, who they were and whence they came, will be solved. Egypt and the Nile, and Arizona and the Colorado will be linked by a historical chain running back to ages which staggers the wildest fancy of the fictionist.

A Thorough Investigation

Under the direction of Prof. S. A. Jordan, the Smithsonian Institute is now prosecuting the most thorough explorations, which will be continued until the last link in the chain is forged. Nearly a mile underground, about 1480 feet below the surface the long main passage has been delved into, to find another mammoth chamber from which radiates scores of passageways, like the spokes of a wheel. Several hundred rooms have been discovered, reached by passageways running from the main passage, one of them having been explored for 854 feet and another 634 feet. The recent finds include articles which have never been known as native to this country, and doubtless they had their origin in the orient. War weapons, copper instruments, sharp-edged and hard as steel indicate the high state of civilization reached by these strange people. So interested have the scientists become that preparations are being made to equip the camp for extensive studies, and the force will be increased to thirty or forty persons.

"Before going further into the cavern, better facilities for lighting will have to be installed, for the darkness is dense and quite impenetrable for the average flashlight. In order to avoid being lost, wires are being strung from the entrance to all passageways leading directly to large chambers. How far this cavern extends no one can guess, but it is now the belief of many that what has already been explored is merely the 'barracks,' to use an American term, for the soldiers, and that far into the underworld will be found the main communal dwellings of the families. The perfect ventilation of the cavern, the steady draught that blows through, indicates that it has another outlet to the surface."

Mr. Kinkaid's Report

Mr. Kinkaid was the first white child born in Idaho and has been an explorer and hunter all his life, thirty years having been in the service of the Smithsonian Institute. Even briefly recounted, his history sounds fabulous, almost grotesque.

"First, I would impress that the cavern is nearly inaccessible. The entrance is 1,486 feet down the sheer canyon wall. It is located on government land and no visitor will be allowed there under penalty of trespass. The scientists wish to work unmolested, without fear of the archaeological discoveries being disturbed by curio or relic hunters. A trip there would be fruitless, and the visitor would be sent on his way. The story of how I found the cavern has been related, but in a paragraph: I was journeying down the Colorado river in a boat, alone, looking for mineral. Some forty-two miles up the river from the El Tovar Crystal canyon, I saw on the east wall, stains in the sedimentary formation about 2,000 feet above the river bed. There was no trail to this point, but I finally reached it with great difficulty. Above a shelf which hid it from view from the river, was the mouth of the cave. There are steps leading from this entrance some thirty yards to what was, at the time the cavern was inhabited, the level of the river. When I saw the chisel marks on the wall inside the entrance, I became interested, [secured] my gun and went in. During that trip I went back several hundred feet along the main passage till I came to the crypt in which I discovered the mummies. One of these I stood up and photographed by flashlight. I gathered a number of relics, which I carried down the Colorado to Yuma, from whence I shipped them to Washington with details of the discovery. Following this, the explorations were undertaken.

The Passages

"The main passageway is about 12 feet wide, narrowing to nine feet toward the farther end. About 57 feet from the entrance, the first side-passages branch off to the right and left, along which, on both sides, are a number of rooms about the size of ordinary living rooms of today, though some are 30 by 40 feet square. These are entered by oval-shaped doors and are ventilated by round air spaces through the walls into the passages. The walls are about three feet six inches in thickness. The passages are chiseled or hewn as straight as could be laid out by an engineer. The ceilings of many of the rooms converge to a center. The side-passages near the entrance run at a sharp angle from the main hall but toward the rear they gradually reach a right angle in direction.

The Shrine

"Over a hundred feet from the entrance is the cross-hall, several hundred feet long in which are found the idol, or image, of the people's god, sitting cross-legged, with a lotus flower or lily in each hand. The cast of the face is oriental, and the carving shows a skillful hand and the entire is remarkably well preserved, as is everything is this cavern. The idol almost resembles Buddha, though the scientists are not certain as to what religious worship it represents. Taking into consideration everything found thus far, it is possible that this worship most resembles the ancient people of Tibet. Surrounding this idol are smaller images, some very beautiful in form; others crooked-necked and distorted shapes, symbolical, probably, of good and evil. There are two large cactus with protruding arms, one on each side of the dais on which the god squats. All this is carved out of hard rock resembling marble. In the opposite corner of this cross-hall were found tools of all descriptions, made of copper. These people undoubtedly knew the lost art of hardening this metal, which has been sought by chemists for centuries without result. On a bench running around the workroom was some charcoal and other material probably used in the process. There is also slag and stuff similar to matte, showing that these ancients smelted ores, but so far no l trace of where or how this was done has been discovered, nor the origin of the ore.

"Among the other finds are vases or urns and cups of copper and gold, made very artistic in design. The pottery work includes enameled ware and glazed vessels. Another passageway leads to granaries such as are found in the oriental temples. They contain seeds of various kinds. One very large storehouse has not yet been entered, as it is twelve feet high and can be reached only from above. Two copper hooks extend on the edge, which indicates that some sort of ladder was attached. These granaries are rounded, as the materials of which they are constructed, I think, is a very hard cement. A gray metal is also found in this cavern, which puzzles the scientists, for its identity has not been established. It resembles platinum. Strewn promiscuously over the floor everywhere are what people call 'cats eyes,' a yellow stone of no great value. Each one is engraved with the head of the Malay type.

The Hieroglyphics

"On all the urns, or walls over doorways, and tablets of stone which were found by the image are the mysterious hieroglyphics, the key to which the Smithsonian Institute hopes yet to discover. The engraving on the tablets probably has something to do with the religion of the people. Similar hieroglyphics have been found in southern Arizona. Among the pictorial writings, only two animals are found. One was of the prehistoric type.

The Crypt

"The tomb or crypt in which the mummies were found is one of the largest of the chambers, the walls slanting back at an angle of about 35 degrees. On these are tiers of mummies, each one occupying a separate hewn shelf. At the head of each is a small bench, on which is found copper cups and pieces of broken swords. Some of the mummies are covered with clay, and all are wrapped in a bark fabric. The urns or cups on the lower tiers are crude, while as the higher shelves are reached, the urns are finer in design, showing a later stage of civilization. It is worthy of note that all the mummies examined so far have proved to be male, no children or females being buried here. This leads to the belief that this exterior section was the warriors' barracks.

"Among the discoveries no bones of animals have been found, no skins, no clothing, no bedding. Many of the rooms are bare but for water vessels. One room, about 40 by 700 feet, was probably the main dining hall, for cooking utensils are found here. What these people lived on is a problem, though it is presumed that they came south in the winter and farmed in the valleys, going back north in the summer. Upwards of 50,000 people could have lived in the caverns comfortably. One theory is that the present Indian tribes found in Arizona are descendants of the serfs of slaves of the people which inhabited the cave. Undoubtedly a good many thousands of years before the Christian era a people lived here which reached a high stage of civilization. The chronology of human history is full of gaps. Professor Jordan is much enthused over the discoveries and believes that the find will prove of incalculable value in archaeological work.

"One thing I have not spoken of, may be of interest. There is one chamber in the passageway to which is not ventilated, and when we approached it a deadly, snaky smell struck us. Our light would not penetrate the gloom, and until stronger ones are available we will not know what the chamber contains. Some say snakes, but other boo hoo this idea and think it may contain a deadly gas or chemicals used by the ancients. No sounds are heard, but it smells snaky just the same. The whole underground installation gives one of shaky nerves the creeps. The gloom is like a weight on one's shoulders, and our flashlights and candles only make the darkness blacker. Imagination can revel in conjectures and ungodly daydreams back through the ages that have elapsed till the mind reels dizzily in space."

An Indian Legend

In connection with this story, it is notable that among the Hopi Indians the tradition is told that their ancestors once lived in an underworld in the Grand Canyon till dissension arose between the good and the bad, the people of one heart and the people of two hearts. Machetto, who was their chief, counseled them to leave the underworld, but there was no way out. The chief then caused a tree to grow up and pierce the roof of the underworld, and then the people of one heart climbed out. They tarried by Paisisvai (Red River), which is the Colorado, and grew grain and corn. They sent out a message to the Temple of the Sun, asking the blessing of peace, good will and rain for people of one heart. That messenger never returned, but today at the Hopi villages at sundown can be seen the old men of the tribe out on the housetops gazing toward the sun, looking for the messenger. When he returns, their lands and ancient dwelling place will be restored to them. That is the tradition. Among the engravings of animals in the cave is seen the image of a heart over the spot where it is located. The legend was reamed by W.E. Rollins, the artist, during a year spent with the Hopi Indians. There are two theories of the origin of the Egyptians. One that they came from Asia, another that the racial cradle was in the upper Nile region. Heeren, an Egyptologist, believed in the Indian origin of the Egyptians. The discoveries in the Grand Canyon may throw further light on human evolution and prehistoric ages.

21 Introduction to Bread From Stones

Dr. Raymond Bernard (A.B., M.N., Ph.D.)

Dr. Julius Hensel was the greatest figure in the history of agricultural chemistry even if his powerful enemies, members of the octopus chemical fertilizer trust, have succeeded in suppressing his memory, destroying his books and getting his Stone Meal fertilizer off the market. But eventually the truth comes to the fore, and its enemies are vanquished. Julius Hensel's pioneer work in opposing the use of chemicals in agriculture, a half a century later, found rebirth in the Organic Movement which has swept through the world. But Hensel is more modern than the most modern agricultural reformer, for he claimed, on the basis of theoretical chemical considerations, and supported by practical tests, that his Stone Meal can replace not only chemical fertilizers but all animal ones as well.

It was the German agricultural chemist Liebig who first put forward the phosphorus-potash-nitrogen theory of chemical fertilization. This false doctrine Hensel bitterly attacked and in so doing, won the ire of the financial interests behind the sale of chemical fertilizers, which used agricultural authorities and university professors to denounce poor Hensel as a charlatan and his Stone Meal as worthless.

Though his fight against chemical fertilizers was a losing battle and he died as a defeated hero, it took a generation for Hensel's efforts to bear fruit in the modern Organic Movement, which has not given its founder the credit due him.

The fight between Liebig, advocate [of] one-sided chemical fertilization, and Hensel, who advocated a more balanced form of plant nutrition, including the trace minerals which Liebig completely overlooked, was a battle between an opportunist, who sought to further the sales of chemical fertilizers, and a true scientist, interested in humanity's welfare. Though

From *Bread From Stones: A New and Rational System of Land Fertilization and Physical Regeneration* by Dr. Julius Hensel (Agricultural Chemist). Translated from the German (1894).

Liebig, with the Chemical Trust behind him, won the battle, Hensel's ideas finally triumphed . . . several decades after his passing.

Liebig claimed that plants require three main elements—nitrogen, phosphorus and potash—the basis of which conception chemical fertilizers were manufactured that supplied these elements. On the other hand, Hensel claimed that plants need many more than these three major elements, stressing the importance of the trace minerals, which at that time were ignored. In place of chemical fertilizers, supplying only three elements in an unnatural, caustic form, Hensel recommended the bland minerals of pulverized rocks, especially granite, a primordial rock which contains the many trace minerals that meet all needs of plant nutrition.

Hensel first made his discovery of powdered rock fertilization when he was a miller. One day, while milling grain, he noticed that some stones were mixed with it and [he] ground [them] into a meal. He sprinkled this stone meal over the soil of his garden and was surprised to note how the vegetables took on a new, more vigorous growth. This led him to repeat the experiment by grinding more stones and applying the stone meal to fruit trees. Much to his surprise, apple trees that formerly bore wormy, imperfect fruit now produced fine quality fruit free from worms. Also vegetables fertilized by stone meal were free from insect pests and diseases. It seemed to be a complete plant food, which produced fine vegetables even in the poorest soil.

Encouraged by these results, Hensel put his "Stone Meal" on the market, and wrote extensively on its superiority over chemical fertilizers, while at the same time opposing the use of animal manure, and the nitrogen theory on which it is based, claiming that when plants are supplied with Stone Meal, plenty of water, air and sunshine, they will grow healthfully even if the soil is poor in nitrogen, since it was his belief that plants derive their nitrogen through their leaves, and do not depend on the soil for this element.

In opposing this use of chemical fertilizer, Hensel awoke the ire of a powerful enemy, which was resolved to liquidate him—the Chemical Trust. Through unfair competition, Hensel's "Stone Meal" business was destroyed and his product was taken off the market. However, the chief object of attack was his book, *Bread From Stones*, in which he expounded his new doctrines of Liebig on which the chemical fertilizer business was based, as well as the "Liebig meat extract." (For Hensel advocated vegetarianism, just as he advocated natural farming without chemicals or manure.) Accordingly, his enemies succeeded in suppressing the further publication of this book and in removing it from libraries, until it became extremely rare and difficult to obtain. It is more fortunate that a surviving copy came into the writer's possession.

Dr. Julius Hensel was not only a student of agricultural chemistry, but also biochemistry and nutrition, and he related all these sciences, and united them into a composite science of life, which he labeled "Makrobiology." His theory was that the chemistry of life is basically determined by the chemistry of the soil, and that chemicals unbalance and pervert soil chemistry while powdered rocks help restore normal soil mineral balance, producing foods favorable to health and life. His discoveries concerning the value of powdered rocks as soil conditioners and plant foods, though rejected and ridiculed when he first proposed them, were adopted by agricultural science nearly a century later, when the application of powdered limestone, rock phosphate and other rocks became standard agricultural pratice. Granite, which Hensel recommended as the most balanced of all rocks as source of soil minerals, was first rejected as worthless, but later appreciated and used as a soil mineralizer.

During the course of his researches, Dr. Hensel found that in the primeval rocks, as granite, lie a potentially inexhaustable supply of all minerals required for the feeding and regeneration of the soil, plants, animals and man. All that is required is to reduce them to finely a pulverized form, so that their mineral elements may be made available to plants. Hensel wrote a book describing his discovery of a new method of creating more perfect fruits and vegetables, rich in all nutritional elements and immune to disease and insect pests, with the result that it produced worm-free fruit without the need of spraying. The foods so produced by rock-meal fertilization were true Organic Super Foods, far superior in flavor and value than those produced under the forcing action of manure or chemical fertilizers.

Hensel was the first to put up a fight against the then-growing new chemical fertilizer industry—a struggle that was continued in the next century by Sir William Howard in England and J. I. Rodale in America. The use of chemical fertilizers, claimed Hensel, leads to the following evil consequences:

1. It poisons the soil, destroying beneficial soil bacteria, earthworms and humus*.

2. It creates unhealthy, unbalanced, mineral-deficient plants, lacking resistance to disease and insect pests, thus leading to the spraying menace in an effort to preserve these defective specimens.

3. It leads to diseases among animals and men who feed on these abnormal plants and their products.

4. It leads to a tremendous expense to the farmer, because chemical fer-

---

* Decayed vegetable or animal matter that provides nutrients for plants.

tilizers, being extremely soluble, are quickly washed from the soil by rainfall and needs constant replacement. (Powdered rocks, on the other hand, being less soluble, are not so easily washed from the soil, but keep releasing minerals to it for many years).

The use of various pulverized rocks, [such] as granite, limestone, rock phosphate, etc., in place of chemical fertilizers, will lead, claimed Hensel, to permanent restoration of even poor soils to the balanced mineral content of the best virgin soils; and the rock dust thus applied will remain year after year and not be washed away by rains or irrigation water, as is the case with highly soluble chemical fertilizers. This will be an economic saving to the grower and enable him to sell foods at a lower price than when he must spend large sums on chemical fertilizers. Also, since foods thus mineralized are healthy and immune to plant diseases and insect pests (as Hensel experimentally demonstrated), there is no need for the expense and dangers of spraying.*

Foods raised by Hensel's followers, including many German gardeners and farmers, who were enthusiastic in praise of his method, were found to possess firmer tissue and better shipping and keeping qualities than those raised with animal manure or chemicals. And most important among the advantages of Hensel's agricultural discovery is that foods grown on mineralized soil are higher both in mineral and vitamin content and so produce better health and greater immunity to disease than those grown by the use of chemical fertilizer sprays.

To kill insects by poisons applied to plants does not remove the cause of their infestation, and poisons both the insect as well as the human consumer of the sprayed plant. Only correct feeding of the soil, and consequently of plants by trees, by proper methods of fertilization, thereby keeping them well-nourished, vigorous and free from disease, will accomplish this, for insects do not seem to attack healthy plants. It appears that insects, like scavengers, attack chiefly unhealthy and demineralized plants, not healthy ones. Dr. Charles Northern has performed experiments in which he raised two tomato plants, entwined with each other, in different pots, one being supplied with an abundance of trace minerals, derived from colloidal phosphate, and the other just chemical fertilizer. The tomato plant grown with chemical fertilizer alone was attacked by insects, while the other one given trace minerals was not.

Hensel pointed out that animal manure and chemical fertilizers produce a forced, unnaturally rapid growth of large-sized produce which fail to acquire the minerals normally secured during a slower, longer development. The result is the production of demineralized, unbalanced plants,

---

* **Editor's Note:** Rock phosphate from some sources contains a high level of the toxic mineral cadmium. It's wise to purchase rock phosphate from a supplier who's able to provide an analysis.

which are weak and unhealthy, falling prey to disease and insect pests. This explains why, coincident with the increased use of chemical fertilizers, during the past century, insect pests steadily increased. So did cancerous conditions among plants, animals and humans, as shown by Keens, an English soil chemist, who presents statistics to show that the increased use of chemical fertilizers is a major cause of the greater incidence of cancer during that last hundred years.

The modern Organic Farming movement has accepted and propagated one of Hensel's theories—his opposition to chemical fertilizers and recommendation of powdered rocks in their place—but has failed to appreciate his other main doctrine—his opposition to the use of animal excrements as plant foods. In this respect, Hensel, though he lived in the last century, [was] far ahead of the Organic Movement and more modern than the most modern agricultural reformer.

Hensel had a great admirer and disciple in England, one Sampson Morgan, who founded his "Clean Culture" doctrine on Hensel's philosophy of soil and biological regeneration by the avoidance of chemical or animal fertilizers. While Hensel was more of a theorist, Morgan was a practical farmer and agricultural experimenter, who proved the truth of Hensel's theories by winning the first prize at all agricultural exhibits at which his super-sized, super-quality, disease- and blight-free rock-dust fertilized fruits and vegetables were displayed. In Sampson Morgan's *Clean Culture*, Morgan's views are presented. In reality they are Hensel's doctrines transplanted to English soil. The reading of Morgan's book will be a valuable supplement to [the reading] of this, to give one a thorough understanding of the subject of Natural Agriculture (i.e., a system of soil culture definitely in advance of Organic Gardening by the compost method).

Practical experience with Hensel's Stone Meal and his non-animal method of soil regeneration, has proven the following:

1. That Stone Meal creates healthier, tastier, more vitaminized and mineralized foods.

2. That Stone Meal creates immunity to insect infestation, worms, fungi and plant diseases of all kinds.

3. That Stone Meal improves the keeping and shipping quality of foods, so that they keep a long time, in contrast to the rapid deterioration of foods given abundant animal manure.

4. That Stone Meal helps plants to resist drought and frost, enabling them to survive when those fed on manure and chemicals perish.

5. That Stone Meal produces larger crops which are more profitable because the farmer is saved the expense of buying chemical fertilizers which are rapidly leached from the soil by rainfall, whereas Stone Meal, being less soluble, is gradually released during the course of years and remain in the soil, being the most economical of fertilizers.

6. That foods raised with Stone Meal are better for human health and the prevention of disease than those grown with chemicals or animal manure.

7. That use of Stone Meal, in place of chemical or animal fertilizers, helps to end the spraying menace (by removing its cause) is proven by the fact that plants and trees grown with Stone Meal are immune to pests and so require no spraying.

22 Scientist With an Attitude: Wilhelm Reich

Jeane Manning

Federal employees worked with a vengeance when instructed to destroy the work of scientist Wilhelm Reich, M.D., at his laboratory in the state of Maine. Their 1956 court injunction said that construction materials in Reich's boxlike "orgone accumulator" could be salvaged, but the workers slashed the Celotex panels into useless junk.

Down the coast in New York, Reich's associates Dr. Michael Silvert and Dr. Victor Sobey were forced to load the literature in the Orgone Institute stock room into a large truck. The freight truck dumped the papers at a Lower Manhattan incinerator, for an FDA-ordered book burning.

The American Civil Liberties Union stepped in when it was too late, with a press release saying that the court order was a violation of free speech because only one of the torched books could be considered [an aid] to promote or explain the controversial orgone accumulator. (Orgone is the name Reich gave to a life force which he discovered to be within and around all living organisms, including the earth.) The civil liberties press release said, in part, "It is a serious challenge to freedom of the press, principles of free thought on which our democratic government is based, for an agency of government to take advantage of such a dragnet injunction to thwart the dissemination of knowledge, however eccentric or unpopular that knowledge may be."

MEDIA LOOKED THE OTHER WAY

No major newspaper used the press release. Furthermore, six scientists and educators sent all major papers in England a letter of protest about the book burning and Reich's sentencing. All the papers remained silent on the topic.

What was the suppression of Reich's scientific work really about? It was apparently about more than just the FDA's responsibility to protect supposedly gullible consumers from spending money on devices which the FDA decreed were useless. Granted, Wilhelm Reich was brought to court because another physician transported Reich's orgone accumulators across state lines in defiance of a federal FDA order. Reich believed the courts had no mandate to judge basic scientific research nor to order him to destroy his life's work, so he returned to his laboratory and continued his writings. As a result of his attitude, he was fined heavily for contempt of court, sentenced to two years and four months, jailed despite his heart condition, and later died in prison just before he was eligible for parole.

So far, the story is comprehensible even though it is tragic. But why do writers even today deflect attention from Reich's most important discoveries related to a cosmic life force—weather control, and the Oranur demonstration of the dangerous effects of atomic radiation? Even academics who present themselves as researchers publish distortions of Reich's ideas. Reich's biographers, W. Edward Mann and Edward Hoffman, point to a 1980 textbook which has sixteen factual errors in two pages on Reich. "Most of them are vicious distortions . . . feeding the notion that Reich was either a quack or a nut."

Other writers, aiming for popular publications, look for ridiculous if not lewd material. Out of the remains of forty years of published opinions, personal correspondence and spontaneous statements from a prolific, courageous freethinker, it is not hard for a skeptic to find a few items which can be presented as amusing. A continual barrage of such ridicule put Reich down in popular history mainly as a psychiatrist and "sex doctor."

The labels do not do justice to Reich. His well-known studies of orgasmic potential (measurements of bio-electric charge correlated with emotions reported by patients) were only a part of the evolution of his work. Each step of his career—from being Sigmund Freud's most promising disciple who worked out how neuroses show up in the human body, to uncovering the pathology of fascism, to discovering entities under the microscope which he claimed were links between the non-living and the living—led him toward wide-ranging findings about the primordial substratum that he called orgone. He found it moving in living organisms and everywhere, saw it pulsating in "bions" under the microscope and glowing in the dark of an orgone accumulator. In the unpolluted oceans and atmosphere the energy could be seen in the blue colour and lively sparkle. It is attracted to water but recoils from certain manmade factors. His later discoveries about the pre-atomic atmospheric substratum, and their implications for health and the environment, dwarfed any of his earlier work which led up to them.

In addition to the question of why detractors still try to diminish Reich, another nagging question remains: Why did the United States government burn his soft cover books and papers wholesale? The fires destroyed piles of copies of twenty books and journals. Crate after crate of his life's work was rounded up wherever it could be found, and hauled away into the furnaces. Decades of scientific journals and publications on politics, psychiatry, education, sociology, sexology, microbiology, meteorology and other disciplines were reduced to ashes.

WHO FELT THREATENED?

Some observers wonder if his free-energy invention played a part in the squashing of his scientific writings and the obliteration of his reputation. Reich claimed that he could power an electric motor with concentrated atmospheric energy. Did economic interests want to crush that possibility?

Or was he correct in seeing the opposition to be more psychological—a gut-level reaction by what he called anti-life "armoured" people who are in denial of his life-affirming discoveries? Did mechanistic-minded people, in positions of power, fear being shown that they and the earth itself and the universe are filled with streamings of a vibrant, pulsating unpredictable life force? Reich's experiments indicated that this living force could actually be measured in terms such as heat or movement, and that it is present in varying degrees depending on sickness or health of the organism. And that this life-giving substratum is bothered by high-voltage power grids and is in effect irritated into a frenzy by unnatural levels of nuclear radioactivity.

A third possibility is that the unprecedented opposition came from the orthodox medical community. The orgone accumulator, central to Reich's legal troubles, was a simple medical-treatment box which concentrated the previously-unknown energy by a certain layering of absorbent organic and reflective inorganic materials. Experiments showed an anomalous rise in temperature inside the box, and even Albert Einstein had experienced this phenomenon under Reich's tutoring. Although Reich himself never claimed that the accumulator cured cancer, patients of a number of physicians reported that they were helped with various conditions by sitting in an accumulator or being treated with a smaller accumulator called a shooter.

The FDA had worked for years on the case before sheriff's officers finally led Reich in handcuffs to a small courthouse in Maine. At his trial for contempt of court, he defended himself but was not allowed to bring testimony about the medical effectiveness of the orgone accumulator nor even to explain "orgone." Myron R. Sharif, Ph.D., later wrote about the trial and said the moment when fundamental issues stood out searingly was when FDA agent Joseph Maguire scornfully referred to Reich's discovery of a primal energy:

> They talk about pre-atomic energy! What's that? We've moved way beyond that—we've got A-energy and now we are getting H-energy [the H-bomb].

Sharif and others knew that when atomic bombs were being tested, Reich's orgone experiments would become disturbed. Measurements inside the accumulators would swing strangely, which he said showed a seething reaction in the life-field of the earth after atomic testing.

Apparently sick at heart over what he saw as its tragic outcome, Sharif reported that the trial discussed meaningless secondary issues while it avoided Reich's scientific evidence. Probably the judge and jury were not capable of grasping a radically different world view—new understanding of a universal force—during the short span of time of a court battle.

Reich had plenty of time afterward to reflect on how his life reached such a distressing low point.

REICH IN DANGER

Born in the Ukrainian part of Austria, Reich's interest in biology began on his father's farm, where he lived until the First World War drew him into the Austrian army for three years. He began his formal education by studying law, switching to medicine and then specialising in psycho-analysis. He was one of Sigmund Freud's inner circle in Vienna in the 1920s, seen as Freud's most brilliant pupil and perhaps successor. About the time he became a political activist, he edged away from tradition-al Freudian methods of psychoanalysis. Revealing the independence of thinking that he kept all his life, he began to develop his own systems of therapy.

He worked in Berlin in the early 1930s. Still resisting Fascism, he had joined the German Communist Party and was a member of a cell block of brave writers and artists. They met in secret while Nazi storm troopers marched the streets. As the decade went on and the Nazis took over Germany, Reich was in increasing danger from Hitler's officers. He had been born of Jewish parents, was a psychiatrist and a Communist—three identities which Hitler hated.

At the same time Reich was studying Fascism and concluding that worsening social situations did not make people swing to the left politi-cally. Instead, he noticed that fear of freedom led people to cling to authority figures who promised a better life.

The same year that Hitler came to power in Germany, 1933, Reich courageously published *The Mass Psychology of Fascism*. In February a student organization invited him to Copenhagen, Denmark, to lecture on Sexual Reform and Social Crisis. When he returned to Berlin on February

28, a conflagration broke out and was followed the next morning by arrest of more than a thousand left-wing intellectuals. Reich's friends either went underground, or were arrested or shot. Disguised as a tourist on a ski holiday to Austria, Reich escaped to Austria.

The psycho-analytic society there was hostile to Reich's views, however, and after two months he emigrated to Denmark. Later that year he was excluded from the Danish Communist party, which he had never joined. One reason for the exclusion was that an article of Reich's on sex education caused a furore. Then he had argued with party officials who were supposed to help immigrants and who turned away a suicidal young immigrant who lacked the proper papers, and Reich made a scene in his protest against the inhumane episode. A third strike against Reich was that he had started a publishing house without the permission of the Communist Party. Fourthly, his book *The Mass Psychology of Fascism* was considered counter-revolutionary.

WITHOUT A COUNTRY

Despite such experiences, Reich continued to be intellectually honest throughout his life, regardless of consequences. Through no fault of his own, much of his vision of a sexual revolution—toward a maturity in people—was lost in what actually happened in society. He would be opposed to pornography, with its emphasis on perverse, infantile and destructive elements. Biographer David Boadella said Reich wanted to take away barriers to "re-emergence of a truly personalized sexuality that could deepen and enrich people's lives so fully that 'trips' to a heightened consciousness on drugs would be . . . irrelevant . . ."

In studying the relation between sexuality and anxiety, Reich the psychoanalyst developed a theory which considered the orgasm in terms of increase of surface electrical tension followed by a decrease. This avenue of study led him to look at plasma movements in one-celled animals. They too followed rhythms of reaching out toward the world and then retreating.

In the last month of that year the Danish Minister of Justice refused to renew Reich's residence permit, because of accusations by psychiatrists who did not agree with Reich's unorthodox writings. He relocated across the three-mile strait to Malmo, Sweden, and many of his Danish students began to commute by boat. But two Copenhagen psychiatrists contacted their counterparts in Sweden, and Swedish and Danish police co-operated in keeping watch on Reich and his students. City police searched his home in Malmo without a warrant. No charges were laid against Reich or his students, but again his residence permit was not renewed. On advice from a friend, Reich re-entered Denmark as an illegal immigrant for a time.

During that time his unorthodox views were co-opted by some psychoanalysts but they did not have the courage to present them in the frank manner which he did. In 1934 the 13th International Congress of Psycho-analysis expelled Reich, the man whom Sigmund Freud had titled "the founder of the modern technique in psycho-analysis."

In the mid- and late 1930s Reich was a refugee in Norway, after accept-ing the invitation of a professor he knew in Oslo. As a psychoanalyst Reich continued to develop new techniques for releasing blocked emotions. The human potential movement and today's bodywork therapies can be traced back to Reich.

BIO-ELECTRIC ORGANISMS

While in Norway he first discovered what he called "bions," a micro-scopic form of particles which Reicheans say are a transitional form between non-living material and living organisms. The scientific commu-nity refused to accept his reports of spontaneous generation of life, nor his contention that as long as medical scientists study dead tissue, their under-standing of living organisms will remain limited.

His previous work led up to the discovery. His professor friend had made facilities at the Psychological Institute of Oslo University available to Reich, and Reich had turned to an assistant there for help on measuring electrical charges of the skin. He wanted to confirm his bio-electric con-cepts. Again he was a pioneer.

Out of his earnings from lectures, Reich paid for the building of sensi-tive new apparatus with electrodes and vacuum tubes connected to an oscillograph.* Mainly, Reich confirmed his tensions-charge theory and the theory that the organism worked like an electrolytic system, and that it has a continuous bio-electric field of excitation between nerve centers in the middle of the body and the skin surface.

The holistic aspect of his work was important; for the first time in this way a scientist showed the organism to be a whole in which disturbance of one part affects it all. The bio-electric experiments showed the presence of one bio-psychological energy. His earlier work had indicated the ener-gy being dammed up and then released in the body, and now his instru-ments showed pleasure causing an increase of measurable charge and dis-pleasure causing decrease of bio-electric charge.

The prolific researcher was about to master yet another area of science. He wanted to study processes of expansion and contraction and corre-sponding bio-electric charges in protozoa—primitive forms of life. Did currents of a biological force work the same in all living creatures?

---

* A device for producing a geographical record of the variation of an oscillating quantity, such as an electric current.

LIFE UNDER THE MICROSCOPE

Loyal friends helped Reich buy equipment for microphotography, sterilization, and detecting electrical charges, as well as to hire assistants. In 1936 time-lapse photography of protozoa was a new idea, but Reich never let that stop him. His critics could not understand why he wanted high-magnification microscopes, since there was an upper limit above which the subject would become increasingly blurry. But he wanted to study movement within the protozoa, not the fine details of form.

A series of accidental or experimental changes in procedures led to his amazing discovery of moving lifelike forms which could be grown in cultures and developed from a variety of apparently non-living materials put in solutions which caused microscopic particles to swell. Artificially-created tiny blue-green vesicles (sacks) which he named "bions" grew in sterilized preparations of materials such as coal or sand. Under high magnification the vesicles could be seen in rolling, pulsating, rotating and merging movements. In controlled experiments he proved that the bions could not have appeared as a result of infection from the air.

While looking at bion cultures under the microscope, his eyes were burned by a non-nuclear radiation from them that he later found in the atmosphere. It was not the type of radiation known to physics. Instead, it corresponded more to the Hindu concept of *prana* or to the Chinese concept of *chi*. This is when he named it orgone—energy of the organism. It is a biological radiation, not electromagnetic, and an Oslo radiologist confirmed that no standard nuclear radiation was present in the bion culture. In the dark, the cultures glowed with a vague greyish-blue light.

Reich also studied cancer tissue at high magnification and showed a leading cancer researcher some moving cancer cells from living tissue. The researcher took the tissue back to his own laboratory, performed the usual procedures which killed cells by drying and staining them, then in a smug tone reported that he had "controlled" Reich's experiment and found Reich's bions to be "only staphylococci." He apparently did not follow Reich's procedures, however.

Reich continued to follow the path which now leads into research on cancer pathology. Eminent Norwegians started a newspaper campaign against his work in all his fields of interest, and once again influential psychiatrists pressed a government to kick Reich out of their country—this time by changing licensing regulations. By now the furore had nothing to do with his former interest in Communism; he had seen it for what it is and became vehemently anti-communist. In the middle of Reich's intense study of bions, he had to quickly pack up his laboratory equipment. On the last boat out of Norway before World War II, Reich again emigrated to another country.

SIXTH NEW START

After he arrived in the United States, Reich settled with his third wife in a rented house on Long Island, New York. The basement was used for experiments, the dining room transformed into a laboratory and the maid's room into an office/preparation room for laboratory cultures. Psychotherapy took place in what had been an extra bedroom. Reich further made a living by lecturing at the New School for Social Research as associate professor of medical psychology until 1941.

During those years Reich's research focus was on cancer and on radiation properties of his bions. To make certain that it was not only his own perceptions, he had his assistants stand in the dark and pick out test tubes which had a bluish glimmer of radiating bion cultures. Accidentally from a rubber glove incident he had found that organic materials absorbed the radiation.

His next experiment was to design an enclosure of metal to prevent leakage of the radiation from cultures. He lined the experimental boxes on the outside with organic materials—cotton or wood. The experiment was controlled by an identical metal box which was empty of bion cultures. To his surprise, the [empty] control box luminated as if it held radiating cultures itself. It appeared to pull the same type of radiation from the very air.

From the experiments with experiencing a lumination visually, he went on to discover that heat concentrated in the box. It felt like the warmth and prickling which bion cultures produced on skin . . . He then learned that metal attracted the unusual radiation and then reflected it away, to be absorbed by the organic materials.

He then designed an accumulator with a glass window behind which a thermometer could be inserted. An identical thermometer at the same height outside the box measured room temperature. Reich found the accumulator was always about a half a degree Celsius warmer than surrounding air.

What it meant was that the life force he had previously found in bion cultures could be collected from the atmosphere by an orgone accumulator. In its one-layer form, it is a wooden box lined with sheet metal. It works like a one-way grid for the orgone, as in the greenhouse effect where a radiation is allowed to enter but is reflected back inside faster than it exits, and the concentration builds up. He and his associates learned they could sit inside the box, soak up a greater charge of life force than they could by sitting outside, and improve their health.

Among the experiments done with the accumulator, one type showed that an electroscope* discharges more slowly inside it. This could not be explained by the current theory on atmospheric electricity. Other experiments showed body temperature of people sitting inside the accumulator

---

* An apparatus for detecting an electric charge.

rose anomalously. Control experiments eliminated all standard explanations for the temperature rise.

To follow what Reich was doing, he said, a scientist would have to drop all the intellectual baggage that's connected to the Second Law of Thermodynamics. Otherwise, "he will not understand the temperature difference; he will feel inclined to do away with it as only heat convection or . . . this or that. He will fail to see its orgonomic, atmospheric significance." Believers in the hypothesis of empty space likewise would not understand that a vacuum could light up and that the effect can vary with weather changes, Reich said.

In the orgone accumulator, heat is not produced out of nothing, Reich said, but rather the moving orgone within it is stopped by the accumulator's inner wall or the palm of a hand, and is then expressed as heat.

Reich's bion experiments continued, including one type which showed fogging on X-ray plates from the bions' radiation.

ACCUMULATOR TO ORANUR

Over the following years Reich's patients reported that the orgone accumulator was helpful in treating many types of disorders such as arthritis, and especially cancer. He never claimed it was a cure for cancer, but somehow he gained the reputation of having claimed this.

He moved from New York to a small rural community, Rangeley, Maine, and set up an institute he called Orgonon. Throughout the 1940s he researched the orgone as well as kept up a practice and publishing his own journal *The International Journal of Sex Economy and Orgone Research.*

Reich also reported discovering the motor force—he claimed that enough energy was collected in an orgone accumulator to run an electric motor about the size of an orange. Plans for the motor were never published because he said humanity was not ready. As with all orgone phenomena, such as the accumulator, the orgone motor varied with the weather. Today, the Wilhelm Reich museum has a film of the motor.

The saga at Orgonon took a frightening turn when in 1951 Reich tried putting a small amount of radioactive material—radium—in an orgone accumulator. His hypothesis was that powerful orgone would wipe out the bad effects of nuclear radiation. He was wrong. Some unknown force, different and more powerful than the radioactive material itself, went crazy.

The reaction of an area highly charged with orgone and then exposed to radioactivity caused a local disaster; Reich's "oranur" experiment contaminated his laboratory, killed mice which he had in the laboratory for experiments and made everyone at the institute quite sick, including Reich, who fainted several times in the sickening atmosphere caused by the experiment.

One worker nearly passed out when he stuck his head in the accumulator. Rocks on the fireplace crumbled mysteriously. Granite sticking out of the ground several hundred yards away in the infected area blackened.

Dark, dull clouds hung overhead for days. The clouds seemed to be connected with an anti-life effect, and people's health worsened in their personal weak areas. For several weeks, radiation counts measured on Geigercounters in a radius of 300 miles from Orgonon were unusually high. Reich did his best to wash and decontaminate the building and surrounding area, but it took a long time and caused much stress to the people at Orgonon.

CLOUDBUSTER

The disaster had a side effect. In an effort to clear the depressing clouds from the area, Reich invented the device he later called the cloudbuster. It is made simply—from hollow metal tubes pointed at an angle at the sky and grounded at the other end in flowing water, because water attracts the life force. The bundle of pipes is said to draw orgone out of the sky wherever it is pointed.

Why would Reich want to do that? He and his associates would reply that radioactive fallout and other pollutants turn the lively natural-state orgone into a stale, stagnant, dead form of orgone which he called DOR, which stands for Deadly Orgone Radiation. He said DOR is a factor in causing droughts by inhibiting rain and cloud formation. One theory of cloudbusting is that by drawing the DOR out of the sky with a cloudbuster and then getting the healthy orgone moving again, the atmosphere returns to its natural cycles which include rain.

For several years he researched what could be done with a cloudbuster to change weather, and said he learned how to raise the energetic level of the surrounding atmosphere instead of just decrease it. When he took his cloudbusting equipment to Arizona, events became really strange, including alleged experiences with UFOs. Reich's journal of his 1954 journey reveals an unusual ability to sense the natural landscape and its moods, similar to the awareness and sensitivities of aboriginal peoples.

Reich viewed the cloudbusting operation as beneficial, bringing rain to the southwest in January 1955. One morning in Tucson there was so much rain that planes were unable to land at the airport. The previous weeks, he reported prairie grass had sprouted in the desert until in December 1954 the grass was a foot high on land that had been barren as long as anyone living could remember. This work was not as well documented, perhaps because of the distractions of a coming showdown with the government agents.

Meanwhile, the United States Food and Drug Administration gathered a case against his use of the orgone accumulator for therapy. The FDA and medical profession did not believe that it worked, and labelled it quackery. In 1954 the FDA ordered his [Reich's] hardcover books banned from circulation and his softcover books, including all his periodicals, burned. It also ordered him to stop making and distributing orgone accumulators. For refusing to obey the injunction against publishing, Reich was sentenced to two years in jail. He died in prison in 1957, shortly before he would have been eligible for parole.

STILL IGNORED BY MAINSTREAM

Years later, mainstream science has not accepted bions or Reich's more important findings regarding the atmosphere. An orgone accumulator sits in the St. Louis Museum of Quackery. However, small groups in several countries carry on the work. Some European health practitioners openly use orgone accumulators. A scientist and former weather forecaster, Dr. Charles R. Kelley, wrote *A New Method of Weather Control* in 1960 and published the only periodical related to Reich's work in the years just after Reich's death, up until 1965. Another of Reich's students, the late Elsworth F. Baker, M.D., founded the American College of Orgonomy and began the *Journal of Orgonomy* about a decade after Reich's death. Headquarters of the small college are now in Princeton, New Jersey. It consists of a group of academics—mostly psychiatrists. The Wilhelm Reich museum at Rangeley, Maine, is open to the public in summer. Unfortunately, his will specified that his archives be sealed in a vault until the year 2007. He hoped that a new generation would seriously look at his work without feeling the need to squash it.

Over the years, some of Reich's publicly-stated views, such as his McCarthy-era accusations that certain government agents were Red Fascists, his claims of UFO-related experience, or his advocacy of adolescent sexual freedom, have been an embarrassment to followers who otherwise want to carry on his work. Some of them claim that, in his last few years of his life, Reich's loneliness and the cumulative effects of his experiences became too heavy. From around 1955 until his death in 1957, says biographer Boadella, "the paranoid ideas ran alongside perfectly rational concepts and insights."

The best of Reich's discoveries live on, although not publicized in mainstream media. A handful of individuals in various countries have continued to learn about "etheric weather modification." Such experimentation with atmospheric processes is not to be taken lightly, according to practitioners. In fact, they say that irresponsible cloudbusting operations can contribute to destructive weather instead of restoring the weather's natural rhythms.

There had been a unifying thread spun by Reich's varied research; most of his findings related to a central discovery. The growing thread of evidence pointed to reality of Life Force which can be scientifically demonstrated. It led to Reich's findings that, when the atmospheric life force over a large area has been assaulted too much, it locks into an immobile state of drought-causing stagnant air.

A Reichean-oriented scientist in Michigan and others add a sombre note about degradation of the atmosphere. Herman Meinke of the Detroit area estimated in 1993 that the life force in the atmosphere is only about one/fortieth of the strength which it was during Reich's experiments. He has been repeating the experiments for many years, and found that they no longer show results which they did previously. He blames the proliferation of nuclear testing and nuclear power plants for weakening the planet's atmospheric life force.

REICH PARALLELS SCHAUBERGER

Reich's biographers hint that squashing of writings about a dynamic atmospheric force in the 1950s was related to the fact that the atomic power industry was emerging at that time; it would not do for the public to debate whether atomic fission and its byproducts turns life force in the environment into a destructive presence which Reich called Deadly Orgone Radiation. Nor would the atomic power industry want people to connect droughts and anomalous weather with atmospheric DOR. Reich's contemporary and fellow Austrian, Viktor Schauberger, also had an advanced understanding of what he saw as an energy whose life could be blown apart by proliferation of atomic radiation in the atmosphere.

Like Schauberger, Reich also learned from observing nature. When Reich published photographs of trees dying from the tops downward because of poisoning of the biosphere—what he called the falling of DOR onto the trees—he was one of the first scientists to warn that the planet could become a lifeless wasteland. His work indicated that the life force within an organism is stimulated by outer orgone in the atmosphere. Has the weakening of the life force in the atmosphere by pollution, been reflected within humanity and other species? The many weakened immune systems—from cancers in sea life to AIDS in humans—presents a strong clue.

Reich's followers today say that Reichean methods to break up blockages in the atmosphere can help save the day, if the causes of Deadly Orgone Radiation are also removed. (Reich said the causes include treatment of babies and children which perpetuates an emotional desert in humanity.) His followers describe a scenario of atmospheric medicine, including cloudbusters, renewed vitality in the air and in organisms, and greening of deserts.

To the end of his life, Reich was close friends with English educator A. S. Neill of the famous Summerhill school, who also pioneered a life-affirming approach to children. Neill wrote in 1958, "If the anti-life men in charge of our lives do not destroy the world, it is possible that people as yet unborn will understand what Reich was doing and discovering."

Former student of Wilhelm Reich, Dr. Charles R. Kelley of 13715 SE 36 St., Steamboat Landing, Vancouver, WA 98685, USA teaches a correspondence course titled Science and the Life Force.

REFERENCES

Blasband, Richard A. "Orgone Energy as a Motor Force," *New Energy Technology*. Planetary Association for Clean Energy. Ottawa, Ontario, 1988.

Boadella, David. *Wilhelm Reich: The Evolution of His Work*. Arkana, London, 1985.

Boadella, David. Appendix One, "The Trial of Wilhelm Reich" by Sharif, Myron R. first published by Ritter Press, 1958.

Burr, Harold Saxton. *Blueprint for Immortality: The Electric Patterns of Life*. Essex, England: C.W. Daniel, 1972.

Eden, Jerome. *Orgone Energy*. Hicksville, New York: Exposition Press, 1972.

Einstein, Albert. Correspondence with Wilhelm Reich 1941-1944, from the Archives of the Orgone Institute.

The History of Orgonomy, "Wilhelm Reich on the Road to Biogenesis," author unknown; this author has only part of this manuscript from the Archives of the Orgone Institute.

Ind, Peter. *Cosmic Metabolism and Vortical Accretion*. Self-published manuscript, England, 1964.

Kelley, Charles R. *A New Method of Weather Control*. Westport, Connecticut: Radix, 1960.

Mann, W. Edward and Hoffman, Edward, *The Man Who Dreamed of Tomorrow*. Los Angeles: J.P. Tarcher, 1980.

Manning, Jeane. "A Cause of Droughts? Interview with Dr. James De Meo," *Explore! Magazine* Vol. 4, No. 1, 1993.

Manning, Jeane. "Travels Across the Continent," *Explore! Magazine*, Vol. 5, No. 3, 1994.

Pulse of the Planet Journal. Orgone Biophysical Research Laboratory, California, 1991.

The Wilhelm Reich Foundation. *The Orgone Energy Accumulator*, Orgone Institute Press 1951.

"A Motor Force in Orgone Energy, Preliminary Communications," *Orgone Energy Bulletin*, Vol. 1, No. 1.

(The last two items are reprinted in *A History of Free Energy Discoveries* by Peter A. Lindemann, Borderland Sciences, Garberville, CA 1986.)

23 The AMA's Charge on the Light Brigade

Stuart Troy

The evils of some men have a karmic momentum that extends beyond the grave, undiminished by their deaths. If you could somehow quantify and accurately ascribe human pain and needless suffering, then the pernicious legacy of Morris Fishbein, M.D. (1889-1976) of the American Medical Association (AMA) would exceed in villainous ignominy the legacies of Hitler and Stalin combined. While a more subtle and quiet offence which may pass unnoticed in the historical moment, ideocide is ultimately, in its continually expansive accumulative enormity, a far more pernicious crime against *all* humanity than any "simple" genocide. When a genocide indictment is finally issued, it contains specifics: dates of onset, locations, duration, victim identity lists. But who can name the victims or measure the pain that marks Fishbein's ideocidal career? Indeed, when can we even end the tally?

If the only adduced instance of Fishbein's ideocide were the persistent, obsessive persecution of Colonial Dinshah Ghadiali, M.D., D.C., Ph.D., L.L.D., from 1924 to 1958 and the attempted eradication of his Spectro-Chrome Therapy (SCT) both from practice and from print, it would tragically suffice to make my point.

Popular history would have us believe that the (now scandalously) shocking FDA-instigated incineration of the printed works of Dr. William Reich was an unprecedented and isolated event in these United States of alleged First Amendment protections.

However, the dubious distinction of having been the first Federal book-burn victim belongs to Dinshah. Ten years previously, in 1947, in compliance with a Federal Court order, he had to "surrender for destruction" his unique library and all printed material pertaining to coloured light therapies to U.S. marshals in Camden, New Jersey. All during those years he remained steadfastly dedicated to truth in the healing arts, and to his personal vision of an earnest, energetic, open America (a vision he formed some fifty years earlier on his first visit). The source of this resiliency is found in part in his often-repeated motto: "Truth can be defeated, *never* conquered."

In the better-known case of Dr. Reich, the very barbarity of the assault itself added to his mystique, impairing a legendary martyrdom and ensuring an elevated niche in history independent of the content or validity of his science. In contrast, very few, even among practitioners of the alternative disciplines, know the story of SCT despite the uninterrupted efforts of the Dinshah Health Society, established and run by his son Darius Dinshah on the original 23-acre Malaga, New Jersey estate. Operating under the strict confines of the final 1958 Food and Drug Administration (FDA) injunction, which is still in effect, SCT has somehow survived to enjoy the modicum of legitimacy conferred by the 1994 recognition and listing (as an information source only) of SCT by the United States Office of Alternative Medicine.

Fortunately for us, the core of the system (any projected light source except fluorescent, plus 12 coloured filters) is so low-tech, and the "tonation" application formulations—laboriously determined and charted by Dinshah—are so simple that the ease of home assembly and utilisation allows for convenient accessibility. (See diagram of the Spectro-Chrome Therapeutic System on p. 262.) Unfortunately for Dinshah (the "Ghadiali" was dropped in America), it was precisely this low-tech accessibility and therapeutic efficacy which made him an irresistible and inevitable target for Fishbein and the healing-for-money establishment.

Born in Bombay, India, in 1873 to a Parsee watchmaker of Persian descent (the Zoroastrian faith to which he adhered is often referred to as "the Faith of Light"), Dinshah's special genius and industry soon became apparent. He began primary school at age three, and high school at eight. By his eleventh year he was an assistant to the Professor of Mathematics and Science at Wilson College, Bombay. His father did not encourage his early fascination with electricity, and Dinshah told of sneaking downstairs to study through the night, retiring for a few hours of sleep shortly before dawn when he and his father would arise together. He took his university exams at fourteen, winning proficiency awards in English, Persian and religion. (In his spare time, he was to achieve competence in eight oriental and eight occidental languages.)

The following year he divided his time between giving demonstrations in physics and chemistry and meeting the demands of running a successful electric doorbell/burglar alarm installation business. It was also the year he began his medical studies.

At eighteen, having mastered the practice of Yoga Shastra and having been awarded a fellowship by The Theosophical Society, he added spiritual subjects to his oratorial repertoire. His reputation and experience as an electrical engineer earned an appointment as Superintendent of Telephone and Telegraph for Dholar state. Three years later found him serving as Electrical Engineer of Patiala state and Mechanical Engineer for the Umbala Flour Mill.

His medical studies completed, in 1896 Dinshah made his first trip to America, where he lectured on X-rays and radioactivity, meeting Tesla, Edison and other scientific notables. A darling of the press, Dinshah was affectionately referred to by *The New York Times* as "the Parsee Edison."

The freedoms, the opportunities, the stimulating intellectual energies he perceived in pre-war America left him with an inspired, compassionate optimism that future events could not dilute. Upon his return to India he became a social reformer and the first publisher/editor of *The Impartial*, a weekly founded "to further the cause of freedom in speech and writing."

The year 1897 was to prove pivotal, for it was the year Dinshah became the first person in India to apply and thus effect a cure for disease in accordance with the hypotheses of Dr. Edwin D. Babbitt (as in his book, *The Principles of Light and Color*, University Books, New Hyde Park, NY, 1876, reprinted 1967) and Dr. Seth Pancost *(Blue and Red Light, or Light and Its Rays as Medicine,* 1877).

During the plague years of the early 1900s, Dinshah's eclectic and unorthodox ministrations effected a 60 percent recovery rate, in contrast to the 40 percent recovery expectations of conventional medical practice.

Responding to an influential Theosophist friend's urgent summons, Dinshah, from his supervisory position in a major light installation several hundred miles away from central India, travelled to the bedside of his aunt who was dying from mucosa colitis (dysentery). Upon arrival, Dinshah faced several handicaps. The attending physician of record was a prominent Parsee and the Honorary Surgeon of no less than a personage than the Viceroy of India. The old woman revered him as a demigod, but contemptuously referred to Dinshah as "that kid doctor."

For three days he had to watch silently as her health continued to fail rapidly under a brutal but conventional medicinal regimen. Although the regimen was well thought out and in conformity with the best recommendations of *The British Pharmacopaedia*, Dinshah saw that the opium administered for the pain was stressful to the heart; the catechu, although a good astringent, was a peristalsis inhibitor; the chalk, intended as a binder, was an intestinal irritant; the bismuth subnitrate, a local antiseptic, choked the alimentary canal; and the anti-flatulent chloroform was escharating damaged tissue.

As Dinshah noted:

Thus she stayed two days more, drinking the poisonous concoction. On the third day she was in such a condition that she lifted her hands to me and implored me, "O, Dinshah, save me!" Medically she was beyond recovery and I said with a sigh, "Call on the Almighty to save you. Dear girl, I have no power, no medicine of which I know I can be of service to you, but if you let me I shall endeavor to do the best otherwise." She nodded her consent and I promptly threw out the drug mixture . . . I brought [indigo-] coloured pickle bottles to act as the slides . . . Within 24 hours the [100 daily bowel evacuations] were reduced to four a day; within 48 hours they came down to two; the third day Jerbanoo was out of bed!

Reflecting an Eastern patience and restraint, and reflecting the slower technological pace of a pre-electronic age, Dinshah did not rush impetuously into print. However, before he could publicly promote SCT he had to be satisfied that he could exercise confident control of the procedure. Thus Dinshah embarked on a lengthy theoretical research project, producing remarkably precise and accurate tonation formulations.

By 1904, at Ajmer and Surat, he had established "Electro-Medical Halls" for the promotion of colour therapy research, magneto- and electro-therapeutic approaches as well as orthodox medicine. However, early on he was forced to abandon the otherwise promising electric modalities due to frequent episodes of nerve anastomosis and the inherent and insurmountable problems he encountered with "unmanageable and freaky currents."

In 1908 he left India to promote his inventions through Europe, eventually, in 1911, dropping anchor in the United States with his (first) wife and two children. He loved America and vigorously embraced the principles and politics of an open democracy. However, the same cannot be said of his wife who, in reaction to her early years of impoverishment and perhaps more than a little culture-shocked, returned alone to India.

Dinshah was so taken by his vision of a Walt Whitman/Horatio Alger America as perceived in those pre-war years that in 1914 he turned down a private offer of $100 thousand for his Internal Combustion Engine Fault Finder which he developed while serving as Professor and Chief Instructor at the New York College of Engineering Science and Automobile Instruction. Instead, he gave all rights to the United States military for aviation application. (Amongst his patented inventions are: #983,703, Electrical Wiring Device, 1911; #1,144,898, Automobile Internal Combustion Engine Fault Finder, 1915; #1,544,973, Color Wave Projector, 1925; #1,724,469, Electric Thermometer, 1929; #2,038,784, Color Wave Projection Apparatus, 1936.)

Dinshah was granted U.S. citizenship in 1917. The following year he was given a commission as Captain in the New York Police Department Reserve, and in recognition of his wartime civilian aeronautic harbour-patrol activities he was promoted to Colonel, awarded the Liberty Medal by New York City Mayor John Hyland and appointed head and principal instructor of the NYPD Aviation School. The banner year of 1919 found him a member in good standing with the American Association of Progressive Medicine, and the elected Vice-President of the Allied Medical Association of America and the National Association of Drugless Practitioners.

This all seems a strange background for a "charlatan" and "huckster," as he was to be labelled and pilloried by Fishbein's AMA, in conjunction with the equally acquiescent, ever-spineless FDA in the decades that followed.

Something insidious and unnoticed had happened during the twenty-four years since Dinshah's first visit to America. The robber barons of the nineteenth century had discovered value in technology and proceeded to exercise the same piratical control over intellectual property as they previously had over the traditional sources of material wealth. Lights dimmed all over the short-lived Age of Enlightenment as new acquisitional and inquisitional institutions became empowered and entrenched.

Social historian and author David Lindsay (*Magnificent Possibilities*, Koodansha America, Fall 1996) notes that with the change of the century there was a change in the perception of the technical man, the inventor. New social forces coalesced, mediating direct contact between people and technology. The control and credibility that had been the scientist's were co-opted by agencies of industry working with agencies of government.

Canadian political scientist Andrew Michrowski fixes the date with even more precision: "It was possible for Nikola Tesla, Alexander Graham Bell, and George Westinghouse to make their mark because in their time, before 1913, the retardant forces were not yet organized enough to totally counterweigh their innovations."

Suffice it to note that 1912 was the year that young Morris Fishbein, MD [sic] (neither passing anatomy nor completing his residency), entered the employment of the already disreputable American Medical Association, without ever practising medicine. In 1913 he became Assistant Editor of the AMA's journal, *JAMA*. A prolific writer of articles, editorials and, later, books crusading for the medical profession, Fishbein became Editor of *JAMA* and *Hygeia* in 1924, holding these two posts for 25 years.

So it was against this background that Dinshah, ignorant of or indifferent to this dawning of a New Age of Darkness, innocently went public with SCT in April 1920 in New York City. The first formal instruction, in

December that year, was attended by 27 students. During the next four years (as Morris Fishbein consolidated his political power within the AMA) Dinshah held 26 classes, training over 800 students, predominantly physicians but including many lay trainees as well.

It was the very ease with which the correct tonation could be determined and applied by laymen in the privacy of their homes (as much as, if not more than, mere efficacy) which constituted the true threat. If Dinshah had kept the SCT technology arcane, the equipment expensively overdesigned and within the preserve of the professional health community, events would have played out quite differently. Unquestionably it was this accessibility and the consequent commercial threat which SCT represented that made Dinshah an early target for an eager Fishbein.

Fortunately for Dinshah, an early attendee was the twenty-three-year-tenured Chief Surgeon of the Woman's Hospital of Philadelphia, the highly credentialled Kate W. Baldwin, M.D., F.A.C.S., a member of the AMA and the Pennsylvania Medical Society, and the first woman in the American Academy of Ophthalmology and Ololaryngology. Until her death in 1937, she remained a private SCT practitioner and a vociferous advocate in public, frequently defending SCT and Dinshah against the dark forces of repression.

Dr. Baldwin enjoyed sufficient status and seniority that the initial anti-light-therapy onslaught could only incommode but not intimate or destroy her. Indeed, so forceful was her presence and so unequivocal her defence testimony at Dinshah's first trial in 1931, that the government refrained from any prosecution of Dinshah—on the basis of science—during her lifetime.

In 1921 Dr. Baldwin arranged for Dinshah to lecture in Philadelphia. Eventually her brother, the equally eminent surgeon L. Grant Baldwin, M.D., F.A.C.S. (Mayo Clinic), was to produce several SCT units to his Brooklyn, New York, practice.

Some of the social (political) history of SCT is to be found reserved in the meagre, but reliable, regular Minutes of the Board of Managers of the Woman's Hospital in Philadelphia. Over the next five years these records were to suggest even more than they revealed.

Dr. Baldwin's request to address the Board on her in-patient SCT work was granted, and on 21 December 1923, according to the Minutes, she "gave an illustrated account of the wonderful work done in the Hospital with the spectrochrome [sic]. She described a remarkable case of a child . . . so badly burned that there seemed no hope of her recovery. With the use of the Spectrochrome [sic] [primarily using the colour turquoise, i.e., blue plus green], the child is almost entirely cured. It is such an unusual case that the Board feels it should be written up for publication by Dr.

Baldwin . . . The Spectrochrome is used in no other hospital and *credit should be given to Dr. Baldwin for developing its use here. There are four instruments in the Hospital and more could be used if the room were larger."* [Author's emphasis in italics.]

It was but a short five weeks later, in the 26 January 1924 issue of the *Journal of the American Medical Association (JAMA)* (which had just recently fallen under the editorial control of Fishbein), that the first salvo was fired: a lengthy, baseless denunciation of SCT, complete with a defamatory attack on the character of Dinshah and, by associative implication, all SC therapists—with explicit reference to Dr. Baldwin, who, among numerous physicians, had been regularly contributing case histories to Dinshah's *Spectro-Chrome* monthly journal (which he published from 1922 to 1947). The *JAMA* article concluded:

Some physicians, after reading this article, may wonder why we have devoted the amount of space to a subject that, on its face, seems so preposterous as to condemn itself. When it is realized that helpless but incredulous patients are being treated for such serious conditions as syphilis, conjunctivitis, ovaritis, diabetes mellitus, pulmonary tuberculosis and chronic gonorrhea with colored lights, the space will not be deemed excessive.

While it took four years for Fishbein finally to bring Dinshah before his first magistrate, the first blood had been drawn much earlier. Two months after the *JAMA* article appeared in print, the Woman's Hospital Board of Managers' Minutes of 28 March 1924 report the receipt of a letter from the staff, requesting that Dr. Baldwin discontinue the use of SCT. The only ground offered for this initiative was the *JAMA* article. The Board's time-tested response was the classic bureaucratic reflex: an ad hoc committee was established to evaluate the situation for later discussion.

Not all the Board's problems conveniently faded during this interval and it was forced to address the issue head on. According to the minutes of 23 May 1924: ". . . the question had been considered from every viewpoint and . . . *the Committee recommended the continuance of present conditions.* This report of the Committee was accepted." [Author's emphasis added.]

Almost a year later, the Minutes of 27 March 1925 record that, "Dr. Baldwin in a letter spoke of her need of more room for the Spectrochrome. She asked to have two cubicles made; she is getting many cases . . ." Subsequently Dr. Baldwin was permitted to install additional treatment cubicles.

Notwithstanding all of these initial successes, buttressed by the consistent clinical evidence, official affirmations and institutional support,

SCT was soon to suffer the first of a nearly unbroken string of reversals. There is a subtle but interesting peculiarity to this sudden, decisive turnabout spontaneously appearing in the Minutes without warning. True, the Minutes give no picture of day-to-day hospital politics, and given their narrow purpose and focus, especially as the sole historical source, they would of course tend to conceal more of a general contextual circumstance than they could reveal.

So we are left to speculate on the strangeness of the impudence of a letter from the *hospital interns*, received by the Board and reported in the Minutes of 24 September 1926, expressing their objections to Dr. Baldwin's presence on the surgical staff.

Traditionally interns, from a socio-political perspective, constitute the least vocal and effective participants in hospital policy formation. However, one can perhaps understand, even sympathise with the interns' position, with their cumulative daily frustration as endless streams of serious surgical candidates and other diseased patients were regularly being sent home without ever seeing a knife or pill. What is less understandable is the effectiveness of their one letter.

The September meeting moved to request Dr. Baldwin's resignation from the surgical staff, but also moved that she "be granted the privilege of practising Spectro-chrome Therapy with her private patients in the Woman's Hospital." Both motions were carried. The Board passed on the request to Dr. Baldwin, and by the meeting of 22 October 1926, without record of internal debate or explanation, the board accepted "with regret" Dr. Baldwin's resignation from the surgical staff. Just before Dinshah's first trial in 1931 in Buffalo, Erie County, New York state, Dr. Baldwin received a letter from the Secretary of the Erie County Medical Society specifically soliciting her comments about the 1924 article and the impending criminal action. The letter read:

> According to [the *JAMA*] article, Susie T., age 9, who was admitted to the Woman's Hospital with a sloughed appendix and peritonitis, developed a pneumonia which was treated by Dr. Baldwin with lemon, turquoise and magenta colored lights. Susie went home well and happy.
>
> Dinshah P. Ghadiali, using the title M.D., is the publisher of *Spectro-Chrome*. He is under arrest in Buffalo charged with grand larceny for selling a course of lectures and leasing a colored light apparatus of alleged curative value for human ailments.
>
> We are wondering if the article in which your name is given is a correct statement. Our Society is somewhat interested in the outcome of this case and we will very much appreciate your telling us if your name was used with authority.

Dr. Baldwin's ringing endorsement was but faint indication of the eye-opening testimony she would soon deliver under oath.

Her reply read:

Your letter of June 9th is just received. The statement printed in the *Journal of the American Medical Association* of January 26th, 1924 is practically as written by me for *Spectro-Chrome* magazine. Susie's was an emergency operation at nine o'clock at night. There was nothing left of the appendix to remove. There were quantities of pus. The wound could not be closed, free drainage was provided and the child put to bed with little hope that she would live until morning. For some days, an enema would simply pass through and out of the abdominal opening. Susie did develop pneumonia. I did use Spectro-Chrome and eventually she did leave the hospital in good condition.

In the Woman's Hospital, I used Spectro-Chrome for many things to the satisfaction of the patients, the staff and the Board. The results were approved by all interested, until the article cited came out in the *Journal.* Then the staff turned traitor. The Board appointed a Special Committee of five to investigate, and a copy of its report I am enclosing. After this investigation I was granted a large space for the work of Spectro-Chrome. The American Medical Association continues to rate me as a Fellow in good standing. *Not the slightest effort to prove the truth has ever been made by the AMA or the doctors. The simple fact that the AMA made the statement against Spectro-Chrome was sufficient to condemn.* At the time I wrote to the *Journal* stating facts. The courtesy of a reply was not granted. The letter was sent by registered mail and a return card showed that it was delivered. Eventually this article was the cause of my losing my position on the surgical staff of the Woman's Hospital.

The AMA has not been just to one of its members or to humanity; within the year of 1929, communications have been sent by the AMA to several of my patients in the shape of a reprint of the article published in the 1924 *Journal* and a letter ridiculing Spectro-Chrome and me.

Spectro-Chrome has more value as a therapeutic measure than all the drugs and serums manufactured. I would close my office tonight, never to reopen, if I could not use Spectro-Chrome. [Author's emphasis added.]

Dinshah was to face tribunals eight times, winning vindication only twice and having to serve a total of eighteen months in prison. His first victory, at Buffalo, NY, in 1931, was the last time the anti-light forces dared expose themselves to a decision rendered by a jury allowed to hear medical evidence and expert scientific testimony. His second victory (the first Camden trial, 1934) rested on the judicial reasoning that, being of Parsee descent, Dinshah was a "white man" and therefore, 17 years after his naturalisation, he was ruled to be not deportable.

Just as, in retrospect, the involvement of the normally non-influential intern role in efficaciously precipitating Baldwin's predicament seems to be more than meets the eye, so too is the circumstance that led to Dinshah's arrest in May 1930. The indictment charged that he "did feloniously steal $175 from one Houseman Hughes by falsely representing and pretending that a certain instrument and machine [Spectro-Chrome] would cure any and all human diseases and ailments."

Again, the reliable Court record here gives scanty insight. Looking back at the actual ascertainable, there is reasonable inference that Hughes was only a point man—but for whom? He was a layman who had admittedly never received, witnessed or administered a tonation treatment. Affidavits and testimony from official records show only that he leased a unit (subsequently defaulting on the payment) and promptly pressed charges. There is no indication that he even removed it from the box. Someone on the prosecution side did turn it on, although did not take advantage of the exercise to attempt a tonation.

The core of the embarrassingly underprepared prosecution (what could they actually say?) was the "expert" testimony of a physicist who testified to the fact that the unit used an ordinary light, projected through ordinary coloured glass filters, producing no spectral alterations nor new rays of any sort. This was rather extraordinary testimony, considering that Dinshah had never claimed otherwise!

(This 1931 trial was also the last time the government was to base any indictment on its "ordinariness." Later, when the FDA, in a convenient about-face, proclaimed the Spectro-Chrome a "medical device" (though unauthorised), it provided the cloak of legality under which they conducted, unopposed, hundreds of warrantless, confiscatory and non-compensatory raids through the living rooms, basements and converted garages of otherwise innocent, non-complaining citizens after 1947.)

Dinshah, despite facing a looming 10-year/$10,000 adverse judgement, chose to defend himself with a five-witness defence which included three M.D.s. He reasoned: "The judge knows the law and I know my science so I can defend it better than any lawyer. Truth can be defeated but *never* conquered."

Unprepared for an impregnable defence, the state produced in its rebuttal its only medical witness: Albert Sy, M.D., a practitioner of the high-tech, expensive and generally inaccessible treatment modalities of radium, X-ray and ultraviolet radiation. This prosecution witness, also testifying to his zero experience with SCT, was forced to admit under oath that he had *no* evidence at all for his "expert" opinion that there "could be no therapeutic value of colored light or other appreciable effect on animals."

Dinshah's first witness, Dr. Welcome Hanor, an early SCT student and enthusiastic proponent, had posted the $1,500 bail. His modest credential was his reputation as a local general practitioner of thirty years' experience. He gave unresolved credit to SCT for his successes with cancers, diabetes, gonorrhea, syphilis, ulcers, neuritis, meningitis, heart conditions *and many other disorders*.

Dr. Martha Peebles had a distinguished twenty-four-year private practice in New York where she had also held public office with the City Department of Health before serving with General Pershing's expeditionary forces (attending up to sixty-one operations a day). Invalided by crippling arthritis and neuritis, her health was restored one month after receiving her first tonation treatment from Dr. Baldwin, and she subsequently re-established her medical practice. In court she recounted her success with cancers, hypotropic arthritis, poliomyelitis, mastoiditis, sciatica, heart disorders, goitres, ulcers, neuritis *and many other disorders*.

Dr. Kate Baldwin's testimony was extensive, forcefully unequivocal and unshakeable. The worst nightmare of a prosecuting cross-examiner, she repeated affirmatively as to SCT's efficacy in the treatment of cataracts, glaucoma, acute eye infections and hemorrhaging; mastoid and middle-ear problems; tonsillitis and adenoidal disorders; tuberculosis, bronchitis and pleurisy; functional and organic heart disorders; ulcers, hemorrhoids, boils, drug addictions, asthma, laryngitis, mouth disorders, rheumatism, lumbago, syphilis, cancer, radiation burns, appendicitis, strangulated hernia *and many other disorders*.

The trial lasted for four days before the jury returned a "not guilty" verdict in ninety minutes. Subsequently (in addition to his previous loss in Portland in 1928), Dinshah was to lose actions in Cleveland, Wilmington, Washington, D.C., Brooklyn (the decisive FDA ruling) and, finally, in Camden in 1947.

By 1941, mail sent to Dinshah's institute was being returned by the local postmaster, marked "Fraudulent: Mail to this address returned by order of the Post-Master General." No doubt this purely postal administration, not the result of a judicial proceeding, contributed measurably to the recognition of the AMA as a para-governmental agency of intimidation, and hastened the final discontinuance of SCT by the dwindling numbers of loyal M.D. practitioners.

Through an internal restructuring and reorganising at the Spectro-Chrome Institute, Dinshah was for a while able to circumvent and neutralise the mail blockade. Suffice it to note that Dinshah's manoeuvre was a short-lived expediency, lasting six years until 1947.

While looking back on Dinshah as an exemplar of indefatigable, persevering resiliency and inner strength of character, in balance we must also note with all due respect the persistence, long memory and vindictive, single-minded purposiveness of the private professional associations, in concert with government regulatory agencies, now legislatively armed to dispose of the inconvenience of the evidences and protocols of both court and clinic.

No narrative, historical reportage or creative mythology of suppression and censure, however factual or fantastical, is complete without the obligatory, timely, lab or library fire under suspicious circumstances. To paraphrase Cervantes very loosely, a tale of "intellectual inquisition" without arson is like a meal without wine.

The 1945 fire that destroyed Dinshah's main building, just ninety days before the Brooklyn trial, caused inestimable damage not just to his defence but to all of us through the destruction of demonstration prototypes and the irreplaceable case histories of twenty-five years.

Losing the second Camden trial (the FDA-driven action on "mislabeling") in 1947, Dinshah was fined $20,000 and sentenced to a five-year probation period, a condition of which was to "surrender for destruction *all printed material* (save some personal notes) pertaining to coloured light therapy—a singular collection valued then at $250,000. He was further ordered to disassociate himself from any research in the field.

Probation completed in 1953, Dinshah again restructured his institute, this time as an educational institution, Visible Spectrum Research Institute, and resumed the dissemination of information and equipment—but with the disclaimer asserting that "in accordance with the current conventional medical view, there is no curative, therapeutic value" to these projection systems. Independent of any SCT/Dinshah data, this scientific "edict" was already known to be false. (In 1958 the FDA finally obtained the permanent injunctions, still in effect today, under which Dinshah was to operate until his death in 1996, at the age of ninety-two.)

Contrary opinions such as those of the medical establishment-respected A. J. Ochsner, M.D., F.A.C.S., author of still-classical surgical textbooks, could again be made weightless by edict. Writing to no apparent effect in those days, he reported:

In a personal experience with septic infection, the pain was so severe that it seemed unbearable. When the use of electric light was suggested, it seemed unlikely that this could act differently from the other forms of therapy that had been employed. Upon applying the light, however, the excruciating pain disappeared almost at once, and since this experience we have employed the light treatments in hundreds of cases of pain caused by septic infection, and quite regularly with results that were eminently satisfactory, not only in relief of pain but also because the remedy assists materially in reducing the infection.

There is little comfort to be taken from the fact that half a century separates us from those distant, dark ages of book-burning in America. The relation of phototherapies to the medical mainstream has not much improved. The dynamics of the relationship, the rules of combat are unchanged. Large and loud, unapologetic denial; unembellished, unforgivable, inexplicable and dangerously erroneous counter-factual utterances are still being recycled by the usual bunch of high-prestige suspects.

For example, the highly regarded *Cancer Journal for Clinicians (CJC)* in 1994 (44:225–127), in an *anonymous* diatribe, characterised the World Research Foundation (WRF) (41 Bell Rock Road #C, Sedona, AZ 86351, USA) as "helping people locate questionable cancer cures . . . . [and has] touted the Spectro-Chrome device." It then astonishingly concluded: "There is no scientific evidence that shining colored lights on the body will produce any biological effects."

This must come as quite a shock—to the generations of the paediatric health community who, for half a century, have routinely been treating the jaundice (neonatal bilirubin imbalance) of premature babies with the spectrally rebalanced, blue-enhanced Westinghouse maternity bulb; to the generations of commercial breeders of chickens, chinchillas and fish who, for half a century, have been using the monochromatic reformulation work of photo-pioneer John Ott (the original champion of full-spectrum light) to manipulate fertility, gender and even behaviour; and to the readers of the respected *American Teacher* (71[6]:16, March 1987), who were gullible and naive enough to believe the account of H. Wohlfrath of the University of Alberta, Canada, who in 1982 *replicated the nearly fifty-year-old* work of Soviet researcher E. I. Kritvitskya, in which high-frequency-restored classroom light reduced absenteeism, eye strain, dental caries, etc. as it increased attention, retention, etc.

When Dr. Sy expressed his "disbelief" in 1931, he could do so with a certain innocent honesty. But would the editors of *CJC* have us dismiss vol. 453 (1985) of the *Annals of the New York Academies of the Sciences*, on "The Medical and Biological Effects of Light" (an entire conference on the subject) as so much chopped liver? Or was the then nine-year-old *Annals* too recent to have come to the attention of the *CJC* editors, or too old for their consideration? Unlike Dr. Sy, they are at least guilty of criminal paucity of scholarship.

Responding to this anonymous *CJC* article, Dr. Steve Ross, writing in the WRF International Health and Environment Network Journal, *World Research News* (2nd quarter, 1995), goes succinctly to the core.

The *Cancer Journal for Clinicians* is sent to virtually all the physicians in the United States dealing with cancer. Could this sort of stupidity and misinformation be one of the reasons why the answer to the cancer problem has not come as quickly as suspected?

> During the Inquisition, individuals were burnt at the stake for believing that the Earth revolved around the Sun. The same Inquisition takes place today when the bastion of the medical community persecutes and removes those individuals who attempt to discuss and utilize therapies that are different than the therapeutic system that is being touted by the pharmaceutical industry.

In all fairness to Fishbein, he did not create the Torquemada mentality—a mindset untroubled by the subtle (or not so subtle) distinction and easy interchangeability between a science of data and a science of dicta. After all, the *JAMA* before, during and after Fishbein was never the arena to seek the open Lockean dialogue in "the free marketplace of ideas" in which truth would always emerge as the best value. Ridicule as retort and censure by consensus pre-date even Galileo. It could be argued that all of this is part of our collective hardwiring.

The real and ongoing legacy of Fishbein—the apotheosis of the peer review, the institutionalisation (professional, academic, corporate and political) of entities that perpetuate and fuel the reactionary, counter-evolutionary potentials of the human intellect—is not a simple, single bequest. It is rather an annuity that pays out incrementally in pain, indefinitely.

Today, a century and a quarter after Dr. Babbitt and 100 years after Dinshah's empirical confirmation, in most Western-style hospitals all over the world you will find the seriously traumatised post-surgical patient routinely maintained under the arbitrarily bizarre and randomly unbalanced spectra from cool and allegedly "white" fluorescence, while meticulously sustained on FDA-determined *minimum* daily nutritive requirements. You may *sneak* in a full-spectrum light; you may *sneak* in anti-oxidant vitamin megadoses. But here in the United States—the Land of the Litigious where the unholy AMA/FDA annuity is issued—take great care to call the light "*only* cheerful," the co-enzyme pills "*only* food." The operative words are "sneak in" and "only." Otherwise apprehension constitutes an inference with the conventions of established (hence, ossified) medical practice; and the consequent shifting of criminal, civil and professional responsibilities (especially monetary liabilities) is quicker than 186,270 miles per second in a vacuum.

This confusion, this melding of the professional proclamations with the proof of the pudding, may, in some Hegelian antithetical manner (the ". . . and one step back" of the historical process), provide some sort of intellectual brake to the evolutionary inevitable. However, as bleak a picture as this is, the flip side of the Hegelian paradigm promises a net gain of one forward step. This could be the philosophical principle that makes Dinshah's motto about defeated truth remaining unconquerable, a feature of the universe rather than mere personal mantric expedient.

At any rate, the work of Dinshah P. Ghadiali, the light of spectrochromology and related phototherapies, although deliberately dimmed for decades has not been extinguished. In fact, SCT endures and modestly thrives under the diligent, dedicated tutelage of Dinshah's son, Darius Dinshah—the accessible and gentle prime-mover for the active work being continued by the Dinshah Health Society.

The Society serves as an active information-clearing centre, holding annual meetings, publishing a newsletter and archiving relevant literature available to an increasingly interested public. Especially recommended, both for historical background as well as for its simple, utilitarian instructional material, is Darius Dinshah's book, *Let There Be Light*.

Note: The Dinshah Health Society is a non-profit, scientific, educational, membership-based corporation. For further information, contact the Society at: P.O. Box 707, Malaga, NJ 08328; telephone: (609) 692–4686; web: http://www.wj.net/dinshah

About the Author: Stuart Troy, a native of New York City, is a researcher and writer. His first foray into print was in 1978 with a critical article, "Sigmund Freud and the Relevance of a Newtonian Scientist in Post-Einsteinian/Heisenberg Age," for *APERION: A Journal of Philosophical Inquiry and Opinion*, for which he was also an associate editor. His first book, co-authored with Jonathan Eisen, was *The Nobel Reader* (Clarkson, N. Potter, NY, 1987).

Stuart is currently working with noted historian Monroe Rosenthal on a history of women warriors of the Jews. His other driving passion is the preparation of a detailed report on the various iniquitous activities of Dr. [sic] Morris Fishbein, and he requests *Nexus* readers with any documentation or anecdotal material on Fishbein to communicate with him care of P.O. Box 5027, Fort Lauderdale, Florida 33310, or e-mail him care of bick@earthling.net.

SPECTRO-CHROME THERAPEUTIC SYSTEM

ATTRIBUTES OF ATTUNED COLOR WAVES

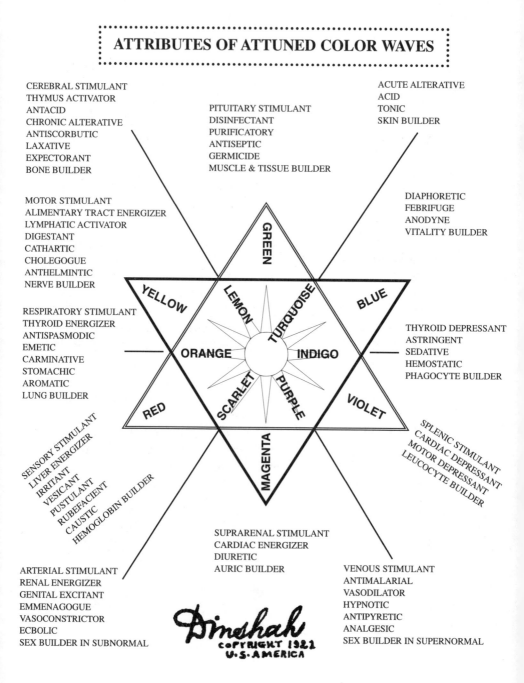

CEREBRAL STIMULANT
THYMUS ACTIVATOR
ANTACID
CHRONIC ALTERATIVE
ANTISCORBUTIC
LAXATIVE
EXPECTORANT
BONE BUILDER

PITUITARY STIMULANT
DISINFECTANT
PURIFICATORY
ANTISEPTIC
GERMICIDE
MUSCLE & TISSUE BUILDER

ACUTE ALTERATIVE
ACID
TONIC
SKIN BUILDER

MOTOR STIMULANT
ALIMENTARY TRACT ENERGIZER
LYMPHATIC ACTIVATOR
DIGESTANT
CATHARTIC
CHOLEGOGUE
ANTHELMINTIC
NERVE BUILDER

DIAPHORETIC
FEBRIFUGE
ANODYNE
VITALITY BUILDER

RESPIRATORY STIMULANT
THYROID ENERGIZER
ANTISPASMODIC
EMETIC
CARMINATIVE
STOMACHIC
AROMATIC
LUNG BUILDER

THYROID DEPRESSANT
ASTRINGENT
SEDATIVE
HEMOSTATIC
PHAGOCYTE BUILDER

GREEN

YELLOW LEMON TURQUOISE BLUE

ORANGE INDIGO

RED SCARLET PURPLE VIOLET

MAGENTA

SENSORY STIMULANT
LIVER ENERGIZER
IRRITANT
VESICANT
PUSTULANT
RUBEFACIENT
CAUSTIC
HEMOGLOBIN BUILDER

SPLENIC STIMULANT
CARDIAC DEPRESSANT
MOTOR DEPRESSANT
LEUCOCYTE BUILDER

SUPRARENAL STIMULANT
CARDIAC ENERGIZER
DIURETIC
AURIC BUILDER

ARTERIAL STIMULANT
RENAL ENERGIZER
GENITAL EXCITANT
EMMENAGOGUE
VASOCONSTRICTOR
ECBOLIC
SEX BUILDER IN SUBNORMAL

VENOUS STIMULANT
ANTIMALARIAL
VASODILATOR
HYPNOTIC
ANTIPYRETIC
ANALGESIC
SEX BUILDER IN SUPERNORMAL

Dinshah
COPYRIGHT 1921
U.S.AMERICA

24 The Neurophone

Patrick Flanagan and Gael-Crystal Flanagan

In the early 1960s, while only a teenager, Life *magazine listed Patrick Flanagan as one of the top scientists in the world. Among his many inventions was a device he called the Neurophone—an electronic instrument that can successfully programme suggestions directly through contact with the skin. When he attempted to patent the device, the government demanded that he prove it worked. When he did, the NSA (National Security Agency) confiscated the Neurophone. It took [Flanagan] years of legal battle to get his invention back.*

When I was fifteen years old, I gave a lecture at the Houston Amateur Radio Club, during which we demonstrated the Neurophone to the audience. The next day we were contacted by a reporter from the *Houston Post* newspaper. He said that he had a relative who was nerve-deaf from spinal meningitis and asked if we might try the Neurophone on his relative. The test was a success. The day after that, an article on the Neurophone as a potential hearing aid for the deaf appeared and went out on the international wire services.

The publicity grew over the next two years. In 1961, *Life* magazine came to our house and lived with us for over a week. They took thousands of photographs and followed me around from dawn to dusk. The article appeared in the 14 September 1962 issue. After that, I was invited to appear on the *I've Got a Secret* show hosted by Gary Moore. The show was telecast from the NBC studios in New York. During the show, I placed electrodes from the Neurophone on the lower back of Bess Meyerson while the panel tried to guess what I was doing to her. She was able to "hear" a poem that was being played through the Neurophone electrodes. The poem was recorded by Andy Griffith, another guest on the show. Since the signal from the Neurophone was only perceived by Bess Meyerson, the panel could not guess what I was doing to her.

History of the Neurophone

The first Neurophone was made when I was 14 years old, in 1958. A description was published in our first book, *Pyramid Power.* The device was constructed by attaching two Brillo pads to insulated copper wires. Brillo pads are copper wire scouring pads used to clean pots and pans. They are about two inches in diameter. The Brillo pads were inserted into plastic bags that acted as insulators to prevent electric shock when applied to the head.

The wires from the Brillo pads were connected to a reversed audio output transformer that was attached to a hi-fi amplifier. The output voltage of the audio transformer was about 1,500 volts peak-to-peak. When the insulated pads were placed on the temples next to the eyes and the amplifier was driven by speech or music, you could "hear" the resulting sound inside your head. The perceived sound quality was very poor, highly distorted and very weak.

I observed that during certain sound peaks in the audio driving signal, the sound perceived in the head was very clear and very loud. When the signal was observed on an oscilloscope while listening to the sound, the signal was perceived as being loudest and clearest when the amplifier was overdriven and square waves were generated. At the same time, the transformer would ring or oscillate with a dampened wave form at frequencies of 40–50 kHz.

The next Neurophone consisted of a variable frequency vacuum tube oscillator that was amplitude-modulated. This output signal was then fed into a high frequency transformer that was flat in frequency response in the 20–100 kHz range. The electrodes were placed on the head and the oscillator was tuned so that maximum resonance was obtained using the

As a result of the *Life* magazine article and the exposure on the Gary Moore Show, we received over a million letters about the Neurophone.

The patent office started giving us problems. The examiner said that the device could not possibly work, and refused to issue the patent for over twelve years. The patent was finally issued after my patent lawyer and I took a working model of the Neurophone to the patent office. This was an unusual move since inventors rarely bring their inventions to the patent examiner. The examiner said that he would allow the patent to issue if we human body as part of the tank circuit. Later models had a feedback mechanism that automatically adjusted the frequency for resonance. We found that the dielectric constant of human skin is highly variable. In order to achieve maximum transfer of energy, the unit had to be retuned to resonance in order to match the "dynamic dielectric response" of the body of the listener.

The 2,000 volt peak-to-peak amplitude-modulated carrier wave was then connected to the body by means of two-inch diameter electrode discs that were insulated by means of mylar films of different thicknesses. The Neurophone is really a scalar wave device since the out-of-phase signals from the electrodes mix in the non-linear complexities of the skin dielectric. The signals from each capacitor electrode are 180 degrees out of phase. Each signal is transmitted into the complex dielectric of the body where phase cancellation takes place. The net result is a scalar vector. Of course I did not know this when I first developed the Neurophone. This knowledge came much later when we learned that the human nervous system is especially sensitive to scalar signals.

The high frequency amplitude-modulated Neurophone had excellent sound clarity. The perceived signal was very clearly perceived as if it were coming from within the head. We established quite early that some totally nerve-deaf people could hear with the device. But for some reason, not all nerve-deaf people hear with it the first time.

We were able to stimulate visual phenomena when the electrodes were placed over the occipital region of the brain. The possibilities of Neurophonic visual stimulation suggest that we may someday be able to use the human brain as a VGA monitor!

I wrote my own patent application with the help of a friend and patent attorney from Shell Oil Company and submitted the application to the patent office.

could make a deaf employee of the patent office hear with the device. To our relief, the employee was able to hear with it and, for the first time in the history of the patent office, the Neurophone file was reopened and the patent was allowed to issue.

After the Gary Moore Show, a research company known as Huyck Corporation became interested in the Neurophone. I believed in their sincerity and allowed Huyck to research my invention. They hired me as a consultant in the summer months. Huyck was owned by a very large and powerful Dutch paper company with offices all over the world.

At Huyck I met two friends who were close to me for many years, Dr. Henri Marie Coanda, the father of fluid dynamics, and G. Harry Stine, scientist and author. Harry Stine wrote the book, *The Silicon Gods* (Bantam), which is about the potential of the Neurophone as a brain/computer interface.

Huyck Corporation was able to confirm the efficacy of the Neurophone but eventually dropped the project because of our problems with the patent office.

The next stage of Neurophone research began when I went to work for Tufts University as a research scientist. In conjunction with a Boston-based corporation, we were involved in a project to develop a language between man and dolphin. Our contracts were from the U.S. Naval Ordinance Test Station out of China Lake, California. The senior scientist on the project was my close friend and business partner, Dr. Dwight Wayne Batteau, Professor of Physics and Mechanical Engineering at Harvard and Tufts.

In the Dolphin Project we developed the basis for many potential new technologies. We were able to ascertain the encoding mechanism used by the human brain to decode speech intelligence patterns, and were also able to decode the mechanism used by the brain to locate sound sources in three dimensional space. . . . These discoveries led to the development of a 3-D holographic sound system that could place sounds in any location in space as perceived by the listener.

We also developed a man-dolphin language translator. The translator was able to decode human speech so that complex dolphin whistles were generated. When dolphins whistled, the loudspeaker on the translator would output human speech sounds. We developed a joint language between ourselves and our two dolphins. The dolphins were located in the lagoon of a small island off of Oahu, Hawaii. We had offices at Sea Life Park and Boston. We commuted from Boston to Hawaii to test out our various electronic gadgets.

We recorded dolphins and whales in the open sea and were able to accurately identify the locations of various marine mammals by 3-D sound-localization algorithms similar to those used by the brain to localize sound in space.

The brain is able to detect phase differences of two microseconds. We were able to confirm this at Tufts University. The pinnae or outer ear is a "phase-encoding" array that generates a time-ratio code that is used by the brain to localise the source of sounds in 3-D space. The localization time ratios are run from two microseconds to several milliseconds. A person with one ear can localize sound sources (non-linear) to a 5 degree angle of accuracy anywhere in space. You can test this by closing your eyes while having a friend jingle keys in space around your head. With you eyes closed you can follow the keys and point to them very accurately. Try to visualize where the keys are in relation to your head. With a little practice, you can accurately point directly at the keys with your eyes closed. If you try to localize a sine wave, the experiment will not work. The signal must be non-linear in character. You can localize the sine wave if the speaker has a nonlinear or distortion in the output wave form. A sine wave cannot be localized because phase differences in a sine wave are very hard to detect. The brain will focus on the distortion and use it to measure time ratios. Clicks or pulses are very easy to localize.

If you distort your pinnae by bending the outer ears out of shape, your ability to localize the sound source is destroyed. The so-called cocktail party effect is the ability to localize voices in a noisy party. This is due to the brain's ability to detect phase differences and then pay attention to localized areas in 3-D space. A favorite "intelligence" trick is to have sensitive conversations in "hard rooms" with wooden walls and floors. A microphone "bug" will pick up all the echoes and this will scramble the voice. Almost all embassies contain "hard rooms" for sensitive conversations. If you put a microphone in the room with a duplicate of the human pinnae on top of it, you will be able to localize the speakers and tune out the echoes—just like you were at a party.

In order to localize whales and dolphins under water, we used metal ears 18 inches in diameter that were attached to hydrophores. When these ears were placed under water, we were able to accurately localize underwater sounds in 3-D space by listening to the sounds by earphones. We used this system to localize whales and dolphins. Sound travels five times faster under water, so we made the "pinnae" larger to give the same time-ratio encoding as we find in the air. We also made large plastic ears that were tested in Vietnam. These ears were of the same proportions as real ears but were much larger. They enabled us to hear distant sounds with a high degree of localization accuracy in the jungle. It seems that we can adapt to ears of almost any size. The reason we can do this is because sound recognition is based on a time-ratio code.

We were able to reverse the process and could take any sound recording and encode it so that sounds were perceived as coming from specific points in space. Using this technique, we could spread out a recording of an orchestra. The effect added reality as if you were actually listening to a live concert. This information has never been used commercially except in one instance when I allowed The Beach Boys to record one of their albums with my special "laser" microphones.

We developed a special Neurophone that enabled us to "hear" dolphin sounds up to 250 thousand Hertz. By using the Neurophone as part of the man-dolphin communicator, we were able to perceive more of the intricacies of the dolphin language. The human ear is limited to a 16 kHz range, while dolphins generate and hear sounds out to 250 kHz. Our special Neurophone enabled us to hear the full range of dolphin sounds.

As a result of the discovery of the encoding system used by the brain to localize sound in space and also to recognize speech intelligence, we were able to create a digital Neurophone.

When our digital Neurophone patent application was sent to the patent office, the Defense Intelligence Agency slapped it under a secrecy order. I was unable to work on the device or talk about it to anyone for another five years.

This was terribly discouraging. The first patent took twelve years to get, and the second patent application was put under secrecy for five years.

The digital Neurophone converts sound waves into a digital signal that matches the time encoding that is used by the brain. These time signals are used not only in speech recognition but also in spatial recognition for the 3-D sound localization.

The digital Neurophone is the version that we eventually produced and sold as the Mark XI and the Thinkman Model 50 versions. These Neurophones were especially useful as subliminal learning machines. If we play educational tapes through the Neurophone, the data is very rapidly incorporated into the long-term memory banks of the brain.

HOW DOES IT WORK?

The skin is our largest and most complex organ. In addition to being the first line of defence against infection, the skin is a gigantic liquid crystal brain.

The skin is piezoelectric. When it is vibrated or rubbed, it generates electric signals and scalar waves. Every organ of perception evolved from the skin. When we are embryos, our sensory organs evolved from folds in the skin. Many primitive organisms and animals can see and hear with their skin.

When the Neurophone was originally developed, neurophysiologists considered that the brain was hard-wired and that the various cranial nerves were hard-wired to every sensory system. The eighth cranial nerve is the nerve bundle that runs from the inner ear to the brain. Theoretically, we should only be able to hear with our ears if our sensor organs are hard-wired. Now the concept of a holographic brain has come into being. The holographic brain theory states that the brain uses a holographic encoding system so that the entire brain may be able to function as a multiple-faceted sensory encoding computer. This means that sensory impressions may be encoded so that any part of the brain can recognize input signals according to a special encoding. Theoretically, we should be able to see and hear through multiple channels.

The key to the Neurophone is the stimulation of the nerves of the skin with a digitally encoded signal that carries the same time-ratio encoding that is recognised as sound by any nerve in the body.

All commercial digital speech recognition circuitry is based on so-called dominant frequency power analysis. While speech can be recognised by such a circuit, the truth is that speech encoding is based on time ratios. If the frequency power analysis circuits are not phased properly, they will not work. The intelligence is carried by phase information. The frequency content of the voice gives our voice a certain quality, but frequency does not contain information. All attempts at computer voice recognition and voice generation are only partially successful. Until digital time-ratio encoding is used, our computers will never be able to really talk to us.

The computer that we developed to recognize speech for the man-dolphin communicator used time-ratio analysis only. By recognising and using time-ratio encoding, we could transmit clear voice data through extremely narrow bandwidths. In one device, we developed a radio transmitter that had a bandwidth of only 300 Hz while maintaining crystal clear transmission. Since signal-to-noise ratio is based on band width considerations, we were able to transmit clear voice over thousands of miles while using milliwatt power.

Improved signal-processing algorithms are the basis of a new series of Neurophones that are currently under development. These new Neurophones use state-of-the-art digital processing to render sound information much more accurately.

ELECTRONIC TELEPATHY?

The Neurophone is really an electronic telepathy machine. Several tests prove that it bypasses the eighth cranial nerve or hearing nerve and transmits sound directly to the brain. This means that the Neurophone stimulates perception through a seventh or alternate sense.

All hearing aids stimulate tiny bones in the middle ear. Sometimes when the eardrum is damaged, the bones of the inner ear are stimulated by a vibrator that is placed behind the ear on the base of the skull. Bone conduction will even work through the teeth. In order for bone conduction to work, the cochlea or inner ear that connects to the eighth cranial nerve must function. People who are nerve-deaf cannot hear through bone conduction because the nerves in the inner ear are not functional.

A number of nerve-deaf people and people who have had the entire inner ear removed by surgery have been able to hear with the Neurophone.

If the Neurophone electrodes are placed on the closed eyes or on the face, the sound can be clearly "heard" as if it were coming from inside the brain. When the electrodes are placed on the face, the sound is perceived through the trigeminal nerve.

We therefore know that the Neurophone can work through the trigeminal or facial nerve. When the facial nerve is deadened by means of anaesthetic injections, we can no longer hear through the face.

In these cases, there is a fine line where the skin on the face is numb. If the electrodes are placed on the numb skin, we cannot hear it but when the electrodes are moved a fraction of an inch over to skin that still has feeling, sound perception is restored.

This proves that the means of sound perception via the Neurophone is by means of skin and not by means of bone conduction.

There was an earlier test performed at Tufts University that was designed by Dr. Dwight Wayne Batteau, one of my partners in the U.S. Navy Dolphin Communications Project. This test was known as the "Beat Frequency Test." It is well known that sound waves of two slightly different frequencies create a "beat" note as the waves interfere with each other. For example, if a sound of 300 Hz and one of 330 Hz are played into one ear at the same time, a beat note of 30 Hz will be perceived. This is a mechanical summation of sound in the bone structure of the inner ear. There is another beat phenomenon known as the binaural beat. In the binaural beat, sounds beat together in the corpus callosum in the center of the brain. This binaural beat is used by Robert Monroe of the Monroe Institute to stimulate altered states. That is, to entrain the brain into high alpha or theta states.

The Neurophone is a powerful brain-entrainment device. If we play alpha or theta signals directly through the Neurophone, we can entrain any brain state we like. In a future article we will tell how the Neurophone has been used as a subliminal learning device and also as a behavior modification system.

Batteau's theory was that if we could place the Neurophone electrodes so that the sound was perceived as coming from one side of the head only, and if we played a 300 Hz signal through the Neurophone, if we also played a 330 Hz signal through an ordinary headphone we would get a beat note if the signals were summing in the inner ear bones.

When the test was conducted, we were able to perceive two distinct tones without a beat. This test again proved that Neurophonic hearing was not through the means of bone conduction.

When we used a stereo Neurophone, we were able to get a beat note that is similar to the binaural beat, but the beat is occurring inside the nervous system and is not a result of bone conduction.

The Neurophone is a "gateway" into altered brain states. Its most powerful use may be in direct communications with the brain centers, thereby bypassing the "filters" or inner mechanisms that may limit our ability to communicate to the brain.

If we can unlock the secret of direct audio communications to the brain, we can unlock the secret of visual communications. The skin has receptors that can detect vibration, light, temperature, pressure and friction. All we have to do is stimulate the skin with the right signals.

We are continuing Neurophonic research. We have recently developed other modes of Neurophonic transmission. We have also reversed the Neurophone and found that we can detect scalar waves that are generated by the living system. The detection technique is actually very similar to the process used by Dr. Hiroshi Motoyama in Japan. Dr. Motoyama used capacitor electrodes very much like those we use with the Neurophone to detect energies from the various chakras.

An example of the secrecy order that enables a government to confiscate a patent.

U.S. DEPARTMENT OF COMMERCE
PATENT OFFICE
WASHINGTON, D.C.   20231

Serial No.   756,124   Filed   Aug. 29, 1968

Applicant   Gillis P. Flanagan

Title   METHOD AND SYSTEM FOR SIMPLIFYING SPEECH WAVEFORMS

SECRECY ORDER
(Title 35, United States Code (1952), sections 181-188)

NOTICE: To the applicant above named, his heirs, and any and all his assignees, attorneys and agents, hereinafter designated principals.

You are hereby notified that your application as above identified has been found to contain subject matter, the unauthorized disclosure of which might be detrimental to the national security, and you are ordered in no wise to publish or disclose the invention or any material information with respect thereto, including hitherto unpublished details of the subject matter of said application, in any way to any person not cognizant of the invention prior to the date of the order, including any employee of the principals, but to keep the same secret except by written consent first obtained of the Commissioner of Patents, under the penalties of 35 U.S.C. (1952) 182, 186.

Any other application already filed or hereafter filed which contains any significant part of the subject matter of the above identified application falls within the scope of this order. If such other application does not stand under a secrecy order, it and the common subject matter should be brought to the attention of the Security Group, Licensing and Review, Patent Office.

If, prior to the issuance of the secrecy order, any significant part of the subject matter has been revealed to any person, the principals shall promptly inform such person of the secrecy order and the penalties for improper disclosure. However, if such part of the subject matter was disclosed to any person in a foreign country or foreign national in the U.S., the principals shall not inform such person of the secrecy order, but instead shall promptly furnish to the Commissioner of Patents the following information to the extent not already furnished: date of disclosure; name and address of the disclosee; identification of such part; and any authorization by a U.S. Government agency to export such part. If the subject matter is included in any foreign patent application, or patent this should be identified. The principals shall comply with any related instructions of the Commissioner.

This order should not be construed in any way to mean that the Government has adopted or contemplates adoption of the alleged invention disclosed in this application; nor is it any indication of the value of such invention.

EDWIN L. REYNOLDS
First Assistant Commissioner

POL-86 (7-68)   USCOMM-DC 64772-P68

An example of the secrecy order that enables a government to confiscate a patent.

Section III

The Suppression of UFO Technologies and Extraterrestrial Contact

The UFO question carries with it the baggage of centuries of speculation as to the very nature of who we are and how we got here. It calls into question conventionally accepted theories of evolution, and has the potential to unravel and invalidate many of our most cherished beliefs in the supremacy of the human species on the evolutionary ladder. Reports of sightings and abductions have unearthed new quandaries concerning government involvement with alien visitors, and the extensive cover-up of these stories by various governments. Clearly the human race has a propensity for avoiding and/or denying uncomfortable information. We attempt to support our fractured and failing paradigms with what, at best, can be considered an amazing display of obstinacy.

It is no wonder that the Brookings Institute was commissioned in the 1950s to study and report on the implications to the social fabric of the revelation that we are "not alone." Sadly, the possibility of extraterrestrial life has highlighted the persistence of the various scientific establishments in cozying up to military money and perquisites at the expense of their science, their ethics, and in the end, their self-respect. Those skeptics curious enough to investigate may find that the UFO information unfolding in the past few decades has been very, very carefully managed so as to achieve the desired results.

And what might these results be? The logic behind any institution seems to be self-perpetuation, and it comes as no surprise that the institutions charged with managing UFO and extraterrestrial information are tasked with a virtual information hot potato. Despite the supposed existence of an alien culture making its presence known and felt, our dominant terrestrial institutions would make it appear as though we are alone in an otherwise lifeless universe. It would not be presumptuous to assume that there has in fact been some sort of bargain or treaty negotiated between the alien visitors and the major Earth powers to perpetuate this appearance. Such a pact might allow the present power hierarchies to continue to operate (with some new limitations and guidelines, of course) while at the same time essentially doing the work they've been assigned. It seems incredible that the fate of all beings of earth may already have been decided without our knowledge or consent, due to the censorship and top secret classification of sensitive information.

Stanton Friedman, UFO researcher and nuclear physicist, has publicly stated that "whoever controls alien technology rules the world." While there is abundant evidence that the U.S. government has had access to some alien technology for several decades, there is no doubt that the ones who really control alien technology are the aliens themselves. Regardless of this, much time and effort has gone into the research and development of these same innovations right here on earth. T. Townsend Brown was succeeding with his antigravity work in the 1940s and 1950s and, in fact, experimentation of this sort has been ongoing for at least seventy years. Indeed, there are many serious researchers who now believe that the U.S. moon landing program was accomplished with the aid of antigravity machines, and there is more than a little reason to believe them. (The lack of a blast-off exhaust from the lunar lander on Apollo 11, for example, is one intriguing bit of evidence for this assertion.) Yet it is only recently that the news of a successful antigravity breakthrough is being allowed to be published in mainstream physics journals.

UFO research is big news these days, and the news is coming fast and thick. I have chosen some of the most revealing stories I have been able to find to illustrate the extent of the cover-up of information concerning UFOs and the principles on which they operate. Avoidance and outright denial have made it possible for our governments to hide the truth thus far. It is interesting to speculate how this knowledge, made public, is likely to alter our perspective.

25 Breakthrough as Boffins Beat Gravity

Scientists in Finland are about to reveal details of the world's first antigravity device. Measuring about 30 cm across, it is said to reduce significantly the weight of anything suspended over it.

The claim—which has been rigorously examined by scientists and is due to appear this month in a physics journal—could spark a technological revolution. By combating gravity—the most ubiquitous force in the universe—everything from transport to power generation, could be transformed.

NASA, the United States space agency, is taking the claims seriously and is financing research into how the antigravity effect could be turned into a means of flight.

The researchers at the Tampere University of Technology in Finland, who discovered the effect, say it could form the heart of a new power source in which it is used to drive fluids past electricity-generating turbines.

Other uses seem limited only by the imagination:

- Lifts in buildings could be replaced by devices built into the ground. People wanting to go up would simply activate the antigravity device—making themselves weightless—and with a gentle push ascend to the floor they want.

- Space travel would become routine, as all the expense and danger of rocket technology is geared towards combating the Earth's gravitational pull.

- By using the devices to raise fluids against gravity and then conventional gravity to pull them back to Earth against electricity-generated turbines, the devices could also revolutionize power generation.

According to Dr. Eugene Podkletnov, who led the research, the discovery was accidental.

From *The New Zealand Star Times,* September 22, 1996

It emerged during routine work on so-called "superconductivity," the ability of some materials to lose their electrical resistance at low temperatures.

The team was carrying out tests on a rapidly spinning disc of superconducting ceramic suspended in the magnetic field of three electrical coils, all enclosed in a low-temperature vessel called a cryostat.

"One of my friends came in and he was smoking his pipe," said Dr. Podkletnov. "He put some smoke over the cryostat and we saw that the smoke was going to the ceiling all the time. It was amazing—we couldn't explain it."

Tests showed a small drop in the weight of objects placed above the device, as if it were shielding the object from the effects of gravity—an effect deemed impossible by most scientists.

The team found that even the air pressure above the device dropped slightly, with the effect detectable directly above the device on every floor above the laboratory.

What makes this claim different from previous "antigravity" devices scorned by the establishment is that it has survived intense scrutiny by skeptical, independent experts and has been accepted for publication by the *Journal of Physics-D: Applied Physics*, published by Britain's Institute of Physics.

Even so, most scientists will not feel comfortable with the idea of antigravity until other teams repeat the experiments.

The Finnish team is already expanding its programme to see if it can amplify the antigravity effect.

In its latest experiments, the team has measured a 2-percent drop in the weight of objects suspended over the device—and double that if one device is suspended over another.

If the team can increase the effect substantially, the commercial implications are enormous.

26 Antigravity on the Rocks: The T. T. Brown Story

Jeane Manning

T. Townsend Brown was jubilant when he returned from France in 1956. The soft-spoken scientist had a solid clue which could lead to fuelless space travel. His saucer-shaped discs flew at speeds of up to several hundred miles per hour, with no moving parts. One thing he was certain of—the phenomena should be investigated by the best scientific institutions. Surely now the science establishment would admit that he really had something. Although the tall, lean physicist—handsome, in a gangly way—was a humble man, even shy, he confidently took his good news to a top-ranking officer he knew in Washington, D.C.

"The experiments in Paris proved that the anomalous motion of my disc airfoils was not all caused by ion wind." The listener would hear Brown's every word, because he took his time in getting words out. "They conclusively proved that the apparatus works even in high vacuum. Here's the documentation . . . "

Anomalous means unusual—a discovery which does not fit into the current box of acknowledged science. In this case, the anomaly revealed a connection between electricity and gravity.

That year *Interavia* magazine reported that Brown's discs reached speeds of several hundred miles per hour when charged with several hundred thousand volts of electricity. A wire running along the leading edge of each disc charged that side with high positive voltage, and the trailing edge was wired for an opposite charge. The high voltage ionized air around them, and a cloud of positive ions formed ahead of the craft and a cloud of negative ions behind. The apparatus was pulled along by its self-generated gravity field, like a surfer riding a wave. *Fate* magazine writer Gaston Burridge in 1958 also described Brown's metal discs, some up to 30 inches in diameter by that time. Because they needed a wire to supply electric charges, the discs were tethered by a wire to a Maypole-like mast. The double-saucer objects circled the pole with a slight humming sound. "In the dark they glow with an eerie lavender light."

Instead of congratulations on the French test results, at the Pentagon he again ran into closed doors. Even his former classmate from officers' candidates school, Admiral Hyman Rickover, discouraged Brown from continuing to explore the dogma-shattering discovery that the force of gravity could be tweaked or even blanked out by the electrical force.

"Townsend, I'm going to do you a favor and tell you: Don't take this work any further. Drop it."

Was this advice given to Brown by a highly-placed friend who knew that the United States military was already exploring electrogravitics? (Recent sleuthing by American scientist, Dr. Paul La Violette, uncovers a paper trail which leads from Brown's early work, toward secret research by the military and eventually points to "Black Project" air craft.)

HARASSMENT

Were the repeated break-ins into Brown's laboratory meant to discourage him from pursuing his line of research?

Brown didn't quit, although by that time he and his family had spent nearly $250,000 of their own money on research. He had already put in more than thirty years seeking scientific explanations for the strange phenomena he witnessed in the laboratory. He earlier called it electrogravitics, but later in his life, trying to get acknowledgement from establishment scientists, he stopped using the word "electrogravitics" and instead used the more acceptable scientific terminology "stress in dielectrics."

No matter what his day job, the obsessed researcher experimented in his home laboratory in his spare time. Above all he wanted to know "Why is this happening?" He was convinced that the coupling of the two forces —electricity and gravity—could be put to practical use.

An arrogant academia ignored his findings. Given the cold-shoulder treatment by the science establishment, Brown spent family savings and even personal food money on laboratory supplies. Perhaps he would not have had the heart to continue his lonely research if he had known in 1956 that nearly thirty more years of hard work were ahead of him. He died in 1985 with the frustration of having his findings still unaccepted.

The last half of his career involved new twists. Instead of electrogravitics, at the end of his life he was demonstrating "gravitoelectrics" and "petrovoltaics"—electricity from rocks. Brown's many patents and findings ranged from an electrostatic motor to unusual high-fidelity speakers and electrostatic cooling, to lighter-than-air materials and advanced dielectrics. His name should be recognized by students of science, but instead it has dropped into obscurity.

Too late to comfort him, some leading-edge scientists of the mid-1990s are now resurrecting Brown's papers. Or what they can find of his papers.

EXTRAORDINARY CURIOSITY

Thomas Townsend Brown was born March 18, 1905, to a prominent Zanesville, Ohio, family. The usual child-like "Why?" questions came from young Townsend with extraordinary intensity. For example, his question "Why do the (high voltage) electric wires sing?" led him later in life to an invention.

His discovery of electrogravitics, on the other hand, came through an intuition. As a sixteen-year-old, Townsend Brown had a hunch that the then-famous Coolidge X-ray tube might give a clue to spaceflight technology. His tests, to find a force in the rays themselves which would move mass, lead to a dead end. But in the meantime the observant experimenter noticed that high voltages applied to the tube itself caused a very slight motion.

Excited, he worked on increasing the effect. Before he graduated from high school, he had an instrument he called a gravitator. "Wow," the teenager may have thought. "Antigravity may be possible!" World-changing technological discoveries start with someone noticing a small effect and then amplifying it.

Unsure of what to do next, the next year he started college at California Institute of Technology. Even then his sensitivity was evident, because he saw the wisdom of going forward cautiously—first gaining respect from his professors instead of prematurely bragging about his discovery of a new electrical principle. He was respected as a promising student and an excellent laboratory worker, but when he did tell his teachers about his discovery they were not interested. He left school and joined the Navy.

Next he tried Kenyon College in Ohio. Again, no scientist would take his discovery seriously. It went against what the professors had been taught; therefore it could not be.

He finally found help at Dennison University in Gambier, Ohio. Townsend met Professor of physics and astronomy Paul Alfred Biefeld, Ph.D., who was from Zurich, Switzerland and had been a classmate of Albert Einstein. Biefeld encouraged Brown to experiment further, and together they developed the principle which is known in the unorthodox scientific literature as the Biefeld-Brown Effect. It concerned the same notion which the teenager had seen on his Coolidge tube—a highly charged electrical condenser moves toward its positive pole and away from its negative pole. Brown's gravitator measured weight losses of up to one percent. (In 1974 researcher Oliver Nichelson pointed out to Brown that before 1918, Professor Francis E. Nipher of St. Louis discovered gravitational propulsion by electrically charging lead balls, so the Brown-Biefeld Effect could more properly be called the Nipher Effect. However,

Brown deserves credit for his sixty years of experimentation and developing further aspects of the principle.)

Brown's 1929 article for the publication *Science and Inventions* was titled bluntly, "How I Control Gravity." The science establishment still turned its back. By then he had graduated from the university, married, and was working under Professor Biefeld at Swazey Observatory.

His career in the early 1930s also included a post at the Naval Research Lab in Washington, D.C.; staff physicist for the Navy's International Gravity Expedition to the West Indies; physicist for the Johnson-Smithsonian Deep Sea Expedition; and soil engineer for a federal agency and administrator with the Federal Communications Commission.

As his country's war effort escalated, he became a Lieutenant in the Navy Reserve and moved to Maryland as a materials engineer for the Martin aircraft company. Brown was then called into the Navy Bureau of Ships. He worked on how to degauss (erase magnetism from) ships to protect them from magnetic-fuse mines, and his magnetic minefield detector saved many sailors' lives.

PHILADELPHIA EXPERIMENT

The "Philadelphia Experiment" which Brown may or may not have joined in 1940 is dramatized in a popular movie as a military experiment in which United States Navy scientists are trying to demagnetize a ship so that it will be invisible to radar. According to the account, the ship and its crew dematerialized and rematerialized—became invisible and later returned from another dimension.

Whatever the Project Invisibility experiment actually was, Brown was probably an insider, as the Navy's officer in charge of magnetic and acoustic mine-sweeping research and development. However, later in life, Brown was said to be mute on the topic of the alleged Philadelphia Experiment, except for brief disclaimers. He told friend Josh Reynolds of California, who made arrangements for Brown's experiments in the early 1980s, that the movie and the controversial book *The Philadelphia Experiment*, by William L. Moore and Charles Berlitz, were greatly inflated. He apparently did not elaborate on that comment.

Reynolds spoke on a panel discussion at a public conference (dedicated to Townsend Brown) in Philadelphia in 1994, along with highly-credentialed physicist Elizabeth Rauscher, Ph.D. Rauscher theorized that the Philadelphia Experiment legend grew out of the fact that certain magnetic fields can in effect "degauss the brain"—cause temporary memory loss. If the huge electrical coils involved in degaussing a ship were mistuned, the sailors could have felt that they "blinked out of time and back into time."

Blinking this account back to 1942: Townsend Brown was made commanding officer of the Navy's radar school at Norfolk, Virginia. The next year he collapsed from nervous exhaustion and retired from the Navy on doctors' recommendations. More than his hard work caused his health to break down, he had suffered years of deeply-felt disappointments because his life's work—the gravitator—had not been recognized by scientific institutions which could have investigated it. The final precipitating factor for his collapse was an incident involving one of his men.

BREAK-IN AT PEARL HARBOR

After he recuperated for six months, his next job was as a radar consultant with Lockheed-Vega. He later left the California aircraft corporation, moved to Hawaii and was a consultant at the Navy yard at Pearl Harbor. An old friend who was teaching calculus there had opened some doors, and in 1945 Brown demonstrated his latest flying tethered discs to a top military officer—Admiral Arthur W. Radford, commander-in-chief for the U.S. Pacific Fleet, who later became Joint Chief of Staff for President Dwight Eisenhower.

Brown was treated with respect because of who he was, but again no one signed up to help investigate his discovery. His colleagues in the Navy treated it lightly because it was anomalous.

When he returned to his room after the Pearl Harbor demonstration, however, the room had been broken into and his notebooks were gone. A day or so later, as Josh Reynolds remembers Brown's account of the incident, "they came to him and said 'we have your work; you'll get it back.' A couple of days later they gave him back his books and said 'we're not interested.'"

"Why?" Brown was given the answer that the effect was a result of ion propulsion, or electric wind, and therefore could not be used in a vacuum such as outer space. The earth's atmosphere can be rich in ions (electrically-charged particles), but a vacuum is not.

He was disgruntled, but not stopped. Later a study funded by a French government agency would prove the effect was not caused by "electric wind." But even before that, Brown knew that it would take an electric hurricane to create the lifting force he saw in his experiments.

Project Winterhaven was his own effort for furthering electrogravitic research. He began the project in 1952 in Cleveland, Ohio. Although he demonstrated two-feet-diameter disk-shaped transducers which reached a speed of 17 feet per second when electrically energized, he was again met with lack of interest. Alone in his enthusiasm, he watched the craft fly in a 20 foot diameter circle around a pole. According to the known laws of science, this should not be happening. And he went on to make spectacular demonstrations.

When La Societe Nationale de Construction Aeronautique Sud Ouest

(SNCASO) in France offered him funding, he went to France and built better devices as well as had them properly tested. Those tests convinced his backers that it could mean a feasible drive system for outer space, he told Reynolds. SNCASO merged with Sud Est in 1956 and funding was cut, so Brown had to return to the United States.

Brown was eager to show the French documentation to all those officials who had raised the wall of indifference in the past. But after his discouraging visit to Washington, D.C. in 1956 and what felt like a put-down from Admiral Rickover, he apparently decided "if the military isn't interested, the aerospace companies might be." Friends say it did not occur to him to ask if the defense industry was already working on electrogravitics, unknown to him. In 1953 he had flown saucer-shaped devices of three feet in diameter in a demonstration for some Air Force officials and men from major aerospace companies. Energized with high voltage, they whizzed around the 50 foot diameter course so fast that the reports of the test were stamped "classified."

Independent researcher Paul LaViolette, Ph.D., traces the path which these impressive results led to—toward the Pentagon, the military hub of the United States. "A recently declassified Air Force intelligence report indicates that by September of 1954 the Pentagon had launched a program to develop a manned antigravity craft of the sort suggested in Project Winterhaven," writes LaViolette.

Meanwhile, Brown went practically door-to-door in Los Angeles to try to rouse some interest in his work. One day he returned to his laboratory to find it had been broken into and much of his belongings were missing.

CHARACTER ASSASSINATION

Then the nasty rumours started. The type of rumours which can discredit a man's character, upset his wife and children, and overall cause deep distress to a sensitive man.

Another tragedy in Brown's life was the sudden death of his friend and helpful supporter, Agnew Bahnson, who funded him to do anti-gravity research and development beginning in 1957 in North Carolina. Did they make too much progress? In 1964 Bahnson, an experienced pilot, mysteriously flew into electric wires and crashed. Bahnson's heirs dissolved the project.

The authors of the book *The Philadelphia Experiment* wrote that in spite of his numerous patents and demonstrations given to governmental and corporate groups, success eluded Townsend Brown. "Such interest as he was able to generate seemed to melt away almost as fast as it developed—almost as if someone . . . was working against him."

Today's researchers looking at Townsend Brown's life have noticed that he went into semi-retirement some time in the 1960s. Tom Valone of Washington, D.C., who in 1994 compiled a book on Brown's work, speculates that the work was classified and Brown was bought off or somehow persuaded to stop promoting electrogravitics. Valone told the April, 1994, meeting in Philadelphia that Dr. LaViolette's detective work sheds new light on what happened to Brown in the 1950s. The speculation of these scientists is that "this project was taken over by the military, worked on for 40 years, and we now have a craft that's flying around." Valone speculates that Brown was de-briefed and told what he could talk about.

In the later 1960s to 1985, Brown turned his attention to other research, although related. He mainly did basic research to try to understand strange effects he saw. As did T. H. Moray, Townsend Brown had decided that waves coming from outer space are not only detected on Earth, but also the waves build up a charge in a properly built device. Instead of making increasingly-complex devices, however, Brown toward the end of his life in the 1980s was getting a charge—voltage to be exact—out of rocks and sand. It was all in search for answers.

If his work had been accepted instead of suppressed by seeming disinterest, he would be known to science students. His work would fill more than one science book; an encyclopedia set could easily be filled with T. T. Brown's experiments and discoveries.

For example, his childhood fascination with the singing wires led him to investigate how to modulate ionized air like that which had carried the high-voltage current. Could this be used for high-fidelity sound systems? Eventually he did invent rich-sounding Ion Plasma Speakers which incidently had a built-in "fac"—a cool breeze of health-enhancing negative ions. Would this discovery have been commercialized if his main interest, electrogravitics, had not been suppressed by ignorance or been co-opted?

He searched for better dielectrics, endlessly trying new combinations. (A dielectric is any material which opposes the flow of electric current while at the same time can store electrical energy.) This search led him to study, when working with Bahnson, the lighter-than-air fine sand, in certain dry river beds, which could be used to make advanced materials. The anomalous sands were first discovered by his hero Charles Brush early in the century. Brush also found that certain materials fell slower in a vacuum chamber than others. He called it gravitational retardation and said they were slightly more interactive with gravity. These materials also spontaneously demonstrated heat. Brush believed that the "etheric gravitational wave" interacted with some materials more richly than with others. Brush's findings were swept under the rug of the science establishment.

Brown followed his idol's lead and did basic research in a number of areas. Gravito-electrics—how neutrinos or gravitons or whatever-they-are converted into electricity. This led him to conduct experiments in various locations, from the ocean to the bottom of the Berkeley mine shaft.

When entrepreneur Josh Reynolds became interested in Brown's work in the last five years of the inventor's life, Brown was able to do the work he loved the most—petrovoltaics. No one else was putting electrodes on rocks to measure the minute voltages of electricity which the rocks somehow soaked up from the cosmos. Brown and Reynolds made artificial rocks to see what various materials could do and how long they would put out a charge.

Their efforts in a number of areas led toward what they called a ForeverReady Battery—a penny-sized piece of rock which put out a tiny amount of voltage indefinitely because they had learned how to "soup-up" the effect. After Brown died, Reynolds carried on the research until funding ran out. He estimates that it would have taken up to $10 million of advanced molecular engineering research to take the discovery to another stage of development. The high-power version of the battery remains on paper—only theory until developed farther.

This discovery alone should have put Brown into science history books. In all his years of experiments with the periodic variations in the strip-chart recordings of the output from the materials, he found that the patterns had a relationship to position of the stars. And orientation toward the centre of the universe seemed to make a difference too. This resulted in further unconventional thinking that only made Brown more of an outcast in the world of sanctioned science.

While he was coming up with the cosmic findings, the military researchers had a different agenda. One of the reports dug up by researcher LaViolette came from a London think tank called Aviation Studies International Ltd. In 1956 the think tank wrote a classified "confidential" survey of work done in electrogravitics. LaViolette says the only original copy of the document, called Report 13, was found in the stacks at Wright-Patterson Air Force Base technical library in Dayton, Ohio. It is not listed in the library's computer.

Excerpts from Report 13 paint a picture of heavy secrecy. A 1954 segment says that infant science of electro-gravitation may be a field where not only the methods are secret, but also the ideas themselves are a secret. "Nothing therefore can be discussed freely at the moment." A further report predicted bluntly that electrogravitics, like other advanced sciences, would be developed as a weapon.

A couple of months later, another now declassified Aviation Report said it looked like the Pentagon was ready to sponsor electrogravitic propulsion devices and that the first disc should be finished by 1960. The report anticipated that it would take the decade of the 60s to develop it properly "even though some combat things might be available ten years from now."

Defense contractors began to line up, as well as universities who get grants from the U.S. Department of Defense.

After he came across Report 13, LaViolette put his knowledge of physics to work and began to piece together a picture of what may have happened in the past thirty years. It includes "black" projects—work which the military decides should be so secretive that even Congress does not get reports about its funding.

A breakthrough in LaViolette's quest for the pieces of the picture came when a few establishment scientists gave out tidbits of formerly-secret information about a "black funding" project—the Stealth B-2 bomber. (The B-2A is described as the world's most expensive aircraft at $1.2 billion.) Their description of the B-2 gave LaViolette and others a number of clues about the bomber—softening of the sonic boom as Brown had talked about in the 1950s, a dielectric flying wing, a charged leading-edge, ions dumped into the exhaust stream and other clues. The B-2 seems to be a culmination of many of Brown's observations made more than forty years ago.

Townsend Brown fought an uphill battle all his adult life, at great cost to himself and to family life. His cause included getting the science of advanced propulsion out into public domain, not hidden behind the Secrecy Act and a wall of classified documents. He died feeling that he had lost the battle.

27 Did NASA Sabotage Its Own Space Capsule?

From *NASA Mooned America!* by René

THE RIGHT STUFF

The Seven Samurai is a 1954 Japanese cult movie about a poverty stricken village that hired seven magnificent warriors to help them fight the bandits. In 1960 Hollywood filmed *The Magnificent Seven* which was effectively the same story set in Mexico as a western. Someone in the hierarchy of NASA had undoubtedly seen one or both movies and decided that seven space samurai was a psychologically appropriate number to start with. We were told that these men represented the nation's finest and that they possessed what was later called that elusive quality: the "Right Stuff."

Virgil Grissom certainly had the "Right Stuff." He was one of the original seven, culled from the first batch of military test pilots almost a decade before. Grissom was not the type of man who "went along to get along." Men who spend their lives seeking the wild hairs on a new airplane's ass seldom are. He was a professional test pilot, a mechanical engineer and had flown 100 combat missions in Korea. But he was dead before his flight to the Moon could fulfill his dream.

ACCIDENTS

Compared to civilian test pilots the astronauts were underpaid. However, their perks were impressive. Their celebrity status instantly conferred upon them all the bonuses usually associated with show business stardom. Each night on the town provided them with all the young women they could handle, plus free drinks in every bar in the country. They were also given a government jet trainer as a personal toy.

Test pilots have a hazardous occupation which probably sees as many fatalities per unit of time as do men in combat. However, before the first Apollo manned flight ever cleared the launching pad, eleven astronauts died in accidents. Grissom, Chaffee, and White were cremated in an Apollo capsule test on the launching pad during a completely and suspiciously unnecessary test. Seven died in six air crashes: Freemen, Basset and See, Rogers, Williams, Adams and Lawrence. Givens was killed in a car crash.

When you reflect on their deaths in the light of the three-man-instant crematorium one wonders. Add the fact that there were eight deaths in 1967 alone. One wonders if these "accidents" weren't NASA's way of correcting mistakes and saying that some of these men really didn't have the "Right Stuff."

After 1967, only Taylor died in another plane crash in 1970. An actuarial statistician would probably go berserk over these numbers considering how small the group was. Another weighty factor, even though they were "hot" pilots, the astronauts flew their trainer jets only part time. And add to that the fact that trainers are usually inherently safer than other planes in the same class. It would raise his eyebrows to find how few of these men would ever enter space.

I can't help but wonder what technicians serviced their ships—because what we have here is an appalling "accident" rate. They were the finest professional pilots in the world, operating government planes where costs have little meaning. Yet they died. Even if we call the cremation an accident we still have five more "accident" deaths in one year. Very interesting! I also wonder what the death rate was among the other NASA employees who were in position to know too much?

THE PRELIMINARIES

The first American in space was Alan Shepard, followed by Grissom and then Glenn. I'm convinced that every Mercury flight was real and that the phony missions only started after Grissom's Gemini 3. And even some of the later Gemini flights were real which leaves most of the original astronauts smelling like a rose. Unfortunately, Wally Schirra and NASA General Tom Stafford's Gemini 6A flight, with its miracle of an undamaged antenna, turned the rosy aroma into real toilet water. So did Alan Shepard's little golf game on the Moon during the Apollo 14 mission.

All of these men barely entered near space (near-Earth-orbit) which I define as any altitude less than 500 miles. Far space I reserve for those interstellar journeys that may come during the next millennium. That is, if we can solve our planetary problems before we dissolve in the stew created by the Four Horsemen of the Apocalypse: War, Famine, Plague, Pestilence. And add a fifth "horseman," Religious Fanaticism, which frequently causes the other four.

Every other "race" involving aircraft, from hot air balloons through rocket planes, entailed serious efforts to go higher and faster than the other guy. For good technical reasons neither we or the Russians played that game. To this day our Shuttle flights are limited to very near space usually well under 200 miles in altitude.

Most writers on the Apollo Program either totally ignored, or played down, the fact that by early January 67, Grissom, was no longer a happy camper. He was very disenchanted with both NASA and the prime capsule contractor, North American Aviation. This company had a phoenix-like ability to weather every storm, including the fire on Pad 34. It ultimately combined with Rockwell Engineering to become North American Rockwell.

GRISSOM'S LEMON

North American Rockwell's first Apollo capsule had been delivered and accepted by NASA in August 66, with a flight date set for November. But time after time the date had to be reset because of problems with the craft. "Grissom, a veteran of two test flights in Mercury and Gemini, normally quiet and easy-going, a flight pro, could not hide his irritation. 'Pretty slim' was the way he put his Apollo's chances of meeting its mission requirements."[1]

According to Mike Gray, "Grissom had a sense of unease about this flight. He told his wife, Betty, 'If there ever is a serious accident in the space program, it's likely to be me.'"[2] We will never know if this statement was the result of a psychic premonition or a burgeoning fear of our government.

Early in January 67, Grissom, probably unaware that NASA had other internal critics, hung a lemon on the Apollo capsule. He was threatening to go public with his complaints.[3] He was already a popular celebrity, especially with the press. He would have had no problem in getting his story out. In a case like this even NASA's censors would have had little control over the news. Headlines like "Popular Astronaut Rips Into NASA!!" couldn't be easily squelched.

SPACE RADIATION

NASA also had another serious problem, besides being in a space race with the Russian Bear. This problem derived from our first answer to the Sputniks. On January 31, 1958, Explorer 1 lifted into orbit. It weighed a mere 18.3 pounds and carried a geiger counter which dutifully reported that a belt of intense radiation surrounded the Earth.

The belt was subsequently named after the Explorer Project Head, James A. Van Allen. However, the radiation was first predicted by Nikola Tesla around the beginning of this century as the result of experimental and theoretical work he had done on electricity in space in general and the electrical charge of the Sun in particular. He tried then to tell our academic natural philosophers (scientists) that the Sun had a fantastic electrical charge and that it must generate a solar wind. But to no avail. The experts knew he was crazy. It would take almost sixty years to prove him right.

However, predicting something is not the same as discovery so the discovery of our magnetic girdle of radiation rightfully belongs to the man who was suspicious enough to put a geiger counter on board the satellite, whichever technician actually thought of it.

Subsequent study showed that this belt, or belts, began in near space about 500 miles out and extends out to over 15,000 miles. Since the radiation there is more or less steady it obviously must receive as much radiation from space as it loses. If not it would either increase until it fried the Earth or decay away to nothing. Van Allen belt radiation is dependent upon the solar wind and is said to focus or concentrate that radiation. However, since it can only trap what has traveled to it in a straight line from the Sun there remains a dangerous question: how much more radiation can there be in the rest of solar space?

The Moon does not have a Van Allen belt. Neither does it have a protective atmosphere. It lies nakedly exposed to the full blast of the solar wind. Were there a large solar flare during any one of the Moon missions massive amounts of radiation would scour both the capsules and the Moon's surface where our astronauts gamboled away the day. The question is worse than dangerous—it's lethal!

In 1963 the Russian space scientists told the famous British astronomer, Bernard Lovell that they "could see no immediate way of protecting cosmonauts from the lethal effects of solar radiation."[4] This had to mean that not even the much thicker metal walls used on the Russian capsules could stop this radiation. How could the very thin metal—almost foil—we used on our capsules stop the radiation? NASA knew that. Space monkeys died in less than ten days but NASA never revealed their cause of death.

Most people, even those interested in space, are still unaware that killer radiation pulses through space. I believe our ignorance was caused by the people who sell us space sagas. Sitting in front of me is a 9-x-12-inch coffee table book titled *The Illustrated Encyclopedia of Space Technology*, printed in 1981. The words "Space Radiation" just do not exist on any of its almost 300 pages. In fact with the dual exceptions of Bill Mauldin's *Prospects for Interstellar Travel* published in 1992 and *Astronautical Engineering and Science* written by early NASA experts, no other book I have read even begins to discuss this extremely serious impediment to space flights. Do I detect the fine hand of my democratic government at work?

The Russians were in a position to know because as early as the spring of 61 their probes had been sent to the backside of the Moon. Upon his return to England Lovell sent this information to NASA's deputy administrator, Hugh Dryden. Dryden, representing NASA obviously ignored it!

Collins spoke of space radiation in only two places in his book. He said "At least the moon was well past the earth's Van Allen belts, which promised a healthy dose of radiation to those who passed and a lethal dose to those who stayed."[5]

In speaking of ways to dodge problems he wrote, "In similar fashion, the Van Allen Radiation belts around the earth and the possibility of solar flares require understanding and planning to avoid exposing the crew to an excessive dose of radioactivity."[6]

So what does "understanding and planning" mean? Does it mean that after the Van Allen Belts are passed that the rest of space is free of radiation? Or did NASA have a strategy for dodging solar flares once they were committed to the trip?

It seems to imply that back in 1969 it was possible to predict solar flares. My astronomy text has this to say on that subject "It is accordingly possible to predict only approximately the date of the future maximum and how plentiful the groups will then become."[7] This text was ten years old by 1969. Later in this book I will show that nothing had changed during the years of Apollo Moon missions.

To continue with the Apollo Program after receiving this information implies that NASA knew something the Russians didn't. Either we had developed an effective extremely light weight radiation shield or NASA already knew that no one was going any where near the Moon.

Could the cloth in our space suits stop the radiation? I doubt that because more than fifteen years have passed since the partial core meltdown at TMI (Three Mile Island) and workers still can't enter the containment dome. We don't yet have the technology to create light weight flexible radiation shielding. High velocity could get the capsule through the Van Allen belt but what could they do about solar flares during the rest of the trip to the Moon? And if we didn't go, why didn't the Soviets, our arch enemies, rat us out?

While I was thinking about this something rang a bell. Around the time we were fighting communism in Vietnam (and other countries in southeast Asia) we began to sell Russia, later to be called the Evil Empire, wheat by the mega-ton at an ultra-cheap price.

On July 8, 1972 our government shocked the entire world by announcing that we would sell about one-fourth of our entire crop of wheat to Russia at a fixed price of $1.63 per bushel. According to these sources we were about to produce another bumper crop while the Russian crop would be 10–20 percent less. The market price at the time of the announcement was $1.50 but immediately soared to a new high of $2.44 a bushel.[8]

Guess who paid the 91 cents difference in price for the Russians? Our bread prices and meat prices were immediately inflated reflecting the suddenly diminished supply. It was the beginning of the high inflation of the 70s. Now how much did the Moon cost us? Would our government be a party to blackmail? Nah!

However, if NASA knew that Kennedy's dream was impossible in the time frame given, they should have reported this to the President. We are civilized now and no longer cut off the right arm of the messenger who brings bad news. Now we cut off budgets! That's safer for the messenger but fatal to the bureaucracy in question.

NASA must have decided if they couldn't make it they would fake it. Big bucks were at stake here, to say nothing of American prestige. Those bucks, properly funneled, would buy a lot of southeast Asia, at least for awhile. And with proper prestidigitation some of the same could wind up in numbered accounts handled either by the "gnomes of Zurich" or offshore Caribbean banks.

NASA'S OTHER PROBLEM

NASA's second problem was magnified as a result of the first. If they were really going to land on the Moon they would be able to take great quantities of real photos and pick up genuine Moon rocks. Such pictures should include the Earth rising or setting against a background of a bona fide starry sky.

However, if they weren't actually going to the Moon, the evidence would have to be synthesized. Credible proof was vital to the continued high rate of funding and to NASA's very survival. NASA's labs could create "Moon rocks" to the specifications of an educated, or rather an expected, guess that would pass any inspection, because there wasn't anything else to compare them to.

Or they could have used rock samples picked up in Antarctica during the intensive exploration of that continent during the International Geophysical Year in 1957. They would do as well provided there were no fossils in them. These rocks could be slowly doled out, but only to those geologists who could be counted on to agree with anything the government said. And most of academia can be relied on to do just that!

Strangely enough rocks were later found in Antarctica that closely resemble "Moon rocks." In point of fact, some geologists are now positive that these rocks were blasted from the Moon to Earth during immense meteoric impacts.

However, true-to-the-Moon photos posed a bit more of a problem. Because the twentieth century is the age of increasingly sophisticated photography, huge amounts of tape and film had to be expended. NASA seemed to do precisely that. As Harry Hurt put it, ". . . Project Apollo was one of the most extensively documented undertakings in human history . . ."[9]

Despite this alleged fact and the fact that the NASA Apollo mission photo numbers seem to indicate that thousands of pictures were taken, we keep seeing the same few dozen pictures in all the books on space.

Using the well developed art of Hollywood style special effects (FX) the astronauts could be photographed "on the Moon" in the top secret studio set up near Mercury, Nevada. Of course, there is a bit more to great FX than having the best equipment. As in any art form, the artists are always more important than their tools. The backbone of superb FX is lodged in the Hollywood professionals who devote their lives to it. Lacking access to these relatively liberal experts NASA was forced to use CIA hacks . . . relative amateurs.

Nevertheless, they did their job well enough to pass casual inspection for many years. It worked only because we wanted to believe! As long as we had something to hang our hats on we could continue to have faith and ignore the anomalies in the evidence the photos provided. It worked . . . for a while!

GRISSOM'S FINAL MISTAKE

At the time of his death Grissom was one of NASA's old-timers. He was the man who, a few short years before, certified that the astronauts had been involved in every step of the program and had been free to criticize at will, and even suggest ideas for improvements. He was the man whose fatal error was no more than in being who he was; an independent thinker . . . a free spirit who seemed to be completely unaware that NASA had wholeheartedly opted to enact the second part of the old saying, "If you can't make it, fake it!"

He had been selected as Commander of Apollo I, the first manned flight of the Apollo series. Grissom's crew included Edward H. White and Roger B. Chaffee. White flew on Gemini 4 but Chaffee was a newcomer who had not as yet been in space, or verified the NASA rite of passage by condemning the visibility of stars and planets.

THE HANDICAP

Right from the beginning, NASA was operating under a tremendous handicap. They were in a space race with a nation who, they knew, had operational rockets that made ours seem like tinker toys by comparison. The Soviets started their space program in capsules that were 50 times heavier than those we were launching six months later.

Russian capsules were closer to being compressed air tanks than flimsy space capsules. Their ships had sufficient wall strength to maintain normal atmospheric pressure inside the craft against the zero pressure outside in space. However, since we didn't have rockets to lift that sort of weight, we couldn't afford this luxury. We had to make light, [almost] tin foil, capsules just to get into the ball game.

The differential in pressure between the 14.7 psi (our normal atmospheric pressure) and the zero pressure of space amounts to 2116 pounds per square foot of outward loading on the enclosing wall of a capsule. Compare this figure with the floor of a house—which is designed to be safely loaded to only 30 pounds per square foot—and you will realize that relatively heavy metal is vital for skin and skeleton if you want to enjoy normal pressure. It is wall strength that prevents catastrophic and explosive depressurization of small capsules. The LEM's walls will be discussed in more detail later in the book.

BREATHING MIXTURES

The greater lifting capacity of their rockets allowed the Russians the luxury of using a mixture of 20 percent oxygen and 80 percent nitrogen—the equivalent to regular air. Naturally it wasn't stored on board as bulky "compressed air." It was stored separately as liquids in cryogenic tanks. However, the nitrogen supply was smaller since the gas is inert to the human body and additional nitrogen is required only to help reestablish pressure when the cabin is vented to space. Oxygen tanks were larger because the only oxygen used is that small portion converted into CO_2, by the necessity of breathing and this is immediately removed from the cabin by chemicals. A great deal is also lost when the cabin is vented to space during depressurization.

PURE OXYGEN

Lacking strong walled capsules, NASA decided right from the beginning to use 50 percent oxygen and 50 percent nitrogen at 7 psi. This specification was changed in August 1962, into the use of pure oxygen at 5 psi.[10]

A policy shift of this nature indicates that approved design of the capsules being manufactured was weaker than expected. The amazing thing is that NASA made this deadly decision despite testing that usually ended in disaster. One would think that after testing showed disaster that one would never implement a dangerous policy. But NASA was in a race with destiny. They had no time for common sense.

NASA TESTS

Here is a list of all government sponsored testing that resulted in oxygen fires. This information was extracted from Appendix in *Mission To The Moon* written by Kennan & Harvey:

September 9, 1962—The first known fire occurred in the Space Cabin Simulator at Brooks Air Force Base in a chamber using 100 percent oxygen at 5 psi. It was explosive and involved the CO_2 scrubber. Both occupants collapsed from smoke inhalation before being rescued.

November 17, 1962—Another incident using 100 percent oxygen at 5 psi in a chamber at the Navy Laboratory (ACEL). There were four occupants in the chamber, but the simple replacing of a burned-out light bulb caused their clothes to catch on fire. They escaped in 40 seconds but all suffered burns. Two were seriously injured. In addition an asbestos "safety" blanket caught fire and burned causing one man's hand to catch fire.

July 1, 1964—This explosion was at an AIResearch facility when they were testing an Apollo cabin air temperature sensor. No one was injured. The composition of the atmosphere and pressure isn't listed, but we have to assume 100 percent oxygen (and possible pressure equal to atmospheric).

February 16, 1965—This fire killed two occupants at the Navy's Experimental Diving Unit in Washington, D.C. The oxygen was at 28 percent and the pressure at 55.6 psi. The material in the chamber apparently supported extremely rapid combustion, driving the pressure up to 130 psi.

April 13, 1965—Another explosion as AIResearch was testing more Apollo equipment. Again, neither pressure or atmospheric composition is given but a polyurethane foam cushion exploded.

April 28, 1966—More Apollo equipment was destroyed as it was being tested under 100 percent oxygen and 5 psi at the Apollo Environmental Control System in Torrance, CA.

January 1, 1967—The last known test was over three weeks before Grissom, Chaffee and White suffered immolation. Two men were handling 16 rabbits in a chamber of 100 percent oxygen at 7.2 psi at Brooks Air Force Base and all living things died in the inferno. The cause may have been as simple as a static discharge from the rabbits' fur . . . but we'll never know.

Of course, NASA's moronic decision to use pure oxygen would play a crucial part in the deadly fire on Pad 34 a few years later. Never mind that the test was classified as "non hazardous" by NASA. Only after Grissom, White and Chaffee died in that fire, would NASA again change the specs to either 60–40 or 50–50 oxygen/nitrogen mixes at 5 psi, depending on what source I've read.[11]

In pure oxygen at normal pressure even a piece of steel wool will burn rapidly. In fact, Michael Collins claims that even stainless steel will burn.[12] As mentioned already an asbestos blanket, normally classed as fireproof, was consumed when used to smother flames during an oxygen fire.[13] Pure oxygen is extremely hazardous!

To successfully switch to reduced pressure breathing of pure oxygen one must first purge the body of nitrogen. This prevents residual nitrogen left in the body from forming small bubbles which expand from the decreasing pressure. To deep sea divers this is known as "the bends." To avoid this lethal hazard astronauts must spend some period of time breathing 100 percent oxygen—which is medically dangerous— at full atmospheric pressure just before the mission.

The pressure problem in a space capsule is [analogous] to those encountered in a submarine. Submarine hulls are deliberately strong, to resist the increasing pressure at depth. If a submarine hull was as thin as our space crafts—at 200 feet deep it would require an internal pressure of 100 psi—at 300 feet a pressure is 150 psi.

PRESSURE TESTING

The Apollo Program command capsules must be regarded as flimsy, even though they were built of titanium which has the strength of steel and weighs half as much. I reason that if our capsules were too weak to withstand normal pressure they must also have been too weak to keep the atmosphere from crushing the capsule on the launching pad. If this was true they had to be using 100 percent oxygen at normal pressure during the launch.

It was found out that this is precisely what NASA did on all their launches. It is obvious that the present Shuttles, with 50 tons of cargo capacity, could use normal pressure and regular air. However, the designers may still begrudge the few pounds of extra material in the cabin that it takes to do this. By the same token our large diameter commercial airliners are able to maintain almost regular atmospheric pressure, and don't have to resort to pure oxygen, even when flying over 40,000 feet. Neither does the SST which hits altitudes of 60,000 feet.

To insure the integrity of the capsule NASA subjected it to their pressure test. One would assume that they would use compressed air for this test because the electric panels had power and live men were inside the unit. However, when it came time to test the 012 capsule on Pad 34 it was decided to use pure oxygen at a pressure somewhat above our atmospheric pressure of 14.7 psi. What the actual pressure was is confusing. It was either 16.7 psi according to Michael Collins, or 20.2 psi as reported by Frank Borman.[14]

One would think that intelligent men with the "Right Stuff" would precisely know the pressures used. But either way, there were astronauts locked inside—practicing for their first Apollo mission. After the accident NASA claimed the test was SOP (Standard Operating Procedure). In either case an idiot was in charge.

If it was SOP, then the idiot was the official who instituted and approved this test program. If not, then it was the low level idiot in direct charge of the test who gave the order to proceed. I have no fear of a libel suit because of this accusation. The only legal defense in a libel suit is whether what you said was the truth, as determined by a jury. If you were on a jury and watched steel wool explode in a 16.7 psi 100 percent oxygen atmosphere what would you decide?

I find it hard to believe that this test was SOP. In fact, I suspect that it wasn't, simply because two men with the "Right Stuff" can't agree. NASA telling us after the fire that it was always done that way, doesn't prove a thing. NASA, like all political organizations, can always be counted on to say anything to better their position. Using pure oxygen at this pressure, once the panels were live, means that every launch was always one small spark away from disaster. Combustion in 100 percent oxygen even at low pressures, is extremely rapid. At higher pressures it becomes explosive!

HIGH PRESSURE OXYGEN

Consider this standard procedure: Burning a substance using high pressure oxygen is precisely the method used to determine the number of calories in that substance. The test procedure requires placing the sample in a strong steel pressure vessel called a "Calorimeter Bomb." The "Bomb" is placed in an insulated container of water holding a known quantity of water at a known temperature. There is an electrical sparking device inside the bomb and sufficient high pressure oxygen is added to insure complete combustion of the material.

Even relatively wet foodstuffs are quickly reduced to ashes once the electric spark initiates combustion. This process produces high pressures in the steel chamber. That's why it's called a Calorimeter Bomb. The heat transfers to the surrounding water and the rise in temperature using known parameters results in the quantity of calories (energy) in the substance tested.

To get back to the discussion, every time an electric switch is thrown the induction of the electric current causes a tiny spark to jump between the two switch contacts. If the unit is explosion proof (like the switches motors, and lighting fixtures used in hazardous or explosive locations), that spark is safely enclosed in a hermetically sealed container. If not anything near it that is combustible can burn.

In standard electrical switches the electrical insulation is some form of plastic (hydrocarbon). All hydrocarbons can be oxidized if there is sufficient oxygen and heat to raise the temperature of some small portion of that substance beyond the flash point. Bear in mind that an electric spark is a plasma. Indeed the temperature at the core of a large spark can be so high it is indeterminable.

SPONTANEOUS COMBUSTION

The phenomena we call spontaneous combustion is also oxidation. Under normal conditions oxygen in the air begins to oxidize almost any material. In fact what we call rust on metal is supposed to be very slow oxidation. If the material is insulated to any degree, the heat created by the process cannot escape as fast as it is generated. So the entrapped heat creates a small temperature rise which increases the rate of oxidation. If some or all of that increased heat cannot escape there is a self-escalating "loop." The temperature continues to rise until the flash point is reached. At that point the material concerned bursts into flame. That's "spontaneous" combustion.

In an atmosphere containing a higher percentage of oxygen or a higher pressure the oxidation rate is greatly increased. It is well known that a pile of oily rags in an oxygen environment will burst into flame. In 100 percent oxygen any hydrocarbon or carbohydrate becomes potential fuel needing only a small spark or increase in heat to set it off.

THE TEST

On January 27, 1967 astronauts Grissom, White and Chaffee approached Pad 34 where an obsolete model of the command capsule had been installed on top of an unfueled Saturn 1B rocket.[15] This was the same type of rocket that had carried the smaller and lighter Gemini capsules. The capsule itself was already outmoded and would be replaced before any Apollo missions were launched.

However this was a full "dress" rehearsal. But somebody neglected to tell the maintenance people to clean out all the extremely combustible extraneous construction materials. The urgency of this test was simply that they were scheduled for a manned mission that had been repeatedly postponed. As we will see later, NASA had every intention of sending Apollo I, Grissom's mission, into space even though neither that Saturn V (actual moon rocket), nor the Apollo capsule, had ever actually been tested in space.

Would you not have smelled a rat? Perhaps Grissom was a bit worried because he got Wally Schirra to ask Joe Shea, NASA's chief administrator, to go through that with him. *"Grissom still wanted Shea to be with him in the spacecraft.*"[16] Shea refused because NASA couldn't patch in a fourth headset in time for the test. Is that likely?

It is difficult to believe that this couldn't have been done in the 24 hour time frame available. If I had a crew of technicians who couldn't install another headset-jack in that amount of time I'd fire the whole damn crew.

The original Apollo capsule had different hatches, but by 1300 hours all three astronauts were strapped in their acceleration couches with the new hatches sealed behind them. It was later revealed that these hatches were so poorly designed that even with outside help and in a non-emergency situation, it took seven or eight minutes to open them. They were originally supposed to spend a few hours practicing throwing the proper switches at the right time in sequential response to computer simulations. However, with delay piled upon delay and everyone in a hurry, each time a switch was thrown, unnoticed by any, tiny sparks jumped.

During the test of the Apollo capsule on Pad 34 Grissom and his crew were in 100 percent oxygen simulating the real thing. In fact they reported a burning smell a few times earlier that day. When that happened technicians would come with "sniffers," open the hatches, but find nothing. One wonders if the review board considered that these hatch openings flushed out the smell with the fresh air admitted by opening the hatch. These incidents delayed the test and time was running out.[17] The extraneous combustible materials may have been combining with the pressurized oxygen each time pure oxygen refilled the cabin. Oxidation makes heat, and if you stop the process that heat remains in the material. Each time you repressurize the craft the combustible material will be at a slightly higher temperature. I sense that the board of review missed this angle.

I also feel that spontaneous combustion would have been much too subtle for the CIA. If it was a CIA hit they would have done it with an electric squib or incendiary device wired to a switch programmed to be thrown toward the end of the test.

While the testing was going on, some mastermind in mission control decided to save some time. In his wisdom that unknown leader made the decision to speed up the testing. As the board of inquiry later noted, *"To save time, the space agency took a short cut."* What he did was simply order the capsule to be pressurized with 100 percent oxygen at either 16.7 or 20.2 psi. Notice, that no name was used. The entire agency takes the blame.

I have great difficulty in believing that apparently not one of these rocket scientists in Control, nor the astronauts themselves, knew that a Calorimeter Bomb consists of a combustible material, pressurized oxygen and a spark. These were highly educated men, men with technical degrees, men who had taken chemistry courses, and men who must have spent some time around welding and cutting torches that used oxygen.

Also I cannot understand why Grissom et al entered that capsule in the first place if they knew it was to be pressurized with oxygen over 14.7 psi. For example in a hospital no one is allowed to smoke in a room where oxygen is in use. In this situation we have only a small section of a room with tiny amounts of low pressure oxygen being used. Yet everyone seems to know of the danger. Grissom was a test pilot and engineer while both White and Chaffee had degrees in aeronautical engineering. Apparently not one of them complained. Didn't anyone know about Calorimeter Bombs? Didn't NASA send them copies of the fire reports? Or maybe no one told them they were jacking up the pressure!

At 1745 hours (5:45 P.M.) Grissom was getting angry with the communication people for a static filled on again-off again communication system. At one point he ragged them *"How do you expect to get us to the moon if you people can't even hook us up with a ground station? Get with it out there."*[18]

In the meantime around 6 P.M. Collins had to attend a general meeting of the astronauts. Let Collins tell you about it in an incredible single paragraph: [19]

On Friday, January 27, 1967, the astronaut office was very quiet and practically deserted, in fact. Al Shepard, who ran the place, was off somewhere, and so were all the old heads. But someone had to go to the Friday staff meeting, Al's secretary pointed out, and I was the senior astronaut present, so off I headed to Slayton's office, note pad in hand, to jot down another week's worth of trivia. Deke wasn't there either, and in his absence, Don Gregory, his assistant presided. We had just barely gotten started when the red crash phone on Deke's desk rang. Don snatched it up and listened impassively. The rest of us said nothing. Red phones were a part of my life, and when they rang it was usually a communications test or a warning of an aircraft accident or a plane aloft in trouble. After what seemed like a very long time, Don finally hung up and said very quietly, "Fire in the spacecraft." That's all he had to say. There was no doubt about which spacecraft (012) or who was in it (Grissom, White, Chaffee) or where (Pad 34 Cape Kennedy) or why (a final systems test) or what (death, the quicker the better). All I could think of was My God, such an obvious thing and yet we hadn't considered it. We worried about engines that wouldn't start or wouldn't stop; we worried about leaks; we even worried about how a flame front might propagate in weightlessness and how cabin pressure might be reduced to stop a fire in space. But right here on the ground, when we should have been most alert, we put three guys inside an untried spacecraft, strapped them into couches, locked two cumbersome hatches behind them, and left them no way of escaping a fire. Oh yes, if a booster caught fire, down below, there were elaborate if impractical, plans for escaping the holocaust by sliding down a wire, but fire inside a spacecraft itself simply couldn't happen. Yet it had happened, and why not? After all, the 100 percent oxygen environment we used in space was at least at a reduced pressure of five pounds per square inch, but on the launch pad the pressure was slightly above atmospheric, or nearly 16 psi. Light a cigarette in pure oxygen at 16 psi and you will get the surprise of your life as you watch it turn to ash in about two seconds. With all those oxygen molecules packed in there at that pressure, any material generally considered "combustible" would instead be almost explosive.

Here Collins reported that the pressure was 16 psi. Other authors went higher. A staff meeting at 6 P.M. on Friday night? Do you have a feeling that this Friday night staff meeting was the first and last in the long history of our government bureaucracies?

THE FIRE

At 6:31:03 P.M., one of the astronauts smelled smoke and yelled fire. The capsule had suddenly turned into a Calorimeter Bomb. They tried their best to open the hatch. Without panic the triple hatch that sealed them in usually took about nine minutes to open. They didn't have nine minutes. In fact, they barely had ninety seconds before their suits burned through and the deadly poisonous gasses released from the plastics silenced them forever.

The capsule's internal pressure soared from the great quantity of hot gasses created by the quasi-explosive burning of all the combustible material. This short term fire was so intense that it melted a silver soldered joint on the oxygen feed pipe pouring even more oxygen into the conflagration.

At 6:31:17, fourteen seconds from the first smell of smoke, the pressure reached 29 pounds and the capsule ruptured, effectively releasing the heat and damping the fire. But it was too late. They were already dead.

Let me put in some additional questions here. If this was not murder and just an example of extreme stupidity in governmental slow motion why did government agents in rapid action, raid Grissom's home before anyone knew about the fire? Why did they remove all his personal papers and his diary? Why didn't they bring his diary, or any other paper with the word "Apollo" on it back, when they returned some of his personal papers to his widow? And if it really took 29 psi to blow the cabin why didn't they use regular air at higher pressure?

Also was it really the vicissitudes of life that the outward opening hatch was coincidentally changed that very morning to one that opened inward? An inward opening hatch meant that any inside pressure, acting outward, would prevent it from being opened—even if someone was standing by, which they weren't. It was also bolted up from the outside and lacked explosive bolts.[20]

THE AFTERMATH

NASA should have known better. And they did! You have read earlier of the men injured in flash explosive fires in their own tests. NASA had even commissioned a report by Dr. Emanuel M. Roth which was published in 1964. Dr. Roth cited difficulties with 100 percent oxygen atmospheres even under low pressures. Any competent engineer should have known the dangers of oxygen at 16.7 or 20.2 psi. This is why I cannot believe that this was "standard operating procedure," or that Grissom and his crew knew that about it. NASA not only ignored their own tests on pure low pressure oxygen but upped the ante by increasing the pressure above atmospheric.

Kennan and Harvey had this to say, "Most U.S. scientists could not believe their ears when they learned that fact. Oxygen at such pressure comes in the category of an 'oxygen bomb.'"[21]

A Board of Inquiry termed "The Apollo 204 Review Board" was quickly convened to investigate the fatal fire by appointing astronaut Frank Borman as the chairman. In effect, NASA sent the fox into the chicken house to investigate mysterious disappearances of the occupants. The board's final report was about what you might expect when an inhouse investigation investigates itself: "One key to the caution which reveals itself on every page of the Board's report is that it was written by government employees. Thompson himself was director of the space agency's Langley Research center, and no fewer than six of the eight Board members were NASA officials."[22]

The pressure of 16.7 psi is quoted from *Journey to Tranquility* where the authors wrote that they learned the pressure of the pure oxygen in the capsule was 2 psi over atmospheric. Collins reported it as nearly 16 psi. It seems strange that NASA told two insiders, Borman and Collins, plus the authors of *Tranquility* three different capsule pressures? Apparently NASA, like the rest of us find it almost impossible to keep all the little white lies straight. And if it's a group lie we get the results shown in this book.

Borman writes that, "We brought in every learned mind we could enlist—including a chemistry expert from Cornell . . ."[23] Didn't this *expert* know that oxygen has a deep and forceful desire to breed little oxides by passionately mating with hydrocarbons and carbohydrates? Didn't this "so-called" expert tell them that?

Borman, played dumb when he was called before Congress. In testifying under oath he said, "*None of us were fully aware of the hazard that existed when you combine a pure-oxygen atmosphere with the extensive distribution of combustible materials and a likely source of ignition . . . and so this test . . . was not classified as hazardous.*"[24] And if Borman was as unaware of all the dangerous fires that erupted during NASA's own safety tests over the years why did he later write about 20.2 psi oxygen in this manner, *"That is an extremely dangerous environment, the equivalent of sitting on a live bomb, waiting for someone to light the fuse."*[25]

Aldrin in his 1989 book, *Men From Earth* written twenty-two years after the cremation has this to say, "As every high school chemistry student learns, when a smoldering match is put into a beaker of oxygen, it blazes into a spectacular flame."[26]

He (Aldrin) continues by telling us how there was a multitude of switches and miles of electrical wiring all of which were easy to short and could act as a match. "But the risk was considered acceptable because, in space, the astronauts could instantly depressurize their cabin . . ."[27] Hey Buzz, didn't you claim that the reason your EVA [extravehicular activity] on the Moon was late in starting because it took so long to vent the last of the oxygen from the LEM?

Say what? Borman, who held a Masters in engineering and taught thermodynamics at West Point claims nobody was aware of the danger! After all these years Aldrin now claims he knew. Obviously, either Borman is lying or Aldrin didn't have the guts to open his mouth.

When Deke Slayton was asked about the pressure test he reportedly blurted out, "Man, we've just been lucky. We've used the same test on everything we've done with the Mercury and the Gemini up to this point, and we've just been lucky as hell."[28]

Why do I doubt that? I suspect that everything about the pressurization test is a lie. I think that it was a one time only occurrence specially configured to suit the job at hand.

Borman contended that Ed White and his wife Pat were friends of his and that he listened to the audio tapes of the fire over and over again. Then he states, "The only comfort derived from listening to the tapes was the knowledge that the agony hadn't lasted long; that death had come from noxious fumes before the flames reached them."[29]

Borman's acumen might be judged by the fact that Eastern Airlines played submarine when he was at the helm as CEO. Nobody dies in 14 seconds from noxious fumes. Ed White died inhaling super heated oxygen which set fire to his lungs, throat and skin the same way that technician's hand burned in the test years before. The chances are that they survived for minutes and were conscious for a good part of that time. However, death was definite after the first breath.

Borman then writes about "nuts" and disgruntled employees who tried to give his committee information:

As the investigation progressed, all sorts of nuts came out of the woodwork with their own theories. There also were some serious allegations directed against North American Aviation, most of them coming from former employees with large axes to grind. They charged the the company with criminal neglect and mismanagement, and we investigated each accusation thoroughly. We found that in every case we were getting input from people who simply had personal grievances against the company, with no evidence to back them up.[30]

That's odd! One of Borman's superiors, General Phillips, had also made a report in November, 1966 that shredded North American Aviation. He could hardly be classified as a disgruntled employee. Speaking of classified, Michael Gray in his book disclosed the fact that Phillip's report was classified.[31] Borman apparently ignored that report.

Time and time again, NASA has bragged about how open NASA was. One wonders, then, who classified this report? What could it possibly have had to do with national security? No wonder that Bill Kaysing was never able to obtain a copy. To paraphrase an old saying, the "TOP SECRET" stamp, because it reflects patriotism, has always been the last refuge of scoundrels.

On April 27, 1967 the 204 Board was still in the process of almost learning new things. A low level employee named Thomas Baron had already testified in Washington and now was a target for NASA's ire. His voluminous reports were day by day accounts of North American's screwups and were written years earlier. It seems very strange that both Baron's and Phillips' reports disappeared. After accepting his reports, the 204 Board wrote off his testimony.

By the very next evening Baron, his wife and stepdaughter would all be dead. The two women were totally innocent but, maybe, that's what they get for associating with a NASA whistle-blower.

One of the common accidents to governmentally sensitive folks in Florida is the old railroad crossing gambit. There are lots of semi-deserted country roads and active railroad tracks in Florida. Usually after the grisly event, the bodies are found by someone so powerful that he can have them immediately cremated, frequently before an autopsy can be performed—which is contrary to Florida state law. And they used to tell us horror stories about the KGB! I no longer live in Florida so if they come after me for writing these words they will have to think up a new method.

And please note: I am not suicidal. I say that because suicide is a common cause of death in this context. For instance there is a suspicion that another casualty of NASA is Mrs. Pat White, who committed suicide a few years after her husband's cremation. From post-mortem reports—she wasn't suicidal either. Low level whistle blowers die like flies and yet, General Phillips, goes on to head NASA after he told basically the same story.

Borman also complained about the windows that kept fogging up on his Gemini 7 mission and again on Apollo 8. North American, for four straight years failed to find a solution for such a simple problem as window fogging yet he couldn't find anything seriously wrong with them. That's about par, isn't it?

Borman was stationed at Clark Air Force base in Manila during 1952 and part of his duty was to inspect a huge warehouse that stored heavy equipment, supposedly ready to roll on an instant's notice. His inspection revealed that, "there wasn't a vehicle or a piece of equipment that wasn't in deplorable shape—most of it unusable without major overhauls. The stuff had been there since the end of the war and obviously hadn't been touched since."[32]

The Captain in charge asked Borman to certify that it was in good condition and he refused. The code of West Point of "Duty and Honor" took precedence. However, when a Colonel insisted that he sign-off as in good condition he caved in. "Honor" be dammed. The new moral code is apparently totally dependent upon the rank of the officer who gives the order? Go along to get along.

Next Borman, still the politician that Collins first pegged him for, tells perhaps the greatest lie of his life. He concludes, *"We didn't sweep a single mistake under the rug, and to this day I'm proud of the committee's honesty and integrity."*[33] Presumably Mr. Borman, had his fingers crossed when he wrote that!

The committee was still in the middle of its stately review process when on April 7, 1967, a House subcommittee was also convened to investigate the fire. The next day a dismayed *New York Times* fired off a lead editorial. They used the words, *"Even a high school chemistry student" (knows better than to play with 100 percent oxygen).* The editorial went on to accuse those in charge of the Apollo project of "incompetence and negligence."[34]

The 204 Board concluded with a real wrist spanker of a statement against NASA:

A sealed cabin, pressurized with a pure oxygen atmosphere without thought of fire hazard; an overly extensive distribution of combustible materials in the cabin; vulnerable wiring carrying spacecraft power; leaky plumbing carrying a combustible and corrosive coolant; inadequate escape provisions for the crew, and inadequate provisions for rescue or medical assistance.

Both committees would prove about as useful as a screen door in space (and about as effective as the politicians who manned the Warren Commission's investigation of the Kennedy assassination a few years before). Like all government inquisitions they use a method best described as "let's all gang-bang the whistle-blower."

At the beginning of the Mercury Program, NASA tests on pure oxygen proved that the safe pressure limit for breathing was between 2.9 and 6.67 psi. But they also concluded that pressures, *"outside these limits would cause severe, if not permanent damage."*[35] In plain English, *murder begins at 6.7 psi!*

Kennan & Harvey have this to say about the fatal test on the capsule, "The day of the plugs-out test, the TV camera inside the space-craft, which was an important piece of flight and test equipment, was absent; its retaining brackets had some how been bent during installation."[36]

These authors never called it murder but they continued with this statement, "It is of the greatest significance that the fire extinguishers were located in that (008) spacecraft during its testing. Not only were fire extinguishers included but fire resistant teflon sheets were draped over wire bundles and the astronaut's couches. These particular items, non flight items, were conspicuously absent in command module 012 during the fatal plugs-out test on January 27, 1967."[37]

They also summed up the test with these statements:

It was the *first* and only use of the new three piece hatch.

It was the *first* plugs-out test in which as many as three hatches were closed on a crew in an oxygen atmosphere at a pressure of sixteen pounds per square inch . . .

It was the *first* occasion of the Apollo emergency escape drill under all-out pre-launch conditions.

It was the *first* occasion when certain non flight flammable materials, such as two foam rubber cushions—were placed in the cockpit.[38]

Later NASA would rule out the use of any material which could be ignited by spark at 400°F in pure oxygen at 16.7 psi.[39] "They included the couch padding, to which astronaut White's body was welded by the heat: this, it emerged, could be ignited by a spark at 250°F."[40] Notice they still had every intention of using 16.7 psi oxygen. Or was it 20.2 psi?

If a civilian corporation killed three men by extreme stupidity there would be criminal proceedings, trials and fines. But because the government is the suspected culpable party nothing happens. To repeat: I cannot believe that in such a highly technical field as space that even the lowest paid technician would not have questioned the moronic decision to use 100 percent oxygen to try a pressure test on a capsule with live electric panels, and which contained locked-in and strapped down astronauts. Especially, on a capsule that would never fly.

At the time, there was talk the Apollo Program would be scratched. But even if 50 people had been killed the operation would have continued, with no more than a brief pause, because the bucks were too big. As Collins points out, "I don't think the fire delayed the first lunar landing one day, because it took until mid-1969 to get all the problems solved in areas completely unrelated to the fire."[41]

According to the newspapers, NASA committed another unequivocal example of utter stupidity on March 19, 1981. They had a chamber on the Space Shuttle Columbia filled with nitrogen and seven people entered it. Two died and five were injured.

I believe that the cremation was mass murder. If not that it was unconscionable stupidity. We may never know for sure. What I am sure of is that the entire Apollo Program was a show, a simulation produced by the CIA, directed by NASA, invested in by Congress, and paid for by Mr. and Mrs. American Taxpayer! As shown, I also believe that, to protect their multi billion dollar income, the CIA murdered three astronauts on Pad 34, plus four more on plane rides, and one in a car.

REFERENCES

1. Barbour, *Footprints on the Moon* (The Associated Press, 1969), p. 117.

2. Gray, *Angle of Attack* (Norton, 1992), p. 218.

3. Barbour, *Footprints on the Moon* (The Associated Press, 1969), p. 117.

4. Young, Silcock and Dunn, *Journey to Tranquility* (Doubleday, 1969), p. 173.

5. Collins, *Carrying the Fire* (Ballentine Books, 1974), p. 62.

6. Ibid, p. 101.

7. Baker, *Astronomy* (Van Nostrand, 1959), p. 291.

8. "Economics Of Wheat Deal," *National Review* (1972), p. 1168.

9. Hurt, *For All Mankind* (Atlantic Monthly Press, 1988), p. 323.

10. Young, Silcock and Dunn, *Journey to Tranquility* (Doubleday, 1969), p. 193.

11. Lewis, *Voyages of Apollo* (Quadrangle, 1974), p. 163.

12. Collins, *Carrying the Fire* (Ballentine Books, 1974), p. 275.

13. Young, Silcock and Dunn, *Journey to Tranquility* (Doubleday, 1969), p. 194.

14. Borman and Serling, *Countdown* (Morrow, 1988), p. 175.

15. Wilford, *We Reach the Moon* (Bantam, 1969), p. 101.

16. Murray and Cox, *Apollo: The Race to the Moon* (Simon & Schuster, 1989), p. 187.

17. Young, Silcock and Dunn, *Journey to Tranquility* (Doubleday, 1969), p. 186.

18. Wilford, *We Reach the Moon* (Bantam, 1969), p. 96.

19. Collins, *Carrying the Fire* (Ballentine Books, 1974), p. 270.

20. Kennan and Harvey, *Mission to The Moon* (Morrow, 1969), p. 32.

21. Ibid. p. xi

22. Young, Silcock and Dunn, *Journey to Tranquility* (Doubleday, 1969), p. 192.

23. Borman and Serling, *Countdown* (Morrow, 1988), p. 174.

24. Kennan and Harvey, *Mission to the Moon* (Morrow, 1969), p. 146.

25. Borman and Serling, *Countdown* (Morrow, 1988), p. 175.

26. Aldrin and McConnell, *Men From Earth* (Bantam, 1989), p. 162.

27. Ibid. p. 163.

28. Gray, *Angle of Attack* (Norton, 1992), p. 233.

29. Borman and Serling, *Countdown* (Morrow, 1988), p. 174.

30. Ibid. p. 178.

31. Gray, *Angle of Attack* (Norton, 1992), p. 241.

32. Borman and Serling, *Countdown* (Morrow, 1988), p. 51.

33. Ibid. p. 178.

34. Murray and Cox, *Apollo: The Race to the Moon* (Simon and Schuster, 1989), p. 220.

35. Baker, *The History of Manned Space Flight* (Crown, 1982), p. 39.

36. Kennan and Harvey, *Mission to the Moon* (Morrow, 1969), p. 21.

37. Ibid. p. 57.

38. Ibid. p. 20.

39. Young, Silcock and Dunn, *Journey to Tranquility* (Doubleday, 1969), p. 195.

40. Ibid. p. 198.

41. Collins, *Carrying the Fire* (Ballentine Books, 1974), p. 276.

28 Extra-Terrestrial Exposure Law Already Passed by Congress

On 5 October 1982, Dr. Brian T. Clifford of the Pentagon announced at a press conference (*The Star*, New York, 5 October 1982) that contact between U.S. citizens and extra-terrestrials or their vehicles is strictly illegal.

According to a law already on the books (Title 14, Section 1211 of the Code of Federal Regulations, adopted on 16 July 1969, before the Apollo moon shots), anyone guilty of such contact automatically becomes a wanted criminal to be jailed for one year and fined U.S. $5,000.

The NASA administrator is empowered to determine, with or without a hearing, that a person or object has been "extra-terrestrially exposed" and impose an indeterminate quarantine under armed guard, which could not be broken even by court order.

There is no limit placed on the number of individuals who could thus be arbitrarily quarantined.

The definition of "extra-terrestrial exposure" is left entirely up to the NASA administrator, who is thus endowed with total dictatorial powers to be exercised at his slightest caprice, which is completely contrary to the Constitution.

According to Dr. Clifford, whose commanding officers have been assuring the public for the last thirty-nine years that UFOs are nothing more than hoaxes and delusions to be dismissed with a condescending smile, "This is really no joke; it's a very serious matter."

This legislation was buried in the 1211th subsection of the 14th section of a batch of regulations very few members of government probably bothered to read in its entirety—the proverbial needle in the haystack—and was slipped onto the books without public debate.

Thus from one day to the next we learn that without having informed the public, in its infinite wisdom, the government of the United States has created a whole new criminal class: UFO contactees.

Source: MUFONET Network, March 23, 1993

The lame excuse offered by NASA as a sugar coating for this bitter pill is that extra-terrestrials might have a virus that could wipe out the human race. This is certainly one of the many possibilities inherent in such contact, but just as certainly not the only one, and in itself not a valid reason to make all contact illegal or to declare contactees criminals to be jailed and fined immediately.

It appears the primary effect of such a law would not be to prevent contact: it would be to silence witnesses.

According to NASA spokesman Fletcher Reel, the law as it stands is not immediately applicable, but in case of need could quickly be made applicable. What this means is that it is ambiguously worded, so that it can be interpreted either one way or the other, as the government desires.

It is certainly not a coincidence that Dr. Clifford held his press conference during the period when the popularity of the film *E.T.* was at its peak. As *E.T.* portrayed a type of extra-terrestrial that was benevolent and lovable, the inference is that the press conference was intended to discourage attempts to communicate or fraternize with UFO occupants. However, instead of having the intended effect, it backfired, causing public furore.

There may be some relationship between this fiasco and the next semi-officially endorsed attempt to deal with the subject of extra-terrestrials, the TV series *V* which was featured with repeat performances and maximum publicity by major networks worldwide. The aliens portrayed in *V* are the most horrifying and repulsive nightmares imaginable, but are defeated thanks largely to a CIA hit man specializing in covert operations.

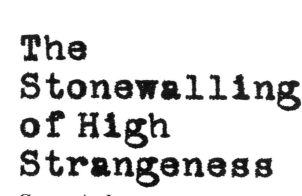

29 The Stonewalling of High Strangeness

George Andrews

A large crack appeared in the stonewalling of high strangeness on October 14, 1988. On that date, there was a semi-official admission of extra-terrestrial intervention in human affairs, in the guise of a two-hour TV special entitled UFO COVER-UP? . . . LIVE! Participation by U.S. and Soviet officials was so extensive that the broadcast could not have occurred without the consent of both governments.

The program was presented to the public simultaneously in the United States and the Soviet Union, the first time in history that any TV program had received such preferential treatment. However, in spite of this clearly implied U.S.-U.S.S.R. seal of approval (or perhaps because of it?), the program contained a clever mix of information and disinformation.

The valid information was that we are not alone, and that the government has made a hitherto secret agreement with short gray humanoids (the "Grays") who say they are from Zeta Reticuli.

Although it was not specifically stated that the Grays were the only E.T. group our government has made contact with, that implication was made. To the extent the implication was made, it was a falsehood.

UFO COVER-UP? . . . LIVE!, hosted by Mike Farrell, implied that Uncle Sam had made a smart deal. However, the truth of the matter is that this deal was the most disastrous mistake, not only in the history of our nation, but in the history of our entire civilization. There were other E.T. groups that we could have made incomparably better arrangements with. Although this scandal is similar in nature to the Iran-Contra deal, it is a tragedy on a scale of such unprecedented magnitude that in comparison Iran-Contra seems like very small change indeed.

The case of contactee Edouard "Billy" Meier in Switzerland, which Mike Farrell stigmatized as "an obvious hoax," is not without its ambiguities, absurdities and contradictions. Nevertheless, the physical evidence is so strong that this is one of the best-substantiated cases on record. Dr. James Deardorff has speculated that the E.T.s may be making deliberate use of absurdity as a form of strategic camouflage, in order to ensure gradual rather than sudden realization of their presence. By cloaking their activities in an aura of absurdity, they repel serious investigation and make people hesitate to endorse the reality of the phenomena, keeping it all borderline and marginal, which distracts human attention from their presence.

If this is the game that is going on, the E.T.s may be feeding Meier a mixture of truth and falsehood, which he relays to the public in the sincere belief that it is all true.

For example, Meier believes himself to be the only genuine UFO contactee, an obvious absurdity, because there have been thousands of genuine human-alien contact cases during recent years. He also states that the many thousands of abduction and crop circle incidents reported during recent years were all hoaxes, another obvious absurdity. I concur with Deardorff that Meier is sincere in his beliefs, but that he is being fed information mixed with disinformation.

One of the most curious aspects of the Meier case is the veritable obsession displayed by the goverment's plainclothes media agents (such as William Moore who has publicly admitted his links with government agencies) in over-zealous attempts to discredit the evidence. An all-out media vendetta has been waged against the Meier case that seems totally out of proportion with the investgation of any one single UFO case. Is this because Meier's contact was not with the Zeta Grays but with a rival E.T. group that in appearance is almost indistinguishable from us, and which shares a common ancestry with us, known among researchers as the Blonds or Swedes? This is the same group that was known in antiquity to the Scandinavians as the Aesir, to the Irish as the Tuatha Te Danaan, and by other names in other cultures. One of the modern Soviet cases mentioned on the Mike Farrell program, in which the UFO opened up like a flower, was an encounter with the Blonds, a group with whom I think we could work out a valid alliance.

Dr. Jean Mundy is a Professor Emeritus in Psychology at Long Island University and has been in private practice in psychotherapy and hypnotherapy for over twenty years. She is a member of a long list of professional and honorary societies, and she has published articles on a wide variety of subjects in both the academic and the popular press. During recent years, she has focused her expertise on the subject of UFO phenomena. Dr. Mundy has written the following analysis of the public reaction to the 1988 TV program especially for this book:

The most astonishing UFO TV documentary aired to date, on October 14, 1988, with Mike Farrell as the host, had all the ingredients of a blockbuster. Note just some of the cast of characters: 1) A Russian scientist, revealing that aliens have landed in Russia and contacted their military personnel. 2) Two CIA agents, indentities disguised, revealing that extra-terrestrial aliens are now the "guests of the U.S. Government." 3) About a hundred residents of Gulf Breeze, Florida, who have witnessed and photographed the extensive UFO activity in their home town. 4) Abductees telling their own stories under hypnosis. 5) Expert witnesses from the military saying that sightings they reported while in service 'disappeared' from the records. 6) A psychiatrist reporting that abductees she has treated were sane and suffering from trauma as a result of their contact with aliens. 7) Evidence by investigators of crashed UFO sites. 8) Testimony from Budd Hopkins, who has investigated hundreds of abductee cases for over 12 years, that genetic experiments are being performed by aliens on humans on a large scale. 9) Other witnesses, each expert in his/her own field, testifying to the validity of photographic or other tangible documentation of UFOs. 10) Paintings of aliens drawn by artists from the testimony of witnesses.

Surely, even by Hollywood standards, here are all the ingredients of a box-office hit. Was it a hit? Well, during the two-hour show, 75,000 viewers did pay one dollar each to have their phone-in vote about their belief in human-alien contact recorded. What other signs of success made the front page? Not a one! The show was a flop in terms of arousing public interest. How can this be? How could such a carefully concocted combination of ingredients, culminating in the most newsworthy disclosure in history, not elicit a bang, nor even a whimper?

I played a videotape of the show to my adult education class on alien-human contact. I watched the reactions of my class members. Thanks to the review/playback capability, I was able to conduct some experiments.

The adult students who signed up for this course were believers in the possiblity of human-alien contact. Some even had personal experience of contact, yet after the screening they were more doubtful than before! I then recognized that UFO COVER-UP? . . . LIVE! made brilliant use of the best propaganda techniques. If you have a tape of the show, watch it again, and listen carefully to the soundtrack. Look at the painted backdrops and, most important, look at Mike Farrell's reactions.

You will see and hear some interesting contradictions. The astute viewer cannot dismiss these contradictions as sloppy production. The production had a huge budget and was planned many months in advance. In fact, Budd Hopkins reported to me that each witness was interviewed at length before the show, then certain words of that interview were selected to be used. The words were put on cue cards, and on the nights of the show, the witnesses were allowed to say only what was on the cards. UFO COVER-UP? . . . LIVE! was carefully rehearsed. Reading from cue cards, unless one is experienced in doing so, gives words a stilted sound that many listeners take as phony. Score one for not believing the witnesses.

Score two: When words are spoken over background music, the music makes an unconscious impression that flavors the emotional impact of the words. For example, the relentless "approach," louder and louder in intensity, of the orchestration of the *Jaws* soundtrack as the shark nears its victim builds up a feeling of dread even before any action takes place. The background music for UFO COVER-UP? . . . LIVE! is light, pleasant dance music. While the Russian is speaking of terrifying abductions, the music is a rousing polka!

Score three: When the Russian says, "The aliens communicated with the terrified soldiers by mental telepathy," Mike Farrell responds, "Oh, they used music, just like in the movie *Close Encounters*." We depend very much on the reaction of others to set our own reaction. Obviously, if someone tells you a story and they are laughing, you think the story is funny. If they are in a panic state while telling it to you, you think of it as tragic. Throughout of the entire show, Mike Farrell reacted to all the UFO information with puzzlement, but in an amused and light-hearted manner. So, for the most part, the audience did too.

Score four: Through the skillful but unobtrusive use of another media technique, the painted backdrop against which the alien Grays are portrayed is a bright sunny room overlooking a garden! Nothing to take seriously there, certainly nothing to worry about.

Also the depictions of the Grays were idealized and did not conform to witness descriptions, giving the impression of benign Disneyesque animated characters.

Score five: Another powerful media technique is to trivialize something important by focusing on a tiny detail. For example, telling the audience that the aliens "prefer strawberry ice cream" reduces these mysterious life forms to a childish format.

Score six: We are so accustomed to Hollywood movies that we are entertained by a "historical" film, but never for a moment truly believe that the "action" taking place on the screen is real. We know it is rehearsed and played by actors. When Mike Farell puts the UFO reports in the same category as the movie *Close Encounters*, the audience automatically thinks of the TV documentary UFO COVER-UP? . . . LIVE! as just another Hollywood movie. No one writes a letter to the editor of their local newspaper or phones their Congressman to demand action about a crime committed in a Hollywood movie.

There is more. One of the women presented in the documentary who has been abducted by aliens is an acquaintance of mine. Her entire life has been affected by her terrifying alien contact. The one and only comment highlighted on the cue card for her to use on the supposedly "LIVE!" Mike Farrell show was, "Oh, he's so ugly." When my students heard this, they giggled! They took it to be comical.

Neither my students nor the majority of the viewing audience took the UFO COVER-UP! . . . LIVE! program seriously. That is why there was no audience response. One of my students said he thought, "The man portraying the Russian couldn't have been a real Russian, because he spoke English too well." No doubt he was trying to account for why the show did not have the ring of truth to it.

The question remains: were the producers of the show, whoever they really are, pleased that their show was a "flop," in that few viewers took the reality of alien-human contact (or invasion) seriously? Or were they disappointed by the lack of audience response? What, indeed, was the real purpose of this carefully staged show?

The format in which UFO COVER-UP? . . . LIVE! was presented tranquilized the general public, as did the movies *E.T.* and *Close Encounters of the Third Kind*, in the comfortable belief that there is absolutely nothing to worry about, as far as any reports of alleged "aliens" are concerned.

As Dr. Mundy has pointed out, this comfortable belief system—so assiduously maintained by the authorities through their constant insistence on relegating the subject of alien intervention in human affairs to the realm of media trivia—relies for its support on some very thin ice indeed.

This ice is so thin that it can be shattered by featherweight random events, such as the arrival in my mailbox of a letter from Rev. John E. Schroeder of the UFO Study Group of Greater St. Louis, saying:

> You mentioned William Moore's TV program hosted by Mike Farrell. Is it any surprise that the November 1989 issue of prestigious *Millimeter* magazine for film and TV producers listed the CIA as paying producer for that show? I wonder what happened to the response requests? Who effectively collected whose paying phone numbers? Why? How are they to be used? Was any data ever given congressional members? The plot thickens!

The plot does indeed thicken. A search of the November 1989 issue of *Millimeter* magazine did not locate the item which Rev. Schroeder is nevertheless sure that he and his wife saw in *Millimeter*, though he may have been mistaken as to the issue. This does not necessarily mean that the item does not exist, but as we go to press the question remains unresolved. In spite of this uncertainty, the item seems worth retaining, because both the possibility and the questions raised by Rev. Schroeder remain relevant, even if neither confirmed nor invalidated.

The ability to face phenomena of high strangeness with an open mind is a rare trait, not shared by the vast majority of contemporary humanity. The average person feels threatened or terrified by any unprecedented divergence from conventionally accepted norms of reality, and may react with dangerous violence.

Let us consider the case of Herbert Schirmer, who in 1967 held the distinction of being the youngest Chief of Police in Nebraska, and who was one of the most prominent citizens in his home town of Ashland, Nebraska. One night while on duty in his patrol car, he encountered a large disc-shaped UFO, which had landed not far from the highway. When a glowing humanoid figure appeared in his headlights, he tried to draw his gun, but found himself inexplicably paralyzed. When the humanoid opened the door of his car, Schirmer felt a cold hard instrument being applied to the back of his neck, and he blacked out. Upon regaining consciousness, he drove straight back to the police station and reported the incident to his fellow officers, who noticed that he was unable to account for about half an hour, There was an unusual welt on the back of his neck, which later left a permanent scar. A qualified hypnotist, Dr. Leo Sprinkle, was brought in to regress him back to the period of missing time. Under hypnosis, the regression indicated that he had been taken aboard the UFO and had communicated with a group of alien humanoids during the period of missing time. A search of the location where the incident occurred revealed physical landing traces.

How did Chief of Police Schirmer's old friends react to his adventure? By firing him from his job, dynamiting his car, and hanging and burning him in effigy in the town square. His wife divorced him, and he was driven out of town. He has since moved frequently, contacting various UFO investigators, trying to make sense of his dilemma.

There is a remarkable resemblance between Herbert Schirmer's ordeal and the Jeff Greenhaw case. In 1973, Jeff Greenhaw was the youngest Chief of Police in Alabama, and was one of the most prominent citizens in his home town of Falkville, Alabama. One night he received an anonymous phone call from a woman who said that a UFO had just landed in a field not far from town. Greenhaw got into his patrol car and drove toward that area. As he approached the field, he encountered a humanoid about six feet tall, covered from head to foot in metallic clothing, standing in the middle of the road. Greenhaw pulled up near him and said: "Howdy, stranger." There was no response. Greenhaw reached for a Polaroid camera he happened to have with him and took four pictures. Then he turned on the flashing blue light on the top of his patrol car. The humanoid began to run, but not in normal fashion, moving sideways instead of forward, taking large leaps of about ten feet at a time very quickly, traveling at extraordinary speed. Greenhaw began to pursue him, but his patrol car suddenly went out of control and into a spin as he reached 45 miles per hour, obliging him to give up the chase. He returned to town and reported the incident.

How did Chief of Police Greenhaw's old friends react to his adventure? By firing him from his job, dynamiting his car, and burning down the trailer in which he lived. His wife divorced him, and he was driven out of town. He has since disappeared.

These two cases are not isolated. One could easily fill a book with the many cases of UFO contactees who have been obliged to leave their homes and change their names because of hostile social pressures. As our history clearly indicates, such witch hunts are nothing new. The average citizen's tolerance of diversity has increased only minimally since medieval times. What sets the Schirmer and Greenhaw cases apart from the hundreds, if not thousands, of other cases of contactee harassment is the secure social position they both enjoyed, until they reported their UFO encounters.

Being the Chief of Police in a small town one has grown up in implies widespread respect and trust from a closely knit group of people who have known you since childhood, and is about as secure a social position in the hierachy of the American system as it is possible to attain. The fact that one encounter with the unknown could in one day transform the role model for an entire community into a despised outcast demonstrates the extreme extent to which the average citizen feels insecure, and therefore fears the unknown.

This is true on the national level as well as on the individual level. The thousands of documents that the government has been obliged to release under the Freedom of Information Act demonstrate that its internal policy concerning UFOs and extra-terrestrials is extremely different from its publicly stated policy. To put it bluntly, our government has been lying to its citizens about UFOs and extra-terrestrials for over 40 years.

To point out merely one example of this duplicity: Air Force Regulation 200-2, JANAP-146* provides a penalty of ten years in prison plus a $10,000 fine and a forefeiture of pay and pension for any member of the Armed Forces who makes an unauthorized statement about UFOs. If you write to the Library of Congress and ask for a copy of this regulation, you will get an answer stating that no such regulation exists. Air Force spokesmen blandly deny that any such regulation exists. However, if you write to the Library of Congress and ask for a copy of *The UFO Enigma* by Marcia Smith and David Havas, which was published by none other than the Library of Congress itself in 1983, you will find in it a statement that Major Donald Keyhoe was the first to make public reference to the previously secret JANAP-146.

Further details are to be found in a book by a well-known French

---

* Joint-Army-Navy-Air Force Publication

researcher, Aime Michel, published in 1969. The preface to a previous book by Michel was written by General L. M. Chassin, General Air Defense Coordinator, Allied Air Forces, Central Europe, NATO. General Chassin commends Michel's ability and integrity in strong terms. The statement Michael made in 1969 translates as follows: "However, if it is so certain that for the American authorities this subject is no more than crazy stories that are completely without interest, how does one explain the extraordinary precautions in Air Force Regulation 200-2, and the ten years in prison plus $10,000 fine of JANAP-146, all of which are still being enforced fifteen years later, and more vigorously than ever?"

Since Michel stated in 1969 that the regulation had been enforced for fifteen years, it must have originated in 1954, which just happens to have been a year during which an exceptional amount of UFO activity took place.

According to Ralph and Judy Blum, who received assistance from government sources while compiling their excellent book *Beyond Earth* (Bantam, 1974), the text of JANAP-146 is contained in an official publication of the Joint Chiefs of Staff, entitled Canadian-United States Communications Instructions for Reporting Vital Intelligence Sightings (CIR-VIS/MERINT, JANAP 146).

As the vast majority of UFO incidents get coverage only in local newspapers, if they even get that, there is no general awareness of the persistent UFO activity occurring in other parts of the country and other parts of the world. Some stories in the media even state that UFO activity is now almost non-existent and was just a passing fad of the 50s and 60s, when the truth is that there are just as many incidents as there ever were, but that the covert censorship is being enforced more effectively.

Major wire services operate in collusion with governmental intelligence agencies in perpetrating this devious form of camouflaged censorship for one basic reason, as clearly expressed by nuclear physicist Stanton Friedman: "No government of Earth would want its citizens to pledge allegiance to the planet rather than to itself, and to think of themselves first as Earthlings, rather than as Americans, Canadians, Russians, etc."

On pages 188–191 of *Extra-Terrestrials Among Us*, I described the adventures of an officer using the pseudonym of "Toulinet,"who had been assigned to write an analysis of the top secret "Grudge 13" report, after which he has been summarily discharged from military service. During the summer of 1989, this man took the courageous step of publicly indentifying himself as former Captain William English of the Green Berets, who had been working at the RAF Security Services Command, RAF Chicksands, England, at the time that he was assigned to write the analysis of "Grudge 13." I have been unable to either confirm or invalidate the rumor that no sooner had he publicly identified himself than his residence was firebombed, as had been the residences of police chiefs Schirmer and Greenhaw in 1967 and 1973.

Attempts have been made to destroy English's credibility, such as the U.S. Army stating that it has no military records concerning him. However, it is standard operating procedure for military commanders to delete, either partially or in totality, the service records of subordinate personnel who have become security risks, as is examplifed in "The Cutolo Affidavit," published in *Erase and Forget* by Paragon Research, P.O. Box 981, Orlando FL 32802, in 1991.

Records are also systematically destroyed by the highly paid defense contractors engaged by the Pentagon for secret projects. In November 1989, physicist Robert Lazar went public with disclosures concerning alien discs and related activities at Area 51 of the Nevada Test Site. From one day to the next, all records of his previous employment, his education, and even of his birth, vanished as if by magic. This would have effectively destroyed his credibility, if there had not been certain items that survived the onslaught of the modern Inquisitors. For example, Lazar stated that he had worked as a physicist for Los Alamos National Laboratories, whose representatives denied that he had ever been employed by them. However, independent investigators found a copy of the telephone directory issued by the Los Alamos Lab in 1982, which listed Robert Lazar among the scientists employed by them. An article in the Los Alamos paper during that same year, 1982, described Lazar's interest in jet cars, mentioning his employment at the Los Alamos Lab as a physicist. Finding themselves unable to destroy Lazar's credibility in any other way, the authorities resorted to a crude but time-tested technique and tarnished his reputation with a sex scandal, which Lazar's lifestyle unfortunately made possible.

Some people have taken issue with my statement that we are about to experience direct confrontation with non-human intelligent beings from elsewhere in the cosmos in the near future, pointing out that UFO intervention in human affairs has been minimal during the last forty years, so why shouldn't that pattern continue indefinitely? I have answered that question at length in my previous book, but would like to extend my response by describing some major incidents that occurred since its publication, which clearly indicate that we have entered a new phase of UFO activity, a phase characterized by deliberate and ostentatious UFO displays over heavily populated areas on an unprecedented scale.

There was a flurry of significant UFO activity both before and after August 12, 1986, but it was on the night of August 12–13 that the climax of the first incident occurred. During that night, reports came flooding in from Lake Huron to Nova Scotia in Canada, and in the United States from Maine, New York, Michigan, Indiana, Illinois, Arkansas, Louisiana, Kentucky, Pennsylvania, South Carolina, and Florida, as well as from the town of Leongatha in the province of Victoria in Australia. There had also been multi-witness sightings of a large UFO in Pennsylvania. Both of the Pennsylvania sightings appeared to involve the same object, described as bright silver and elliptical, the size of three buses in length. On the night of August 12–13, the reports from Arkansas, Louisiana, and Kentucky described a large cloud-like ball of fire. In Clark County, Kentucky, the appearance of an enormous ball of fire that lit up the whole sky was accompanied by a sonic boom that made houses shake, along with an odor described as similar to gunpowder. From Illinois, Indiana, Michigan, and Ontario came a flood of reports about "a spiral cloud with a star-like object beside it." An air traffic controller, Tim Jones, saw it both visually and on his radar screen, and stated that the way it was behaving was unlike any aircraft he had ever seen. On the same night in Leongatha, Victoria Province, Australia, an English teacher and a science teacher at the local high school reported that a UFO with flashing lights had hovered and maneuvered above them for about forty minutes.

On the following day, the scramble for explanations began. Both NASA and NORAD* denied any involvement in the phenomena, and further specified that it did not correspond with any known Soviet space activity. Speculation that a new Japanese satellite might have exploded was squelched by a statement from Japan's Tanageshima Space Center that their satellite was functioning normally.

At this point, as it usually does whenever a UFO incident occurs that is difficult to explain away in conventional terms, the U.S. government brought NASA expert (or establishment hatchet man, according to your point of view) James Oberg into the controversy to make a statement. Mr. Oberg stated authoritatively that what everyone had seen was fuel being dumped from a rocket on the Japanese satellite, as the rocket boosted the satellite into orbit, and that was that. Although this explanation left many questions unanswered (in particular, concerning the incidents in Clark County, Kentucky, and Syracuse, New York), it was accepted tamely without protest by the entire news media of this great nation as the final solution to the mystery of the night.

A lady named Lorraine Whitaker in Lanesboro, Pennsylvania, got a clear photograph of what had been visible in the sky over her area on the night of August 12, which depicts a sharply defined, intensely bright cigar shape, emitting a swirling cloud of luminous gas.

Paul Oles, who is the Planetarium Director at the Buhl Science Center in Pittsburgh, made the following statement: "We know what it wasn't,

---

* North American Air Defense Command

but we have no idea what it was. Our most logical explanations have been totally ruled out. It now falls into the category of an unidentified sighting." However, only one newspaper even mentioned the statement by Mr. Oles. Every other newspaper nationwide that carried the story featured the statement by Mr. Oberg of NASA as definitive.

On the night of August 15, 1986, three days after the incidents I have just described, Angelo and Grazia Ricci of Verona, Italy, were abducted while on a summer vacation camping trip near Belluno, Italy. They were taken aboard a UFO by two humanoids, each about six feet, six inches tall, who were dressed in gray coveralls that left only their heads exposed. Their heads were long and hairless and had very pale skin. Their eyes were phosphorescent. They had pointed ears, a normal nose, and a narrow slit where the mouth should be. Mr. and Mrs. Ricci were subjected to medical examinations and various tests for about three hours before being released.

A series of events comparable in importance to those of August 12 occurred on September 23, 1986. They began at daybreak, when two brothers fishing on a lake near Daventry, England, reported that shortly after dawn they had seen six UFOs flying in formation behind a large UFO. Within the next few hours, thousands of people (including police) reported UFOs flying in formation and performing maneuvers, during which they left behind multi-colored vapor trails over West Germany, Holland, Belgium, Luxembourg, and France. In Paris, whole crowds of people driving to work during the morning rush witnessed a fleet of fifteen UFOs flying in formation. Simultaneously a ball of fire was seen over Amsterdam; what was described as a "flying machine" was reported by a staff member of the Royal Observatory in Belgium; "a very luminous object shaped like a rocket, three times as large as an airplane" as well as "a cluster of five or six luminous green objects" were reported from Luxembourg; and "a bright flying object with a luminous tail" was reported from West Germany. There were also similar reports from Derbyshire and Leicestershire in England.

The fifteen UFOs seen over the Montreuil region of Paris were described as silver-colored, but over the Chatelet region of Paris witnesses perceived them as intensely luminous green and turquoise blue, some of them emitting green flame. Over Paris they were traveling at a leisurely pace, about the speed of an airplane during an air show.

The nearly simultaneous occurrence of such phenomena over six of the nations of Western Europe on the morning of September 23, 1986, has all the characteristics of a carefully orchestrated and deliberately ostentatious display, obviously intended to bring about widespread recognition of the reality represented by UFOs among the intelligent citizens of these key countries.

What was the result?

It is hard to know whether to laugh or cry over the incorrigible hypocrisy with which the news media handled this story. How did the journalists deal with this unprecedented manifestation of high strangeness in the skies of Europe? They understandably requested an explanation from NORAD. However, when NORAD explained that what everyone had seen was debris from a Soviet booster rocket, an explanation that was directly contradicted by the observations the journalists had themselves recorded, this implausible explanation was instantly and uncritically accepted by the news media, which abdicated all pretense of independent reasoning and parroted it ad nauseam as the only rational solution to the enigma of what had happened that morning throughout six nations of Western Europe.

The next example of a deliberate and ostentatious UFO display did not occur over a heavily populated area, but northeastern Alaska is certainly a sensitive military zone. The report did not reach the U.S. news media until January 1, 1987, though the incident happened on November 17, 1986. The time lag between the date of the incident and the date the report was made public supports the hypothesis of covert censorship of the news media.

This case bears a remarkable resemblance to the case officially announced by the Soviet Academy of Sciences in 1985, in which a Soviet airliner was followed by a UFO for approximately 800 miles.

The case of November 17, 1986, involved the pilot, co-pilot, and flight engineer of a Japan Air Lines cargo jet that was making a return trip from Iceland to Anchorage, Alaska. The crew members first became aware of the three UFOs in the vicinity of their jet while over northeastern Alaska. Two of the UFOs were small, but the third was enormous, twice the size of an aircraft carrier. The UFOs followed the cargo jet for about 400 miles, during nearly an hour. They emitted flashing amber, green, and yellow lights. They played games with the jet: disappearing, reappearing, moving at incredible speeds, and hovering. At one point, the two smaller UFOs maintained positions directly in front of the cockpit of the cargo jet at close range, pacing the jet for several minutes at a distance of only a few feet in front of the cockpit, although the jet was traveling at 570 miles per hour at the time.

The large object appeared on the radar screens of Federal Aviation Administration flight controllers, who gave the Japan Air Lines pilot permission to attempt evasive action. Veteran pilot Captain Terauchi carried out evasive maneuvers, but was not able to shake off his pursuers. The UFOs later abandoned the pursuit of their own accord, without having taken any hostile action.

FAA officials interviewed the crew members upon their arrival at Anchorage and issued a statement saying that the crew was "normal, professional, rational, and had no drug or alcohol involvement."

At first the FAA confirmed the sighting, then a few days later decided that one air traffic controller had mistakenly interpreted a split image of the cargo plane as a separate object. Establishment hatchet man Phil Klass was then called in to kill the story by announcing that Captain Terauchi, despite twenty-nine years of experience as a pilot and a hitherto impeccable record, had mistaken the planet Jupiter for a UFO. The fact that the large UFO had been witnessed not only visually by all crew members, but also on the jet's radar screen, and that neither Jupiter nor any other planet appears on radar screens, was ignored by Philip Klass. Hal Bernton, a reporter for the *Daily News* of Anchorage, Alaska, conducted an interview with the air traffic controller in question, Sam Rich, which was printed on January 9, 1987. Sam Rich's testimony contradicted the FAA's version of the event in several important ways.

Rich, who has worked with the FAA for over a decade, denied categorically that he was the only air traffic controller to have seen the radar track of the UFO. The two other controllers who were working that shift also saw it. The track was not very strong, but neither he nor his two colleagues thought that it could be a split image, a possiblity they considered at the time. Right after spotting the track, Rich phoned the Military Regional Operations Control Center, and "they informed me that they had the same radar track."

Rich confirmed that double images often occur on the FAA radar screen but said that the JAL plane was not in the area where these split images usually occur. Also, over the past decade there have been about half a dozen reports by pilots of unidentified lights in the region where the JAL plane sighted the UFOs.

To all this, I can now add the fact that there have been several sightings from the area of the JAL encounter since the incident took place, reported both by airplane pilots and by people on the ground. So who are we to believe—the air traffic controller who was actually on the job at the time of the incident, or the professional disinformation agents?

Another interesting aspect of the Japan Air Lines story is that although the incident occurred over Alaska on November 17, 1986, no U.S. media coverage of it took place until January 1987. When this six-week delay in making the story public was investigated, it turned out that the story never would have been made public at all in the United States if a family member of one of the JAL crew had not leaked the news to journalists in Japan. Once the story had entered the public domain in Japan, the U.S. authorities could no longer pretend that nothing had happened.

Yet another major development in the story of this case, which apparently just refuses to die, occurred at the end of August 1987 when MUFON* researcher T. Scott Crain, Jr. revealed (in an article entitled

---

* Mutual UFO Network

"New JAL Sighting Information," California UFO, vol. 2, no. 3) that there were indications that the images on the radar tapes had been tampered with. The FAA officials in Anchorage, Alaska, had sent the radar tapes to the main FAA office in Washington, D.C.—but they had not sent them directly. The tapes had traveled an indirect route, making an unexplained detour via the FAA Technical Center at Atlantic City, New Jersey. Researchers suspect that it was during this brief sojourn at the FAA Technical Center that the images on the tapes were altered. The Freedom of Information Act request that Mr. Crain sent to the FAA Technical Center was answered evasively.

So once again, the story of the way this case has been handled by the authorities provides a detailed demonstration of how the covert censorship enforces the UFO cover-up.

What this prolonged series of deliberate ostentatious displays appears to add up to is a reinforcement schedule, discreetly but firmly making the presence of extra-terrestrials undeniably obvious, puncturing the balloon of the big lie that has been foisted on U.S. citizens and the world for over 40 years, deflating it gradually in a manner that is calculated to oblige public recognition while avoiding public panic.

Another important case that just refuses to die is that of the "Westchester Wing," which was described in the Appendix of my previous book and which continues to be persistently reported. A major incident occurred on March 17, 1988, when hundreds of reports came in from northern New Jersey, New York City, and up the Hudson River Valley past Ossining to Mahopac, New York. As usual, the same old implausible explanation was spewed forth by the authorities: pranksters in ultralight aircraft. Attempts have been made, presumably by the authorities, to bolster the acceptability of this nonsense by sending up a fleet of ultralights now and then to imitate the Westchester Wing, but the imitations are so obviously different from the genuine sightings that this desperate ploy has been a complete flop.

Scientists at the Jet Propulsion Laboratory in Pasadena, California, who examined a videotape of the Westchester Wing made in 1984, gave their "unofficial" opinion in the form of a letter that the lights are on a single, solid object—thereby ruling out formations of ultralight aircraft. They would not, of course, go on record with an official opinion, being employees of the same authorities who continue to maintain, despite overwhelming evidence to the contrary, that what huge numbers of people in greater New York have been seeing for the last ten years is pranksters in ultralight aircraft. This unprecedented series of sightings over the same area, the first of which took place on December 31, 1982, remains ongoing.

According to UFO researcher Rosemary Decker:

In pointing out that the vast majority of contemporary humanity feels threatened or terrified by any unprecedented divergence from conventionally accepted norms of reality, you are presenting a powerful argument in favor of governmental silence and media low-profiling. In view of the fact that government agencies are already acutely aware of the fear problem, we should be willing to see that some degree of reserve and silence is appropriate. There is no good reason why everyone should be entitled to know all there is to know on an immediate and widespread scale, as most of the population could not handle it, though it would be unwise to try to tell them so directly.

The behavior patterns of our visitors indicate that they also must be aware of the dangers of sudden wide-spread publicity concerning their presence. Otherwise why would they consistently manifest in waves within specfic limited areas for specific periods of time, build gradually to peaks, and then withdraw from these areas for long periods?

Discriminating reserve and caution on the part of officialdom are appropriate. However, blatant lying, deceit, and silencing of witnesses by ridicule or personal threats are deplorable. Such tactics are undermining both national and international security. The population of the entire world has by now received absurd explanations and outright lies from their respective governments for so many years that distrust of governments has reached epidemic proportions on a global scale.

If, during the 1940s or early 1950s, the official agencies had agreed among themselves on a policy of gradual and cautious, but honest presentation of the facts known to them, with the humility to be able to say 'We don't know' at times, the situation would not have gotten so completely out of hand, as it now is. Ever since 1947, officialdom has suffered from disagreements between agencies, between individuals within a given agency, and from differences in direction as key UFO policy personnel were toppled from office and replaced. Part of the problem of inconsistency in policy has been due to varying degrees of fear of public reaction, but is also due to the individual fears of those in office, as office-holders.

What government wants to admit that it does not know everything? Researchers who have been studying this subject full-time for as long as forty years admit that they don't know everything about it, and that there are frequently extreme differences of opinion between even the best informed of the experts.

What government wants its citizens to begin to think of themselves as citizens of the planet rather than as French, Russian, American, etc.? To give their allegiance primarily to the planet and only secondarily to the nation to which they belong? With every year that passes, many more millions of people all over the world are becoming aware that we Earth-folk are being visited from elsewhere, irrespective of our national borders.

With sightings and abductions having escalated to unprecedented and ever-increasing levels, the situation is now completely out of control. The lid of secrecy imposed on the subject by the government for over 40 years is about to blow, no matter how desperately the government may attempt to continue to stonewall the high strangeness. The most effective way to avoid a sudden explosion, traumatic for all concerned, is to decrease the pressure by releasing as much information as possible in forms that the public can assimilate without being excessively traumatized, such as through this and other books and unbiased media coverage, so that there is no longer such a gross disparity between what the public has been conditioned to believe and what is actually going on.

A major development in the release of previously secret information has been the publication of the briefing papers for President Eisenhower by William Moore, Stanton Friedman, and Jaime Shandera in the spring of the 1987. There has been considerable debate over the authenticity of these documents, which describe the circumstances under which President Truman created the top secret Majestic-12 group in order to investigate the national security implications of UFO phenomena. One of the original members of MJ-12 was Admiral Roscoe Hillenkoetter, who was head of the CIA at the time that the incoming Eisenhower administration was installed, and whose signature was appended to one of the controversial documents, dated November 18, 1952.

Critics bent on disparaging the authenticity of the documents were dealt a major blow when Dr. Roger W. Westcott announced the result of his in-depth study on the basis of stylistic analysis. Dr. Westcott graduated from Princeton *Summa Cum Laude* and is Director of the Linguistics Department at Drew University. He has published 40 books in Linquistics, approximately 400 articles, and is considered the most eminent authority on this subject in the United States. Dr. Westcott compared the signature on the controversial document with the signatures on 27 other documents signed by Admiral Hillenkoetter, the authenticity of which is not in question, and with 1,200 pages of personal correspondence and memoranda written by Hillenkoetter. Dr. Westcott concluded that Admiral Hillenkoetter's signature on the controversial document is authentic.

It would seem that such a verdict, combined with the information content of the document whose authenticity was thus confirmed—concerning a crashed UFO and the recovery of four small alien bodies—should be sufficient to deal a final death blow to the credibility of our government's publicly stated official attitude towards UFOs. However, as any psychologist will tell you, deeply entrenched, long held, rigidly assumed, conventionally accepted, blind and fanatical belief systems do not die easily. They tend to be thick-skinned to the point of being almost impervious to logic. I stress that "almost," as it is our only hope of at last achieving a sane and rational approach to the subject.

Edward Mazur made some very relevant remarks about the MJ-12 controversy in the July/August 1989 issue of the *Arkansas MUFON Newsletter:*

> The unauthorized disclosure of a highly classified document is a serious federal crime. The forging of a classified document purported to come from the highest levels of government is perhaps an even more serious crime. Yet in the five years or so since the documents surfaced, there have been no arrests or prosecutions by the Department of Justice. Why?
>
> According to this writer's logic, the FBI could have easily determined, through the issuing agency, whether the document was authentic or forged. If it was a forgery, there wouldn't be great difficulty in finding the forger, prosecuting him, and setting an example.
>
> This action would also discredit and ridicule the gullible UFO community who had 'bought' MJ-12. Why didn't the FBI take advantage of this opportunity if the document was phony? Or wasn't it?
>
> But if the document was genuine, what would the government gain by apprehending its leaker? Prosecution would be a public admission that MJ-12 was authentic and would reveal the very fact that the document's high and sensitive classification was designed to suppress. It would be far better to treat the matter with benign neglect, as is the case now, and to work behind the scenes to thwart any progress the might be made by UFO researchers, while undermining their activity wherever possible.
>
> The fact that there have been no indications of any investigations, arrests, or prosecutions in the past five years in the matter of these documents is of great significance.

According to the *Nevada Aerial Research Newsletter*, P.O. Box 81407, Las Vegas, NV 89180, the black-uniformed elite Delta Special Forces, which carry out their missions in black unmarked helicopters and which act as security for the U.S. government alien-related projects, are selected almost exclusively from soldiers who grew up as orphans or have no close family ties.

Is this because the enemy they are trained to fight is the citizens of the United States? And we are paying for this with our own tax dollars?

The publication of the U.S. edition of *Above Top Secret* by British researcher Timothy Good (Morrow, 1988) was a landmark event that from here on out puts the critics who persist in denying the reality of UFO phenomena on the defensive. Timothy Good employs a similar technique to that of Barry Greenwood and Lawrence Fawcett in their *Clear Intent* (Prentice-Hall, 1984), using contradictions within government documents to demonstrate that the government is perpetrating a cover-up. However, *Clear Intent* was focused mainly on the devious activities of the intelligence community within the United States. The scope of *Above Top Secret* is world-wide. Timothy Good also deals with the United States, bringing up much material that was not included in *Clear Intent*, but the main thrust of his book is a meticulously detailed investigation of what went on concerning UFOs within the intelligence communities and officialdom of England, Canada, Australia, Russia, China, France, Italy, Portugal, and Spain. It is particularly illuminating to compare the information contained in the *Above Top Secret* with information contained in Clear Intent, as they supplement each other in remarkable fashion, and the correlations provide powerful confirmation of their basic hypotheses. Both books are focused primarily on unidentified flying objects as aircraft, dealing only marginally with the subject of UFO occupants, which is of course the subject that my own books have been devoted to investigating. The publication of *Clear Intent* put the cynics who maintain that UFO sightings are all explicable in terms of weather balloons, the planet Venus, swamp gas, mass hysteria, or flocks of geese in a difficult position. The publication of the worldwide evidence presented with such concise, conservatively understated, devastating effectiveness in *Above Top Secret* put these same cynics in an impossible situation, from which there is no way they can recover their lost credibility.

Let us now turn our attention to the aborted attempt to bring the subject of UFOs to the attention of the United Nations.

A highly important figure in this series of events was Major Colman Von Keviczky, whose background was summarized in the following terms by his colleague and long-time research associate, J. Antonion Hunccus, in the *New York City Tribune* of May 19, 1988:

Von Keviczky received his Master of Military Science and Engineering (MMSE) at the historical Ludovica University in Budapest. As a Captain and then Major with the Royal Hungarian Army, he created the Audio-Visual Department of the Hungarian General Staff before World War II. After the war he worked for the U.S. occupation forces in Germany and emigrated to the United States in 1952, the year his interest in UFOs began. Von Keviczky is a member of the American Institute of Aeronautics & Astronautics (AIAA) and his biography appears in *Who's Who in Aviation & Aerospace.*

In the mid-60s, Von Keviczky worked with the United Nations staff audio-visual department, where he became involved in a controversy over UFOs and the UN. In 1966, Von Keviczky was actually commissioned by Secretary General U Thant to work on a preliminary memo on how the UFO problem could be inserted in the UN agenda. However, the Major found he no longer had a job at the UN after he leaked news of this assignment to the press. Yet U Thant confided around that time that "he considers UFOs the most important problem facing the UN next to the war in Vietnam,' as reported in Drew Pearson's syndicated column.

While Von Keviczky was employed as a staff member of the United Nations Secretariat's Office of Public Information, the UFO wave of 1952 over Washington, D.C., occurred. Being an expert in photography, Von Keviczky realized that the photographs were genuine, and became interested in the subject. Private discussions with diplomats, scientists, and old friends who were still military officers convinced him of the subject's importance. In 1966, he undertook the initiative that destroyed his career, which will now be described in his own words:

In February 1966, after a long-scrutinized military study of the UFOs' global operation, as Staff Member of the United Nations Secretariat, I addressed THE FIRST UFO MEMORANDUM to my Secretary General U Thant. Seizing on his constitutional duty regarding the endangered international security, Thant assigned me to elaborate the FIRST UN-UFO PROJECT. This project referred to:

1. A coordinated cooperation amongst the nations to control the UFOs' global operation and activities.

2. Immediate STOP to any HOSTILE CONFRONTATION, which at any time could trigger a fatal Space War.

3. Seek OFFICIAL CONTACT AND COMMUNICATION with the exploring UFO forces, assisted by UNESCO, and by the governments' respective UFO organizations.

4. Declare the 550-mile belt around our Celestial Body under the PROTECTION and JURISDICTION of the United Nations.

Thant's common sense and constitutional duty on the alarming worldwide UFO fever is demonstrated by his remark within the diplomatic corps and his cabinet that: "UFOS ARE THE MOST IMPORTANT PROBLEM FACING THE UNITED NATIONS AFTER THE WAR IN VIETNAM." The Pentagon and the U.S. diplomacy were immediately alerted to stop him!

During the next month, in March, to thwart Thant's UN-UFO Project, the Air Force Scientific Research Board "AD HOC PANEL" was mobilized in haste to find a suitable University to study the UFO phenomena. Evidence: CONDON REPORT, preface, pages 7–9 written by the Vice President of the Colorado University.

Thant was totally silenced! A "mysterious" diplomatic power constrained him to violate the UN Constitution, and confess also toward the public that his interest in UFOs was only "purely academic and personal."

According to the Associated Press, Ambassador Trofimirovich

Fedorenko of the U.S.S.R. comforted him thus: "UFOs are only the nightmares of the imperialist and capitalist countries." But against this, on the other side of the token . . .

"For my honorarily-accepted UN-UFO Project, I became the No. 1 ENEMY OF THE UNITED STATES UFO POLICY. I was awarded with the notorious SECURITY RISK for scientists! This governmental denunciation over my lifetime prevented me from having any career job at reputable firms in the United States. . . ."

United Nations Secretary General U Thant, and the Member Nations in 1966, were convinced that the Colorado University SCIENTIFIC STUDY OF UFOs had been established to supervise the Pentagon UFO file trustworthiness and credibility, which would liquidate the media-generated UFO fever.

Only three years later, the Report's preface, written by the University's vice president Thruston E. Manning, exposed how he was hoodwinked in 1966, because the Committee was assigned to study the UFO phenomena "WHOLLY OUTSIDE THE JURISDICTION OF THE AIR FORCE." Namely, "outside the jurisdiction" meant the scientific study of the hundreds of UFO weekend clubs and news clipping collector hobbyists, thereby discrediting the respected NICAP* and APRO valuable public research. Evidence:

On February 20, 1967, before the Committee started, the CIA gave Dr. Edward U. Condon, Committee Director, the necessary guideline and instruction. THE MEETING WAS SECRET. No comments on the masterly delusion and deception of the nations!

Well, in UFO research all roads lead to the USA's Rome—as we have learned—THE OMNISCIENT AND OMNIPOTENT C! I! A!

The climax of the struggle over whether or not the subject of UFOs should be placed on the agenda of the United Nations will now be briefly described.

It is a matter of historic record that Dr. J. Allen Hynek and Dr. Jacques Vallee prevented Von Keviczky from presenting his evidence that UFOs are space-craft of extra-terrestrial origin before the Special Political Committee of the United Nations, by threatening to boycott the Committee if Von Keviczky was allowed to testify. The other three experts were Dr. David Saunders, Dr. Claude Poher, and astronaut Gordon Cooper. Only Gordon Cooper, who held the rank of Colonel in the U.S. Air Force and was therefore under obligation to obey orders, supported Hynek and Vallee on this issue. Due to the pressure that Hynek and Vallee exerted, Von Keviczky's invitation to testify was canceled.

In documents concerning the Robertson Panel that was convened by the CIA in 1953 to deal with the subject of UFOs, Dr. Hynek was listed not only as an Air Force Project Bluebook consultant, but also as an "OSI"

---

* National Investigations Committee on Aerial Phenomena

consultant. "OSI" is an abbreviation for the Office of Scientific Intelligence, which is a sub-section of the CIA. According to a public statement made by Dr. Hynek, he was not invited to attend all the sessions of the Robertson Panel. However, this statement is contradicted by an official document dated January 27, 1953, and declassified in 1977, which stated that Hynek "sat in on all the sessions after the first day," but did not sign the report as an official group member. Hynek was already an OSI consultant before he became a consultant to Project Bluebook. Did he remain an OSI consultant for the rest of his life? Was he assigned to play the part of the maverick scientist in revolt against the authorities, when the authorities realized the explanations, such as "swamp gas" were no longer credible, in order to more effectively acquire information from and control over the genuinely independent researchers?

Jacques Vallee began his UFO research career in the 1960s as the assistant of Dr. Hynek. Considering this in combination with the result of his close collaboration with Dr. Hynek at the United Nations, is Dr. Vallee also an OSI consultant?

On November 27, 1978, Dr. Vallee stated to the Special Political Committee of the United Nations, which had been convened to decide whether or not to place the subject of UFOs on the UN agenda, that: "although the UFO phenomenon is real and appears to be caused by an unknown physical stimulus, I have so far failed to discover any evidence that it represents the arrival of visitors from outer space." Having said that, he skipped briefly over the physical manifestations of the phenomenon, saying only they should be studied. He then stressed the importance of studying the psychophysiological effects on witnesses, carefully pointing out that: "I do not believe it is within the province or the budget of the United Nations to address such effects directly . . ." He continued by insisting at length on the importance of studying the social belief systems generated by the phenomenon, as well as the emotional factors involved—aspects of the phenomenon so clearly outside the province and budget of the United Nations that it was not necessary for him to repeat this a second time. Thus with a few deft strokes, he effectively sabotaged the placing of the subject of the UFO phenomenon on the agenda of the United Nations, which was exactly what the CIA wanted to prevent. If Von Keviczky had been allowed to present his evidence, there is a strong probability that the Committee would have decided to place the subject on the UN agenda, after which the cover-up would have been impossible to maintain.

I salute the example set by Major Colman Von Keviczky, as I consider him a genuine modern hero. The four propositions he suggested to Secretary General U Thant should be implemented by the United Nations without further delay, since they are as valid now as they were in 1966.

In my opinion, the close encounter and abduction cases constitute the spearhead of UFO research. Of course, it is essential to distinguish between genuine and fraudulent cases, and this is not always easy to do. However, with patient, open-minded, persistent, and alert attention, it can be done. The analogy of sorting out batches of gemstone rough is relevant here. There are ways of detecting whether one is in the presence of the real thing or an imitation.

The genuine contact and abduction cases are the interface between terrestrial humanity and the UFO phenomenon. There are literally thousands of such case histories on record, and perhaps tens or hundreds of thousands, or more, which have not been reported. I would suggest that about two-thirds of the hypothetical budget be allocated to investigating as thoroughly as possible the thousands of case histories already on record: evaluating them for authenticity, conducting follow-up interviews and hypnotic regression sessions when appropriate, and feeding all the information obtained into computers.

The data bank derived from the review of contact and abduction cases should then be correlated with the data bank derived from the lights-in-the-sky and physical nuts-and-bolts manifestations.

If project funds are still available after this procedure has been completed, we should start interviewing the oceanic multitude of contactees and abductees whose stories have only recently begun to surface and are not yet on record. These interviews would be conducted and tabulated along the same lines as those case histories already on record.

From these myriad correlations, certain major patterns should emerge. What I would consider to be of prime importance would be the patterns indicating the characteristics of the different types of extra-terrestrial and/or inter-dimensional humanoids, human-appearing beings, and extremely dissimilar alien entities involved in these manifestations, to which we have been applying the catch-all UFO label. The information derived from such profiles would include not only the physical characteristics and types of craft most frequently used, but also the behavioral characteristics. Typical ways of interacting with us would contain clues as to motivations for making contact with us, as well as to psychological traits and the extent to which communication may be possible. Such questions as superiority, inferiority, or equality of intelligence between them and us would be explored, as well as unusual aspects of their intelligence. Indications of the relative benevolence, malevolence, or neutrality of the various types would show up clearly in such profiles, as well as of their friendliness or hostility to each other. As abductee Ida Kannenberg has so perceptively pointed out: "There are so many different types of extra-terrestrials that it is not possible to make statements that are valid for all of them. Many types are as alien to each other as they are to us."

By establishing reliable profiles of the different types most persistently reported, we would at least know what we are dealing with and be in a far better position to communicate meaningfully.

If there was a war in heaven that is still going on, in which who we give our allegiance to may be a matter of importance, though Earth may be no more than a single sector of a multi-galactic battle zone, at least we would be able to make an informed choice concerning which group we enter an alliance with. That would certainly be preferable to making a decision of such importance in our present state of blind ignorance.

If, on the other hand, peace and harmony reign supreme over the inhabitants of outer space, and it is we humans who must learn to transcend our aggressive bellicose natures in order to become eligible for galactic citizenship, we still need to know who we are dealing with, and be able to communicate with them.

Those who are still arguing about whether or not UFOs are real will continue to do so, until obliged to face the facts with their own eyes. For those of us who are already aware that UFOs are real, the question becomes: What types of beings are piloting them, and what is their motivation for keeping us under surveillance and clandestinely interacting with us?

It is a matter of extreme urgency that an all-out effort be made to find out as much as possible about the different types of non-human intelligent life-forms at present hovering above us and among us. We can no longer continue to pretend that we are dealing with misidentifications of weather balloons, the planet Venus, flocks of geese, or swamp gas, and retain our position as the dominant life-form on planet Earth. It is time to face the fact that outer space is inhabited, and that ever since we exploded the first atomic bombs its inhabitants have been watching us very closely. It is obvious that in comparison to a number of already existing alien civilizations, our space technology is in the kindergarten stage. We must establish open alliance with the groups we can work out mutually beneficial relationships with, and take appropriate measures to defend ourselves against the predatory activities of the groups who have come here to exploit us. It is imperative that we learn to distinguish between extra-terrestrial friends and foes. What is at stake is our survival, not only as individuals, but as a freely evolving species.

30 UFOs and the U.S. Air Force

PART I

Unidentified Flying Objects (UFOs)

What is an Unidentified Flying Object (UFO)?

Well, according to United States Air Force Regulation 80-17 (dated 19 September 1966), a UFO is "any aerial phenomenon or object which is unknown or appears to be out of the ordinary to the observer."

This is a very broad definition which applies equally well to one individual seeing his first noctilucent [luminous] cloud at twilight as it does to another individual seeing his first helicopter. However, at present most people consider the term UFO to mean an object which behaves in a strange or erratic manner while moving through the Earth's atmosphere. That strange phenomenon has evoked strong emotions and great curiosity among a large segment of our world's population. The average person is interested because he loves a mystery, the professional military man is involved because of the possible threat to national security, and some scientists are interested because of the basic curiosity that led them into becoming researchers.

The literature on UFOs is so vast, and the stories so many and varied, that we can only present a sketchy outline of the subject in this chapter. That outline includes description classifications, operational domains (temporal and spatial), some theories as to the nature of the UFO phenomenon, human reactions, attempts to attack the problem scientifically, and some tentative conclusions.

33.1—Descriptions

One of the greatest problems you encounter when attempting to catalog UFO sightings, is selection of a system for cataloging. No effective sys-

From the United States Air Force Academy (Department of Physics) Textbook *Introductory Space Science, Volume II.*

tem has yet been devised, although a number of different systems have been proposed. The net result is that almost all UFO data are either treated in the form of individual cases, or in the forms of inadequate qualification systems. However, these systems do tend to have some common factors, and a collection of these factors is as follows:

[a] Size; [b] Shape (disc, ellipse, football, etc.); [c] Luminosity; [d] Color; [e] Number of UFOs.

Behavior: [a] Location (altitude, direction, etc.); [b] Patterns of paths (straight line, climbing, zig-zagging, etc.); [c] Flight characteristics (wobbling, fluttering, etc.); [d] Periodicity of sightings; [e] Time duration; [f] Curiosity or inquisitiveness; [g] Avoidance; [h] Hostility.

Associated Effects: [a] Electro-magnetic (compass, radio, ignition systems, etc.); [b] Radiation (burns, induced radioactivity, etc.); [c] Ground disturbance (dust stirred up, leaves moved, standing wave peaks of surface of water, etc.); [d] Sound (none, hissing, humming, roaring, thunderclaps, etc.); [e] Vibration (weak, strong, slow, fast); [f] Smell (ozone or other odor); [g] Flame (how much, where, when, color); [h] Smoke or cloud (amount, color, persistence); [I] Debris (type, amount, color, persistence); [j] Inhibition of voluntary movement by observers; [k] Sighting of "creatures" or "beings."

After Effects: [a] Burned areas or animals; [b] Depressed or flattened areas; [c] Dead or "missing" animals; [d] Mentally disturbed people; [e] Missing items.

We make no attempt here to present available data in terms of the foregoing descriptors.

33.2—Operational Domains—Temporal and Spatial

What we will do here is to present evidence that UFOs are a global phenomenon which may have persisted for many thousands of years. During this discussion, please remember that the more ancient the reports the less sophisticated the observer. Not only were the ancient observers lacking the terminology necessary to describe complex devices (such as present day helicopters) but they were also lacking the concepts necessary to understand the true nature of such things as television, spaceships, rockets, nuclear weapons and radiation effects. To some, the most advanced technological concept was a war chariot with knife blades attached to the wheels. By the same token, the very lack of accurate terminology and descriptions leaves the more ancient reports open to considerable misinterpretation, and it may well be that present evaluations of individual reports are completely wrong. Nevertheless, let us start with an intriguing story in one of the oldest chronicles of India—the *Book of Dzyan*.

The book is a group of "story-teller" legends which were finally gathered in manuscript form when man learned to write. One of the stories is of a small group of beings who supposedly came to Earth many thousands of years ago in a metal craft which orbited the Earth several times before landing. As told in the Book:

> These beings lived on Earth while largely keeping to themselves and were revered by the humans among whom they had settled. But eventually differences arose among them and they divided their numbers, several of the men and women and some children settled in another city, where they were promptly installed as rulers by the awe-stricken populace.
>
> Separation did not bring peace to these people and finally their anger reached a point where the ruler of the original city took with him a small number of his warriors and they rose into the air in a huge shining metal vessel. While they were many leagues from the city of their enemies, they launched a great shining lance that rode on a beam of light. It burst apart in the city of their enemies with a great ball of flame that shot up to the heavens, almost to the stars. All those who were in the city were horribly burned and even those who were not in the city—but nearby—were burned also. Those who looked upon the lance and the ball of fire were blinded forever afterward. Those who entered the city on foot became ill and died. Even the dust of the city was poisoned, as were the rivers that flowed through it. Men dared not go near it, and it gradually crumbled into dust and was forgotten by men.
>
> When the leader saw what he had done to his own people he retired to his palace and refused to see anyone. Then he gathered about him those warriors who remained, and their wives and children, and they entered their vessels and rose one by one into the sky and sailed away. Nor did they return.

Could this foregoing legend really be an account of an extraterrestrial colonization, complete with guided missile, nuclear warhead and radiation effects? It is difficult to assess the validity of that explanation—just as it is difficult to explain why Greek, Roman and Nordic Mythology all discuss wars and conflicts among their "Gods." (Even the Bible records conflict between the legions of God and Satan.) Could it be that each group recorded their parochial view of what was actually a global conflict among alien colonists or visitors? Or is it that man has led such a violent existence that he tends to expect conflict and violence among even his gods?

Evidence of perhaps an even earlier possible contact was uncovered by Tschi Pen Lao of the University of Peking. He discovered astonishing carvings in granite on a mountain in Hunan Province and on an island in Lake Tungting. These carvings have been evaluated as 47,000 years old, and they show people with large trunks (breathing apparatus?) or "elephant" heads shown on human bodies. (Remember, the Egyptians often represented their gods as animal heads on human bodies.) Only 8,000 years ago, rocks were sculpted in the Tassili plateau of Sahara, depicting what appeared to be human beings but with strange round heads (helmets? or "sun" heads on human bodies?) And even more recently, in the Bible, Genesis 6:4 tells of angels from the sky mating with women of Earth, who bore them children. Genesis 19:3 tells of Lot meeting two angels in the desert and his later feeding them at his house. The Bible also tells a rather unusual story of Ezekiel who witnessed what has been interpreted by some to have been a spacecraft or aircraft landing near the Chebar River in Chaldea* (593 B.C.).

Even the Irish have recorded strange visitations. In the Speculum Regali in Konungs Skuggsa (and other accounts of the era about 956 A.D.) are numerous stories of "demonships" in the skies. In one case a rope from one such ship became entangled with part of a church. A man from the ship climbed down the rope to free it, but was seized by the townspeople. The bishop made the people release the man, who climbed back to the ship, where the crew cut the rope and the ship rose and sailed out of sight. In all of his actions, the climbing man appeared as if he were swimming in water. Stories such as this makes one wonder if the legends of the "little people" of Ireland were based upon imagination alone.

About the same time, in Lyons, France, three men and a woman supposedly descended from an airship or spaceship and were captured by a mob. These foreigners admitted to being wizards, and were killed. (No mention is made of the methods employed to extract the admissions.) Many documented UFO sightings occurred throughout the Middle Ages, including an especially startling one of a UFO over London on 16 December 1742. However, we do not have room to include any more of the Middle Ages sightings. Instead, two "more recent" sightings are contained in this section to bring us up to modern times.

In a sworn statement dated 21 April 1897, a prosperous and prominent farmer named Alexander Hamilton (Lea Roy, Kansas) told of an attack upon his cattle at about 10:30 P.M. the previous Monday. He, his son, and his tenant grabbed axes and ran some 700 feet from the house to the cow lot where a great cigar-shaped ship about 300 feet long floated some 30 feet above his cattle. It had a carriage underneath which was brightly lighted within (dirigible and gondola?) and which had numerous windows. Inside were six strange-looking beings jabbering in a for-

---

* An ancient region of Mesopotamia.

eign language. These beings suddenly became aware of Hamilton and the others. They immediately turned a searchlight on the farmer, and also turned on some power which sped up a turbine wheel (about 30 feet in diameter) located under the craft. The ship rose, taking with it a two-year-old heifer which was roped about the neck by a cable of one-half inch thick, red material. The next day a neighbor, Link Thomas, found the animal's hide, legs and head in his field. He was mystified at how the remains got to where they were because of the lack of tracks in the soft soil. Alexander Hamilton's sworn statement was accompanied by an affidavit as to his veracity. The affidavit was signed by ten of the local leading citizens.

On the evening of 4 November 1957 at Fort Itaipu, Brazil, two sentries noted a "new star" in the sky. The "star" grew in size and within seconds stopped over the fort. It drifted slowly downward, was as large as a big aircraft, and was surrounded by a strong orange glow. A distinct humming sound was heard, and then the heat struck. A sentry collapsed almost immediately, the other managed to slide to shelter under the heavy cannons where his loud cries awoke the garrison. While the troops were scrambling towards their battle stations, complete electrical failure occurred. There was panic until the lights came back on but a number of men still managed to see an orange glow leaving the area at high speed. Both sentries were found badly burned—one unconscious and the other incoherent, suffering from deep shock.

Thus, UFO sightings not only appear to extend back to 47,000 years through time but also are global in nature. One has the feeling that this phenomenon deserves some sort of valid scientific investigation, even if it is a low level effort.

33.3—Some Theories as to the Nature of the UFO Phenomenon

There are very few cohesive theories as to the nature of UFOs. Those theories that have been advanced can be collected in five groups:

[a] Mysticism; [b] Hoaxes, and rantings due to unstable personalities; [c] Secret Weapons; [d] Natural Phenomena; [e] Alien visitors.

[a] Mysticism. It is believed by some cults that the mission of UFOs and their crews is a spiritual one, and that all materialistic efforts to determine the UFOs' natures are doomed to failure.

[b] Hoaxes, and Rantings due to Unstable Personalities. Some have suggested that all UFO reports were the results of pranks and

hoaxes, or were made by people with unstable personalities. This attitude was particularly prevalent during the time period when the Air Force investigation was being operated under the code name of Project Grudge. A few airlines even went as far as to ground every pilot who reported seeing a "flying saucer." The only way for the pilot to regain flight status was to undergo a psychiatric examination. There was a noticeable decline in pilot reports during this time interval, and a few interpreted this decline to prove that UFOs were either hoaxes or the result of unstable personalities. It is of interest that NICAP (The National Investigations Committee on Aerial Phenomena) even today still receives reports from commercial pilots who neglect to notify either the Air Force or their own airline.

There are a number of cases which indicate that not all reports fall in the hoax category. We will examine one such case now. It is the Socorro, New Mexico sighting made by police Sergeant Lonnie Zamora. Sergeant Zamora was patrolling the streets of Socorro on 24 April 1964 when he saw a shiny object drift down into an area of gullies on the edge of town. He also heard a loud roaring noise which sounded as if an old dynamite shed located out that way had exploded. He immediately radioed police headquarters, and drove out toward the shed. Zamora was forced to stop about 150 yards away from a deep gully in which there appeared to be an overturned car. He radioed that he was investigating a possible wreck, and then worked his car up onto the mesa and over toward the edge of the gully. He parked short, and when he walked the final few feet to the edge, he was amazed to see that it was not a car but instead was a weird egg-shaped object about fifteen feet long, white in color and resting on short, metal legs. Beside it, unaware of his presence, were two humanoids dressed in silvery coveralls. They seemed to be working on a portion of the underside of the object. Zamora was still standing there, surprised, when they suddenly noticed him and dove out of sight around the object. Zamora also headed the other way, back toward his car. He glanced back at the object just as a bright blue flame shot down from the underside. Within seconds the egg-shaped thing rose out of the gully with "an ear-splitting roar." The object was out of sight over the nearby mountains almost immediately, and Sergeant Zamora was moving the opposite direction almost as fast when he met Sergeant Sam Chavez who was responding to Zamora's earlier radio calls. Together they investigated the gully and found the bushes charred and still smoking where the blue flame had jetted down on them. About the charred area were four deep marks where the metal legs had been. Each mark was three and one half inches deep, and was circular in shape. The sand in the gully was very hard packed, so no sign of the humanoids' footprints could be found. An official investigation was launched that same day, and all data obtained supported the stories of Zamora and Chavez. It is rather difficult to label this episode a hoax, and it is also doubtful that both Zamora and Chavez shared portions of the same hallucination.

[c] Secret Weapons. A few individuals have proposed that UFOs are actually advanced weapon systems, and that their natures must not be revealed. Very few people accept this as a credible suggestion.

[d] Natural Phenomena. It has also been suggested that at least some, and possibly all of the UFO cases were just misinterpreted manifestations of natural phenomena. Undoubtedly this suggestion has some merit. People have reported, as UFOs, objects which were conclusively proven to be balloons (weather and skyhook), the planet Venus, man-made artificial satellites, normal aircraft, unusual cloud formations, and lights from ceilometers (equipment projecting light beams on cloud bases to determine the height of the aircraft visual ceiling). It is also suspected that people have reported mirages, optical illusions, swamp gas and ball lightning (a poorly-understood discharge of electrical energy in a spheroidal or ellipsoidal shape . . . some charges have lasted for up to fifteen minutes but the ball is usually no bigger than a large orange). But it is difficult to tell a swamp dweller that the strange, fast-moving light he saw in the sky was swamp gas; and it is just as difficult to tell a farmer that a bright UFO in the sky is the same ball lightning that he has seen rolling along his fence wires in dry weather. Thus accidental misidentification of what might well be natural phenomena breeds mistrust and disbelief; it leads to the hasty conclusion that the truth is deliberately not being told. One last suggestion of interest has been made, that the UFOs were plasmoids from space-concentrated blobs of solar wind that succeeded in reaching the surface of the Earth. Somehow this last suggestion does not seem to be very plausible; perhaps because it ignores such things as penetration of Earth's magnetic field.

PART II

Alien Visitors

The most stimulating theory for us is that the UFOs are material objects which are either "manned" or remote-controlled by beings who are alien to this planet. There is some evidence supporting this viewpoint. In addition to police Sergeant Lonnie Zamora's experience, let us consider the case of Barney and Betty Hill. On a trip through New England they lost two hours on the night of 19 September 1961 without even realizing it. However, after that night both Barney and Betty began developing psychological problems which eventually grew sufficiently severe that they submitted themselves to psychiatric examination and treatment. During the course of treatment, hypnotherapy was used, and it yielded remarkably detailed and similar stories from both Barney and Betty. Essentially they had been hypnotically kidnapped, taken aboard a UFO, submitted to two-hour physicals, and released with posthypnotic suggestions to forget the entire incident. The evidence is rather strong that this is what the Hills, even in their subconscious, believe happened to them. And it is of particular importance that after the "posthypnotic block" was removed, both of the Hills ceased having their psychological problems.

The Hill's description of the aliens was similar to descriptions provided in other cases, but this particular type of alien appears to be in the minority. The most commonly described alien is about three and one half feet tall, has a round head (helmet?), arms reaching to or below his knees, and is wearing a silvery space suit or coveralls. Other aliens appear to be essentially the same as Earthmen, while still others have particularly wide (wrap around) eyes and mouths with very thin lips. And there is a rare group reported as about 4 feet tall, weight of around 35 pounds, and covered with thick hair or fur (clothing?). Members of this last group are described as being extremely strong. If such beings are visiting Earth, two questions arise: 1) Why haven't they attempted to contact us officially? The answer to the first question may exist partially in Sergeant Lonnie Zamora's experience, and may exist partially in the Tunguska meteor.* It was suggested that the Tunguska meteor was actually a comet which exploded in the atmosphere, the ices melted and the dust spread out. Hence, no debris. However, it has also been suggested that the Tunguska meteor was actually an alien spacecraft that entered the atmosphere too rapidly, suffered mechanical failure, and lost its power supply and/or weapons in a nuclear explosion. While that hypothesis may seem far-fetched, sample of tree rings from around the world reveal that, immediately after the Tunguska meteor explosion, the level of radioactivity in the world rose sharply for a short period of time. It is difficult to find a natural explanation for that increase in radioactivity, although the suggestion has been advanced that enough of the meteor's great Kinetic energy was converted into heat (by atmospheric friction) that a fusion reaction occurred. This still leaves us with no answer to the second question: Why no contact? That question is very easy to answer in several ways: 1) We may be the object of intensive sociological and psychological study. In such studies you usually avoid disturbing the test subjects' environment; 2) You do not "contact" a colony of ants, and humans may seem that way to any aliens (variation: a zoo is fun to visit, but you don't "contact" the lizards); 3) Such contact may have already taken place secretly; and 4) Such con-

---

* A massive explosion that occurred in Siberia in 1908.

tact may have already taken place on a different plane of awareness and we are not yet sensitive to communications on such a plane. These are just a few of the reasons. You may add to the list as you desire.

33.4—Human Fear and Hostility

Besides the foregoing reasons, contacting humans is downright dangerous. Think about that for a moment! On the microscopic level our bodies reject and fight (through production of antibodies) any alien material; this process helps us fight off disease but it also sometimes results in allergic reactions to innocuous materials. On the macroscopic (psychological and sociological) level we are antagonistic to beings that are "different." For proof of that, just watch how an odd child is treated by other children, or how a minority group is socially deprived. . . . In case you are hesitant to extend that concept to the treatment of aliens let me point out that in very ancient times, possible extraterrestrials may have been treated as Gods but in the last 2000 years, the evidence is that any possible aliens have been ripped apart by mobs, shot and shot at, physically assaulted, and in general treated with fear and aggression.

In Ireland about 1000 A.D., supposed airships were treated as "demonships." In Lyons, France, "admitted" space travelers were killed. More recently, on 24 July 1957 Russian anti-aircraft batteries on the Kouril Islands opened fire on UFOs.* Although all Soviet anti-aircraft batteries on the Islands were in action, no hits were made. The UFOs were luminous and moved very fast. We, too, have fired on UFOs. About ten o'clock one morning, a radar site near a fighter base picked up a UFO doing 700 miles per hour. The UFO then slowed to 100 miles per hour, and two F-86s scrambled to intercept. Eventually one F-86 closed on the UFO at about 3000 feet altitude. The UFO began to accelerate away but the pilot still managed to get within 500 yards of the target for a short period of time. It was definitely saucer shaped. As the pilot pushed the F-86 at top speed, the UFO began to pull away. When the range reached 1000 yards, the pilot armed his guns and fired in an attempt to down the saucer. He failed, and the UFO pulled away rapidly, vanishing in the distance.

This same basic situation may have happened on a more personal level. On Sunday evening 21 August 1955, eight adults and three children were on the Sutton Farm (one-half mile from Kelly, Kentucky) when, according to them, one of the children saw a brightly glowing UFO settle behind the barn, out of sight from where he stood. Other witnesses on nearby farms also saw the object. However, the Suttons dismissed it as a "shooting star," and did not investigate. Approximately thirty minutes later (at 8:00 P.M.), the family dogs began barking, so two of the men went to the

---

* The Kouril Islands are located north of Japan.

back door and looked out. Approximately 50 feet away and coming toward them was a creature wearing a glowing silvery suit. It was about three-and-one-half feet tall with a large round head and very long arms. It had large webbed hands which were equipped with claws. The two Suttons grabbed a twelve gauge shotgun and a .22 caliber pistol, and fired at close range. They could hear the pellets and bullet ricochet as if off of metal. The creature was knocked down, but jumped up and scrambled away. The Suttons retreated into the house, turned off all inside lights, and turned on the porch light. At that moment, one of the women who was peeking out of the dining room window discovered that a creature with some sort of helmet and wide slit eyes was peeking back at her. She screamed, the men rushed in and started shooting. The creature was knocked backwards but again scrambled away without apparent harm. More shooting occurred (a total of about fifty rounds) over the next twenty minutes and the creatures finally left (perhaps feeling unwelcome?). After about a two hour wait (for safety), the Suttons left too. By the time the police got there, the aliens were gone but the Suttons would not move back to the farm. They sold it and departed. This reported incident does bear out the contention though that aliens are dangerous. At no time in the story did the supposed aliens shoot back, although one is left with the impression that the described creatures were having fun scaring humans.

33.5—Attempts at Scientific Approaches

In any scientific endeavor, the first step is to acquire data, the second step is to classify the data, and the third step is to form a hypothesis. This hypothesis is tested by repeating the entire process, with each cycle resulting in an increase in understanding (we hope). The UFO phenomenon does not yield readily to this approach because the data taken so far exhibits both excessive variety and vagueness. The vagueness is caused in part by the lack of preparation of the observer—very few people leave their house knowing that they are going to see a UFO that evening. Photographs are overexposed or underexposed, and rarely in color. Hardly anyone carries around a radiation counter or magnetometer. And, in addition to this, there is a very high level of "noise" in the data.

The noise consists of mistaken reports of known natural phenomena, hoaxes, reports by unstable individuals, and mistaken removal of data regarding possible unnatural or unknown natural phenomena (by overzealous individuals who are trying to eliminate all data due to known natural phenomena). In addition, those data, which do appear to be valid, exhibit an excessive amount of variety relative to the statistical samples which are available. This has led to very clumsy classification systems, which in turn provide quite unfertile ground for formulation of hypotheses.

One hypothesis which looked promising for a time was that of OR-THOTENY (i.e., UFO sightings fall on "area circle" routes). At first, plots of sightings seemed to verify the concept of orthoteny but recent use of computers has revealed that even random numbers yield "great circle" plots as neatly as do UFO sightings.

There is one solid advance that has been made though. Jacques and Janine Vallee have taken a particular type of UFO—namely those that are lower than tree-top level when sighted—and plotted the UFOs' estimated diameters versus the estimated distance from the observer. The result yields an average diameter of 5 meters with a very characteristic drop for short viewing distances. This behavior at the extremes of the curve is well known to astronomers and psychologists as the "moon illusion." The illusion only occurs when the object being viewed is a real, physical object. Because this implies that the observers have viewed a real object, it permits us to accept also their statement that these particular UFOs had a rotational axis of symmetry.

Another, less solid, advance made by the Vallees was their plotting of the total number of sightings per week versus the date. They did this for the time span from 1947 to 1962, and then attempted to match the peaks of the curve (every 2 years, 2 months) to the times of Earth-Mars conjunction (every 2 years, 1.4 months). The match was very good between 1950 and 1956 but was poor outside those limits. Also, the peaks were not only at the times of Earth-Mars conjunction but also roughly at the first harmonic (very loosely, every 13 months). This raises the question why should UFOs only visit Earth when Mars is in conjunction and when it is on the opposite side of the sun. Obviously, the conjunction periodicity of Mars is not the final answer. As it happens, there is an interesting possibility to consider. Suppose Jupiter's conjunctions were used; they are every 13.1 months. That would satisfy the observed periods nicely, except for every even data peak being of different magnitude from every odd data peak. Perhaps a combination of Martian, Jovian, and Saturnian (and even other planetary) conjunctions will be necessary to match the frequency plot—if it can be matched.

Further data correlation is quite difficult. There are a large number of different saucer shapes but this may mean little. For example, look at the number of different types of aircraft which are in use in the U.S. Air Force alone.

It is obvious that intensive scientific study is needed in this area; no such study has yet been undertaken at the necessary levels of intensity needed. Something that must be guarded against in any such study is the trap of implicitly assuming that our knowledge of physics (or any other branch of science) is complete. An example of one such trap is selecting a group of physical laws which we now accept as valid, and assuming that they will never be superseded.

Five such laws might be:

1. Every action must have an opposite and equal reaction.

2. Every particle in the universe attracts every other particle with a force proportional to the product of the masses and inversely as the square of the distance.

3. Energy, mass and momentum are conserved.

4. No material body can have a speed as great as c, the speed of light in free space.

5. The maximum energy, E, which can be obtained from a body at rest is $E=mc^2$, where m is the rest mass of the body.

Laws numbered 1 and 3 seem fairly safe, but let us hesitate and take another look. Actually, law number 3 is only valid (now) from a relativistic viewpoint; and for that matter so are laws 4 and 5. But relativity completely revised these physical concepts after 1915, before then Newtonian mechanics were supreme. We should also note that general relativity has not yet been verified. Thus we have the peculiar situation of five laws which appear to deny the possibility of intelligent alien control of UFOs, yet three of the laws are recent in concept and may not even be valid. Also, law number 2 has not yet been tested under conditions of large relative speeds or accelerations. We should not deny the possibility of alien control of UFOs on the basis of preconceived notions not established as related or relevant to the UFOs.

33.6—Conclusion

From available information, the UFO phenomenon appears to have been global in nature for almost 50,000 years. The majority of known witnesses have been reliable people who have seen easily-explained natural phenomena, and there appears to be no overall positive correlation with population density. The entire phenomenon could be psychological in nature but that is quite doubtful. However, psychological factors probably do enter the data picture as "noise." The phenomenon could also be entirely due to known and unknown phenomena (with some psychological noise added in) but that too is questionable in view of some of the available data.

This leaves us with the unpleasant possibility of alien visitors to our planet, or at least of alien controlled UFOs. However, the data are not well correlated, and what questionable data there are suggest the existence of at least three and maybe four different groups of aliens (possibly at different states of development). This too is difficult to accept. It implies the existence of intelligent life on a majority of the planets in our solar system, or a surprisingly strong interest in Earth by members of other solar systems.

A solution to the UFO problem may be obtained by the long and diligent effort of a large group of well-financed and competent scientists; unfortunately there is no evidence suggesting that such an effort is going to be made. However, even if such an effort were made, there is no guarantee of success because of the isolated and sporadic nature of the sightings. Also, there may be nothing to find, and that would mean a long search with no profit at the end.

The best thing to do is to keep an open and skeptical mind, and not take an extreme position on any side of the question.

31 UFOs and the CIA: Anatomy of a Cover-Up

Reg A. Davidson

The modern age of UFO phenomena began on a July afternoon in 1947 when private pilot Kenneth Arnold reported nine unidentifiable silvery, crescent-shaped objects that skimmed through the sky at an incredible rate of speed.

Their motion, Arnold said, reminded him of "a saucer skipping over water." A news reporter took up Arnold's description and the phrase "flying saucers" soon became imprinted on the collective consciousness.

When strange objects continued to be reported by competent witnesses, the U.S. authorities began investigating the phenomenon. The task fell under the auspices of the United States Air Force, but few were aware that the CIA took an interest in the strange phenomena soon after the first reports of "flying saucers" emerged.

The Air Force was actually in a state of near panic due to the wave of sightings. UFOs were reported over Maxwell Air Force base in Alabama, then, to the horror of the top military brass, over the White Sands Proving Ground—right in the middle of their atom bomb territory. General Nathan Twining, commander of the Air Material Command, wrote to the commanding general of the Army-Air Force stating that the phenomenon was something real, that it was not "visionary or fictitious," and that the objects were disc-shaped, as large as aircraft, and *controlled*.

The press latched onto the reports and sensationalized stories of alien invasion gripped the population. The press and the Government were demanding answers. The Air Force, worried that the whole situation was getting out of hand, tried to quell public angst by ordering a full investigation.

On December 30, 1947, Major General L. C. Craigie ordered the establishment of Project Sign at what became known as Wright-Patterson Air Force Base in Dayton, Ohio. Operating under auspices of the Air Material Command's Technical Intelligence Division, Project Sign was directed "to collect, collate, evaluate and distribute to interested government agencies and contractors all information concerning sightings and phenomena in the atmosphere which can be construed to be of concern to the national security."

The project was given a 2A restricted classification security rating under a system that acknowledged 1A as the highest, or most secret, designation.

The following year, three men from Wright-Patterson approached Dr. J. Allen Hynek, an astronomer then employed by Ohio State University in nearby Columbus. "They said they needed some astronomical consultation because it was their job to find out what these flying saucer stories were all about," Hynek recalls. Hynek was hired as a consultant with the Air Force and remained in that capacity for over two decades as Sign evolved into Projects Grudge and Blue Book, the last officially ceasing in December of 1969.

According to Hynek, the Air Force had a simple, but effective, method to explain UFOs: *Dismiss all sightings as misidentified astronomical phenomena.* The problem, says Hynek, was the Air Force "regarded it as an intelligence matter" instead of handing the investigation to an academic or university group. Therefore, any serious investigation of the new phenomena was stultified [rendered useless] because top military brass believed it was an "intelligence" matter, another intrigue of the emerging Cold War.

However, military personnel directly involved in Project Sign had a different view. While 96 percent of reports turned out to be misidentified astronomical phenomena (e.g., the planet Venus), the other 4 percent were not so easily discredited or explained, and a minority of military personnel took these seriously.

Minority intelligence opinion then divided into the two camps, namely, those who saw UFOs as evidence of new Soviet technology, and those who thought they might be precursors of an invasion by extraterrestrials.

"FLYING SAUCERS" AND THE CIA

Ever since 1948 the CIA has maintained an interest in UFOs and remains tight-lipped to this very day on the subject, keeping evidence and documents on the phenomena many levels above Top Secret.

A memo sent on January 29, 1952 to the CIA's deputy director of Intelligence from Ralph Clark of the Office of Scientific Intelligence (OSI) states: "In the past several weeks numerous UFOs have been sighted visually and on special UFO group radar. This office has maintained a continuing review of reputed sightings for the past three years and a special group has been formed to review the sightings to date."

Many researchers believe that from the very beginning the CIA was quite certain UFOs were not just Soviet technology. In fact, as evidence accumulated pointing to the possible extraterrestrial origin of UFOs, the CIA became increasingly nervous that other U.S. government agencies might launch their own inquiries into the matter. Secrecy would be an impossibility if everyone investigated UFOs, and in a matter of time, details would leak to the media and the public.

In response to these concerns, the CIA began a process of maintaining a tight rein over the investigations to ensure no public inquiries would ever take place. To discredit the phenomenon, the CIA set up a panel of experts whose job was to explain away UFOs.

The CIA convened on 14 January, 1953, a confab that became known as the Robertson Panel, after its Chairman Dr. H. P. Robertson, then director of the Weapons Systems Evaluation Group in the Office of the Secretary of Defense, and also a CIA employee.

The sequence of events leading directly to the Robertson Panel involved a series of UFO sightings over the nation's capital in the summer of 1952, sightings confirmed by military personnel, including radar operators and scrambled interceptor pilots, and which themselves resulted in the largest post-WWII military press conference to date. At the press conference itself, the repeated radar sightings were put down to "temperature inversions," and the attending Air Force officers made no mention of the scrambled jet fighters.

Besides the esteemed Dr. Robertson, the Panel also included as members physicist Dr. Luis Alvarez, later a Nobel Laureate, Dr. Samuel Goudsmit, another physicist from Brookhaven National Laboratories who was an associate of Einstein's and had discovered electron spin, a former University of Chicago astronomer and then deputy director of the Johns Hopkins Operations Research office, Dr. Thornton Page, and finally Dr. Lloyd Berkner, yet another physicist and one of Brookhaven's directors.

The Panel was addressed by a variety of CIA and Air Force personnel who reviewed some twenty of the better UFO cases and showed two film strips of alleged flying saucers, one of which purportedly portrayed objects characterized as "self-luminous" by no less an authoritative source than the Navy's Photograph Interpretation Laboratory which had spent over 1,000 hours analyzing the particular movie film in question.

Although impressive evidence was presented by the panel, highlighted by detailed reports documented by the Air Force, its recommendations read like they were formulated before the panel even convened. The CIA had already developed a cover story to cloak the real story: UFOs were to be dismissed as just another scientific enigma, a Cold War datum, one that might be cleverly manipulated by the enemy.

In short, the Robertson Panel ruled "that the evidence presented on Unidentified Flying Objects shows no indication that these phenomena constitute a direct physical threat to national security." While this ruling is considered contentious by many UFO researchers, it was the panel's second conclusion that really shocked. The panel decreed there was no national security threat from UFOs, however, its members did see a real and distinct danger posed by UFO reports!

In the panel's own words, it concluded "that the continued emphasis on the reporting of these phenomena, in these perilous times, result in a threat to the orderly functioning of the protective organs of the body politic."

"We cite as an example [of such danger]," the Panel continued, "the clogging of channels of communication by irrelevant reports, the danger of being led by continued false alarms to ignore real indications of hostile action, and the cultivation of a morbid national psychology in which skillful hostile propaganda could induce hysterical behavior and harmful distrust of duly constituted authority." In other words, UFO reports might induce national psychosis that could be subject to manipulation by the Soviets.

In the final list of recommendations, the panel calls for "national security agencies to take immediate steps to strip the Unidentified Flying Objects of the special status they have been given. . . ."

The CIA had effectively halted any serious research into the phenomena, and now controlled all ongoing U.S. military investigations.

RUPPELT VS. THE CIA

The public became aware of the panel a few years later with the publication of "The Report on Unidentified Flying Objects" by Captain Edward J. Ruppelt, former commander of Project Blue Book. Both Ruppelt and his Intelligence Liaison Officer, Major Dewey J. Fournet, gave evidence to the Robertson Panel.

Although the panel relegated UFOs to the dustbin of history, Walter Smith, then director of the CIA, saw fit to keep all evidence classified. The CIA's decision shocked Captain Ruppelt and Major Fournet. Both were part of the minority of intelligence officials that believed the evidence for UFOs was incontrovertible. They also believed the possibility of hysteria would be reduced if the public were told the truth.

Ruppelt had fought hard to keep the Air Force investigations afloat, after joining the Project Grudge team in January 1951, but soon found the CIA constantly interfering and withholding valuable information. Project Grudge evolved into the now famous Project Blue Book in March 1952 with Captain Ruppelt appointed as its chief. All this came in response to a spate of UFO sightings, beginning with the 25 August, 1951 famous sightings at Lubbock, Texas, which caused an enormous stir with the American public. And soon after, on 12 September, 1951, a major UFO sighting above the skies of Fort Monmouth [New Jersey] in clear view of visiting military brass, contributed to the Air Force's new found enthusiasm.

Ruppelt first became aware of the CIA's unwanted presence after the Washington UFO "invasion" of July 1952, when he was hampered from doing his job, and witnesses to the sightings were intimidated into changing their reports or simply remaining silent.

The person who most worried Ruppelt was Chief of Staff General Hoyt S. Vandenberg. It was Vandenberg who had buried Project Sign's official UFO "Estimate" report, caused its incineration, and had the project renamed Project Grudge. It is not clear just how much Vandenberg was influencing top military officials responsible for implementing the Air Force's UFO projects. Vandenberg had been head of the Central Intelligence Group (later the CIA) from June 1946 to May 1947, and his uncle was once chairman of the Foreign Relations Committee, then the most powerful committee in the U.S. Senate. Clearly, Vandenberg still had great influence in those areas— and according to Ruppelt, pressure was always coming from them to suppress the results of official UFO investigations.

Thus, Ruppelt was not surprised when the CIA and other high-ranking officers including General Vandenberg convened a panel of scientists to "analyze" all the Blue Book data. Nor was he too surprised when the Robertson Panel found that no further study was necessary.

The pieces of the jigsaw puzzle started to fall into place. It was clear to Captain Ruppelt and other members of Project Blue Book, that the purpose of the Robertson Panel was to enable the CIA and Air Force to state in the future that an *impartial* body had examined the UFO data and found no evidence for anything unusual in the skies. Subsequently, the Air Force embarked upon a public relations campaign to eliminate UFO reports totally. The CIA decided not to declassify the sighting reports *and* to tighten security even more while continuing to deny "non-military personnel" access to UFO files.

One month later CIA director Walter Smith classified all UFO documentation and all subsequent directors continued to endorse the policy.

INITIATION OF A COVER-UP

In August 1953 Ruppelt left the Air Force out of disgust and because of the limitations placed on his work by the CIA. The same month the Pentagon issued the notorious Air Force Regulation 200-2, that prohibited the release of *any* information about a sighting to the public or media, except when it was *positively* identified as natural phenomenon. The new regulation also ensured that *all* sightings would be classified as restricted.

In December 1953 the much worse Joint-Army-Navy-Air Force Publication 146 made the releasing of *any* information to the public a crime under the Espionage Act. And the most ominous aspect of JANAP 146 was that it applied to anyone who knew it existed, including commercial airline pilots. Any information flow to the public was effectively cut.

By the end of the year Project Blue Book was severely decimated and for all intents and purposes, UFO research plunged into secrecy and under the control of the CIA. In just over six years since Kenneth Arnold's sighting of strange silvery objects, the infamous intelligence agency had secured complete official silence on the subject of UFOs.

The cover-up began and continues today, due to the CIA's indomitable power over all other intelligence groups within the U.S. security establishment. The truth is out there . . . and it just might be somewhere deep inside the secret files of the CIA.

32 NASA

Timothy Good

The National Aeronautics and Space Administration, established in 1958, coordinates and directs the aeronautical and space research programme in the United States. Its budget for space activities alone is larger than the general budgets of a number of the world's important countries.

Although officially a civilian agency, NASA collaborates with the Department of Defense, National Reconnaissance Office, National Security Agency, and other agencies, and many of its personnel have security clearances owing to the sensitive intelligence aspects of its programmes. Research into UFOs is one such programme.

In May 1962 NASA pilot Joseph A. Walker admitted that it was one of his appointed tasks to detect unidentified objects during his flights in the rocket-powered X-15 aircraft, and referred to five or six cylindrical shaped objects that he had filmed during his record-breaking high flight in April that year. He also admitted that it was the second occasion on which he had filmed UFOs in flight. "I don't feel like talking about them," he said during a lecture at the Second National Conference on the Peaceful Uses of Space Research in Seattle, Washington. "All I know is what appeared on the film which was developed after the flight."

Britain's FSR magazine cabled NASA headquarters requesting further information and copies of stills from the film taken by Walker. "Objects reported by NASA pilot Joe Walker have now been identified as ice flaking off the X-15 aircraft," NASA replied. "Analysis of additional cameras mounted on top the X-15 led to identification of the previously unidentifiable objects. . . . *No still photos are available*." [Emphasis added.]

In July 1962 Major Robert White piloted an X-15 to a height of fifty-eight miles at the top of his climb, and on his return reported having seen as strange object. "I have no idea what it could be," he said. "It was grayish in color and about thirty to forty feet away." Then, according to *Time* magazine, Major White is reported to have said excitedly over his radio:

"There are things out there. There absolutely is!"

"Two years ago," a NASA scientist said in 1967, "most of us regarded UFOs as a branch of witchcraft, one of the foibles of modern man. But so many reputable people have expressed interest in confidence to NASA, that I would not be in the least surprised to see the space agency begin work on a UFO study contract within the next twelve months."

One of those who expressed interest was Dr. Allen Hynek, who wanted NASA to use its superlative space-tracking network to monitor and document the entry of unidentified objects into the Earth's atmosphere. The problem then—as now—is that UFO sightings tracked by NASA remain exempt from public disclosure since they are classified top secret. But there have been leaks.

In April 1964 two radar technicians at Cape Kennedy revealed that they had observed UFOs in pursuit of an unmanned Gemini space capsule. And in January 1961 it was reliably reported that the Cape's automatic tracking gear locked on to a mysterious object which was apparently following a Polaris missile over the South Atlantic.

A 1967 NASA Management Instruction established procedures for handling reports of sightings of objects such as "fragments or component parts of space vehicles known or alleged by an observer to have impacted upon the earth's surface as a result of safety destruct action, failure in flight, or re-entry into the earth's atmosphere," and also includes "reports of sightings of objects not related to space vehicles." A rather euphemistic way of putting it, to be sure, but the internal instruction continues: "It is KSC [Kennedy Space Center] policy to respond to reported sightings of space vehicle fragments *and unidentified flying objects* as promptly as possible. . . . *Under no circumstances will the origin of the object be discussed with the observer or person making the call.*" [Emphasis added.]

A 1978 NASA information sheet gives the agency's official policy on the subject:

NASA is the focal point for answering public enquiries to the White House relating to UFOs. *NASA is not engaged in a research program involving these phenomena, nor is any other government agency.* Reports of unidentified objects entering United States air space are of interest to the military as a regular part of defense surveillance. Beyond that, the U.S. Air Force no longer investigates reports of UFO sightings.

In 1978 CAUS (Citizens Against UFO Secrecy) filed a request for information relating to a NASA report entitled *UFO Study Considerations,* which had previously been prepared in association with the CIA. In his response, Miles Waggoner of NASA's Public Information Services Branch denied this. "There were no formal meetings or any correspondence with the CIA," he stated. Following another enquiry by CAUS, NASA's Associate Administrator for External Relations, Kenneth Chapman, explained that the NASA report had been prepared solely by NASA employees but that the CIA had been consulted by telephone to determine "whether they were aware of any tangible or physical UFO evidence that could be analyzed; the CIA responded that they were aware of no such evidence, either classified or unclassified."

NASA's statement in the 1978 information sheet that it was not engaged in a research programme involving UFOs, "nor is any other government agency," is demonstrably false, as is its denial of Air Force investigations.

In a leaked secret document purporting to originate with the Air Force Office of Special Investigations (AFOSI) headquarters at Bolling Air Force Base, DC, there appears an intriguing reference to clandestine government UFO research, led by NASA. The document is dated 17 November 1980, and includes this relevant passage:

SEVERAL GOVERNMENT AGENCIES, LED BY NASA, ACTIVELY INVESTIGATE LEGITIMATE SIGHTINGS THROUGH COVERT COVER. . . . ONE SUCH COVER IS UFO REPORTING CENTER, U.S. COAST AND GEODETIC SURVEY, ROCKVILLE, MD 20852. NASA FILTERS RESULTS OF SIGHTINGS TO APPROPRIATE MILITARY DEPARTMENTS WITH INTEREST IN THAT PARTICULAR SIGHTING.

"We have no information relative to the contents of the document," NASA told me in 1985. "Additionally, we have been informed that [it] is not an authentic AFOSI document." In this case, NASA is right. Although substantially legitimate, the document is a re-typed version containing errors, including the reference to NASA, which should be NSA—the National Security Agency.

PRESIDENT CARTER SEEKS TO RE-OPEN INVESTIGATIONS

During his election campaign in 1976, Jimmy Carter revealed that he had seen a UFO at Leary, Georgia, in 1969, together with witnesses, prior to giving a speech at the local Lions Club. "It was the darndest thing I've ever seen," he told reporters. "It was big, it was very bright, it changed colors, and it was about the size of the moon. We watched it for ten minutes, but none of us could figure out what it was. One thing's for sure; I'll never make fun of people who say they've seen unidentified objects in the sky."

Carter's sighting has been ridiculed by skeptics such as Philip Klass and Robert Sheaffer. While there appear to be legitimate grounds for disputing the date of the incident, Sheaffer's verdict that the UFO was nothing more exotic than the planet Venus is not tenable. As a graduate in nuclear physics who served as a line officer on U.S. Navy nuclear submarines, Carter would not have been fooled by anything so prosaic as Venus, and in any case he described the UFO as being about the same size as the Moon.

"If I become President," Carter vowed, "I'll make every piece of information this country has about UFO sightings available to the public and the scientists." Although President Carter did all he could to fulfill his election pledge, he was thwarted, and it is clear that NASA had a hand in blocking his attempts to re-open investigations. When Carter's science adviser, Dr. Frank Press, wrote to NASA administrator Dr. Robert Frosch in February 1977 suggesting that NASA should become the "focal point for the UFO question," Dr. Frosch replied that although he was prepared to continue responding to public enquiries, he proposed that "NASA take no steps to establish a research activity in this area or to convene a symposium on this subject."

In a letter from Colonel Charles Senn, Chief of the Air Force Community Relations Division, to Lieutenant General Duward Crow of NASA, dated 1 September 1977, Colonel Senn made the following astonishing statement: "I sincerely hope *that you are successful in preventing a reopening of UFO investigations.*" So it is clear that NASA (as well as the Air Force and almost certainly the CIA and National Security Agency) was anxious to ensure that the President's election pledge remained unfulfilled.

DR. JAMES MCDONALD

Dr. James McDonald, senior physicist at the Institute of Atmospheric Physics and Professor in the Department of Meteorology at the University of Arizona, who committed suicide in unusual circumstances in 1971, tried unsuccessfully to persuade NASA to take on primary responsibility for UFO investigations. He reported in 1967:

> Curiously, I have said this both in NASA and fairly widely reported public discussions before scientific colleagues, yet the response from NASA has been nil. . . . Even attempting to get a small group within NASA to undertake a study group approach to the available published effort seems to have generated no response. I realize, of course, that there may be semi-political considerations that make it awkward for NASA to fish in these waters at present, but if this is what is holding up serious scientific attention to the UFO problem at NASA, this is all the more reason Congress had better take a good hard look at the problem and reshuffle the deck. . . . I have learned from a number of unquotable sources that the Air Force has long wished to get rid of the burden of the troublesome UFO problem and has twice tried to "peddle" it to NASA— without success.

While McDonald recognized that there were "semi-political considerations" affecting NASA's reluctance to become publicly involved in UFO investigations, he failed to perceive that UFOs are more an intelligence problem than a scientific one. He was simply unaware of the true extent of NASA's secret involvement.

THE PIONEERS

One of the great pioneers in astronautics is Dr. Hermann Oberth, whom I had the honour of meeting in 1972. In 1955 Oberth was invited by Dr. Wernher von Braun to go to the United States where he worked on rockets with the Army Ballistic Missile Agency, and later NASA at the George C. Marshall Space Flight Center. Oberth's statements on the UFO question have always been unequivocal, and he told me that he is convinced UFOs are extraterrestrial in origin. In the following he elaborated on his hypothesis for UFO propulsion:

> . . . today we cannot produce machines that fly as UFOs do. They are flying by means of artificial fields of gravity. This would explain the sudden changes of directions. . . . This hypothesis would also explain the piling up of these discs into a cylindrical or cigar-shaped mothership upon leaving the earth, because in this fashion only one field of gravity would be required for all discs.
>
> They produce high-tension electric charges in order to push the air out of their path . . . and strong magnetic fields to influence the ionised air at higher altitudes. . . . This would explain their luminosity. . . . Secondly, it would explain the noiselessness of UFO flight. Finally, this assumption also explains the strong electrical and magnetic effects sometimes, but not always, observed in the vicinity of UFOs.

Earlier, Dr. Oberth hinted that there had been actual contact with the UFOs at a scientific level. "We cannot take credit for our record advancement in certain scientific fields alone; we have been helped," he is quoted as having said. When asked by whom, he replied: *"The people of other worlds."* There are persistent rumours that the U.S. has even test-flown a few advanced vehicles, based on information allegedly acquired as a result of contact with extra-terrestrials and the study of grounded UFOs.

In 1959 Dr. Wernher von Braun, another great space pioneer, made an intriguing statement, reported in Germany. Referring to the deflection from orbit of the U.S. *Juno 2* rocket, he stated: "We find ourselves faced by powers which are far stronger than we had hitherto assumed, and whose base is at present unknown to us. More I cannot say at present. *We are now engaged in entering into closer contact with those powers*, and in six or nine months' time it may be possible to speak with more precision on the matter." [Emphasis added.]

There has been nothing further published on the matter. As Dr. Robert Sarbacher has commented, von Braun was probably involved in the recoveries of crashed UFOs in the late 1940s, and it is my opinion that he was constrained from elaborating on the subject owing to the security oath that he must have been subject to. I cannot prove this, of course, any more than I can substantiate information I have received from a reliable source that top secret contacts have been made by extraterrestrials with selected scientists in the space programme. It must be admitted, though, that von Braun's statement comes close to corroborating this. What else could he have meant when he said, "We are now engaged in entering into closer contact with those powers"? The Soviets?

NASA WITHHOLDS PHYSICAL EVIDENCE

That NASA has been engaged in UFO research behind the scenes is alone proven, to my satisfaction at least, by its shady involvement in the analysis of metal samples discovered at the site where Sergeant Lonnie Zamora encountered a landed UFO and occupants at Socorro, New Mexico, in April 1964. On 31 July 1964 Ray Stanford and some members of NICAP* visited NASA's Goddard Space Flight Center at Greenbelt, Maryland, in order to have a rock with particles of metal on it analyzed by NASA scientists. Dr. Henry Frankel, head of the Spacecraft Systems Branch, directed the analysis. The particles had apparently been scraped on to the rock by one of the UFO's landing legs. On first inspection of the rock through a microscope, Dr. Frankel declared that some of the particles "look like they may have been in a molten state when scraped onto the rock," and expressed the desire to remove them from the rock for further analysis. Stanford agreed to this, but said that he wanted to retain half of the particles for his own use.

The researchers were invited to go to lunch while NASA engineers conducted their analysis. After lunch, Stanford and the others (Richard Hall, Robert McGarey and Walter Webb), returned to the laboratory building. A NASA technician brought the rock over to the group. "As he handed it to me," said Stanford, "I was able to carefully observe it in the bright light inside the room. *The whole thing had been scraped clean.* Someone had gone over that rock with the equivalent of a fine-toothed comb. There was nothing, *not a speck* of the metal left . . . even the very few tiny particles that I had known were rather well-hidden had been removed."

When Stanford complained, the technician insisted that half of the samples were still on the rock, as promised, but seeing Stanford's disbelief hastily left the room. Dr. Frankel then returned, and after Stanford had remonstrated with him, explained what had happened. "Well, we tried to leave you

---

* National Investigations Committee on Aerial Phenomena

some," he said, "but we also had to get enough to make an accurate analysis. The sample will be placed under radiation this afternoon, where it will remain the entire weekend. Monday, we will remove it for X-ray diffraction tests. That should tell us the elements it contains . . . if you will call me, say on Wednesday, I should be able to tell you something very definite."

Before contacting Dr. Frankel again, Stanford and McGarey had a meeting with a U.S. Navy captain in Washington who was interested in the Socorro case. The captain told the researchers that they would never get their metal samples back from Frankel. "If that metal is in any way unusual," he said, "he will never give you any documentation to prove it . . . Those boys at Goddard know that they must report any findings as important as a strange, UFO alloy to the highest authority in NASA. Once that authority receives the news, the President will be informed, for the matter is pertinent to national security and stability. A *security* directive will instruct those self-appointed authorities at Goddard as to just whose hands the matter is really in. . . ."

On 5 August 1964 Ray Stanford phoned Dr. Frankel at the Goddard Space Flight Center. "I'm glad you called," the scientist said. "I have some news that I think will make you happy." He went on:

> The particles are comprised of a material that could not occur naturally. Specifically, it consists predominantly of *two metallic elements*, and there is something that is rather exciting about the zinc-iron alloy of which we find the particles to consist: *Our charts of all alloys known to be manufactured on Earth, the U.S.S.R. included, do not show any alloy of the specific combination or ratio of the two main elements involved here.* This finding definitely strengthens the case that might be made for an extraterrestrial origin of the Socorro object.

Dr. Frankel added that the alloy would make "an excellent, highly malleable, and corrosive-resistant coating for a spacecraft landing gear, or for about anything where those qualities are needed." He also said that he was prepared to make a statement before a Congressional hearing to this effect, if necessary.

Frankel went on to say that further analysis would be carried out, and that Stanford should call him again the following week. On 12 August Stanford placed a call to Frankel, but was told by his secretary that he was "not available" and suggested he try contacting him the following day. On 13 August Stanford phoned again. "Dr. Frankel simply is not available today," the secretary announced. "He wonders if you might try him the first part of next week?"

On 17 August Stanford rang Frankel's office, only to be told yet again that he was not available. Ominously, the secretary added: *"Dr. Frankel is unprepared, at this time, to discuss the information you are calling about."* On 18 August Stanford tried again. "I'm sorry," the secretary said, "but *Dr. Frankel is in a top-level security conference.* I doubt that he will be able to talk with you until tomorrow or the next day."

Failing to get hold of Frankel the following day, Stanford left a telephone number with the secretary. On 20 August Thomas P. Sciacca Jr. of NASA's Spacecraft Systems Branch phoned Stanford. "I have been appointed to call you and report the official conclusion of the Socorro sample analysis," he said. "Dr. Frankel is no longer involved with the matter, so in response to your repeated enquiries, I want to tell you the results of the analysis. *Everything you were told earlier by Dr. Frankel was a mistake.* The sample was determined to be silica, SiO_2."

In 1967 Dr. Allen Hynek invited Ray Stanford to a lecture he was giving in Phoenix, and afterwards Hynek asked: "Whatever happened with the analysis at Goddard of that metallic sample from the rock you took from the Socorro site?" Both Hynek and Stanford had been closely involved in investigations at the landing site, but Stanford was puzzled as to how Hynek knew about the NASA analysis. "I was not sure where Hynek had learned of the fact that I had taken the rock which Lonnie Zamora had pointed out to both of us, and which the astronomer had ignored," he said. "I was interested to note that he specifically knew it was analyzed at Goddard. That fact had never been published."

Stanford told Hynek that NASA's "official" analysis had revealed it to be common silica. "That cannot be true!" exclaimed Hynek. "I am familiar with the analysis techniques involved. Silica could not be mistaken for a zinc-iron alloy. They haven't given you the truth! I would accept Frankel's original report and forget the later disclaimer."

Given that the original analysis was accurate it is worth recording NASA Administrator Dr. Robert Frosch's statement in the letter he wrote to President Carter's science advisor, Dr. Frank Press, in 1977: "There is an absence of tangible or physical evidence available for thorough laboratory analysis . . . To proceed [therefore] on a research task without a disciplinary framework and an exploratory technique in mind would be wasteful and probably unproductive."

THE SILVER SPRING FILM

In my first book I devoted a chapter to the controversial 8mm colour movie film taken by George Adamski in the presence of Madeleine Rodeffer and other unnamed witnesses outside Madeleine's home at Silver Spring, Maryland, in February 1965. I have been taken to task for endorsing the authenticity of this "obviously fake" film taken by a "proven charlatan," but I have yet to see any conclusive evidence that it was actually faked. Both my co-author Lou Zinsstag and I exposed as many of the inconsistencies in Adamski's claims that were available to us at the time of writing, but that short piece of film, taken a few months before Adamski's death, remains authentic in my opinion at least.

Sometime between 3 and 4 P.M. on 26 February 1965 an unidentified craft of the famous type photographed by Adamski in 1952 (and others subsequently) described a series of manoeuvres over Madeleine's front yard, retracting and lowering one of its three pods and making a gentle humming and swishing sound as it did so. Adamski began filming the craft with Madeleine's 8 mm camera. "It looked blackish-brown or grayish-brown at times," Madeleine told me, "but when it came in close it looked greenish and blueish, and it looked aluminium: it depended on which way it was tilting. Then at one point it actually stood absolutely still between the bottom of the steps and the driveway." The craft then disappeared from view, but reappeared above the roof and described manoeuvres once more before finally disappearing vertically. Madeleine told me that she could make out human figures at the portholes, but details were obscured.

When the film was developed the following week something was obviously wrong with many of the frames and it was apparent that it had been interfered with. Obviously faked frames had been substituted by person or persons unknown. "They took the original film," Madeleine believes, "and what I think they did was rephotograph portions of the original . . . and then fake some stuff. The film I got back is not the original film at all."

Fortunately enough frames showing the craft as they had remembered it survived out of the twenty-five feet that had been taken, and these were analyzed by William T. Sherwood, an optical physicist who was formerly a senior project development engineer for the Eastman-Kodak Company in Rochester, NY. I spent many hours discussing the film with Bill, and in 1968 he provided me with a brief technical summary of his evaluations as they related to the prints he made from the "original" film.

It's hard to capture the nuances of the original film. None of the movie duplicates are good: too much contrast. The outlines look "peculiar" due to distortions, I believe, caused by the "forcefield." The glow beneath the flange is, I think, significant. Incidentally, the tree [near the top of which the craft manoeuvred] is very high (90 ft?). Roughly, the geometry of imagery is this:

$$\frac{\text{object size}}{\text{image size}} \simeq \frac{\text{distance}}{\text{focal length}} \quad \text{or} \quad \frac{27\text{ft}}{2\text{mm}} \simeq \frac{90\text{ft}}{9\text{mm}}$$

In 1977 Bill Sherwood sent me further details of his evaluations:
The camera was a Bell & Howell Animation Autoload Standard 8, Model

315, with a fl.8 lens, 9-29 mm, used in the 9mm position. . . . As you can measure, the image on the film (original) is about 2.7mm maximum. So for a 90 ft distant object, [the diameter] would be about 27 feet. . . . It was a large tree, and the limb that the saucer seems to "touch" could have been about that distance from the camera . . . but unfortunately I could not find a single frame where the saucer could clearly be said to be *behind* the limb. So it is not conclusive as for distance, and therefore for size. . . . In some of the frames of the original, portholes are seen.

In reply to my query as to whether it was possible to authenticate the film unequivocally, Bill said that there is no absolutely foolproof way of assessing whether a photo is "real" or not. One must just take everything into account, including as much as one can learn about the person involved, and then make an educated guess. In the final analysis, he said, it comes down to this question: "Is this the kind of person whom I can imagine going to all the trouble and expense of simulating what only a well-equipped studio with a large budget could begin to approximate, and defending it through the years with no apparent gain and much inconvenience?"

One of the peculiarities of the film is that the outlines of the craft look peculiarly distorted at times. Bill Sherwood believes this is due to a powerful gravitational field that produces optical distortions, an opinion that is shared by Leonard Cramp, an aeronautical engineer and designer who has worked for De Havilland, Napier, Saunders-Roe, and Westland Aircraft companies. In his pioneering book, *Piece for a Jig-Saw*, Cramp proposed a theory to account for this peculiar effect:

Earlier, when discussing light in terms of the G [gravitational field] theory, we saw how we might expect such a field to form an atmospheric lens, producing optical effects which might be further augmented by other field effects as well as the gravitational bending of light. . . . Now it follows that if there would be a local *increase* in atmospheric pressure due to a powerful G field, then similarly we could expect a *decrease* in atmospheric pressure to accompany a powerful R [repulsion] field, and again we would not be surprised to find optical effects . . . we can now say, while a G field might produce optical magnifying properties, an R field could produce optical *reducing* properties.

Leonard Cramp had not seen the Silver Spring film prior to publishing his book, and was delighted that it seemed to confirm his hypothesis. Like Bill Sherwood and myself, he is in no doubt that the film is authentic.

On 27 February 1967 (two years after it had been taken) the film was shown to twenty-two NASA officials at the Goddard Space Flight Center. Discussion afterwards lasted for an hour and a half, and just before

Madeleine left, one of the two friends with her was allegedly told that it was "a very important piece of film" and that the craft was twenty-seven feet in diameter (the figure calculated independently by Bill Sherwood). Unfortunately, I have been unable to confirm this.

In reply to my queries, NASA scientist Paul D. Lowman Jr., of the Geophysics Branch at Goddard, stated that according to one of those present, Herbert A. Tiedemann, everyone considered the Silver Spring film to be fake. Dr. Lowman, who had helped set up the meeting but was unable to attend, offered the following comments on the color photos from the film that I sent him:

> First, it is not possible to make any precise determination of the object's size from the relationship (which is basically correct) quoted by Mr. Sherwood. Given any three of these quantities, one can calculate the fourth. The focal length and image size are obviously known, but not the distance, which can only be roughly estimated. The equation can be no better than its most inexact quantity, and one might as well just estimate the size of the object directly. My own strong impression is that these frames show a small object, perhaps up to 2 or 3 feet across, a short distance from the camera. Judging from the photo of Mrs. Rodeffer's house, a 27 foot UFO would have occupied most of the cleared area in the front yard, and from such a short distance would have been a very large photographic object.

Although Bill Sherwood readily concedes that his estimate of the precise distance from the camera is arbitrary, he is sure that it is reasonably accurate, and my own tests at the site show that, with the camera lens set on wide angle (as it was at the time), an object of this approximate size and distance would appear exactly as it does on the film. That either Adamski or Madeleine (or both) could have faked the film using a small model, and then have the audacity to show it at NASA, seems far-fetched in the extreme. Moreover, to produce the distortion effects as well as the lowering and retracting of one of the pods with a small model, is out of the question as far as I am concerned. As a semi-professional photographer I can speak with some authority on the matter myself.

Following the death of Adamski, Madeleine Rodeffer experienced a great deal of ridicule and harassment, and nearly all copies of the "faked" film have been stolen—in the United States and elsewhere.

Two photographs of an identical craft were taken by young Stephen Darbishire in the presence of his cousin Adrian Myers in Coniston, England, in February 1954. For the benefit of those who contend that Darbishire had faked the pictures and recanted later, the following statement from a letter he wrote to me in 1986 is illuminating:

... when I said that I had seen a UFO I was laughed at, attacked, and surrounded by strange people. . . . In desperation I remember I refuted the statement and said it was a fake. I was counter-attacked, accused of working with the "Dark Powers" . . . or patronizingly "understood" for following orders from some secret government department.

There was something. It happened a long time ago, and I do not wish to be drawn into the labyrinth again. Unfortunately the negatives were stolen and all the prints gone . . .

THE ASTRONAUTS

In the early 1970s I had the pleasure of several meetings in Britain and the United States with the former U.S. Navy test pilot, intelligence officer, and pioneer astronaut Scott Carpenter, who had reputedly seen UFOs and photographed one of them during his flight in the *Mercury 7* capsule on 24 May 1962. Scott vehemently denied this, and poured scorn on other reports of sightings by fellow astronauts. I noticed that he appeared to be ill at ease when discussing the subject, and whenever I produced documentary evidence for official concern in this area he became visibly nervous. But in November 1972 Scott kindly wrote on my behalf to astronauts Gordon Cooper, Dick Gordon, James Lovell and James McDivitt, asking about reports attributed to them. James Lovell replied as follows:

I have to honestly say that during my four flights into space, I have not seen or heard any phenomena that I could not explain. . . . *I don't believe any of us in the space program believe that there are such things as UFOs.* . . . However, most of us believe that there must be a star like our sun that also has a planetary system [which] must support intelligent life as we know it. . . . I hope this is sufficient information for Tim Good, and I hope he isn't too disappointed in my answer. [Emphasis added.]

But according to the transcript of Lovell's flight on *Gemini 7*, an anomalous object was encountered:

SPACECRAFT:	Bogey at 10 o'clock high.
CAPCOM:	This is Houston. Say again 7.
SPACECRAFT:	Said we have a bogey at 10 o'clock high.
CAPCOM:	Gemini 7, is that the booster or is that an actual sighting?
SPACECRAFT:	We have several, looks like debris up here. Actual sighting.
CAPCOM:	. . . Estimate distance or size?
SPACECRAFT:	We also have the booster in sight . . .

Franklin Roach, of the University of Colorado UFO study set up by the Air Force in 1966, concluded that in addition to the booster traveling in an orbit similar to that of the spacecraft, "there was another bright object [the "bogey"] together with many illuminated particles. It might be conjectured," he said, "that the bogey and particles were fragments from the launching of *Gemini 7*, but this is impossible if they were traveling in a polar orbit as they appeared to be doing."

James McDivitt confirmed that although he did see an unidentified object during the *Gemini 4* flight on 4 June 1965, he does not believe it was anomalous:

During *Gemini 4,* while we were in drifting flight, I noticed an object out the front window of the spacecraft. It appeared to be cylindrical in shape with a high fineness ratio. From one end protruded a long, cylindrical pole with the approximate fineness of pencil. I had no idea what the size was or what the distance to the object was. It could have been very small and very near or very large and very far away.

I attempted to take a photograph of this object with each of the two cameras we had on board. Since this object was only in my view for a short time, I did not have time to properly adjust the cameras and I just took the picture with whatever settings the camera had at that time. The object appeared to be relatively close and I went through the trouble of turning on the control system in case I needed to take any evasive actions.

The spacecraft was in drifting flight and when the sun shone on the duty window, the object disappeared from view. I was unable to relocate it, since the attitude reference in the spacecraft was also disabled, and I did not know which way to maneuver to find it.

After landing, the film from *Gemini 4* was flown back to Houston immediately, whereas Ed White and I stayed on the aircraft carrier for three days. During this period of time a film technician at NASA evaluated the photographs and selected what he thought was the photograph of this particular object. Unfortunately, what he selected was a photograph of sunspots [flares] on the window and had nothing whatsoever to do with the object that I had seen. The photograph was released before I returned and had a chance to point out the error in the selection. I, subsequently, went through the photographs myself and was unable to find any photograph like the object I had seen. Apparently, the camera settings were not appropriate for the pictures.

I do not feel that there was anything strange or exotic about this particular object. Rather, only that I could not identify it. In a combination of both *Gemini 4* and *Apollo 9* I saw numerous satellites, some of which we identified and some of which we didn't. . . . I have seen a lot of objects that I could not identify, but I have yet to see one that could be identified as a spaceship from some other planet. I can't say that there aren't any, only that I haven't seen any. I hope this helps Tim.

Neither Gordon Cooper nor Dick Gordon replied to Scott's letter, it seems, and I have never been able to receive a reply from Cooper, although he has spoken publicly of his interest in the subject. In fact, interest in UFOs was one of the reasons that inspired him to become an astronaut. "I . . . had the idea that there might be some interesting forms of life out in space for us to discover and get acquainted with," he wrote in 1962. "As far as I am concerned there have been far too many unexplained examples of unidentified objects sighted around the earth . . . the fact that many experienced pilots had reported strange sights did heighten my curiosity about space . . . This was one of the reasons, then, why I wanted to become an Astronaut."

In 1978 Cooper attended a meeting of the Special Political Committee United Nations General Assembly in order to discuss UFOs. Later that year a letter from Cooper was read at another UN meeting:

> . . . *I believe that these extraterrestrial vehicles and their crews are visiting this planet from other planets, which are obviously a little more advanced than we are here on earth.*
>
> I feel that we need to have a top-level, coordinated program to specifically collect and analyze data from all over the earth concerning the type of encounter, and to determine how best to interface with these visitors in a friendly fashion.
>
> Also, *I did have occasion in 1951 to have two days of observation . . . flights of them, of different sizes, flying in fighter formation,* from east to west over Europe. [Emphasis added.]

Cooper said that most astronauts were reluctant to discuss UFOs "due to the great numbers of people who have indiscriminately sold fake and forged documents abusing their names and reputations without hesitation." But he added that there were "several of us who do believe in UFOs" and who have had occasion to *see a UFO on, around, or from* an aircraft. "There was only one occasion from space which may have been a UFO," Cooper's letter revealed, without elaborating.

A UFO seen on the ground by an astronaut? The only reference I have to such an incident is contained in an article which the late Lou Zinsstag translated from the French for me in 1973. Unfortunately, I have neither the name of the paper nor the date, but the article was written by J. L. Ferrando, based on an interview with an astronaut at a congress in New York in mid-1973, tape-recorded by Benny Manocchia. The name of the astronaut? None other than Gordon Cooper! The following extracts are highly significant—if true:

> For many years I have lived with a secret, in a secrecy imposed on all specialists in astronautics. I can now reveal that every day, in the USA,

our radar instruments capture objects of form and composition unknown to us. And there are thousands of witness reports and a quantity of documents to prove this, but nobody wants to make them public. Why? Because authority is afraid that people may think of God knows what kind of horrible invaders. So the password still is: we have to avoid panic by all means.

I was furthermore a witness to an extraordinary phenomenon, here on this planet earth. It happened a few months ago in Florida. There I saw with my own eyes a defined area of ground being consumed by flames, with four indentations left by a flying object which had descended in the middle of a field. Beings had left the craft (there were other traces to prove this). They seemed to have studied topography, they had collected soil samples and, eventually, they returned to where they had come from, disappearing at enormous speed. . . . I happen to know that the authorities did just about everything to keep this incident from the press and TV, in fear of a panicky reaction from the public.

I immediately wrote to Cooper at Aerofoil Systems Inc., Cape Canaveral, Florida, asking if there was any truth to these statements. "If the whole story is a hoax," I said, "somebody ought to be sued." But there was no reply from him, even when I sent reminders and a stamped addressed envelope. I then wrote to Scott Carpenter, asking if he would forward it to Cooper, and this he promised to do. To this day, I have heard nothing.

In the same letter to Scott I asked for the complete story of the photograph he took during his flight in *Mercury 7* on 24 May 1962. According to a commentator on BBC TV in 1973, Carpenter had been withdrawn from duties as an astronaut for wasting time taking pictures of "sunrise." I thought this was rather unlikely, especially since Scott's friend, André Previn, told me that Scott had not been allowed in space again owing to a slight heart murmur. The released photograph shows what some have interpreted as a UFO, others as a lens flare, ice crystals, or the fabric and aluminium balloon that was deployed at one stage. I wanted the facts.

When I reminded Scott of my request a year later, he replied that he resented

. . . your continuing implication that I am lying and/or withholding truths from you. Your blindly stubborn belief in Flying Saucers makes interesting talk for awhile, but your inability to rationally consider any thought that runs counter to yours makes further discussion of no interest— indeed unpleasant in prospect—to me. I have sent your letter to Gordon Cooper without comment other than a copy of this letter to you. Let's do be friends, Tim, but let's talk about such things as music and SCUBA diving where maybe both of us can learn something.

I have never insisted that Scott Carpenter photographed a UFO, but because of the rumours surrounding the incident I wanted to know the truth. To me, that seems reasonable. In any event, my friendship towards, and respect for Scott remains undiminished.

An anonymous source with secret clearance claims that Carpenter told him that at no time when the astronauts were in space were they alone: there was constant surveillance by UFOs. And Dr Garry Henderson, a senior research scientist for General Dynamics, has confirmed that the astronauts are under strict orders not to discuss their sightings with anyone. Dr Henderson says that NASA "has many actual photos of these crafts, taken at close range by hand and movie camera."

In November 1979 Lou Zinsstag and I received an unofficial invitation to visit the Lyndon B. Johnson Space Center in Houston. The invitation came from Alan Holt, a physicist and aerospace engineer whose main work at that time centered on the development of the astronaut and flight controller training programs associated with the Spacelab. He is also engaged in theoretical research into advanced types of propulsion for spacecraft, and has long been involved in an unofficial NASA UFO study group called Project VISIT (Vehicle Internal Systems Investigative Team). I asked about photographs and films of UFOs allegedly taken by astronauts and was simply told that the National Security Agency screens *all* films prior to releasing them to NASA.

It may be coincidental that a former Director of the National Security Agency and Deputy Director of the CIA, Lew Allen, was appointed head of NASA's Jet Propulsion Laboratory in June 1982. JPL runs NASA's unmanned planetary space programme, whose phenomenal achievements include the landing on Mars by the Viking probes and, more recently, the Voyagers which transmitted such spectacular pictures of Jupiter, Saturn and Uranus. Allen had also been the USAF Chief of Staff, and as one of the pioneers of aerial espionage served as deputy director for Advanced Plans in the Directorate of Special Projects of the National Reconnaissance Office, and later director of the NRO's Office of Space Systems. NRO—America's most secret intelligence agency—liaises closely with the CIA, NSA—and of course NASA.

In an interview in 1986, Lew Allen stated that up to a third of JPL's work was funded by the Department of Defense, but gave details of various fascinating civilian projects. "One of the most exciting of these future programs, called Cassini," he said, "is an investigation of Saturn's moon Titan. Its atmosphere was too dense for the Voyagers to give us any clues about what lies beneath. The Cassini mission . . . would probe this atmosphere . . . we've concluded that it is very similar to what the earth's must have been at the earliest stages of its evolution."

Maurice Chatelain, former chief of NASA communications systems, claims that all the Apollo and Gemini flights were followed at a distance and sometimes quite closely by space vehicles of extraterrestrial origin, but Mission Control ordered absolute secrecy. Chatelain believes that some UFOs may come from our own solar system—specifically Titan.

During a BBC radio interview in December 1972, astronaut Edgar Mitchell, lunar module pilot on *Apollo 14,* was asked by a listener if NASA had made any provisions for encountering extra-terrestrials on the Moon or nearby planets. He replied in the affirmative. When the interviewer intervened and suggested that, if and when we ultimately come into contact with other civilizations, it would only be via radio-astronomy, Mitchell emphatically disagreed, making a point of recommending Allen Hynek's book, *The UFO Experience*, which contradicted official policy on the subject.

I wrote to Dr. Mitchell and asked him to elaborate on this and another statement he made on the programme, to the effect that there had been no concealment of UFO sightings either in transit to or on the Moon, and that such information was open to all. Mitchell's assistant, Harry Jones, replied: "Dr. Mitchell asked me to write and tell you that to his knowledge there have been no unexplained UFO sightings. *All unexplained sightings have subsequently been explained.* Dr. Mitchell personally attests that there has never been any lid of secrecy placed on any NASA astronaut that he is aware of." [Emphasis added.]

Although puzzled by this contradictory reply I did not pursue the matter further, since the publicity from UFO reports in 1973 led to a number of positive statements by some other astronauts. "I'm one of those guys who has never seen a UFO," said Eugene Cernan, commander of *Apollo 17*, at a press conference. "But I've been asked, and I've said publicly I thought they were somebody else, some other civilization."

In 1979 former Mercury astronaut Donald Slayton revealed in an interview with Paul Levy that he had seen a UFO while test-flying an aircraft in 1951:

I was testing a P-51 fighter in Minneapolis when I spotted this object. I was at about 10,000 feet on a nice, bright, sunny afternoon. I thought the object was a kite, then realized that no kite is gonna [sic] fly that high.

As I got closer it looked like a weather balloon, gray and about three feet in diameter. But as soon as I got behind the darn thing it didn't look like a balloon anymore. It looked like a saucer, a disc.

About that same time, I realized that it was suddenly going away from me—and there I was, running at about 300 miles an hour. I tracked it for a little way, and then all of a sudden the damn thing just took off. It pulled about a 45-degree climbing turn and accelerated and just flat disappeared.

A couple of days later I was having a beer with my commanding officer, and I thought, "What the hell, I'd better mention something to him about it." I did, and he told me to get on down to intelligence and give them a report. I did, and I never heard anything more on it.

DID APOLLO 11 ENCOUNTER UFOS ON THE MOON?

According to hitherto unconfirmed reports, both Neil Armstrong and Edwin "Buzz" Aldrin saw UFOs shortly after that historic landing on the Moon in *Apollo 11* on 21 July 1969. I remember hearing one of the astronauts refer to a "light" in or on a crater during the televized transmission, followed by a request from mission control for further information. Nothing more was heard.

According to former NASA employee Otto Binder, unnamed radio hams with their own VHF receiving facilities that by-passed NASA's broadcasting outlets picked up the following exchange:

MISSION CONTROL: What's there? Mission control calling Apollo 11.

APOLLO 11:   These babies are huge, sir . . . enormous. . . . Oh, God, you wouldn't believe it! I'm telling you there are other spacecraft out there . . . lined up on the far side of the crater edge . . . they're on the moon watching us. . . .

The story has been relegated to the world of science fiction since it first appeared, but in 1979 Maurice Chatelain, former chief of NASA communications systems and one of the scientists who conceived and designed the Apollo spacecraft, confirmed that Armstrong had indeed reported seeing two UFOs on the rim of a crater. "The encounter was common knowledge in NASA," he revealed, "but nobody has talked about it until now."

Soviet scientists were allegedly the first to confirm the incident. "According to our information, the encounter was reported immediately after the landing of the module," said Dr. Vladimir Azhazha, a physicist and professor of mathematics at Moscow University. "Neil Armstrong relayed the message to mission control that two large mysterious objects were watching them after having landed near the moon module. But his message was never heard by the public—because NASA censored it." According to another Soviet scientist, Dr. Aleksandr Kazantsev, Buzz Aldrin took color movie film of the UFOs from inside the module, and continued filming them after he and Armstrong went outside. Dr. Azhazha claims that the UFOs departed just minutes after the astronauts came out on to the lunar surface.

Maurice Chatelain also confirmed that *Apollo 11's* radio transmissions were interrupted on several occasions in order to hide the news from the public. NASA chief spokesman John McLeaish denied that the agency censored any voice transmissions from *Apollo 11*, but admitted that a slight delay in transmission took place, due simply to processing through electronic equipment.

Before dismissing Chatelain's sensational claims, it is worth noting his impressive background in the aerospace industry and space programme. His first job after moving from France was as an electronics engineer with Convair, specializing in telecommunications, telemetry and radar. In 1959 he was in charge of an electromagnetic research group, developing new radar and telecommunications systems for Ryan. One of his eleven patents was an automatic radar landing system that ignited retro rockets at a given altitude, used in the Ranger and Surveyor flights to the Moon. Later, at North American Aviation, Chatelain was offered the job of designing and building the Apollo communication and data processing system.

In his book, Chatelain claims that "all Apollo and Gemini flights were followed, both at a distance and sometimes also quite closely, by space vehicles of extraterrestrial origin—flying saucers, or UFOs . . . if you want to call them by that name. Every time it occurred, the astronauts informed Mission Control, who then ordered absolute silence." He goes on to say:

I think that Walter Schirra aboard *Mercury 8* was the first of the astronauts to use the code name "Santa Claus" to indicate the presence of flying saucers next to space capsules. However, his announcements were barely noticed by the general public. It was a little different when James Lovell on board the Apollo 8 command module came out from behind the moon and said for everybody to hear: "Please be informed that there is a Santa Claus." Even though this happened on Christmas Day 1968, many people sensed a hidden meaning in those words.

I asked Dr. Paul Lowman of NASA's Goddard Space Flight Center what he thought about the *Apollo 11* story. He replied:

Most of the radio communications from the Apollo crew on the surface were relayed in real time to earth. I am continually amazed by people who claim that we have concealed the discovery of extra-terrestrial activity on the Moon. The confirmed detection of extraterrestrial life, even if only by radio, will be the greatest scientific discovery of all time, and I speak without exaggeration. The idea that a civilian agency such as NASA, operating in the glare of publicity, could hide such a discovery is absurd, even if it wanted to. One would have to swear to secrecy not only the dozen astronauts who landed on the Moon but also the hundreds of engineers, technicians, and secretaries directly involved in the missions and the communication links.

Yet the rumours persist. NASA may well be a civilian agency, but many of its programmes are funded by the defence budget, as I have pointed out, and most of the astronauts are subject to military security regulations. Apart from the fact that the National Security Agency screens all films (and probably radio communications as well), we have the statements by Otto Binder, Dr. Garry Henderson and Maurice Chatelain that the astronauts were under strict orders not to discuss their sightings. And Gordon Cooper has testified to a UN committee that one of the astronauts actually witnessed a UFO on the ground. If there is no secrecy, why has this sighting not been made public?

Not all communications between the astronauts and ground control are public, as NASA itself admits. John McLeaish, Chief of Public Information at the Manned Spacecraft Center (now Lyndon B. Johnson Space Center) in Houston, explained to me in 1970 that although there is no separate radio frequency used by the astronauts for private conversations with mission control, private conversations, "usually to discuss medical problems," are re-routed: "When the astronauts request a private conversation, or when a private conversation is deemed necessary by officials on the ground, it is transmitted on the same S-band radio frequencies as are normally used but it is routed through different audio circuits on the ground; and unlike other air-to-ground conversations with the spacecraft, it is not released to the general public."

But is there any truth to the *Apollo 11* story? A friend of mine who formerly served in a branch of British military intelligence has provided me with unexpected corroboration. I am not permitted to reveal the name of my source, nor the location and date of the following conversation that was overheard and subsequently confirmed by my friend, which will inevitably lay me open to charges of fabricating the story or being the victim of a hoax. Yet the story must be told, however apocryphal.

A certain professor (whose name is known to me) was engaged in an earnest discussion with Neil Armstrong during a NASA symposium, and according to my friend's recollection, part of the conversation went as follows:

PROFESSOR: What really happened out there with Apollo 11?

ARMSTRONG: It was incredible . . . of course, we had always known
               there was a possibility . . . the fact is, we were
               warned off. There was never any question then of a
               space station or a moon city.

PROFESSOR: How do you mean "warned off"?

ARMSTRONG: I can't go into details, except to say that their ships were far superior to ours both in size and technology—Boy, were they big! . . . and menacing. . . . No, there is no question of a space station.

PROFESSOR: But NASA had other missions after Apollo 11?

ARMSTRONG: Naturally—NASA was committed at that time, and couldn't risk a panic on earth. . . . But it really was a quick scoop and back again. . . .

Later, when my friend confronted Armstrong, the latter confirmed that the story was true but refused to go into further detail, beyond admitting that the CIA was behind the cover-up.

What does Neil Armstrong have to say about the matter officially? In reply to my enquiry he simply stated: "Your 'reliable sources' are unreliable. There were no objects reported, found, or seen on *Apollo 11* or any other Apollo flight other than of natural origin. All observations on all Apollo flights were fully reported to the public."

33 UFO Phenomena and the Self-Censorship of Science

George C. Andrews

In the field of UFO research, there is a constant tug-of-war between zealot skeptics and zealot true believers, which like a Punch-and-Judy show distracts public attention from open-minded attempts to address the real issues, since both of these groups have their minds made up in advance.

It is unfortunate that a large proportion of the academic community falls into the category of zealot skeptics, insofar as UFO phenomena are concerned. Although regrettable, this is understandable, since any other attitude endangers the grants on which their livelihood depends, as well as their prestige in the hierarchy's pecking order.

The treatment Dr. John Mack received from his colleagues and the trustees at Harvard after his book on UFO abductions was published amply illustrates what happens when a previously respected professor investigates a taboo subject and comes up with unconventional conclusions. However, Dr. Mack emerged from the controversy relatively unscathed, when one compares what happened to him with what happened to Dr. James E. McDonald about a quarter of a century earlier.

Dr. James E. McDonald was Senior Physicist of the Institute of Atmospheric Sciences at the University of Arizona. He thought that the Federal Power Commission was evading the evidence concerning UFO involvement in the total power failure that paralyzed New York on July 13th, 1965, and dared to say so in front of a Congressional committee. His courageous statements on this and other occasions triggered a torrent of derision and abuse, and he was ostentatiously ostracized by his colleagues, in ways reminiscent of the treatment Dr. Mack recently received from his colleagues at Harvard. However, unlike Dr. Mack, Dr. McDonald was shortly thereafter found dead under suspicious circumstances, which to this day have not been satisfactorily elucidated.

Arbitrary denial of the reality of UFO phenomena by the academic community, in spite of the substantial evidence to the contrary which has been surfacing persistently at irregular intervals for the last fifty years, demonstrates a self-censorship that amounts to an abdication of responsibility and is incompatible with the principles on which their work is supposed to be based. No matter what the subject matter, scientific research is supposed to be carried out impartially, following the trail of truth wherever it may lead, without skewing the results one way or another to make them fit preconceived biases. It should make no difference if the results are unpopular or subject to ridicule by the ignorant who have not bothered to examine the evidence themselves, even if some of the ignorant happen to be in positions of authority that control research grants and advancement in the academic hierarchy.

It is the academic research community which sets the tone for so-called serious media coverage, as well as statements made by government representatives. Because it has systematically deprecated, minimized or denied evidence out of fear of ridicule, for a full half-century adopting an attitude of zealous skepticism, the academic community now bears a large part of the responsibility for the catastrophic present situation, in which the population as a whole must adjust to the shock of acknowledging the reality of the alien presence on this planet, although deeply conditioned for fifty years to dismiss it as a laughing matter, as easily controlled as a television set. Of course, the decision made in 1953 by the CIA's Robertson Panel to pursue a policy of systematic ridicule towards civilian UFO reports is also a major factor in the equation. This decision illustrates the extent to which contemporary science is influenced by the military/industrial complex, since that disastrous policy is still being implemented to the present day.

What is the evidence I claim is being arbitrarily denied? An incident witnessed by a single person is always open to question, and an eyewitness report on its own does not constitute substantial evidence. However, in the investigation of a traffic accident or a crime, if there are multiple witnesses who independently give similar descriptions of the event, their cumulative testimony tends to be taken seriously in a court of law. If there are literally hundreds or even thousands of witnesses independently giving similar descriptions of an event, the cumulative weight of their testimony becomes overwhelming. Long-term patterns over periods of several decades that include entire populations of towns and cities making similar reports should be considered scientifically as even more decisively significant, no matter what the subject matter.

The exception is the taboo topic of UFO phenomena. There are literally hundreds of examples I could point to, but one incident illustrates particularly well how this taboo operates.

I'll begin by specifying my sources, which are articles in the following newspapers: *Arkansas Gazette*, Little Rock, AR, January 23, 1988; *Arkansas*

Democrat, Little Rock, AR, January 23, 1988; *Gazette,* Texarkana, TX, January 23 & 24, 1988; *BEE,* Dequeen, AR, January 28, 1988; *Northwest Arkansas Times,* Fayetteville, AR, February 4–8 and March 27, 1988; *McCurtain County Gazette,* Idabel, OK, April 10, 1988.

The magnitude and extent of the incidents that began to be reported on January 19, 1988, from Little River County in Arkansas were on a scale that went beyond any other UFO phenomena that occurred in 1988. The incidents clustered around the towns of Foreman and Ashdown in southwest Arkansas, near the Texas border. A few sporadic sightings had occurred in previous months, including a low-altitude sighting of a UFO as large as a football field in November, 1987, but the witnesses did not dare speak out for fear of ridicule. The local population tends to be quite conservative, and the first witnesses to go public after a UFO chased three women in a car at terrifyingly close range on January 19, 1988, were subjected to persistent harassment and ostracism, until hundreds of citizens began seeing the phenomena simultaneously and its reality became undeniable. A typical report described

> . . . a ball of light that was as big as a hay wagon at first, but which got smaller when as many as 100 people gathered to look at it. The object changed color from red to green to blue. It was first seen near ground level, then flew high into the sky. It got under the moon and it looked just like a star up there until everyone went away, then it came back down. When it was up off the ground, lights were flashing, and you had to see it to believe it.

Witnesses included a professional astronomer, an Air Force veteran with 1,800 hours of flying time who had been a navigator on a B-52, a science teacher who had been selected as a finalist for the NASA "teacher in space" program, and a design engineer familiar with propulsion systems. Photos were taken that neither the Arkansas Sky Observatory, NORAD [North American Air Defense Command] or NASA were able to give plausible explanations for. However, Clay Sherrod, the Director of the Arkansas Sky Observatory, succeeded in insulting everyone's intelligence by maintaining that the extremely mobile metallic objects with multicolored flashing lights being perceived simultaneously by whole crowds of people, hovering at low altitude then suddenly rising straight up at incredible speed, performing maneuvers such as no known aircraft can perform, were either misidentifications of the planet Venus or moonlight reflecting off the bellies of white snow geese flying overhead.

Although newspaper coverage of the incidents ceased on March 27, the incidents continued to occur for approximately one full year well into 1989, without even being mentioned in the local press. They were considered no longer newsworthy, having been persistently disparaged by the authorities and the national news media, which parroted the "planet Venus" and "moonlight reflecting off the bellies of snow geese" explanations made by the Director of the Arkansas Sky Observatory, who was hundreds of miles away from the scene of the action in his office in Little Rock.

Besides the many eyewitness reports of UFO sightings, there have been many cases that involve craft landings, sometimes with physical evidence of landing traces left behind after the craft's departure. These traces of physical evidence have often been carefully investigated, and once again there are literally hundreds of examples I could point to. However, one specific case is outstanding because of the remarkable way these details were supported by the meticulously conducted research of high-level scientists, which backed up the anecdotal eyewitness reports with hard physical evidence.

Trans-en-Provence is a little village near Avignon in France. The incident took place there at 5:10 P.M. on January 8, 1981. Renato Nicolai, aged 55, a retired mason who had become a farmer, saw a strange aircraft land in his garden, where it remained for about one minute. It then took off and disappeared over the horizon.

Mr. Nicolai thought that it was probably some sort of experimental craft being tried out by the French Air Force. He did not believe in flying saucers. That evening when his wife came home from work, he described to her what he had seen. The next morning she went with him to look at the markings on the ground, then told a neighbor about the incident. The neighbor was frightened and informed the police.

A contingent of the Draguignan police came to Mr. Nicolai's farm. He described the craft to them as approximately 6 feet in length and 7 1/2 feet in diameter. The color was a dull gray, like that of lead. The shape was flat and circular, bulging slightly above and below. The craft rested on small telescopic legs. There was no light, and no smoke or flames. There was no sound except for a faint whistling. It first appeared at an altitude of about 150 feet, like a mass of stone falling. However, it came down lightly on the ground. He approached it and could see the craft clearly. He had advanced about thirty paces toward it, when it took off at very high speed. When he saw the object from beneath, it was round, and had four port-holes.

The police reported that there was a circular outline about half to three-quarters of an inch deep and 7 1/4 feet in diameter, with skid marks at two places. The site had the appearance of a circular stain, being darker in color than its surroundings. The police collected samples of soil and vegetation along a straight line through the impact site, writing on each sample taken its distance from the impact site. Upon their return to Draguignan, they transmitted their report and the samples to GEPAN

(Group for the Study of Unidentified Aerial and Space Phenomena), which is a branch of CNRS (National Center of Space Research, the French equivalent of NASA). GEPAN passed the samples on to INRA (National Institute of Agricultural Research) and several other government research institutes for analysis. GEPAN personnel visited the site to take further samples on two other occasions. On June 7, 1983, after two and a half years of analyses, a bulky preliminary report which assembled data from the different laboratories was turned in. The government scientists attributed the circular outline to a soil fracture caused by the combined action of strong mechanical pressure and a heat of about 600°C, which is about 1100°F. Dr. Bounias, who was the Director of the Biochemical Laboratory at INRA, had personally taken charge of the examination of the plant specimens. He carried out the analyses in the most rigorous fashion possible. First he established samples from plants of the same species (alfalfa), taken at different distances from the point of impact. Then he and his assistants meticulously analyzed the photosynthetic pigments (such as carotene, chlorophyll, and xantophyle), the glucides, the amino acids and other constituents. He found differences sufficiently important that the statistical significance of the results is irrefutable. Certain substances that were present in the close-range samples were not present in those taken further away, and vice versa. The biochemical trauma revealed by examination of the leaves diminished as the distance from the UFO impact site increased. Some of the plants had been dehydrated, but were not burned or carbonized. The following year control samples were taken from the site, which confirmed the changes made in the vegetation. After completing the analyses, Dr. Bounias made the following formal statement: "We worked on very young leaves. They all had the anatomic and physiologic characteristics of their age. However, they had the biochemical characteristics of advanced senescence, or old age! This bears no resemblance to anything known to exist on our planet."

Dr. Bounias refused to speculate about the cause of the strange facts he had established, or to propose any explanation at all for them.

Although neither Dr. Bounias nor the French government have followed through on the implications of this evidence, or proceeded any further with it, at least as far as the general public is concerned, the Trans-en-Provence case remains one of the most strongly substantiated investigations of landing traces in the history of UFO research.

Another aspect of UFO research which involves physical evidence is the crop circles, though there has been much dispute over whether they are caused by UFOs or by human hoaxers. I believe that some are made by UFOs, and some are made by human hoaxers. Other theories have been proposed, but at present these are the only ones that have retained their plausibility, since freak whirlwinds and hypothetical plasma vortices cannot by any stretch of the imagination explain geometrically precise pictograms and other complex symbolic formations. From 1978 to 1989, the shapes were for the most part simple circles. However, since 1989, the patterns have become more and more intricate, eliminating the possibility that they could be caused by unusual meteorological conditions.

The summer of 1991 was a quantum leap as patterns of rings and circles became true complex pictograms. Straight bars, or boxes, and arcs, both inside and outside of circles, were combined with circles and rings to form complex pictograms. Some pictograms combined more than thirty elements.

Crop circle developments during the summer of 1991 were well described by Michael Chorost in the October 1991 issue of the *MUFON* [Mutual UFO Network] *UFO Journal*:

> One of the most interesting formations was a representation of the Mandelbrot set, a two-dimensional graph made famous by chaos theory . . . the last two seasons of crop circles have clustered densely in a tiny area containing Europe's most remarkable ancient constructions: Avebury, Silbury Hill, Windmill Hill, Barbury Castle, Adam's Grave, the White Horses, and the East and West Kennet Long Barrows. . . .
> I invite my readers to consider that the mystery of the crop circles is very much like the mystery of the megaliths. Each consists of compelling geometric forms. No one knows why they were made, nor why they are where they are. Nor do we know how either were made. Perhaps the two mysteries are deeply intertwined. Not that either one "caused" or "inspired" the other, but that the two phenomena somehow "talk" about the same thing, a thing still unknown to us, or "do" a single thing, taken together as a total system. It could be that solving one mystery will automatically solve the other.

Chorost goes on to describe the research results of Marshall Dudley, a systems engineer for Tennelec/Nucleus of Oak Ridge, Tennessee, as well as those of Michigan biophysicist Dr. W. C. Levengood. Dudley detected significant isotope changes in the soil samples from crop circles he had been provided with, and Levengood found that cell pits in plant cells in the affected formations had been subjected to rapid heating that had separated the cell pits. He found this to be true in samples from England, the United States and Canada.

Another major breakthrough was made by Gerald Hawkins, the author of *Stonehenge Decoded*, who discovered that in eighteen photographs of crop circle formations, there was a repeated pattern of frequency ratios that are equivalent to the diatonic scale (the white keys on a piano). In addition to that finding, he has outlined four new theorems about relationships of triangles to circles to squares that he finds in the crop circle formations, and these theorems do not exist in any known academic text.

That is a very brief condensation of a large amount of highly complex technical research. In light of the fact that there is strong and abundant evidence in support of these results, one would think that the news media would eagerly leap upon so thoroughly substantiated a sensational story, and proclaim it to the world in banner headlines and TV special features.

What actually happened?

The world news media instead leaped eagerly on a flimsy story full of holes: that two British senior citizens had "confessed" to hoaxing the circles with no equipment except some planks. This was triumphantly proclaimed to the world as the final and definitive solution to the mystery of the crop circles, in spite of the obvious fact that two men with planks cannot produce significant isotope changes in the soil, nor heating so rapid that it separates the cell pits without leaving burn marks on the outside of the plants. Other obvious impossibilities deliberately ignored were how these two senior citizens had managed to make so many hundreds of circles without having once been detected, or how they managed to make patterns of such precision and size and complexity with planks while working in the dark. All the factual evidence was deliberately ignored in order to convince the public that the mystery of the crop circles had now been at least definitely solved: Doug and Dave did it. The public was bombarded with ten-second TV shots of Doug and Dave flattening some wheat with some planks, until finally the vast majority was conditioned into accepting this absurdity as the proven explanation. The minority of those who persisted in trying to point out flaws in this explanation was then subjected to scathing ridicule and social ostracism. Vast numbers of copy-cat imitators followed the example set by Doug and Dave, and have ever since devoted themselves to muddying the water and confusing the research picture, egged on with the full collaboration of the news media, intent on trivializing the subject.

In spite of the sabotage and harassment, the research haltingly continues. An intriguing development that occurred recently in England is that a group of hoaxers busily at work making yet another faked crop formation noticed several balls of light hovering above them, which seemed to be under intelligent control. This frightened them to the point that they abandoned their work and fled from the field. There are now quite a few eyewitness reports of small white discs and grapefruit-sized balls of light seen in the vicinity of the crop formations, and the small white discs have been captured twice on videotape.

Some of the crop formation patterns resemble traditional geometric artwork of indigenous tribal cultures from all over the world so closely as to be identical. Without exception the religious traditions of these indigenous cultures describe contacts with celestial beings in deep antiquity at the time of their tribal origins. According to researcher Colin Andrews, *all but a few of the symbols on the panels from the wreckage in the so-called Roswell film have been clearly and precisely reproduced in the crop circle glyphs.*

34 Mars—The Telescopic Evidence

Daniel Ross

When the spaceships appeared in the late 1940s, and sightings reports began to number in the thousands, scientific specialists advising government and military authorities believed that Venus and Mars were the origin of the spacecraft. They were more certain after recovering a few ships that had crashed near our atomic test sites. Then an almost impenetrable security lid came down, to censor any evidence from official sources that life existed beyond the earth. A Silence Group, working for those in entrenched worldly positions, infiltrated secret departments and intelligence agencies to insure that confirmation would never come from official sources or government. Public or private institutions, being generally conservative in matters of science, were unlikely to speculate on the UFO evidence, but in any event, those institutions would not have the means to confirm the origin of the visiting spaceships.

Complete, uncontestable confirmation was strictly the domain of a government space agency, and the official results of any achievements in space exploration were under the sole control of the National Security Agency. Public disclosures regarding planetary environments were carefully slanted to coincide with long-held orthodox views, and with theories that had become rigid and dogmatic with the scientific establishment. That Venus and Mars have not been shown as having earthlike environments, is not due to a lack of technology in our space probe exploration, but due to secretive censoring by intelligence agencies directing operations from behind the scenes. So many false ideas on space have been promoted through official channels, and then become solidified in scientific journalism, that one may reasonably wonder if in today's world it can ever be straightened out.

There is no grand conspiracy by science writers to deceive, nor by scientific spokesmen with their speculations on space conditions. They actually believe what they write or say, because these are widely-shared and firmly held perceptions which have been taught for a long time. Their ideas have also been reinforced by the false disclosures publicized through the media by those in control of past space ventures.

Likewise, this article is not in direct opposition to general astronomy. In fact, a lot of information in this present work is based on the observations and lifetime work of expert astronomers. But in establishing the truth about our solar system, it will be noted that there is little agreement with orthodox thinking in the astronomical field. And if one were to restrict himself to one field—any field—one would have very limited knowledge. Determining the reality behind UFOs requires a complete study involving the whole scope of space sciences.

The problem with all planetary research and common speculation to date, must be defined here at the beginning. It is this: Official presentations regarding planetary space conditions have been made to coincide with (complement) the long standing suppression, and censorship, of the real UFO evidence by our government. This is why the truth about Mars has never been known, publicized, or accepted, up to the present. Yet it is an important correlation, that eighteen years of UFO sightings, with reports numbering into the thousands, predated the first U.S. space probe to reach Mars on a flyby in July, 1965. Of course, it was never officially admitted that UFOs were a major stimulus for us to investigate the planet. Now, in this present work, it will be established that the Martian environment is very similar to earthly conditions, by a review of the early telescopic record, and then through a logical analysis of the more recent space probe developments.

The early history of telescopic observation of Mars has been recounted in numerous books. It began in 1877, when Giovanni Schiaparelli observed through his 8.75 inch reflecting telescope, a number of long lines on the Martian surface that connected up to larger dark areas. He described the lines as "canali," which in his native language meant channels. But the translation quickly became "canals," and his discovery of them led to the idea that intelligent beings on Mars must have constructed artificial waterways. While Schiaparelli didn't publicly suggest that conclusion himself, he didn't really discourage others who were promoting the idea, because he had found 113 different canali that were long, straight, and neatly defined. He intricately mapped the planet from years of observation. His maps were the standard for many years, and he gave ancient names from Biblical and classical mythology, along with names from the old geography of the Middle East, to the large surface areas and distinct markings of the planet. The names he gave to the surface features are still existent on maps today.

A distinguished American astronomer, Percival Lowell, decided to dedicate his life to studying Mars. In 1894, he built the Flagstaff Observatory in Arizona, which housed a 24-inch refracting telescope. By 1915, he and his staff had charted nearly 700 canals—a precise network of large-scale construction on Mars that channeled water from the polar ice caps. They were straight, narrow, sometimes parallel, and at numerous locations the canals intersected geometrically. These latter areas were noted to become seasonally dark, and Lowell named them oases, indicating that vegetation and crop growing were abundant. He naturally concluded that there would be attendant cities for the Martian people at these oases.

Lowell understood that the actual waterways could not be seen from Earth, if it were not for the broad areas of seasonal growth lining both sides. It was the combination of both factors that made it possible to see the network of geometric lines on Mars' surface with clarity. Some of the channels were approximately 3000 miles long, and from 15 to 25 miles wide.[1] In 1915 Lowell stated to the scientific world, "Mars is inhabited, and we have absolute proof." He proclaimed that the Martian civilization had an intricate and highly advanced irrigation system that could be seen and photographed through Earth-based telescopes. A few pictures had been taken as early as 1907. Lowell's position was so revolutionary to the orthodox views of the scientific establishment, that it received harsh contempt from many, and went virtually ignored by others.

Once every twenty-six months, Earth and Mars are at their closest distance from each other in their orbits around the sun, and in astronomy this is called being in opposition. But because the orbits are elliptical, the most favorable opposition occurs only once every fifteen to seventeen years, and at this time the two planets are at their closest, about 35 million miles distant. To view the extensive canals and markings, an astronomer had to have unlimited patience and determination, and more importantly, an open mind. Like the establishment scientists today, Lowell's contemporaries often lacked such traits. Studying the distant features on Mars through the telescope was difficult and tricky, and could only be done at the large observatories when the local atmospheric conditions and other visibility factors were exceptionally coordinated. But even during the brief periods of favorable opposition, the disk-like image showed a blurring of detail almost continuously, due to the ever-present atmospheric turbulence around both the Earth and Mars.

Our atmosphere is constantly in molecular motion due to thermal activity. To the naked eye, the sky might seem so clear and calm, that a person would assume there is perfect seeing conditions. For looking at stars and nebulae, that would be true, but it's not the same when we view our neighboring planets with a large telescope. Through the high power magnification of a telescope, the barely perceptible dynamics of heat (wind) movement in the atmosphere causes a slight shimmering effect, and while the broad features of a planetary image may be easily recognizable, any fine detail is lost in an almost continual slight blurring. Ever so momentarily, our atmospheric unsteadiness will cease for a second or two. At that precise moment an astute telescopic observer will have a perfect seeing condition, and be able to see in fine detail the planetary image 35 million miles away. Yet these views last but a few seconds, making it extremely difficult to obtain a distinct photograph. Furthermore, the only way that the photographic evidence of the canals can be obtained is when the planet is viewed directly overhead. These observations must be made from the best suitable locations in our southern hemisphere.

Lowell made a special expedition to Chile in 1907 and obtained the first photographic evidence of the canals. His successor, Dr. E. C. Slipher, had better success in later years with observations from South Africa, when camera equipment had improved considerably. The Martian canals are seen on plates VI and XLVII in the book, *The Photographic Story of Mars*, by E. C. Slipher. The edition I obtained was published by Northland Press, Flagstaff, Arizona, in 1962.

The quality of photographs can always be debated by the establishment scientist who denies everything he has not seen for himself. In reality, the eye is superior in viewing telescopic images in detail, compared to the photographic results when taking telescopic pictures of a planet 35 million miles distant. Dr. Slipher stated in 1962, "The history of the canal problem shows that every skilled observer who goes to the best available site for his observations has had no great difficulty of seeing and convincing himself of the reality of the canals. I am not aware of a single exception to this." A fellow astronomer, Dr. Pettit, confirmed this visual documentation, by reporting in 1953 that "there are moments when the whole canal pattern can be seen on Mars."

Today's literature never fails to mention that the early Mariner probes during the 1960's proved that the canals are non-existent, and that the controversy over the Schiaparelli and Lowell evidence has been laid to rest. It is true that no actual evidence of canals was released by NASA, but it should be realized, that if the picture-taking cameras on those early probes did photograph certain areas showing canals with sufficient clarity, the evidence would not have been released anyway. The fact is, that until Mariner 9, only a very small and unrepresentative fraction of the Martian surface was photographed, and most of that, very poorly. Mariners 4, 6, and 7 never even found the huge 2300-mile-long Valles Marineus canyon on Mars, which is a natural formation. The fuzzy black and white photos that were released to the public lacked any clarity whatsoever. We can get better telescopic photographs of the Moon 240,000 miles away, than those camera pictures taken only a few thousand miles from Mars. An important point to realize is that the probes carried cameras, not telescopes. Even NASA admitted that the cameras aboard the Mariners could not have provided evidence of a Martian civilization from their photographic distance. However, the publicly-released photos were quickly interpreted as disproving the canal controversy.

What are space photographs in reality? The "picture" is relayed back to Earth in the form of numerous dots, contained within a radio signal. The picture has to be reconstructed from this electronic message, by computer imaging each dot into a shade of gray. The first image processing is considered the raw picture, and is basically a washed-out, blurry gradation of gray. Then the imaging team can reassign the gray levels by computer, in order to better distinguish any identifiable spots or features on the raw picture. A slightly improved image is given to the public.

American astronauts have said that the only visible man-made construction on Earth that they could see from their high orbit around our planet was the Great Wall of China. If there were a Great Wall on Mars and it turned up on one of the photographs, the space agency could still release the picture, but without the slightest trace of a wall. With computer imaging, it is easy to fade out features and erase contrast, to the point of an unidentifiable gray blotch. By starting with the original raw, washed-out picture, it is only a matter of re-assigning the gray levels so that the wall never appears during processing. On the other hand, if another photograph shows a natural landmark or feature, that picture can be electronically sharpened and focused to show great detail. We have reached a new state of the art: we can increase or decrease picture quality by subtle electronic brushing.

Now to clarify the situation regarding the canal evidence first discovered by Schiaparelli and Lowell through their telescopic studies. It was only "laid to rest" because authorities withheld official confirmation. Mariner 4 did photograph some straight-line canals, and this was finally admitted some time later by Dr. William Pickering, the head of Jet Propulsion Laboratory.[2] (JPL conducts all the planetary projects for NASA.) Dr. Clyde Tombaugh, the scientist who discovered Pluto, also confirmed that the canals were photographed by the 1965 probe. But officially, this type of evidence has never been released. The public was shown computer-enhanced photographs, but the detailed originals were in the hands of the authorities. And if the canals were filmed by that first probe, it is a certainty that they were filmed by later Mariner and Viking probes, yet that information has always been withheld. We'll discuss the censoring aspect relative to the later space missions thoroughly, but first let's continue with the telescopic record.

Early in this century, expert astronomers recorded several anomalies during their observations of Mars. On one occasion, a long series of blinking lights lasting seventy minutes was observed, leading one observatory director to describe the incident as "absolutely inexplicable."[3] In 1937 and again in 1949, Japanese experts witnessed a brilliant glow on the surface of Mars, that was as bright as a 6th magnitude star. To be visible from the Earth, these "flares" had to be tremendous. Any type of volcanic activity couldn't possibly be seen from our distance, and so the cause of the brightness remained a mystery. Other strange lights were seen on different occasions.

There was a cloud-like object observed and photographed in 1954, that was in the perfect shape of a W, or an M if we consider that a telescope inverts an image. It was 1100 miles across and remained in a fixed position above the planet for more than a month. (Natural atmospheric clouds will change shape and dissipate within a few days.) At the three intersections of the W, were intense bright spots, or "knobs." Speculation was running high, even at the Carnegie Institution at Washington. It was such a rigid and unusual shape, that there was a strong suggestion of artificial origin.

Throughout the 1920s and 30s, recurring radio signals were picked up coming from the direction of Mars. The spacing and pattern of the radio waves ruled out the possibility that these cryptic signals were random radio noise or electrical disturbances in space, because there was an intelligent coding system to these radio waves. That much was certain, even though they remained indecipherable on our end. Even the famous scientist Marconi, the man who invented the "wireless," picked up these interplanetary radio waves with his advanced experimental equipment in 1921, and later stated that he believed he had intercepted messages from Mars. He emphasized that the transmission wavelength of the coded signals was 150 kilometers, whereas the maximum wavelength used by our transmitting stations at the time was about 14 kilometers.[4]

Many others had come to the same conclusion over the next few years when intercepting these signals, especially when Mars was in orbital proximity to Earth. And speaking to the British Association for the Advancement of Science in 1931, the late Bishop Barnes stated his belief that many other inhabited worlds exist, and that many must certainly be able to propagate interplanetary radio communication. It was such messages that were being picked up now, he said. And when these interplanetary signals were recognized and acknowledged by our Earth, it would be the dawn of a new era of humanity. But at this beginning, he added, there would be opposition between those who welcome the new knowledge and those who deem it dangerous for that information to be known and accepted. And is this not what happened two decades later, when UFOs demonstrated the very existence of life on other worlds? Was it not the beginning of an era of opposition between those who were open and accepting of the new knowledge about space, and those who worked to prevent the truth from coming out?

Along with the later observations of mysterious clouds and lights, the cryptic radio signals led some independent astronomers to conclude that we were being given rudimentary signals from Mars to challenge our thinking about life beyond the Earth. Regarding habitability, there was even more scientific certainty in other telescopic studies. As early as 1926, photographs were taken in ultraviolet light that clearly showed a substantial atmosphere on Mars. Compared with infrared photographs taken at the same time, the pictures proved that there is a dense atmosphere, possibly 40 miles in depth. There are undoubtedly more rarefied layers above this altitude, much like the upper, tenuous atmosphere around the Earth, that would be too thin to be recorded by photography. It has been suggested that the top of the Martian atmosphere might reach 400 miles, by the British scientist-writer Earl Nelson, author of *There Is Life on Mars* (1956).

The early photographs showing the Martian atmosphere were taken by G. E. Hale of the Mount Palomar Observatory [Southern California] and are reproduced in Nelson's book. There are two immediate and important conclusions that can be drawn from these observations. The surface gravity on Mars must be substantially higher than has been taught, for a low gravity would not be sufficient to retain such a sizable atmosphere. Secondly, with such a dense atmosphere, the sun's energy would interact much differently than orthodox theories suggested, and the temperatures on Mars would be considerably warmer, more moderate, and more Earth-like.

Although the length of the Martian year is nearly double our 365-day year, the seasons on Mars vary and alternate just like on Earth. When the northern hemisphere is in its summer cycle, the southern hemisphere has its winter. The length of the Martian day is 24 hours and 37 minutes, and the inclination of its axis is 25 degrees, which is very close to Earth's 23 degrees.

Both the northern and southern polar caps extend nearly half way to the Martian equator during their respective winters. With the onset of spring in either hemisphere, its ice cap recedes and a wave of darkening over broad areas spreads slowly towards the equator. This cyclic surface darkening was widely considered to be seasonal vegetation growth as water was liberated from the polar caps. Each polar cap will shrink considerably during its respective summer cycle. Sometimes the southern polar cap melts completely.

The broad areas near the equator, such as Mare Serpentis, Mare Sirenium, and Syrtis Major, change from their winter shade of brown, to light green and then to dark green. This latter stage has often been described as a dark blue-green. Astronomers also noted that as the seasons changed to autumn, the colors would gradually turn to yellow and gold, finally returning to brown in winter. (The surface color of Mars is not dark red, as I will prove later.)

The parade of colorful seasons was interpreted by open-minded astronomers as the seasonal growth and ripening of vegetation. Cyclic growth coincided regularly with the natural climatic changes on the planet, just as we have here on Earth. I am not discussing the canals and their irrigation for crop growing, at the moment. These seasonal changes showing cyclic plant life would be taking place even if man were not there on Mars.

The presence of vegetation on Mars was held to be a certainty in some quarters, but hotly debated by others. But the way to end all argument was to prove the existence of carbon dioxide, oxygen, and the water in the Martian environment, which would indicate that photosynthesis (the life process) of plants was in fact taking place. Carbon dioxide was there in abundance—even conservative scientists agreed on that, for it was commonly speculated that the atmosphere's chief constituent was carbon dioxide. Oxygen seemed likely, though it could not be detected in the atmosphere from earth-based studies. The evidence for oxygen was indicated by some regional soil colors, which indicated that certain areas contained a large amount of ferrous oxide, or limonite. We have some tropical regions on Earth where the soil is reddish-brown limonite, and two things are necessary for its formation: abundant oxygen and extreme humidity in the air. Apparently, oxygen was in the atmosphere of Mars, as the natural product of plant photosynthesis.

To briefly explain photosynthesis, it is the biological process by which green plants containing chlorophyll use the energy of sunlight to synthesize carbohydrates from carbon dioxide and water. Six molecules of water and six molecules of carbon dioxide are transformed with the aid of solar energy into one molecule of glucose and six molecules of oxygen. The oxygen is then liberated into the atmosphere. We breathe in the oxygen and exhale carbon dioxide, which in turn the green plants use in photosynthesis, and oxygen is returned to the atmosphere. This is Nature's perfect cycle. If all the green plants were suddenly removed from the Earth, all human and animal life would die, because the oxygen we breathe would not be replenished.

The last thing that needed to be confirmed in order to prove the seasonal vegetation on Mars, was the existence of water. For this evidence it is easiest to jump ahead for a moment to the U.S. Viking project of 1976. The Viking I orbiter photographed extensive ground fog, mists, and cloud cover in the northern hemisphere, and from readings taken by sensitive instruments on the orbiting probe, it was proven once and for all, that the polar caps were frozen water.[5] If the polar caps were completely melted, it was estimated that the water produced would cover the entire planet to a depth of about 20 feet.

Along with the early ultraviolet photographs showing a substantial atmosphere, it has been shown that the environmental constituents for life exists on Mars. The three basic parameters are carbon dioxide, water, and oxygen—the ferrous oxide soil being the indirect evidence for oxygen. It is necessary to point to the indirect evidence for oxygen, since NASA refuses to confirm the presence of oxygen in the Martian atmosphere. That is the single remaining ace in their hand. And they keep it, because they know that only the process of photosynthesis by living plants can account for the presence of oxygen in any planet's atmosphere. During the Viking mission, NASA admitted finding nitrogen, argon, carbon dioxide, and water vapor, although they kept the relative percentages and overall density out of proportion to the true conditions. But NASA is holding out on the oxygen and will not admit finding it with the Viking probes, because *atmospheric oxygen* would be recognized by scientists as positive proof that life exists on Mars. But the remaining evidence to be discussed will prove the case.

Before the space agency came into existence on October 1, 1958, scientific astronomers at the large observatories were still the experts and authorities on the planets. It seems as though it was preordained in the heavens that the independent thinkers would have one last chance to probe the mystery of our neighboring planet, as Mars swung by in favorable opposition in 1954 and 1956. In its first approach, Mars came within a distance of 39,800,000 miles. The second time, in 1956, the planet was only 35,120,000 miles away. It would not be that close again until 1971, when planetary exploration and pronouncements were in the hands of NASA.

But in 1954 the excitement ran high in astronomical circles, because an international Mars Committee had been formed, to plan an around the world "Mars patrol." Prominent scientists from seventeen countries would be coordinating telescopic studies from the world's largest observatories, as Mars made its closest approach in July. Some of the countries involved included the United States, France, Italy, Turkey, India, Japan, Australia, South Africa, Java, Egypt, and Argentina.[6]

The international team of scientists was headed by the world's greatest Mars expert, Dr. E. C. Slipher, then the Director of the Lowell Observatory [Flagstaff, Arizona]. He and most of the committee members were well aware of all the previous astronomical records—the mysterious clouds, flares, markings, radio signals, and the evidence for canals and vegetation. Some privately believed that there was an intelligent civilization on Mars, for in 1938, it had been announced that the Lowell Observatory found evidence of changes in the canal system, and the changes appeared to have been altered by design. This 1954 Mars "expedition" was primarily planned to settle the question. It is quite possible that some members linked the numerous flying saucer sightings that had been widely reported since 1947, to the renewed and intense interest in Mars.

Because the government was heavily guarding the UFO evidence, the National Security Agency made it a top priority to use its influence to keep check on the developments of the Mars patrol study. It was imperative that planetary speculations and press statements be kept in a totally ambiguous light. The censors were especially concerned about the Mars patrol because of the caliber of open-minded men who were involved with the project. They included Dr. Seymour Hess, a meteorology expert who was on record as having sighted a UFO; Dr. Harold C. Urey, a prominent astrophysicist who was genuinely curious about life on other planets; and Dr. Slipher, who was following in the footsteps of the pioneer Percival Lowell. Dr. Slipher assigned himself to make observations from the best location possible—the Lamont Hussey Observatory in South Africa. It had the largest refracting telescope in the southern hemisphere, and Mars would be passing directly overhead each night during opposition. And before the project got underway, Slipher publicly stated that if he found proof of life on Mars, he would announce it to the world.

The Mars Expedition took 20,000 photographs and confirmed the presence of both the canals and vegetation. The canals did not meander at all like a river would; they followed great-circle courses, which are the shortest distance between two points on a globe. Many planetary astronomers had speculated previously, that if photographs showed that the canals were along great circle paths, it could be concluded that they were the work of intelligent beings. The scientists were getting exceptional pictures also, because the Lowell Observatory was using a new electronic camera that could amplify faint markings, and photograph in one-tenth of a second to prevent atmospheric turbulence from blurring the details. One canal was found to run straight as an arrow for 1,500 miles, something that no natural water channel could do.

Dr. Slipher brought enough photographs back from South Africa to prove that the canals were real, and man-made. While providing abundant vegetation growth alongside their straight-line courses, the canals also proved to be the common link between the green oases. An intricate pumping system seemed to be the only explanation when considering the distances involved. More than 40 canals and 15 oases were photographed in the first week.[7] But the Mars Committee reports never became public, and they were therefore unknown outside a very limited part of the astronomy community. The new findings were privately logged at the observatories, and sparing details were barely covered in only a few astronomical journals. But everything was kept out of the newspapers.

The government's intelligence agency had succeeded in blocking the Committee's early plans for public reports and press conferences. Then they firmly executed their plans for a blackout of real information about Mars. The government keeps itself in control by keeping a world-wide opinion in control, especially with regard to sensitive and dramatic issues. Allowing an announcement by an international team of scientists suggesting that Mars was inhabited, would be tantamount to the government confirming that UFOs are visiting our planet. So the censors knew what they had to do.

Pressure was put on those who headed the project to furnish no reports to the public press. Though the astronomers studied Mars for five months, only one little statement was given to the public at the beginning. Dr. Slipher had announced that some new and interesting changes were observed on Mars with their photographic study. Following that report, there was only silence. All plans for further publicity were blocked, and no worthy Mars Patrol bulletins were ever released.[8] The excuses given out were in the category of "difficulties in communication and coordination, disagreement as to what had been seen and photographed, months of studies and review were necessary to properly analyze, and so on."

How can any silencing agency of the government achieve such suppression of this, or any other, kind of dramatic information? It is difficult to determine for each case just what methods are employed, but their forceful persuasion does escalate until the cooperation is achieved. Presumably, they start out with the position that such information is related to the national security, and that the government is the entrusted agency to best handle the social implications of confirmed announcements. They imply that the public isn't quite ready for this information, that the world isn't ready for this information. That the economy isn't prepared for this type of information. They fear that there would be an upheaval in thinking (although I am certain that it would be an "upliftment" in thinking, and this is the real problem that threatens the censors).

They will say that the public might panic, or they could offer the excuse that there might be an attack from Mars. The possibilities for persuasive argument are endless, but the only end requirement is that planetary evidence be shown as inconclusive, vague, and debatable. It has always been maintained officially, that known life does not exist beyond the Earth, unless possibly it is light years away from us in another part of the galaxy. In which case, the distance is so great that our civilizations will never meet.

After being persuaded to withhold the significant findings, including the discovery of the great-circle paths of the canals, the Mars Committee only issued a simple press release. Dr. Slipher made a statement to the effect that Mars is alive. That certainly satisfied the censors' insistence in keeping things nebulous. (Alive—how? Geologically with volcanos, dust storms, and polar cap shrinkage? Or alive in the sense of intelligent constructions?) He noted that there were color changes in the Martian geography that were more interesting than in his previous observations over the years. But the tiny report was essentially meaningless, and obviously did not affect public or scientific opinion. The question of Mars might have still been left open, but the orthodox theories of inhability were not threatened in the least.

It was not until eight years afterwards that notable documentation of the 1954 Mars observations was published, in a book titled *The Photographic Story of Mars*. Recently, I obtained a copy of this book, and it appears that the publication had a relatively small printing, and was mainly published to be a reference type of book for science libraries. Certainly, few in the public would have been inclined to buy such a costly book, and take it on their own to study an involved scientific text. Yet the answers are there if one wishes to read through complex analyses and carefully worded discussions. The book was written by Dr. Slipher in 1962, and the full text is based on fifty years of telescopic studies, and thousands of photographic images taken at the world's largest observatories. The conclusions also referenced the last major finding by astronomy regarding the Martian environment.

During the November 1958 opposition, Dr. William Sinton conducted studies at the Smithsonian Astrophysical Observatory. The scientist-astronomer performed careful infrared scans of the bright desert areas and the dark green oases, and found that the sun's energy was absorbed in certain wavelengths over the dark areas, but not over the desert regions. The absorption wavelengths were at 3.43, 3.56, and 3.67 microns*, and these are exactly the same wavelengths absorbed by hydrocarbon compounds. His study proved that there is green plant life on the broad oases of Mars, and that it is organically composed of carbon-hydrogen compounds, the same as our own terrestrial vegetation. In other words, his scientific evidence showed that Martian plant life is based on the same carbon cycle as we find on Earth.

But new experimental evidence is never accepted that quickly. It is always challenged, and subject to much debate, because old established theories are very hard to change. The old theories had predicted that there was no appreciable water or atmosphere on Mars, and that the surface temperatures were too extreme for vegetation life. (The canal evidence also, carried too many implications to be considered acceptable, and was rejected out-

---

* A micron is equivalent to one millionth of a meter.

right as incompatible with respected establishment theories.) Dr. Sinton's experiments with infrared scans were viewed as inconclusive, and any such results would have to be confirmed over and over before conservative science would budge. The scientific community much rather preferred to wait until future space probes settled the questions about Mars. The scientific arguments lingered in limbo, until the government formed a space age bureaucracy, called NASA, that could preempt all discussions on matters of space. The days of independent astronomy speculating on planetary conditions were soon over. While representing the government in its authoritative role, NASA's position was unassailable—almost.

Initially, NASA had three functions:

1. To launch artificial satellites into orbit around the Earth.

2. To put men into space.

3. To explore the other planetary members of our solar system, including the Moon, with remotely controlled space probes.

The first two they did admirably well, and mankind was on the threshold of becoming a space civilization. But with the third, NASA did not advance our knowledge towards an age of enlightenment. In fact, there is a bitter irony to our space age developments, in that our authorities led thinking back to the Dark Ages, through distortion and suppression of actual space findings.

Long ago, Earth was isolated from the rest of the system, through its ignorance and superstitious thinking. By the twentieth century, man's intelligence had progressed to where he could rationally understand and accept that advanced civilizations do travel space and have home planets similar to our Earth. Space visitors traveling in ships which we have termed UFOs were making their existence known at the same time we were reaching technological crossroads in science. But the men of war and all their institutions denied it, and the censors would not allow the confirmation of life beyond the Earth, whether in spaceships or on planets. The doors were kept shut by the silence group and vested interests opposed to the truth, and NASA then turned out the lights. NASA made out space to be an uninteresting wasteland, devoid of life or recognizable purpose. The end result was that mankind on Earth reverted back to an extreme thinking of self-importance, alone in his own egotistical sphere of a world.

Had our authorities left a few questions open for balanced speculations, it would be easier to be less critical. But instead, they determinedly set out to present a completely negative picture of the planet. A living environment was totally negated without qualification, in order to complement the suppression of UFO evidence. It was apparent that our planetary probing was not conducted with any objectivity, right from the start. *Being a government bureaucracy*, NASA had no choice but to serve the hands of the most powerful economic interests of our present-day world. NASA censors cooperated with the corporate interests that demanded the continual coverup and suppression regarding UFOs and their origins, and therefore publicly presented an unrealistic picture of the planets.

The first official flyby of Mars was achieved by Mariner 4 in July 1965. The probe radioed back twenty-two pictures of the Martian surface, and NASA initially claimed that there were no canals. Lifetimes of telescopic studies were casually obliterated with that one statement. A radar occultation reading provided a basis for NASA to declare that the atmosphere density on Mars was less than one percent of Earth's, and another type of signal allowed experts to suggest that the planet had no magnetosphere. At its closest distance, Mariner 4 was 6,000 miles from the planet, yet NASA spokesmen claimed readings showing that the average surface temperature on Mars was 170°F.

The censors may have had a tough time back in the Mars Patrol days curbing speculation, but this was a brand new ballgame. NASA was the perfect vehicle to paint a lifeless picture beyond the Earth. Who could possibly challenge statements coming from the U.S. space program? Telescopes or not. UFOs or not. Anybody who still wanted to claim intelligent life existed on Mars would be considered a lunatic.

The Mariner 4 flyby did not have the capability to realistically confirm habitability. That much can be conceded. But likewise, the space probe could not realistically confirm those alleged planetary conditions that were put out as flat statements by the authorities, either. With future planetary probes, it became apparent how the censorship was orchestrated, and by whom.

The real problem, however, is not with NASA specifically. The space agency had practically no choice but to follow the dictates of the powerful economic interests that control governments and their subordinate agencies. It is these international cartels that have been behind all censorship regarding planetary space. NASA has only been the publicly-identifiable distorter, regarding space pronouncements.

So NASA can be partly excused for not being in a position to objectively conduct space probe explorations. But false values can never be changed by anything but the truth. And we are at the critical crossroad of time. Either we become a space civilization, or we will be a nuclear extinction.

REFERENCES

1. Max Miller, *Flying Saucers—Fact or Fiction?* (Trend Books, 1957), p. 54.

2. Donald Keyhoe, *Aliens From Space* (New York: Doubleday, 1973), p. 171.

3. Maz Miller, *Flying Saucers—Fact or Fiction?* (Trend Books, 1957), p. 54.

4. Ibid. p. 43.

5. David Chandler, *Life on Mars* (New York: Dutton, 1979), p. 61.

6. Donald Keyhoe, *The Flying Saucer Conspiracy* (New York: Holt, 1955), p. 122.

7. Ibid. p. 209.

8. Ibid. pp. 209–211.

35 Never a Straight Answer: A Book Review of NASA Mooned America!

Thomas J. Brown

Long has Earth's Moon been a source of mystery and puzzlement, as well as an inspiration for love and art. It is also the source of vital [terrestrial] life rhythms. It has been mankind's dream to touch this strange world, for some simply to discover and explore, for others to exploit. It is now pretty much accepted as common knowledge that the U.S. government's NASA has sent manned craft to the Moon, and that they have landed thereon. But wait, not everybody is buying the official story! What's up?

There have been many books questioning the official story of a dead Moon, visited only by a few handpicked humans catapulted there in fancy tin cans. Not necessarily in order of appearance, some of these titles are: *Somebody Else is on the Moon* by George H. Leonard, who claims huge mining machines are moving about on the lunar surface; *Our Mysterious Spaceship Moon* and *Secrets of our Spaceship Moon* both by Don Wilson, who claims that the Moon is a giant artificial spaceship and is still inhabited; *The Moon: Outpost of the Gods* by Jean Sendy, who claims that extraterrestrials used the Moon as an Earth observation post and became the gods of old as they interfered with human development; *Flying Saucers on the Moon* by Riley Hansard Crabb, who claims that the Moon is a flying saucer base, and goes on to describe moving lights and changing craters recorded by orthodox astronomers in the 1700s and 1800s; *Moongate: Suppressed Findings of the U.S. Space Program* by William L. Brian II, who claims that the Moon has a heavy gravity (75 percent of Earth's) and atmosphere, and that a top secret antigravity propulsion system was necessary to get on and off the Moon; *We Discovered Alien Bases on the Moon* by Fred Steckling, which shows quite a number of startling NASA photos indicating vegetation, clouds and domed structures on the Moon. Steckling claims we discovered aliens already there when we got there, and that NASA just couldn't bear to tell us poor, common mortals this astounding news; *Extraterrestrial Archaeology* by David Hatcher

Childress, who claims that the Moon is long inhabited and that Mercury, Venus, Mars and some of the moons of the outer planets show signs of current or past inhabitation; *We Never Went to the Moon: America's Thirty Billion Dollar Swindle!* by Bill Kaysing and Randy Reid; and lastly, the subject of this review, *NASA Mooned America!* by René, the last two books dealing with a mass of discrepancies in NASA's public output which the authors take to mean that the Moon shots were faked. All these books are well worth acquiring to broaden one's outlook on this subject.

It is a big charge to claim that NASA never went to the Moon, that it was all a fake, yet this man René has come up with a large body of information that has to be seriously considered. He is obviously not writing this book for the fun of it, I doubt he's making any money at it, and is sure to be scorned and ridiculed simply for asking good questions which go against the common belief.

EVIDENCE OF PICTURES

René begins with the photographic evidence. The more one looks at photos of the Apollo landings, the more one begins to wonder. No blast craters exist under the lunar modules (LEMs), no dust arose from their rocket-softened landings, though the lunar rovers toss dust into the air as though there were an atmosphere acting on the particles. Questions, questions. One important early faked photo is shown here in sequence. Photo 1 is from the book *Carrying the Fire* by Astronaut Michael Collins. It shows Collins in a no-gravity test inside an airplane. Photo 2 is from the same book and is allegedly of a Gemini 10 space walk. René noticed something fishy about these photos and reversed #1 and sized it so he could overlay it on #2. They match . . . they are the same photo . . . and this is official NASA output!

René's book shows several other interesting photos which indicate various anomalies. On a splashdown photo of Gemini 6A there is a whip antenna in excellent condition clearly shown, with no burn marks or scorching (5000°F on reentry). No other Gemini had this antenna, and simple logic indicates that it would have burned off during reentry. Such an antenna is designed for frequencies not used in space.

The cover photo on the book shows two lunar astronauts (or astro-nots as René calls them), one reflecting in the other's visor. The reflected astronaut is not holding a camera, so who took the picture with only two [men] on the Moon? Also, in the same photo a piece of what appears to be scaffolding with a spotlight on it appears on the left edge of the photo. I've seen this same photo in several other places, but it is always cropped so the scaffolding is missing.

My favorite of the photo anomalies in the book is shown here as Photo

3, which René has titled "Mutt and Jeff." The anomaly in this photo is obvious. This is a photo of Armstrong, holding the staff, and Aldrin, holding the flag. While the two astronauts are basically the same height, the shadow of Armstrong is about 75 percent the length of Aldrin's. The shadows are not parallel as they should be, but converge, indicating two sources of light. René used trigonometry to discover that Aldrin's personal source of illumination is at 26.4 degrees of altitude, while Armstrong's is at 34.9 degrees. The sun was at 13.5 degrees altitude on the real Moon, so where were these guys? Certainly not where we have been led to believe. Perhaps a soundstage in the American desert?

NASA allegedly shot tens of thousands of pictures of the lunar landings, yet it is very difficult to procure even a decent percentage of these, and the same ones show up in most publications. The television footage of the first Moon landing was very poor. While having access to the finest of technology, NASA would not allow a direct feed of the footage, but forced networks and news services to film through an optically enlarged television screen, adding quite a bit of distortion. During the Apollo 16 lunar lift-off the camera followed the ship up off the surface. No one was left on the Moon, so who panned the camera? NASA later claimed that the camera was radio controlled from Earth, but how could they have followed the ship so closely given the transmission time lag . . . ?

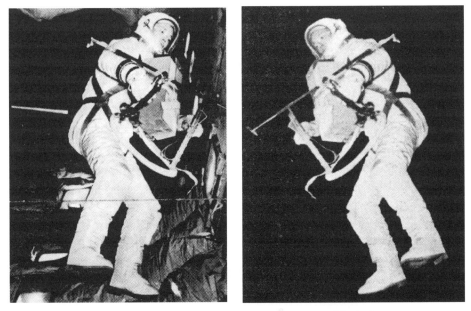

Photo 1   Photo 2

The Spacey Twins. NASA used the same photo twice, perhaps to cut expenses?

Photo 3. Mutt & Jeff

However, other lunar photos not mentioned by René, but appearing in some of the other previously mentioned books, indicate movement and structure on the Moon. One well known photo (Photo 4), shot from an unmanned orbiter, which has even appeared in *National Geographic*, shows "boulders" rolling, allegedly from a moonquake. However, basic scrutiny shows that they are rolling up and down hill. Lesser known photographs show these same "boulders" on other areas of the Moon making identical track marks. Photo 5 is from *National Geographic*, September 1973, and was shot by the crew of Apollo 17. A close-up shot of this same boulder has appeared in a recent issue of *Nexus* magazine, October–November 1995, which shows it to be the same cylindrical shape with arms as the larger rolling boulder of Photo 4.

Domed and pointed building-like structures appear in craters. These have been written off as lava bubbles or geological responses to the impacts causing the craters, even though the physical evidence goes against them being impact craters (the rim heights are similar regardless of crater diameter). Photo 6 is an enlarged section of a photo of the crater Kepler which appeared in the February 1969 *National Geographic*. It looks to me like there is an artificial complex in Kepler. Fred Steckling has a blowup of this structure in his book.

While René has shown some serious problems with the lunar photos, mostly those of the astronauts on the Moon, he doesn't show any of the boulder or dome photos. I would have to concur that many of the astronaut photos are faked, it becomes obvious after a while. However, I don't think this necessarily brings us to the conclusion that NASA didn't go, but

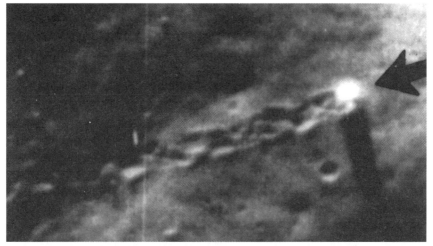

Photo 4. Rolling Boulder

Photo 5. Rolling Boulder gets around

it certainly means we're being lied to about something. But let us go on to further discrepancies:

STAR LIGHT, STAR DARK!

If one were to add up all the astronauts' stated observations of the appearance of space above the atmosphere one would come to the conclusion that they were either crazy, incompetent, or they never went, or, perhaps some of them were lying? Alan Sheppard, first American to be catapulted up reported seeing no stars, ditto for Virgil Grissom. John Glenn reported seeing some brighter stars only (and he saw [what NASA claimed were] "fireflies").

To quote some astronauts on the subject:

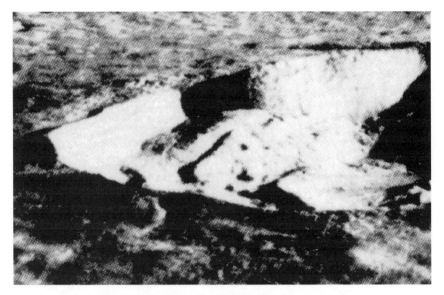

Photo 6. A "city" in the center of Crater Kepler?

Neil Armstrong: "The sky is black, you know." ; "It's a very dark sky."

Mike Collins on Gemini 10: "My God, the stars are everywhere: above me on all sides, even below me somewhat, down there next to that obscure horizon. The stars are bright and they are steady." This was written 14 years later, and remember that the Gemini 10 space walk photo shown here has now been proven fake

Mike Collins on Apollo 11: "I can't see the earth, only the black starless sky behind the Agena [rocket]. . . . As I slowly cartwheel away from the Agena, I see nothing but the black sky for several seconds. . . ." ; "What I see is disappointing for only the brightest stars are visible through the telescope, and it is difficult to recognize them when they are not accompanied by the dimmer stars. . . ."

Gene Cernan on Apollo 17: "When the sunlight comes through the blackness of space, it's black. I didn't say it's dark, I said black. So black you can't conceive how black it is in your mind. The sunlight doesn't strike on anything, so all you see is black."

Yuri Gagarin, first Russian cosmonaut: "Astonishingly bright cold stars could be seen through the windows."

Professor August Piccard on his high altitude balloon flight circa 1938 (many miles up with special heated suit) said that the sky turned from blue to deep violet to black. It is said that he claimed the sun disappeared as he got to the higher altitudes, though I have been unable to locate this exact reference.

My own investigations of NASA, circa 1987, revealed people who claimed that the stars could not be seen in space, but that special diffraction gratings were being developed to attempt to see them. This was from the period from Sheppard on to Skylab. I later spoke with John Bartoe who was up on an early shuttle flight and he laughed at this, said he couldn't believe that anyone in NASA would say that because he was in space and the stars were brighter than they are on Earth! (They must have slipped him a working diffraction grating.) I called back my contact in NASA and he told me, "Sir, the astronaut is a trained observer and is reporting what he saw, but the information I gave you (about the blackness of space) was essentially correct." I spoke with the man who developed the film for NASA for twenty-five years and he told me that the astronauts weren't even sure if they could see the sun, *that it may have been the appearance of the sun on their windows!*

The fact is that there are no visible light photographs of the sun, the stars, or any planets (other than the Earth and Moon, and not including specific probes sent to those planets) available in any NASA photo catalog. The fact that no stars appear on any photos was one of the main pillars of evidence for Bill Kaysing's book. René is the source of the astronauts' quotes as above and feels that there must be some serious problem with this selective star-blindness. As there is no definite answer available to us right now as to whether or not we can see the stars in space, I would have to say that we cannot base our conclusion as to the validity of the Apollo flights on the evidence of the appearance (or nonappearance) of stars in NASA photos.

THERMAL PROBLEMS

Space is supposed to be at absolute zero. Anything directly in the sunshine heats up incredibly. Skylab overheated when one of its solar panels failed to deploy properly, yet Apollo 13, in direct sunlight and in a lethal radiation zone, supposedly got cold! On the launch pad the ship is air conditioned from ground services. In space the ship is air conditioned (powered by fuel cells), if you turn off that air conditioner the ship gets cold! At least that is what NASA's line of logic dictates. Apollo 17's LEM sat on the Moon in the direct sun for 75 hours straight. Without massive power and refrigeration units the only way to cool the LEM would have been with the explosive cooling of water. Many tons would have been necessary for that time period. The astronauts reported that the LEMs were "too cold to sleep in." How cold would your closed car be after 75 hours of direct sunlight (or even 1 hour)?

The life support backpacks that the astronauts wore were supposed to COOL them on the lunar surface by discharging water from a blowhole. Conservative calculation of the water necessary to accomplish this, given standard metabolic heat and solar radiation, indicates that the backpacks had to be filled 40 percent with water, allowing room also for an oxygen bottle, carbon dioxide scrubber, dehumidifier, water bladder for the cooling circuit and one for dump water, a heat exchanger, a radio monitoring bodily function, a communications radio with power to reach Houston, and a battery to power all this. Also, for the cooling to be functioning, the water had to be ejected from the blowhole regularly. This would have created the effect of a fountain spewing minute crystals of water, quite a tremendous photo opportunity NASA seemed to have missed on tens of thousands of photos. NASA's own cutaway drawing of the backpack shows a water storage capacity of about .43 gallons, almost enough for 27 minutes of operation at the impossible efficiency of 100 percent. NASA claims 4 hours of operation.

ODD FIT

With the backpacks on, the astronauts would need about 35 inches of clearance to crawl through the 30-inch hatch on the LEM (lunar [excursion] module) in the manner claimed by NASA. Awful tight fit!

After getting back in the LEM the astronauts "repressurized their cabin." Then [according to NASA] "they removed their boots, slipped out of the backpacks heavy with life-support equipment that had kept them alive on the Moon, reopened the hatch, and dumped them along with crumpled food packages and filled urine bags onto the surface" (Apollo 11). There is no airlock on the LEM, how did they open the door after repressurization and dump their suits and garbage without dying from the supposed vacuum and heat (or was that cold)?

SPACE RADIATION

Van Allen radiation belts and solar flares create deadly radiation in space.* NASA spacecraft were not shielded against this. Apollos 8, 9, 10, 11 and 12 flew during the peak of solar cycle 20, with large flares occurring during the flights. All those astronauts would have received many hundreds or thousands of times the LIFETIME radiation limits for nuclear energy workers. A Supersonic Transport (SST) must drop altitude when it gets a dosage of 10 millirems, at 100 millirems it must alter its flight plan. 170 millirems is dangerous and almost guarantees cancer in the future.

---

* . . . the radiation was first predicted by Nikola Tesla around the beginning of this century as the result of experimental and theoretical work he had done on electricity in space in general and the electrical charge of the Sun in particular. He tried then to tell our academic natural philosophers (scientists) that the Sun had a fantastic electrical charge and that it must generate a solar wind. But to no avail. . . .

Subsequent study showed that this belt, or belts, [begins] in near space about 500 miles out and extends out to over 15,000 miles. . . . Van Allen belt radiation is dependent upon the solar wind and is said to focus or concentrate that radiation.

During Apollo 14 and 16 the solar flares would have given the astronauts approximately 75 rem (not millirems!). In an article in *National Geographic*, "Chernobyl—One Year Later" it says: "In general, 5 rem is considered acceptable for a nuclear-plant employee in a year, with 25 rem (the total countenanced for Chernobyl cleanup workers) an allowable once-in-a-lifetime dose." The walls of the spacecraft were "paper thin" and the fabric suits had no radiation shielding built in, anyway only very thick lead or a large measure of water (approximately 6 feet of shielding mass) will reduce the radiation of solar flares to anywhere close to safe levels. How did NASA protect the astronauts against this deadly radiation? The words "Space Radiation" appear extremely rarely, if at all, in books about manned space flights. Russian scientists told astronomer Bernard Lovell that they know of no way to shield from radiation outside the Van Allen radiation belts.

OUTRIGHT MURDER?

Shortly before the 1967 test-pad fire that killed three astronauts (Grissom, White and Chaffee) Virgil Grissom told his wife, Betty, "If there is ever a serious accident in the space program, it's likely to be me." He had become a critic of the Space Program and had expressed unease about the success of actually getting men on the Moon. The decision to run this test with pure oxygen at pressure was nothing short of moronic—it created a calorimetric bomb which was set off by the astronauts being told to flip switches that caused tiny sparks. Immediately after the test-pad fire, before anyone was notified, government agents raided Grissom's home and took all his personal papers. When they returned his papers to his widow his personal diary and all papers containing the word "Apollo" were missing!

Five other astronauts died in "accidents" that same year. Before the first Apollo manned mission left the launch pad eleven astronauts had died in "accidents," Grissom, White and Chaffee in the capsule fire, Freemen, Basset, See, Rogers, Williams, Adams and Lawrence died in airplane crashes (remember these were the world's best pilots flying their private aircraft, government supplied [jet] trainers—very safe craft) and Givens was killed in a car crash. In 1970 Taylor died in a plane crash.

CONCLUSION

There is no doubt that René has opened a can of worms with *NASA Mooned America!*. It is a challenge to us researchers to check out his information. Only a small portion of his research appears in this review. René has done his homework well and he is an intelligent man. I agree with most of his information, some of his figuring is beyond my technical grasp, and some things I don't agree with, such as his claim that the lunar footprints could only be made in wet soil. I have made very clear footprints in fine, dry desert dust, and I have used them as a guide on returns from the wilderness. I agree that many photos have been faked, that the information doesn't add up, and that NASA will go to great lengths, even murder, to cover up whatever is really happening. I think, and have so for a long time, that there is and has been for centuries, perhaps millennia, intelligent activities occurring on, and perhaps below, the surface of the Moon. I may be wrong, and I admit that possibility, but there is a tremendous body of information to back up my belief.

As far as NASA is concerned, I think that it is just a dog-and-pony show, while the real space program goes on behind the scenes. All of the authors mentioned in this review can be likened to the six blind men and the elephant. They all have an important piece of the puzzle which is being hidden from us. Let's put those pieces together and work towards a clear picture.

It is hard to make a definite conclusion with all this conflicting information. What is being covered up? It is possible, given the light of René's information on radiation and thermal problems that NASA never went to the Moon. However, there has been quite a bit of activity noticed on the Moon since the discovery of the telescope, and unmanned missions have sent back photos of boulders, obelisks and domes. Perhaps robot craft were landed and sent the photos necessary to fake the backgrounds of the manned missions, or perhaps there is another answer: That there is an advanced technology being used in space that us mere mortals have no access to. We can speculate that antigravity drives would create a protective field (like a personal Van Allen belt) which would shield those inside the craft from deadly radiation. If so, then manned missions may well have been undertaken, but for some reason or another NASA still felt the need to fake some of the informational output. All we can say for sure is that we Never Get a Straight Answer!

Section IV

The Suppression of Fuel Savers and Alternate Energy Resources

A chemical war has been declared on our planet. As a species, we are at the end of our grace period, and we can no longer afford to spew out toxins from our industrial plants, and filth from our cars and trucks. If we are to somehow survive and carry on into the next century, to preserve a healthy planet for the future generations, we must conserve our resources. Better still, we must rely on alternate, clean forms of energy.

You need to look only as far as your driveway to find evidence of our abuse of existing energy resources. Conventional carburetion and fuel injection introduce a fine mist of gasoline droplets into the combustion chamber of your car. Some of this mist is vaporized, and that is what propels the pistons down their cylinders, driving the car along. But droplets merely burn—a waste of rapidly depleting fuel resources—and hydrocarbon and carbon monoxide emissions are the result. True vaporization is the answer to ridding the air of these poisons.

Charles Pogue knew that the most explosive part of gasoline is its vapor, and so invented a system that would induct the vapor from the air space above the fuel in a gasoline tank. He was thus able to get more than 200 miles per gallon on two-ton cars with eight-cylinder engines. Pogue held three basic patents for vaporizing carburetors he developed for General Motors in the 1930s. With such an outstanding outcome, one would think that these devices would be standard on today's automobiles.

Unfortunately for us, Pogue's facilities were destroyed in the late 1930s and he was wounded by gunfire in incidents that persuaded him to forego further development of his invention. However, the fundamental concept lives on in various forms. Honda cars, for example, now have

407

sophisticated vaporizing technology enabling high mileage performance. In 1998 Mitsubishi announced the introduction of similar "lean burn" technology that involved vaporizing the fuel. Although none of these modern adaptations go as far as Pogue and some of his contemporaries, Future Perfect Ltd. of Auckland, New Zealand, is currently marketing an aftermarket vapor device that reputedly cuts hydrocarbon pollution by 60 percent.

So it *is* possible for us to use our fuel resources responsibly and economically. But this does not alter the reality that, despite our best efforts at conservation, our energy supplies are finite. Thus we must seek out abundant resources. And, truth be known, we have the technology to harness these resources for a cleaner, safer environment.

However, since the early years of this century, power and petroleum companies have been resolute in their denial that alternate energy resources exist. When faced with irrefutable proof, they have been even more resolute in their efforts to suppress devices that would allow us to harness this energy. The power of these corporations is such that, today, neither free energy nor hydrogen have a place in a world reliant on fossil fuels.

Science, far from being value-free or disinterested, as it likes to portray itself publicly, has always been an advocate for the dominant system in which it has an important stake. If the current fossil-fuel economy were to be disbanded, funds for pet research projects would rapidly disappear. Because alternate energy researchers are a perceived threat to the organizations which provide research funds, unconventional energy devices do not and cannot work, or are doomed as quackery—"in scientific opinion."

Take, for example, the free energy idea, which contends that the earth is floating on (or in) a sea of energy. Space is not the vacuum that some would have our students believe in science classes. *It is a veritable sea of energy.* Indeed the very term "space" is almost a dead giveaway for the message that establishment science wishes to create. In reality, it is not as though matter has been created *ex nihilo*, it is that matter is created out of this sea of energy.

Nikola Tesla and Henry Moray were inventors who actually designed and built machinery that tapped into the free energy of space and harnessed it to drive electric motors, operate radios, and light electric light bulbs. Lester Hendershot invented a generator powered by the earth's magnetic field that achieved similar results. Like so many other inventors trying to make alternate energy a viable option, these men lacked "scientific" training, and worked independently in their small workshops. Scientists working for the establishment endeavored to discredit them, rationalizing along the lines that, as the inventors were not learned in academic science, they could not know what they were doing.

But suppression has not been limited only to free energy inventions. It extends to cover research into hydrogen as an energy resource as well. Francisco Pacheo successfully powered a car and a boat with the hydrogen energy of seawater. Martin Fleischmann and Stanley Pons shocked the scientific community when they announced that they had fused two hydrogen nuclei in a jar of heavy water. So it seems that water can become the major fuel for the world.

That's because when water's two constituent gases, hydrogen and oxygen, are combined with a spark they explode with tremendous force, producing super-abundant quantities of energy that is *totally non-polluting*. Then they recombine to form water. They can also be made to burn with a controlled flame for welding torches; for cooking; for steam generation; *for power*.

The media never shows you an exploding gasoline storage tank and editorializes that gasoline is an extremely dangerous fuel. But with hydrogen the dominant association foisted on the people is the Hindenberg "disaster." We are never told about the probable sabotage of a dirigible that was built with German thoroughness for detail and safety by people who were knowledgeable about the properties of hydrogen; nor it is ever really stressed that thanks to the fact that hydrogen is so incredibly light most of the explosion went up instead of engulfing the gondola . . . and most of the passengers actually survived the blast.

We are at the most crucial time in recorded history in terms of not only our survival but our fulfillment as an intelligent, compassionate, and creative species. It is only now—when the need for a global transformation in energy usage is dire—that this technology is absolutely necessary. If necessity is the mother of invention, perhaps we can hope that it is also the mother of the invention's general acceptance.

36 Nikola Tesla: A Brief Introduction

Jonathan Eisen

Nikola Tesla was arguably the greatest inventive genius of the twentieth century, perhaps the greatest at least as far back as Leonardo da Vinci. What a shame and an indictment on our educational institutions that his name enlists barely a mention here and there in the hallways of learning.

When pressed, electrical engineers, who in fact owe their livelihood to Tesla, will tell you that Tesla invented alternating current and the "Tesla coil" which they play around with every so often when they have to. But they will probably not be able to tell you anything about his other 700 basic patents.

Or his ability to fetch electricity from the ambient atmosphere. Or his conclusion that the earth itself is a capacitor, and his experiments with transmitting electricity around the globe to virtually anywhere. Or his invention of the radio (well before Marconi), X-rays, the transistor and countless other inventions so far ahead of the times that even today they are still virtually unknown.

Even so, his bladeless turbine does seem to be making something of a comeback. And the Tesla Society is trying to interest the world in reviving some of his other lost inventions. He was convinced that "free energy" is a fact, rather than mere speculation, and over the years he has become something of a magnate for people working in the field.

Tesla's legacy is well known to a small-but-growing group of interested scientists and researchers. His astonishing story is recounted still: How he tore up his contract with Westinghouse in order for his alternating current electrification of America to proceed; how he had the rug pulled out from under him by J. P. Morgan when it looked as though his Colorado Springs experiments showed that wireless electricity transmission was feasible; how his Wardenclyffe tower on Long Island was destroyed when it seemed that his new system was about to supplant his old AC system, making free electricity available to everyone.

What a tragedy that a genius of such magnitude should die broke in his room at The New Yorker Hotel. For sometime afterward the FBI was quite interested in his papers, some of which dealt with new kinds of torpedoes, "death rays," and other inventions too numerous to mention here. As Chaney writes in her biography of Tesla:

> Like Einstein he had been an outsider and, like Edison, a wide ranging generalist. As he himself had said, he had the boldness of ignorance. Where others stopped short, aware of what could not be done, he continued. The survival of such mutants and polymaths as Tesla tends to be discouraged by modern scientific guilds. Whether either he or Edison could have flourished in today's milieu is conjectural.
>
> The example set by Tesla has always been particularly inspiring to the lone runner. At the same time, however, his legacy to establishment science is profound for his research, although sometimes esoteric, was almost always sweeping in its potential to transform society. His turbine failed in part because it would have required fundamental changes by whole industries. Alternating current triumphed only after it had overcome the resistance of an entire industry.

We must consider ourselves fortunate to have benefited from Tesla's alternating current technology, without which the world as we know it would not exist. How else might our lives differ today if formidable opposition had not halted his free energy research? Clearly, humanity would no longer operate according to a fossil fuel economy.

37 Tesla's Controversial Life and Death

Jeane Manning

*Electric power is everywhere, present in unlimited
quantities and can drive the world's machinery without the need
of coal, oil, gas or any other fuels.*

Nikola Tesla (1856–1943)

Colorado Springs, International Tesla Symposium, July, 1988—The man sitting next to me was in tears, shaking with quiet hiccuping sobs as if trying to be unobtrusive. He was rotund and wore thick glasses, but otherwise there was little to distinguish his appearance from that of two hundred other electrical engineers and other Tesla fans in the convention hall, still attentive to the scientist who had addressed them so eloquently and was leaving the podium.

It was not difficult to figure out why the man beside me was moved emotionally. The guest speaker, astrophysicist Adam Trombly, seemed to have choreographed his talk to lead to the moment. First he warmed up his audience by praising their hero. He reminded them that Nikola Tesla was the turn-of-the-century genius who fathered alternating current technologies, radar, flourescent tubes, and bladeless turbines. Tesla also presented the first viable arguments for robots, rockets, and particle beams. If society had followed up on the inventions Nikola Tesla envisioned at the turn of the century as he rode in a carriage near what is now this hotel, said Trombly, "we wouldn't have a fossil-fuel economy today. And J. P. Morgan, Rockefeller and a number of others wouldn't have amassed extraordinary fortunes on the basis of that fossil fuel economy."

FREE ENERGY FROM "VACUUM" OF SPACE

Trombly added that if Tesla's vision had prevailed, we would be dipping into a clean and abundant energy, like taking water from the well of space.

After all, the theoretical basis for vacuum energy is now part of the physics literature:

> . . . Not just in the literature of the fringe; it's been in *Physical Review* since 1975, *Review of Modern Physics* since 1962, and in European physics literature since the early 50s. Harold Puthoff in his May 1987 article in *Physical Review D* pointed out that in order for the hydrogen atom in its ground state not to collapse, it had to be absorbing energy from the vacuum.

The astrophysicist saw this scientific work as further vindication of Tesla. Trombly said that in the nineteenth century Tesla prophesied that people would someday hook their machinery up to "the very wheelworks of nature"—the energy of vacuum space.

Trombly noted that electrons themselves must spontaneously appear out of the background field of energy, or "we would have to invoke a rather Neanderthal concept that everything had its start in a certain moment." The speaker paused as if to let the audience catch his sarcasm, then added, "because we have embraced this [Big Bang] cosmology for the last couple of decades, we have some real problems."

In contrast, Trombly said, a more advanced cosmology sees everything as a modification of an energy-rich background field. Our physical bodies are relatively insignificant modifications of that field. The field itself has a potential energy equivalence, in grams, of 10-to-the-94th power grams per cubic centimetre. The human body, in comparison, has a gram equivalent of only about one gram per cubic centimeter. That means that the background energy is 10 (wish 94 zeros after the ten) times more energy-rich than our physical bodies.

THE PLAN: TELL ROOSEVELT

It's a lot of energy, Trombly said. Why not invent a pocket size device which could tap a kilowatt of this space energy? It could "just kind of scrape the surface, ever so slightly" of the 10-to-the-94th-power grams per cubic centimeter supply of energy.

"That's what Nikola Tesla was scheduled to tell Franklin Delano Roosevelt in 1943. In 1943 he proposed to FDR that perhaps we should look carefully at the fact that we can get all the energy we need from any space we happen to be in.

"He didn't show up for the meeting; he was found dead in his apartment— 'natural causes.'"

The speaker added quietly that despite the official statement on the cause of death, there is some suspicion that Tesla's paranoia about what he ate was more premonition than paranoia. Trombly then related an incident which fueled this suspicion. He had given a speech at the University of Toronto, Canada, for the 1981 conference on Non-Conventional Energy. Afterward, an older gentleman with a heavy New York accent came up to Trombly and said he had been a detective at the time Nikola Tesla was found dead, and had been involved in the investigation. The old man had produced vintage credentials to show Trombly that he had indeed been a detective. The man appeared to be old enough to have been an adult in 1943.

In a soft voice Trombly said that the old man had said that "for national security reasons no one was to know that the coroner's report showed that Tesla was poisoned."

A shocked silence descended on the Colorado Springs meeting room when the Tesla Society heard this, coming from a physicist who would not lightly risk his reputation by relating such a story. The silence lifted as the audience honored Trombly with applause at the end of his speech.

To understand why Tesla's story—the life of a dead inventor—can so grip the emotions of yet another generation of technophiles, we need to look at some highlights.

TESLA'S LEGACY

Tesla was a witty, elegantly-dressed loner, at the height of his fame in the late 1800s when the world knew he had invented the whole system of alternating current (AC) electrical generation and distribution which lit up the cities. But that was barely the beginning of his productivity.

Born in 1856 in the rural village of Smiljan in what became Yugoslavia, Nikola Tesla in his boyhood went from the highs of mystical communion with nature to the lows of suffering with cholera and the loss of his older brother. His father was a minister who wrote poetry and his mother a storyteller with a photographic memory. She was also an inventor of domestic laborsaving devices.

Nikola showed his true direction from an early age; at the age of five he invented a unique bladeless waterwheel and placed the little model in a creek. The child also built a motor powered by sixteen live June bugs. His father was not impressed. He insisted that Nikola would follow family tradition and be a clergyman, so he began his son's education at a young age with rigorous mental exercises.

When he was of legal age, Nikola managed to get his father's permission to study engineering instead of the ministry. After he completed his studies at the Austrian Polytechnic School in Graz and then in 1880 at the University of Prague, he worked for a European telephone company and upgraded their technology.

Meanwhile, a more difficult challenge which he had shouldered in his college days was always with him; he was determined to improve the electrical motor and dynamo. Dynamos naturally make alternating current, the type of electric flow which continually changes directions. Tesla intuitively felt that it should be possible to run a motor on AC electricity and eliminate the inefficient sparking of brushes from a commutator. His theory went against textbook knowledge in those early days of electrification, when direct current (DC) was considered the only type of current that would run motors.

MAGNETIC WHIRLWIND

Despite ridicule from his engineering professor, Tesla maintained that there had to be a better way. He worked so intensely on this and other engineering problems that his health broke down. While Tesla recuperated, a friend who was a master mechanic and an athlete took him for long walks through Budapest. In February of 1882 one day while they walked in a park, Tesla was inspired by the setting sun. To his amazement, that is when he made a breakthrough to answering the technical challenge of making a workable AC electrical system to turn a motor. He was reciting lines from the German poet Goethe's *Faust:*

> The glow retreats, done is the break of toil;
> It yonder hastes, new fields of life exploring.
> Ah, that no wing can lift me from the soil,
> Upon its track to follow, follow soaring!

Tesla was stopped in his tracks by a vivid vision. It was as if a 3-D holographic picture of a rotating magnetic field was in motion in front of his eyes and he could reach out and put his hands into it. He saw how the field—a magnetic whirlwind—was produced by alternating currents out of step with each other. He saw separate coils of wire, arranged as four segments of a circle. The first alternating current would energize a coil creating an electromagnetic field which attracted the magnet and then faded. The second overlapping current would feed the next coil and drag the magnet around further and then fade and so on. He saw it as a process similar to the sun traveling around and "giving life wherever she goes."

Speechless, Tesla waved his arms in excitement. His buddy tried to lead him to a nearby bench, but Tesla grabbed a stick to draw a diagram in the dust.

"See my motor here! Watch me reverse it," Tesla blurted out. His friend was afraid that Tesla had lost his mind. Tesla was indeed in another world at that moment. As he watched his vision move, he saw the electrical principle that later made the twentieth century operate.

His rotating magnetic field would not only mean a better motor, it would revolutionize the electrical industry. He mapped out refinements of the idea with several or even five overlapping currents at a time—the basis of a polyphase transmission system. But first he had to convince someone to finance the development of these world-changing inventions. A stepping-stone to that goal was a job in Paris later that year, where he attracted the attention of the Continental Edison Company by his successes as a trou-bleshooter who fixed their dynamos. Another step was to demonstrate the first induction motor for the mayor of Strassburg. The mayor had invited wealthy potential investors to the demonstration, but they failed to compre-hend Tesla's vision of a future for the brushless motor.

DITCHDIGGER TO MILLIONAIRE

Surely it would be welcomed in America, Tesla thought. At twenty-eight years of age he was ready to make his move to the land of opportunity, where he expected that his great discovery would be quickly developed for humanity's use. Before Tesla left Paris, one of his bosses at Continental Edison handed him a letter introducing him to the famous inventor Thomas Alva Edison.

"I know two great men and you are one of them; the other is this young man," the letter read.

When Tesla stepped off the ship in New York on June 6, 1884, he only had four pennies in his pocket, because he had been robbed on the way to the ship. But he did not at all resemble the stereotypical impoverished immigrant; he wore a bowler hat and stylish coat, and his posture was aris-tocratic. He still had the letter of introduction to Thomas Edison.

Edison, then age thirty-seven, had already proven his ability as a busi-nessman as well as inventor. He was a hero to Tesla at first. The polite European admired Edison's accomplishments—discoveries made by trial-and-error and with only grade-school level of formal education. Tesla ignored his rough manners. But Edison on the other hand repudiated Tesla's theory on how to work with AC electricity; Edison used DC in his electric lamps and had invested all his efforts in DC technologies.

Tesla was put to work repairing and improving Edison's DC dynamos and motors on board a ship. He also won Edison's grudging respect by working eighteen-hour days in Edison's Manhattan workshop, seven days a week, and by conquering difficult technical problems.

One day Tesla described how he could improve the efficiency of Edison's dynamo, and Edison reportedly replied, "There's fifty thousand dollars in it for you if you can do it." The European immigrant worked tirelessly—thirty-two hours in one stretch. After months of work, the new machines were tested and found to measure up, and Edison prepared to profit from his improved dynamo. When Tesla went to the boss and asked for the promised $50,000 bonus, however, Edison would not pay.

"Tesla," he said, "you don't understand our American humor."

Nikola Tesla had a well-developed sense of humor, but when someone reneged on a verbal deal he was not amused. He walked out, and into a job on a crew digging ditches with pick and shovel.

Two years later Tesla's luck changed; he had the opportunity to develop his "polyphase system" of AC and patented the AC motor, generator and transformer. By 1891, Tesla had forty patents on his AC induction motor and polyphase system.

An industrialist and inventor of the railroad air brake, George Westinghouse of Pittsburgh, helped Tesla to change history. Westinghouse, a stocky, adventurous man with a walrus mustache, shared Tesla's vision of a power system that could harness hydroelectric resources such as Niagara Falls and could send high-voltage electricity on wires over vast distances. He bought all of Tesla's patents on the polyphase AC system, and signed a contract to pay Tesla a million dollars cash, plus royalties of $2.50 per horsepower produced by the system. Tesla thought he would never have to worry about money again; he could invent to his heart's content.

HIGH STAKES

One of the first challenges that Westinghouse and Tesla faced together was what was called the War of the Currents—the AC/DC battle. It was a time when America's power grid had not yet been built but DC proponents were nevertheless becoming an entrenched interest group stubbornly fighting the use of alternating current (AC) for generating, sending and using electricity. Thomas Edison led the opposition. His own inventions used direct current (DC). However, DC does not travel well. To give people electrical lights, heat and other uses of the current, a power plant had to be built for every square mile served. At the end of a mile of DC power line, light bulbs barely glowed. Skyscrapers and their elevators would have been impossible to build if Edison's views had won.

Tesla knew that AC was the better system for electrical distribution; it could easily travel for hundreds of miles down very slender wires at high pressures (high voltage) and then transformers could reduce the voltage for household use.

In the War of the Currents, most of the casualties were animals. During the time that Edison gave speeches defending the merits of DC over AC, the neighborhood around his New Jersey laboratory was mysteriously losing dogs and cats. Throughout 1887 Edison or his staff grabbed animals off the street by day, and at night invited reporters and other guests to watch what happened when an unsuspecting dog was pushed onto a tin sheet and electrocuted with high voltages—using the Tesla/Westinghouse AC current, of course. Edison refered to electrocuting as "Westinghousing."

Carrying on this strategy of linking AC with electrocution and death, the Edison camp distributed scare pamphlets warning that Westinghouse wanted to put this deadly AC current into every American home. However, Edison omitted the fact that the current would first be reduced in voltage. Through this disinformation campaign, Edison was determined to sway the public toward his DC technology, inefficient as it was.

To answer accusations against the safety of AC, Tesla in turn developed showmanship; he proved that he could conduct AC through his own body without ill effects. He stood on a platform in white tie and tails and cork-bottomed shoes. Bolts of electricity crackled and snapped, and he allowed several hundred thousand volts to dance over his body and light the bulbs in his hands. However, although the voltage (pressure) of the electricity was high, he reduced the amperage (quantity) and used high frequencies. That type of electrical current crawls over a body and therefore doesn't reach vital organs. As an argument against Edison it was cheating, because domestic AC switches back and forth on a conductor 60 times a second, not thousands of times as in high frequencies.

Edison, however, played dirtier. He persuaded state prison authorities to kill a death-row prisoner with AC current instead of executing him by hanging. It was a further attempt to popularize the phrase "to Westinghouse" as a replacement for "to electrocute." Prison officials miscalculated the amount of current needed to kill the condemned man, and newspaper reporters witnessed a messy smoky execution.

Despite Edison's efforts, Tesla and Westinghouse won the Battle of the Currents. In 1892 Westinghouse built an AC system for lighting the 1893 world fair in Chicago.

TYCOONS PUT SQUEEZE ON WESTINGHOUSE

A big hydroelectric project was the second major victory for AC supporters; in 1895 Tesla's first generating unit was put into operation at Niagara Falls. Eventually, Tesla's distribution system delivered immense amounts of electrical power across the continent. Since Westinghouse had signed a contract giving Tesla $2.50 per horsepower, Tesla could have died as a multibillionaire.

"Morganization" intervened, however, with cut-throat practices directed against George Westinghouse. Business competitors in the real-life game of Monopoly tried to squeeze him out of the power picture and gave him an ultimatum: "get rid of your contract with Tesla or you're finished." When Westinghouse laid his cards in front of Tesla and admitted to being in financial trouble, Tesla demonstrated his priorities. He remembered that

Westinghouse had believed in him and had invested in the new AC patents when others had not had such courage. Therefore, so that Westinghouse would survive financially and the technology would be developed, Tesla took a cash settlement and walked away from the millions of future dollars assigned to him by the per-horsepower deal. He tore up the lucrative contract in order to help a friend.

Meanwhile, the power monopolists were poised to grab as much money as possible. When Tesla's inventions made it possible to send electrical power from huge waterfalls across the states, tycoons prepared to make fortunes in utility companies. These captains of industry wanted the 60-cycle-per-second AC power system to continue to grow and cover the earth with power poles, transformers and wires. Transmission towers would march up and down mountainsides and across deserts. Power companies would dam rivers for hydro power and make the people pay for every watt sent over the companies' copper wires. The power magnates did not want the inventor to uproot this growing forest of money trees. J. Pierpont Morgan pulled the strings that formed the huge company General Electric, for example, and had already bought up copper mines knowing that transmission wires would eventually crisscross every industrialized continent.

But Tesla was a discoverer, not a business shark. His new plan was wireless transmission of energy—free energy for anyone who sticks a tuned receiver into the ground while Tesla's tuned transmitter was resonating frequencies!

The financiers on Wall Street didn't catch the drift of Tesla's "wireless" talk right away. The plan was so futuristic that it was literally over everyone's head. But he was giving enough clues for anyone who had been ready to catch his vision. In the same year that the lighting of the World's Fair dazzled society, he talked about "earth resonance" at a lecture to the prestigious Franklin Institute. Earth resonance was part of his vision for wireless power. The secret is sending out the correct frequency—speed of vibration—with electrical pulses. Just as a piano string will vibrate when another instrument at a distance hits the same note as its tuned frequency, wireless receivers would resonate with the transmitter frequencies. The power would be tuned in just like you tune in a radio station. Some Tesla researchers also believe that he could have resonated the cavity between the ionosphere and the ground. Just like the cavity within a violin, this spherical Schumann cavity has its own resonant frequency.

Disregarding the danger of making his own previous inventions obsolete, in the next few years he thought up the processes necessary for futuristic wireless transmission. While the business community assumed he was talking about wireless communications signals only, he had a far grander plan—sending power wirelessly in order that anyone at any place on the planet could plug into freely-available electricity. Before his financiers figured out where Tesla's research was leading, it was briefly funded by men such as Colonel John Jacob Astor as well as Morgan.

The same year that Tesla's generator turned on the power from Niagara Falls, he suffered a major setback. One night in March of 1895 his laboratory burned down, with all files and apparatus destroyed. When he returned from a meeting, he discovered the smoking mess of twisted metal that had fallen through two floors to the foundations of the building. Afterward he wandered through the streets in a daze for hours. The loss of his papers meant that he could not document what he had been working on. For example, later that year the discovery of X-rays by German physicist Wilhelm Conrad Roentgen was made public. Tesla's papers could have proven that he had been the first to take pictures by X-ray.

GOD OF LIGHTNING

Next Tesla concentrated on patenting his methods for sending power and messages wirelessly. In 1889 to 1890, Tesla moved his operations to the high country of Colorado Springs, Colorado, to test his new ideas and develop the art of tuned radio frequency. He built a high-voltage laboratory on a hillside cow pasture. Inside his lab was the world's largest Tesla coil, and the building was topped by a flagpole-like structure. While experimenting on a massive scale, toward his new goal of sending electromagnetic vibrations throughout Earth, he predicted that Tesla coils could also be pocket-size message receiving devices.

Tesla's God of Lightning experiments in Colorado Springs were truly dramatic. Thunder reverberated for at least 15 miles when he fired up the electrical discharges. His massive 52-foot diameter Tesla coils discharged more than 12 million volts at a burst, and threw electric sparks of more than a hundred feet in length from the copper ball on top of his pole. The townspeople sometimes thought his laboratory was on fire. The ground under their feet was so highly charged that spectators at a distance from the laboratory would see tiny sparks between their heels and the sandy soil when they walked, according to biographer Margaret Cheney. Half a mile away, horses would get a shock from their metal horseshoes and would bolt in panic.

The inventor did start a fire one day, when his "magnifying transmitter" experiment accidentally burned out the power plant for the town of Colorado Springs. The town went dark and the overloaded dynamo was in flames. It took Tesla's team of technicians a week to repair the town's generator.

WARDENCLYFFE

Satisfied that he knew enough to carry out his magnificent vision of a world telegraphy system and wireless power, Tesla returned to New York. He hired an architect to design a building with a 154 foot high wooden tower, to be used as a huge transmitter. The tower was topped with a doughnut-shaped copper electrode. As the design changed, the structure evolved to the shape of a giant mushroom sprouting above the low hills of Long Island. Tesla named the project Wardenclyffe, envisioning a station to send out power as well as to broadcast communication channels of all radio wavelengths. The tower was nearly finished in 1902, along with the square brick building, 100 feet on each side, built below it for a power-house and laboratory.

Tesla predicted that when people experience wireless transmission of electrical power affecting their everyday lives, "humanity will be like an ant heap stirred up with a stick." The excitement that he anticipated never had a chance to develop, however. Work on the structure halted in 1906 after J. Pierpont Morgan stopped funding it.

Some historians believe that Morgan had been sincerely interested in wireless broadcasting. Others argue that Morgan's motivation for briefly funding Tesla's tower was to gain control over Tesla. As long as Tesla was an uncontrolled loner, a wild card in the industrial world, his inventions could threaten Morgan's investments in the electrical industry. If wireless transmission of power worked, of course, the value of power utilities and copper mines would plummet. Morgan's companies such as General Electric could have toppled.

While Tesla's fortunes went downhill starting in 1906, Morgan would not reply to Tesla's letters, and other financiers on Wall Street also turned their backs on Tesla for the remainder of his life. In a letter begging an associate for financial help, Tesla mentioned one of the tactics used to discredit him. "My enemies have been so successful in representing me as a poet and a visionary . . ."

One of Tesla's biographers is Dr. Marc Seifer, a psychology professor who researched a psycho-biography of Tesla for his doctoral thesis. Seifer believes that Tesla sowed the seeds of his own financial ruin by not making clear to J. P. Morgan, Sr. his intention to broadcast power from Wardenclyffe as well as to send communications. However, Seifer also thinks that Morgan could have transcended his own limitations and given Tesla the money to complete at least the radio portion of the tower "and the world would have evolved in a totally different way."

MORGAN SABOTAGED TESLA DEALS

Instead, from that time onward Tesla was unable to build the technologies which he believed would help humanity. Seifer mentions the influential men whom Morgan paid a visit when they were ready to close a deal with Tesla. "Morgan purposefully scuttled any future ways Tesla could raise money."

He was deeply in debt, having plowed all his resources into his experiments and Wardenclyffe. Having a strong taste for the elegant life, he had run up an outrageous tab in his more than twenty years of living at the Waldorf-Astoria hotel. The hotel took the deed for Wardenclyffe in lieu of payment. Seifer feels that one reason for Tesla handing over the property to the owner of the Waldorf-Astoria is that he thought he could eventually resurrect the project. His plan was to develop an invention that would be a big money-maker, and his hopes were pinned onto his bladeless turbine/pump. Tesla expected the bladeless turbine to replace the gasoline engine in automobiles, ocean liners and airplanes and then he would use the subsequent wealth to complete his project for world-wide wireless power.

Seifer concludes that one of Tesla's motivations for another invention, a beam weapon which was also called a death ray, was to convince his government that the Wardenclyffe tower should be saved for military use. By attaching a beam weapon to it, he could have claimed that the tower was a strategic property for shooting down incoming aircraft or submarines during World War I.

His efforts were further scattered during this time by a lawsuit against Guglielmo Marconi, the Italian who had hung around his laboratory before the fire of March 13, 1885. In 1901 Marconi sent a signal across the Atlantic which in the eyes of the public secured Marconi's claim to be the inventor of radio. When Tesla had heard the news of the transatlantic wireless signal, he reportedly said, "Marconi is a good fellow. Let him continue. He's using seventeen of my patents."

By the time Tesla tried to collect the hundreds of thousands of dollars owed him so he could rescue Wardenclyffe, most of his patents had elapsed. He did resurrect his main radio patent in 1914, Seifer said. Tesla did not win his suit against Marconi, not because of the legal strength of his case but because World War I interfered. The assistant attorney general of the time, Franklin Roosevelt, and President Woodrow Wilson pushed for a law saying there could be no patent disputes during the war. Seifer added that by the time the war was over it was much more difficult for Tesla to sue. (Eight months after his death, the U.S. Supreme Court ruled that Tesla's radio-related patents preceded Marconi's. Even after the court's decision, school history books continue to credit Marconi for inventing radio.)

Tesla was squeezed out of the picture by the force of corporate interests. "David Sarnoff was Marconi's front man, and Sarnoff created RCA and NBC and purposely kept Tesla's patents out of the loop," Seifer said. "So when people like Hammond and Marconi were getting $500,000 at a clip for their wireless patents, Tesla got nothing."

RADIO CORPORATION ELBOWS HIM OUT

The picture of corporate ruthlessness is reinforced by the experience of the late Philo T. Farnsworth, an inventor of television. In Philo's biography, Elma G. Farnsworth told about Sarnoff's treatment of her husband, and about the early 1930s when RCA dominated the radio industry to the point where no one could make broadcasting or receiving equipment without paying patent royalties to RCA. "RCA's policy regarding patents, licenses, and royalties was very simple: the company was formed to collect patent royalties. It never paid them." Elma Farnsworth added that corporations have always been ambivalent toward inventors and patents. "Although they regard patents as a huge bulwark when protecting their own monopolies, they see the patent system as a great nuisance when it upholds the rights of an individual." She gives the example of two pioneers of radio who battled RCA for their rights unsuccessfully. Dr. Lee DeForest died bankrupt and Major Howard Armstrong put on his coat, hat and gloves and walked out the high window of his New York Apartment.

Tesla never threatened suicide, but he did admit to despairing. Before he could make much progress with the bladeless turbine, his dream of saving the Wardenclyffe structure began to crumble. For one thing, the new owner saw no value in the project and did not post guards on the property. Since the businessman believed that Tesla was just a vain dreamer, he did not try to protect the contents of the laboratory and it was vandalized and stripped.

The Wardenclyffe tower was dynamited in 1917, but not by the government as some legends would have it. Instead it was torn down to be sold as scrap metal. After this dramatic turning point in Tesla's career, he began to disappear from public view.

HOPES PINNED ON TURBINE

Perhaps partly to run away from the sight of the ruined Wardenclyffe structure, the inventor travelled to Chicago. That city held memories of earlier, more triumphant, times such as the World's Fair of 1893 which showcased his AC technologies. Now he spent time with biographer Hugo Gernsback as well as worked on technical problems with the round disks in his bladeless turbine. In his day the available steel was not strong enough for anything moving at such a high speed. (Again, he was ahead of his time and in the 1990s engineers are beginning to catch up and even improve on his designs. The Tesla Engine Builders' Association is a cooperative network of researchers doing just what their name says. This is perhaps the most practical Tesla invention at this time, and could be extensively replacing fossil fuel or nuclear power generation.)

From Chicago he moved again, living alternately in Milwaukee and New York for a few years. During this time he sold a speedometer which he invented to a watch company. It was installed in the luxury cars of the day and provided him some income. Among other inventions which earlier had fleetingly provided income was a fountain which he designed in 1915. He figured out how to power a decorative fountain to get aesthetically-pleasing effects with little water.

DESPERATELY SEEKING FUNDS

Was Tesla also a would-be defense contractor? Tesla had a liaison in Germany before World War I and in 1916 to 1917 they planned to put the bladeless turbine in tanks and other war vehicles. This was the reason that J. P. Morgan, Jr. doled out more than $20,000 to Tesla to develop the turbine, Seifer notes.

In a recent book, Dr. Seifer chronicles Tesla's "lost years," from 1915 onward, when the inventor tried unsuccessfully to raise money for resurrecting his wireless project. Seifer encountered correspondence and articles linking Tesla to such shadowy figures as a Nazi propagandist and a German munitions manufacturer from whom the desperate inventor was trying to get funding by selling his death ray concepts. Those attempts ended when war was declared between their two countries. About Tesla's links to warlords during the 1930s, Seifer says "There's a whole secret side here that needs to be explored further. I did the best I could."

Unknown to most Teslaphiles, the inventor was not always based in New York during those hidden years. For example, around the year 1925 to 1926 he was in Philadelphia working on the turbine design, and in 1931 he was in Massachusetts working with the head of U.S. Steel in an attempt to put his turbines in the steel mills.

Seifer says a 300 page book was written about Tesla's turbine, but it has not surfaced since the inventor's death.

CAR RAN ON FREE ENERGY?

Tesla kept a much lower profile regarding another invention. The story—seemingly impossible to document, generations later—is that when he was around sixty-five, Tesla or his helpers pulled the gasoline engine out of a new Pierce-Arrow and stuck in an 80 horsepower alternating current electric motor. But no batteries! Instead, he bought a dozen vacuum tubes, wires and resisters. Soon he had the parts arranged in a box which sat beside him in the front seat of the car. One account says the mysterious box was two feet long, a foot wide and six inches high, with two rods sticking out of it. From the driver's side, Tesla reached over and pushed the rods in, and the car took off at up to 80 miles per hour. He is reported to have test-driven the loaned Pierce-Arrow for a week. If this story is true, the secret of his power source died with him.

There are clues that indicate he could well have driven a car on "free energy." For example, Tesla wrote to his friend Robert Johnson, editor of *Century* magazine, that he had invented an electrical generator that didn't need an outside source of power. In the early 1930s, Tesla announced that he had, more than twenty-five years earlier, harnessed cosmic rays and made them operate a moving device.

Trying to discover what he had been talking about, today's researchers comb through his patents, such as "Apparatus for the Utilization of Radiant Energy," U.S. Patent No. 658,957, 1901. The research indicates Tesla was working on his "free energy" generator before he hammered out a major article for Robert Johnson's June 1900 issue of *Century*, in which he describes sending power wirelessly. He writes that a device for getting energy directly from the sun would not be very profitable and therefore would not be the best solution. Researchers such as scientist Oliver Nichelson of Utah read this to mean that Tesla had learned that a "free energy" device would never be allowed to reach the market, but a system in which someone could still profit by selling power delivered wirelessly had more of a chance of being allowed by the financial tycoons.

Today's creative-edge physicists may be vindicating Tesla's so-called free energy invention with their theories about the possibility of tapping incredibly abundant—estimated to be the energy equivalent of 10-to-the-94th-power grams per cubic centimeter —supply of energy from the vacuum of space that Adam Trombly spoke about.

GOVERNMENT AGENTS TAKE HIS PAPERS

According to his biographers, Tesla died in genteel poverty in a hotel room in 1943 at age eighty-seven. His memory was honored in a funeral service at St. John's cathedral, attended by more than two thousand people including the elite of the day.

Although Tesla had become a United States citizen in 1899 and valued his citizenship highly for the next fifty-nine years, he was strangely treated like a recent immigrant at the end of his life. After his death the public was told that his papers had been shipped back to Yugoslavia, and that authorities in Washington had sent in the Custodian of Alien Property to deal with his belongings. U.S. government agents reportedly had first crack at his safe and other papers. Later a Tesla museum was built in

Zagreb, Yugoslavia, to house whatever Tesla memorabilia survived the events after his death.

When biographer Margaret Cheney looked into the military's possession of Tesla papers taken from the Office of Alien Properties, the trail led to Wright-Patterson Air Force Base, Ohio. The response from Wright-Patterson AFB under the Freedom of Information Act in 1980 was that "The organization (Equipment Laboratory) that performed the evaluation of Tesla's papers was deactivated several years ago. After conducting an extensive search of lists of records retired by that organization, in which we found no mention of Tesla's papers, we concluded that the documents were destroyed at the time the laboratory was deactivated."

Believe that or not, the fact remains that a great discoverer was left out of our history books but is known among researchers of alternative technology. Does the military own Tesla technology information which could be used for cleaning up the planet instead of for destructive purposes? Did those industrialists who have monopolies on coal and oil also try to control Tesla's legacy? Consider his claim of inventing an electrical generator that would not consume any fuel. "Ere many generations pass, our machinery will be driven by a power obtainable at any point in the universe," Tesla said. ". . . Throughout space there is energy." If that energy had been harnessed, those who profit by the myth of scarcity would not have been able to drum up support for their oil wars.

Whether he died of natural causes or was deliberately given arsenic, the story of Nikola Tesla is clouded by the actions of those who lacked his dedication to improving the lot of humanity.

The man softly crying as he sat beside me at the Tesla symposium may have been a finely-tuned receiver for the prevailing mood in the room. His fist clenched when Adam Trombly said, "Thomas Edison was promoted and promoted, but Nikola Tesla was a genius who was orders of magnitude greater."

REFERENCES

Bearden, Tom. *Planetary Association for Clean Energy*, Vol. 8 (1995), p. 10.

Bird, Christopher, and Nichelson, Oliver. "Great Scientist, Forgotten Genius Nikola Tesla," *New Age Magazine* (1967).

Cheney, Margaret. *Tesla: Man Out of Time*. New York: Dell Publishing, 1981.

Farnsworth, Elma G. *Distant Vision: Romance and Discovery on an Invisible Frontier*. Salt Lake City: PemberlyKent Publishers, 1990.

Marvin, Carolyn. *When Old Technologies Were New*. Oxford University Press, 1988.

O'Neill, John J. *Prodigal Genius*. California: Angriff Press, 1978.

Peterson, Gary. "Nikola Tesla, Man with Many Solutions," *Journal of Power and Resonance*. Colorado Springs, 1990.

Quinby, E.J., USN Commander (ret). "Nikola Tesla, World's Greatest Engineer." Proceedings of Radio Club of America Inc., Fall 1971.

Rauscher, Elizabeth. *Planetary Association for Clean Energy*, Vol. 8 (1995), p. 9.

Seifer, Marc. *Nikola Tesla & John Hays Hammond Jr., A History of Remote Control Robotics*. Fall River, Massachusetts.

Tesla, Nikola. "My Inventions: The Autobiography of Nikola Tesla," *Electrical Experimenter Magazine*, 1919. (Vermont: Hart Brothers, 1982).

Tesla, Nikola, "The Problem of Increasing Human Energy," *The Century Illustrated Monthly Magazine* (New York, June 1900).

The Tesla Journal. (Lackawanna, New York, 1989/90).

Wohleber, Curt. "The Work of the World," *Invention & Technology* (Winter, 1992).

Wright, Charles. "The Great AC/DC War," 1988 International Tesla Syposium, Colorado Springs.

38 Transmission of Electrical Energy Without Wires

Nikola Tesla

It is impossible to resist your courteous request extended on an occasion of such moment in the life of your journal. Your letter has vivified the memory of our beginning friendship, of the first imperfect attempts and undeserved successes, of kindnesses and misunderstandings. It has brought painfully to my mind the greatness of early expectations, the quick flight of time, and alas! the smallest of realizations. The following lines which, but for your initiative, might not have been given to the world for a long time yet, are an offering in the friendly spirit of old, and my best wishes for your future success accompany them.

Towards the close of 1898 a systematic research, carried on for a number of years with the object of perfecting a method of transmission of electrical energy through the natural medium, led me to recognize three important necessities: First, to develop a transmitter of great power; second, to perfect means for individualizing and isolating the energy transmitted; and, third, to ascertain the laws of propagation of currents through the earth and the atmosphere. Various reasons, not the least of which was the help proffered by my friend Leonard E. Curtis and the Colorado Springs Electric Company, determined me to select for my experimental investigations the large plateau, two thousand meters above sea-level, in the vicinity of that delightful resort, which I reached in May, 1899. I had not been there but a few days when I congratulated myself on the happy choice and I began the task, for which I had long trained myself, with a grateful sense and full of inspiring hope. The perfect purity of the air, the unequaled beauty of the sky, the imposing sight of a high mountain range, the quiet and restfulness of the place—all around contributed to make the

Communicated to the Thirtieth Anniversary Number of the *Electrical World and Engineer*, March 5, 1904.

Experimental Laboratory, Colorado Springs.

conditions for scientific observation ideal. To this was added the exhilarating influence of a glorious climate and a singular sharpening of the senses. In those regions the organs undergo perceptible physical changes. The eyes assume an extraordinary limpidity, improving vision; the ears dry out and become more susceptible to sound. Objects can be clearly distinguished there at distances such that I prefer to have them told by someone else, and I have heard—this I can venture to vouch for—the claps of thunder 700 and 800 kilometers [roughly 400 to 500 miles] away. I might have done better still, had it not been tedious to wait for the sounds to arrive, in definite intervals, as heralded precisely by an electrical indicating apparatus—nearly an hour before.

In the middle of June, while preparations for other work were going on, I arranged one of my receiving transformers with the view of determining in a novel manner, experimentally, the electric potential of the globe and studying its periodic and casual fluctuations. This formed part of a plan carefully mapped out in advance. A highly sensitive, self-restorative device, controlling a recording instrument, was included in the secondary circuit, while the primary was connected to the ground and an elevated terminal of adjustable capacity. The variations of potential gave rise to electric surgings in the primary; these generated secondary currents, which in turn affected the sensitive device and recorder in proportion to their intensity. The earth was found to be, literally, alive with electrical vibrations, and soon I was deeply absorbed in this interesting investigation. No better opportunities for such observations as I intended to make could be found anywhere. Colorado is a country famous for the natural displays of electric force. In that dry and rarefied atmosphere the sun's rays beat the objects with fierce intensity. I raised steam, to a dangerous pressure, in barrels filled with concentrated salt solution, and the tin-foil coatings of some of my elevated terminals shriveled up in the fiery blaze. An experimental high-tension transformer, carelessly exposed to the rays of the setting sun, had most of its insulating compound melted out and was rendered useless. Aided by the dryness and rarefaction of the air, the water evaporates as in a boiler, and static electricity is developed in abundance. Lightning discharges are, accordingly, very frequent and sometimes of inconceivable violence. On one occasion approximately twelve thousand discharges occurred in two hours, and all in a radius of certainly less than fifty kilometers from the laboratory. Many of them resembled gigantic trees of fire with the trunks up or down. I never saw fire balls, but as a compensation for my disappointment I succeeded later in determining the mode of their formation and producing them artificially.

In the latter part of the same month I noticed several times that my instruments were affected stronger by discharges taking place at great distances than by those nearby. This puzzled me very much. What was the cause? A number of observations proved that it could not be due to the differences in the intensity of the individual discharges, and I readily ascertained that the phenomenon was not the result of a varying relation between the periods of my receiving circuits and those of the terrestrial disturbances. One night, as I was walking home with an assistant, meditating over these experiences, I was suddenly staggered by a thought. Years ago, when I wrote a chapter of my lecture before the Franklin Institute and the National Electric Light Association, it had presented itself to me, but I had dismissed it as absurd and impossible. I banished it again. Nevertheless, my instinct was aroused and somehow I felt that I was nearing a great revelation.

It was on the third of July—the date I shall never forget—when I obtained the first decisive experimental evidence of a truth of overwhelming importance for the advancement of humanity. A dense mass of strongly charged clouds gathered in the west and towards the evening a violent storm broke loose which, after spending much of its fury in the mountains, was driven away with great velocity over the plains. Heavy and long persisting arcs formed almost in regular time intervals. My observations were now greatly facilitated and rendered more accurate by the experiences already gained. I was able to handle my instruments quickly and I was prepared. The recording apparatus being properly adjusted, its indications became fainter and fainter with the increasing distance of the storm, until they ceased altogether. I was watching in eager expectation. Sure enough, in a little while the indications began again, grew stronger and stronger and, after passing through a maximum, gradually decreased and ceased once more. Many times, in regularly recurring

Experimental Laboratory, Colorado Springs.

intervals, the same actions were repeated until the storm which, as evident from simple computations, was moving with nearly constant speed, had retreated to a distance of about three hundred kilometers. Nor did these strange actions stop then, but continued to manifest themselves with undiminished force. Subsequently, similar observations were also made by my assistant, Mr. Fritz Lowenstein, and shortly afterward several admirable opportunities presented themselves which brought out, still more forcibly, and unmistakably, the true nature of the wonderful phenomenon. No doubts whatever remained: I was observing stationary waves.

As the source of disturbances moved away the receiving circuit came successfully upon their nodes and loops. Impossible as it seemed, this planet, despite its vast extent, behaved like a conductor of limited dimensions. The tremendous significance of this fact in the transmission of energy by my system had already become quite clear to me. Not only was it practicable to send telegraphic messages to any distance without wires, as I recognized long ago, but also able to impress upon the entire globe the faint modulations of the human voice, far more still, to transmit power, in unlimited amounts, to any terrestrial distance and almost without any loss.

With these stupendous possibilities in sight, with the experimental evidence before me that their realization was henceforth merely a question of expert knowledge, patience and skill, I attacked vigorously the development of my magnifying transmitter, now, however, not so much with the original intention of producing one of great power, as with the object of learning how to construct the best one. This is, essentially, a circuit of very high self-induction and small resistance which in its arrangement, mode of excitation and action, may be said to be the diametrical opposite of a transmitting circuit typical of telegraphy by Hertzian or electromagnetic radiations. It is difficult to form an adequate idea of the marvelous power of this unique appliance, by the aid of which the globe will be transformed. The electromagnetic radiations being reduced to an insignificant quantity, and proper conditions of resonance maintained, the circuit acts like an immense pendulum, storing indefinitely the energy of the primary exciting impulses and impressions upon the earth and its conducting atmosphere uniform harmonic oscillations of intensities which, as actual tests have shown, may be pushed so far as to surpass those attained in the natural displays of static electricity.

Simultaneously with these endeavors, the means of individualization and isolation were gradually improved. Great importance was attached to this, for it was found that simple tuning was not sufficient to meet the vigorous practical requirements. The fundamental idea of employing a number of distinctive elements, co-operatively associated, for the purpose of isolating energy transmitted, I trace directly to my perusal of Spencer's clear and suggested exposition of the human nerve mechanism. The influence of this principle on the transmission of intelligence, and electrical energy in general, cannot as yet be estimated, for the art is still in the embryonic stage; but many thousands of simultaneously telegraphic and telephonic messages, through one single conducting channel, natural or artificial, and without serious mutual interference, are certainly practicable, while millions are possible. On the other hand, any desired degree of individualization may be secured by the use of a great number of co-operative elements and arbitrary variation of their distinctive features and order of succession. For obvious reasons, the principle will also be valuable in the extension of the distance of transmission.

Progress though of necessity slow was steady and sure, for the objects aimed at were in a direction of my constant study and exercise. It is, therefore, not astonishing that before the end of 1899 I completed the task undertaken and reached the results which I have announced in my article in the *Century Magazine* of June, 1900, every word of which was carefully weighed.

The experimental station at Colorado Springs showing the structure used to determine the rate of incremental capacity with reference to the earth.

Much has already been done towards making my system commercially available, in the transmission of energy in small amounts for specific purposes, as well as on an industrial scale. The results attained by me have made my scheme of intelligence transmission, for which the name of

"World Telegraphy" has been suggested, easily realizable. It continues, I believe, in its principle of operation, means employed and capacities of application, a radical and fruitful departure from what has been done heretofore. I have no doubt that it will prove very efficient in enlightening the masses, particularly in still uncivilized countries and less accessible regions, and that it will add materially to general safety, comfort and convenience, and maintenance of peaceful relations. It involves the employment of a number of plants, all of which are capable of transmitting individualized signals to the uttermost confines of the earth. Each of them will be preferably located near some import center of civilization and the news it receives through any channel will be flashed to all points of the globe. A cheap and simple device, which might be carried in one's pocket, may then be set up somewhere on sea or land, and it will record the world's news or such special messages as may be intended for it. Thus the entire earth will be converted into a huge brain, as it were, capable of response

Tesla's tower at Wardenclyffe for sending messages across the Atlantic and electricity into the atmosphere as it appeared in 1904.

in every one of its parts. Since a single plant of but one hundred horse-power can operate hundreds of millions of instruments, the system will have a virtually infinite working capacity, and it must immensely facilitate and cheapen the transmission of intelligence.

The first of these central plants would have been already completed had it not been for unforeseen delays which, fortunately, have nothing to do with its purely technical features. But this loss of time, while vexatious, may, after all, prove to be a blessing in disguise. The best design of which I know has been adopted, and the transmitter will emit a wave complex of a total maximum activity of ten million horse-power, one percent of which is amply sufficient to "girdle the globe." This enormous rate of energy delivery, approximately twice that of the combined falls of Niagara, is obtainable only by the use of certain artifices, which I shall make known in due course.

For a large part of the work which I have done so far I am indebted to the noble generosity of Mr. J. Pierpont Morgan, which was all the more welcoming and stimulating, as it was extended at a time when those, who have since promised most, were the greatest of doubters. I also have to thank my friend, Stanford White, for much unselfish and valuable assistance. This work is now far advanced, and though the results may be tardy, they are sure to come.

Meanwhile, the transmission of energy on an industrial scale is not being neglected. The Canadian Niagara Power Company have offered me a splendid inducement, and next to achieving success for the sake of the art, it will give me the greatest satisfaction to make their concession financially profitable to them. In this first power plant, which I have been designing for a long time, I propose to distribute ten thousand horse-power under a tension of one hundred million volts, which I am now able to produce and handle with safety.

This energy will be collected all over the globe preferably in small amounts, ranging from a fraction of one to a few horse-power. One of its chief uses will be the illumination of isolated homes. It takes very little power to light a dwelling with vacuum tubes operated by high-frequency currents and in each instance a terminal a little above the roof will be sufficient. Another valuable application will be the driving of clocks and other such apparatus. These clocks will be exceedingly simple, will require absolutely no attention and will indicate rigorously correct time. The idea of impressing upon the earth American time is fascinating and very likely to become popular. There are innumerable devices of all kinds which are either now employed or can be supplied, and by operating them in this manner I may be able to offer a great convenience to the whole world with a plant of no more than ten thousand horse-power. The introduction of this

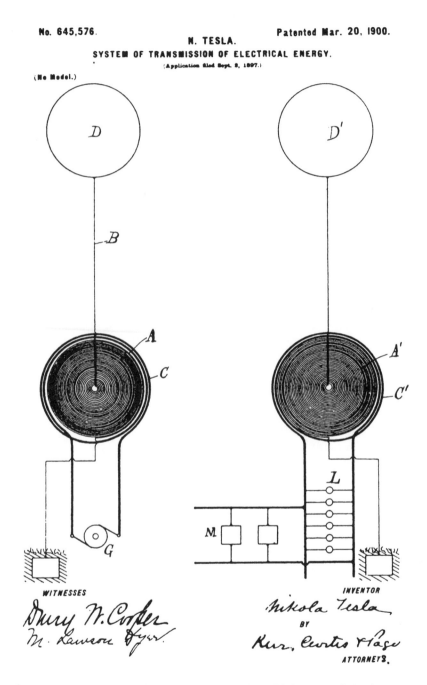

No. 645,576.

N. TESLA.

Patented Mar. 20, 1900.

SYSTEM OF TRANSMISSION OF ELECTRICAL ENERGY.

(Application filed Sept. 2, 1897.)

(No Model.)

WITNESSES

INVENTOR

Nikola Tesla

BY

Kur. Curtis Hase

ATTORNEYS.

Tesla's perfected system of wireless transmission with four tuned circuits was described in U.S. patent numbers 645,576 (March 20, 1900) and 649,621 (May 15, 1900). The applications were filed on September 2, 1897.

system will give opportunities for invention and manufacture such as never been presented themselves before.

Knowing the far-reaching importance of this first attempt and its effect upon future development, I shall proceed slowly and carefully. Experience has taught me not to assign a term to enterprises the consummation of which is not wholly dependent on my own abilities and exertions. But I am hopeful that these great realizations are not far off, and I know that when this first work is completed they will follow with mathematical certitude.

When the great truth accidentally revealed and experimentally confirmed is fully recognized, that this planet, with all its appalling immensity, is to electric currents virtually no more than a small metal ball and that by this fact many possibilities, each baffling imagination and of incalculable consequence, are rendered absolutely sure of accomplishment; when the first plant is inaugurated and it is shown that a telegraphic message, almost as secret and non-interferable as a thought, can be transmitted to any terrestrial device, the sound of the human voice, with all its intonations and inflections, faithfully and instantly reproduced at any other point of the globe, the energy of a waterfall made available for supplying light, heat or motive power, anywhere—on sea, or land, or high in the air—humanity will be like an ant heap stirred up with a stick: See excitement coming!

39 From the Archives of Lester J. Hendershot

Mark M. Hendershot

THE LESTER J. HENDERSHOT STORY
TOLD BY MARK M. HENDERSHOT

My name is Mark Hendershot, Lester J. Hendershot was my Father.

Lester was an inventor and in his many attempts at producing practical items, he had a moderate success a few times with electronic toys, and had sold some of his ideas to small manufacturers. His biggest idea, however, was so revolutionary that it embarrassed the nation's top scientists because they couldn't explain it, and if it could be perfected, it would possibly eliminate the need for public electric utilities in many instances, and it would completely change most of our present concepts of motivation.

His earlier invention was called a "motor" by the newspapers, but it was actually a generator which was powered by the magnetic field of the earth. His later models created enough electricity to simultaneously light a 120 volt light bulb and a table model radio. I witnessed it furnishing the power to run a television set and a sewing machine for hours at a time in our living room.

It was in 1927 and 1928 that my Father began to think seriously about this "fuel-less" generator. He had taken up flying in 1925 and he soon realized that the ultimate development of aviation would be greatly enhanced by the creation of an absolutely true and reliable compass, and his first efforts were to produce such an instrument.

He theorized that the magnetic compass did not point to true north and varies from true north to a different extent at almost every point on the earth's surface. Also, the induction compass has to be set before each flight and at that time was not always reliable. He claimed that with a premagnetized core he could set up a magnetized field that would indicate the true north, but he didn't know just how to utilize that in the compass he had set out to develop.

In continuing his experiments, he found that by cutting the same line of magnetic force north and south, he had an indicator of the true north and that by cutting the magnetic field east and west, he could develop a rotary motion.

With this principle in mind, he switched his plans and began working on a motor which utilized this magnetic power. He built one that would rotate at a constant speed, a speed pre-determined when the motor was built. It could be built for a desired speed, he said, and he felt that a reliable constant speed motor was one of the greatest needs in aviation at that time. The one he built developed 1,800 revolutions per minute.

In the following years, he realized that the idea of a magnetically powered motor was not as practical as a magnetically-powered generator, so his later work was directed toward the generator.

To avoid confusion, it should be pointed out that the early experiments began on a magnetically-powered *motor*, and later a *generator*.

The first significant experiments on the motor version were held at Selfridge Field, Detroit, under the direction of Major Thomas G. Lanphier, commandant of the field and leader of the First Pursuit Group.

The device demonstrated at Selfridge was a small model of what he hoped would be developed into an airplane engine [powered by earth's magnetic field]. Quotes in the newspapers referred to top aeronautical brass of the day and their impressions of what they saw.

One such report was credited to William B. Stout, president of the Stout Air Service, Inc., and designer of the all-metal type plane used by the Ford Motor Company. Stout's comments were: "The demonstration was very impressive. It was actually uncanny. I would like very much to see a large model, designed to develop enough power to lift an airplane."

Major Lanphier's comments to reporters after the demonstrations were: "The whole thing is so mysterious and startling that it has the appearance of being a fake."

"I was extremely skeptical when I saw the first model," he continued, "but I helped to build the second one and witnessed the winding of the magnet. I am sure there was nothing phoney about it."

My Father had first shown the military brass how his model worked, then he supervised army technicians in building their own model, which worked perfectly. Major Lanphier said that the electrical men to whom they had shown the motor ". . . laughed at the way we wired it up and said it wouldn't work. Then it DID work."

It was the Selfridge Field experiment which touched off the series of stories in the national press. Stories with blaring headlines in such papers as the *Detroit Free Press*, *Detroit News*, *Detroit Times*, Pittsburgh papers, *The New York Times*, and many others. Most of them tagged the instrument demonstrated at Selfridge the "miracle motor," and there were pictures of Major Lanphier and Col. Lindbergh, my Father and the motor.

Anything in the news during that period which could be connected with Lindbergh was front page whether he had an active interest, or just happened to be in the area at the time. Headlines in the various papers read, "Gasless Motor Tested for Lindy," "Lindy Inspects Fuelless Motor for Airplanes," and "Lindbergh Tries Motor That The Earth Runs." One story even stated that, at its request, Lindbergh and Lanphier were flying to New York to show the motor to the Guggenheim Foundation for the Promotion of Aeronautics.

Later reports, however, emphasized that Lindbergh actually had nothing whatsoever to do with the experiments, and that he had just witnessed a couple of the demonstrations as the guest of his friend, Major Lanphier.

The Selfridge tests seemed to satisfy Lanphier and his associates, however, and during the period he was there, the model the technicians built obtained as high as 1,800 revolutions per minute and they announced its performance was entirely satisfactory. It was estimated these motors would run for 2,000 to 3,000 hours before the magnet center would have to be recharged.

A man named Dr. F. W. Hochstetter, of the Hochstetter Research Laboratories in Pittsburgh, hastily called a news conference and displayed models of what he said were the "Hendershot Motor." He demonstrated them, and when they wouldn't work, he declared Hendershot was a fake, and that the motors worked only because of power derived from concealed pencil batteries.

After he exhibited his models of the motor, Dr. Hochstetter announced that they wouldn't generate enough electricity to ". . . light a 1-volt firefly" or to ". . . stitch a fairy's britches."

Noting the lavish lecture room in a New York hotel which had been rented by (or for) Dr. Hochstetter for the press conference, Dr. Hochstetter was asked [by one reporter] why he was so interested in the Hendershot demonstrations and in trying to discredit them. He replied merely that he had "come to expose a fraud which would be capable of destroying faith in science for 1,000 years" and he claimed his only motive was that "pure science might shine forth untarnished."

It was obvious to those who were pro-Hendershot that, in view of all the fuss and bother of such a noted scientist as Dr. Hochstetter, somewhere behind it all, someone was anxious for the innovation to be ridiculed.

When approached with the accusations, my Father smiled and told reporters, "Dr. Hochstetter is correct, to a degree. I *have* concealed batteries in a model or two because I found that I could not trust some of my visitors, and I also had evidence that someone had tampered with my work.

So, I put a couple of batteries in on occasion to lead the intruders away from what I was working on."

He added that Major Lanphier and his army technicians were proof enough of his claims. "I didn't build the motor that was demonstrated at Detroit," he pointed out. "That was built by Army men under orders from Major Lanphier and under my direction. I didn't even so much as wind the motor. They built the motor and it works. That's my answer to all the critics—it works."

Dr. Hochstetter and his associates also claimed my Father had signed a contract and received $25,000 for exploitation of the motor, but after a brief period of excitement, the matter was dropped—unproven.

Not long after his demonstrations of the motor, Dr. Hochstetter died under unusual circumstances. He was in a Baltimore and Ohio train wreck, and he was the only passenger on the whole train who lost his life!

My Father was the butt of many jokes and comments at the time of the debates about his invention. An artist, drawing for one of the Pittsburgh papers, depicted him in a cartoon riding a propellorless airplane. The caption made fun of him.

In later years my Father remarked, "Every time I see a jet plane go over now, I think of that cartoon and how everyone laughed at me for suggesting a plane could some day fly without a propeller. Twenty-five years ago I tried to tell them that."

As suddenly as it all started, the publicity and sensationalism of the Hendershot motor stopped. The last news story to appear was on March 10, 1928, when a small article appeared in most papers saying that Lester Hendershot was a patient in Emergency Hospital in Washington.

The personal account he gave was much the same as the newspaper quote, with the minor exception that he was it by a bolt of 220 volts, not the jolting 2,000 the over-eager reporter had written. He was demonstrating the motor in the patent office, and the shock paralyzed his vocal chords, resulting in several weeks of recuperation before he completely recovered.

Something happened during this period that could explain the actions of Dr. Hochstetter and his associates. My Father said that while he was in the hospital, he was approached by a large corporation to stop his activity in connection with the motor or generator.

Until the day he died, he would not reveal the name of the company, only that if he were successful with his generator, it would be a serious threat to their multimillion dollar industry. He named the sum he accepted as $25,000, and the condition was that he was not to build another unit for twenty years. That's when he dropped out of sight.

I've thought about the bizarre events connected with the generator, and feel it is possible the "large corporation" first tried to stop the activities through Dr. Hochstetter. When this failed, they approached my Father personally and bought him off. It's interesting to note that one of the doctor's charges was that he was paid $25,000 to exploit his work. Isn't it odd that this is the same figure actually paid, *but to stop his activities, but was quoted before he was* approached with the offer?

My Father admitted that he and the family lived in constant fear, as we were being contacted every so often by crackpots who had delved into the records and discovered his creation, and had gone to the trouble of searching him out. Some of them, he suspected, were representatives of subversive groups and/or foreign powers.

This latter charge seems a little exaggerated, but was supported by a series of letters he received from a fellow in Ohio in 1952. He had traced my Father by going back to his hometown in Pennsylvania and talking to my uncle about the generator.

The first letter explained that he was a part of a group of scientists who were privately financing their own research on the same phenomena my Father discovered in 1928. He emphasized they would not allow backing by any organization or government since an invention such as the Hendershot Generator should be for "all the peoples and should not be controlled by national governments, but should be given gratis to the World Government when it is ready to assume World Responsibility." He was critical of my Father for allowing the military to look at it in 1928.

That letter was written in April, and in June a postcard came with the following terse message, "Will shortly make public via radio and newspaper, connection your generator with 'Flying Saucer Propulsion.' Request Security Clearance from Security Chief your group within forty-eight hours. Have succeeded in duplication of your Generator."

In July my Father received a four-page hand-written epistle from the Ohio man. To my knowledge, it was the last letter the writer sent on the subject. He discussed information his intelligence had received on flying saucers, modestly admitting his sources were better than the CIA or the FBI, which he claimed had investigated him several times. He intimated that a Pasadena scientist had recently been kidnapped because he was working on an attempt to adapt the generator to aircraft.

Then he went into a long and rambling dissertation on how he got interested in what he called the "Ether Vortex Phenomena" and the generator. He explained that the magnetic field in the earth and volcanic action are related, according to his studies. He had spent two and one-half years in Japan working with Japanese volcanic scientists on the subject.

He mentioned one study he had made, and pointed out that the shift of the strata causing the volcano was due to a rotation of the Electromagnetic

Field of the volcano at high speeds. He urged my Father to write a complete paper on his findings and publish them (preferably send them to the Earthquake Research Institute in Tokyo).

Referring to a particularly bad earthquake which had just occurred in the Los Angeles vicinity a few months before, the writer warned my Father not to operate his generator in the area near the San Andreas (seismic) Fault which runs through the area. He said, "You may not believe it, *but you can cause earthquake activity to increase* if you continue to operate your generator in that district. I am wondering if you were not directly responsible for the recent earthquake near Los Angeles?"

Then he promised that he and his associates would keep the possibility of his involvement in the earthquake to themselves.

Letters such as these, plus occasional phone calls when the callers would not identify themselves, and a threat from an admitted Communist which was turned over to the FBI, caused my Father concern much of the time. If a large organization would take over the generator and its research, all he wanted out of it was enough money to take care of himself and his family in the future years.

One of the most encouraging offers came in September of 1956 when my Father received word that officials of the Mexican government wanted to meet with him and discuss the possibility of using his generator for the rural development program in Mexico.

Government officials flew to Los Angeles and drove out to our house, where our family doctor who spoke Spanish acted as interpreter. Arrangements were made for the family to go to Mexico City, and for my Father to work with Mexican technicians on the generator.

We all flew to Mexico City and were housed in an apartment near the home of the Director of Electricity. My Father supervised the Mexicans in building a model. He had been working with them for several weeks becoming more and more tense as time passed. He confessed to my Mother that he was frightened because he understood no Spanish, and his fellow workers talked constantly in little groups by themselves, often glancing over at him. He couldn't understand a word they were saying, and it worried him considerably.

One morning in February of 1957, the laboratory called and asked where my Father was. My Mother told them he had left for work in the morning, and if he wasn't there, she had no idea where he might be. She became increasingly concerned as the day passed and there was no word from him.

That night he didn't come home, and we were on the verge of hysteria by next morning, then we received a telegram from Los Angeles. My Father's fear had worked itself into a nervous frenzy and he had rushed to the airport the day before and taken a plane for California. To the day of his death, it was a closed subject and he would never explain why he was compelled to leave us so suddenly under such strange circumstances, except that he feared for his life.

The final attempt to promote the generator came in the latter part of 1960 when a Dr. Lloyd E. Cannon convinced my Father that he had the facilities to present the project to the United States Navy for research and development.

Cannon said he was the General Manager of his own company, Force Research of Los Angeles, Palm Springs and the Mojave Desert. Cannon explained that his group was made up of many dedicated scientists of various fields who contributed time and knowledge to Force Research projects. The range of experimentation covered electronics, astronautics, free energies, propulsion, and parapsychology.

Under my Father's supervision, two models were built and 100 copies of a fifty-six-page "proposal" were printed for presentation to the various government agencies and politicians who would have to review the project for its consideration.

After the completion of the proposal and it had been sent to the government with no results, Cannon traveled the southwestern United States with the models trying to raise money for research. His visits were increasingly less frequent to our home until 1961, when a tragic climax to this story occurred.

On April 19, 1961, upon returning home from school, I found my Father dead. It was recorded as a suicide without any further investigation.

For those who might be interested in my Father's analysis of how his generator worked, the following are his theories on the subject:

This field of magnetism surrounding the earth is similar to the field of magnetism in a man-made generator.

The rotor of a generator is revolved by some means of power, cutting the lines of magnetism, creating electric power. The earth is turning inside of a field of magnetism. That, no one contradicts, yet it is claimed there is no power to be derived from it.

Let's say we have a mechanism that will collect, polarize and create a positive and negative connection to this tremendous power that is ever-present on the earth.

Take a survey compass. You can hold the needle east or west, and let go of it, and it immediately goes north and south. This same power, when cut by the proper apparatus as the earth rotates inside this magnetism, will produce power, the amount of which is not calculated at this time.

As long as the earth rotates around the sun, it will create electric power which some scientists claim does not exist. Yet, we dig into the mountains for material that costs us unbelievable sums, to create the same power.

This magnetism surrounding the earth is in the same relation to electric power as uranium is to atomic power. Earth's magnetism is ever-present at any height or depth. It is equal to uranium as a by-product for power, namely electricity.

Magnetism must be cut. The lines of force circling the earth are constant and if this force is broken up, and polarized, you have the equivalent of uranium broken up, which creates a heat and in turn creates power.

Breaking up the forces of magnetism, polarizing them, thereby creating a resistance for power, is the same principal as atomic energy.

Scientists claim it requires friction to generate electricity. I claim the earth rotating as it does, according to scientific theory, creates friction as a generator. The ever-present magnetism is the field, or stator.

We have only to utilize this source of power to light every home, highway, bridge, airplane or any type of thing that cannot now be lighted because of inadequacy of present facilities.

A very small unit composed of wire, a magnet, several especially designed coils, condensors, collector units, and a few other minor items, will cut this force. Another especially designed mechanism will polarize it, giving a positive and a negative connection to any resistance and the result is the generation of electricity.

There you have the theory of how to create electricity from the magnetic force of the earth, as written by a man with only a high school education.

As years went by I've always wanted to continue with my Father's invention, but have worried myself about possibly running into the same problems my Father did.

It would not do my Father justice just to stop all work on it and now I am ready to fulfill his dream. Since childhood I have been fascinated by electricity and have spent over twenty-six years in the electrical trade. Of his three sons, I alone have pursued this fascination and have applied my knowledge and experience to carry on my Father's work.

40 Gunfire in the Laboratory: T. Henry Moray and the Free Energy Machine

Jeane Manning

*Doctors of science . . . are just as involved in industrial
espionage as are their business counterparts. And so
T. Henry Moray's Radiant Energy Device was . . . suppressed
by readiness, suspicion and desire for power . . .*

John Moray, *The Sea of Energy in Which the Earth Floats*

Professional skeptics were stumped, a generation or two ago, by an invention in Utah. Incredulously, people witnessed a working "free energy" device. Men of science mailed impressive credentials ahead to open the inventor's workshop door, then strode in to examine his table top apparatus from all angles, poking it and interrogating him in their search for evidence of fraud. Scientists were allowed to dismantle everything except a delicate two-ounce component, the Radiant Energy detector. When the unit was put back together, they ended up witnessing—but not all believing their eyes—as the self-contained unit converted some unknown energy into usable power, and ran continually for days at a time. Without any moving parts, the device produced a strange cold form of electricity which lit incandescent bulbs, heated a flat iron and ran a motor.

The inventor—T. Henry Moray, D.Sc. of Salt Lake City, Utah—in the late 1920s was a confident thirty-three-year-old engineer with a young family and a gift to give humanity. The gift was his Radiant Energy invention, which as he saw it converted power from the cosmos from rays which, on their eternally-launched flights through space, constantly pierce the earth from all directions.

Despite his self-confidence, there were hints that he might be stopped from mass producing his device. His family was harassed by mysterious threats. "Your husband's life is not worth a plugged nickel unless he cooperates on Radiant Energy," an anonymous caller told Ella Moray over the telephone. Their home was repeatedly broken into when the family was away, as if in warnings of worse to come.

But the young man believed in his dream, and expected that the world would accept his discovery and would eventually have abundant clean energy for homes, vehicles and industry. Many people did arrive at the Moray house in apparent sincerity, and he tuned up the Radiant Energy device for them.

An example of Henry's work in 1926 is described in the book by Henry and John Moray, *The Sea of Energy In Which the Earth Floats*, in a letter from E. G. Jensen to an associate. One October morning in that year, Jensen, another businessman, an attorney, and Henry Moray packed his electrical equipment and a lunch into an automobile and drove into the Utah mountains. Henry kept an eye on the cloudy sky through the car window; he did not like to work in a storm. His spirits rose when the sky lightened occasionally and cheered him with shafts of sunlight.

He sat back and let the other men pick the location; the more they had a hand in the work, the more likely that they would believe it. They chose to drive 26 miles from the nearest power line, to a spot on a little stream which undulated down a grassy flat to Strawberry Lake. After they unloaded the car, the businessmen pounded the six-foot long lower section of his ground pipe into the creek bed, then screwed a four-foot section of the half-inch water pipe onto it. Also without help from Henry, the witnesses to the test put up two antenna poles about 90 feet apart.

Other than the antenna and ground wires, Moray's only equipment was a brown container about the size of a butter box, another slightly smaller box, a fibreboard box about $6 \times 4 \times 4$ inches containing mysterious "tubes" and one other piece—a metal baseboard with what seemed to be a magnet at one end, a switch and a receptacle for an electric light globe as well as posts for connecting wires.

He set these parts on the car's running board and stood on a rubber mat on top of two dry boards to protect against electric shock. Wrong plan; it turned out that the running board was not wide enough to be a workbench. Unruffled by the change of plans, he gently moved his equipment onto the planks on the ground. Snowflakes drifted lightly in the air, so the three spectators hung a tarpaulin over open car doors to protect the electrical equipment.

Before Henry primed and tuned his apparatus, he put a key into the post and showed the men that there was no power flowing. Then he tuned the device by stroking the end of a magnet across two pieces of metal sticking out from what seemed to be another magnet. After tuning for about ten minutes, Moray put the key into the post, and the 100-watt light bulb brought along by one of the men burned brightly for fifteen minutes. Jensen wrote that the light was even, without fluctuations.

While the light was burning, Mr. Moray disconnected the antenna lead-in wire from the apparatus and the light went out. He connected it again and the light appeared. He also disconnected the ground wire and the light went out.

Mr. Moray . . . said he could do the same thing in the middle of the Sahara Desert or in the deepest mine. When the demonstration was over we congratulated Mr. Moray and I felt confident that he had a real invention and that no hoax was being perpetrated.

Where, then, was the dazzling light—the strange electricity which seemed to ignite the entire contents of a light bulb—coming from? Moray's device had no batteries. Was this Utah scientist gifted with advanced intuitive understanding about a previously-unknown source of energy? The answer may be found in Henry's words: "Energy can be obtained by oscillatory means in harmony with the vibrations of the universe . . . the Moray Radiant Energy Device is a high-speed electron oscillating device." He also said that those vibrations continually surged onto the earth like waves onto a seashore.

"The power—the surges—would come in so strongly during the day that it would burn out his detector," Henry and Ella Moray's oldest son, John Moray, told the author. "So he mainly worked at night."

Since the device seemed to go against current "laws" of physics, professional doubters went to ludicrous lengths in attempts to dismiss it as a hoax. Moray's sons remember the family laughing about a visitor who saw the device working in Moray's basement. He insisted that "Mrs. Moray was secretly powering it; she must have been pacing back and forth on a carpet upstairs and generating static electricity!"

Would-be debunkers, sabotage, and lack of funding were only some of the obstacles in the way of further developing the invention. Because of betrayals, Henry Moray himself eventually distrusted people outside his family and he guarded his technical secrets closely—even to the point of losing a potential business deal.

HERITAGE OF WARINESS

Causes for Henry's untrusting nature are outlined by John Moray, in the second edition of *The Sea of Energy*. To begin with, a heritage of wariness was passed on from previous generations. Henry's mother, Swedish immigrant Petronella Larson, had a rather difficult life before she married an American, James Cain Moray. James had been born in Ireland to a family which had to hide from being killed by political enemies. After

Henry's father died (in Salt Lake City) of natural causes, certain individuals—people whom the Morays trusted—swindled his mother out of the family fortune.

She turned to her only son, hoping that Henry would specialize in money matters, and she insisted that he attend a Latter Day Saints (Mormon church) college because it had a good business course.

However, from the age of nine Henry had had a driving interest of his own—radio and electrical science. In his spare time as a boy he searched the garbage dump for scraps of wire and other materials for basement tinkering. By age fifteen, he had a job wiring houses, which taught him more about electricity. Meanwhile, the beginnings of the Radiant Energy concepts were pounding through his mind. In the summer of 1909 he started experimenting with taking electricity from the ground, and by autumn of the next year he had enough power to run a miniature arc light. Thinking about Benjamin Franklin's kite experiment, Henry at first figured he was dealing with static electricity. He later changed his view.

He firmly believed in his energy idea, despite the reigning scientific ideas which would label it as impossible. Even when his experiments only converted enough energy to make a slight click in a telephone receiver, he was sure that he was on the right track. During Christmas holidays of 1911, he became more certain that the mysterious energy was not static, but was oscillating (swinging back and forth) like pendulum upon pendulum across the universe. And he realized that the energy was not coming out of the earth, but instead it was coming to the earth from some outside source. The electrical oscillations pound the earth day and night, "always coming, in vibrations from the reservoir of colossal energy out there in space."

After a correspondence course in electrical engineering, the next step in his education was an extended stay in Sweden; he went on a mission for the Church of Jesus Christ of Latter Day Saints. The young missionary managed to study science at the University of Upsalla and complete a doctoral thesis. Naturally, the thesis related to his idea that there is energy throughout space.

While he was a homesick student/missionary in Scandinavia in the summer of 1913, Henry picked up a soft, white stone-like material out of a railroad car at Abisco, Sweden. He also took some of the material from the side of a hill, tested it and decided the stone might be good to use in a valve-like detector of energy. This led Henry to his research in semiconductive materials; from this stone he developed the "Moray valve" that was used in his early Radiant Energy devices.

After he returned to the United States in 1917 he married Ella Ryser and they later had five children. On his career ladder, Moray worked his way up through various jobs to electrical engineering and positions such as design engineer for the largest oil-cooled electrical switch yard in the world.

An industrial accident at a power substation in late 1920 burned the retina of his eyes and propelled him into legal battles for compensation.

In a way, losing much of his eyesight for years turned out to be a blessing. Although it meant an empty bank account at the time because he was unable to work at his usual profession, being forced away from the drawing table led him back into Radiant Energy research.

SENATOR PROTECTED UTILITIES

Far from being the stereotype of a reclusive basement inventor, Moray was known in his community and was listed in a 1925 *Who's Who in Engineering*. On July 24 of that year, Senator Reed Smoot invited the young inventor to meet with him in the senator's office in the Hotel Utah. Henry Moray made an offer which, if accepted, could perhaps have dramatically changed events in this century. Oil wars, nuclear plant accidents, acid rain were yet to come.

Henry offered his Radiant Energy discovery to the United States government. Free of cost. According to *The Sea of Energy*, the senator thanked Moray but replied that the government would decline such an offer. Why? "On the grounds that the government was not competing with public utilities."

Undeterred, Henry spent countless hours in his basement working on solid state physics with what he called the Moray Valve as a detector for radio frequencies. According to his records, early in the 1930s he made a radio which was no bigger than a wristwatch.

Part of Henry's invention was his pioneering use of semi-conductors. Moray's first germanium solid-state device (a transistor) was sent to the U.S. Patent Office in 1927, and was rejected on the basis that it would not work without a heated cathode. Heated cathodes were commonly used in vacuum tubes of that time. This means that Henry Moray was so far ahead of his time in semi-conductor technology that the patent office had not heard of it, and so the bureaucrats decreed that what he had was impossible. Of course society later learned that cold cathodes are most definitely possible. But when the transistor was officially invented twenty years later, no credit was given to Henry Moray.

The second generation of Moray's radio valve not only picked up radio waves, it also detected a small amount of power. Launched by these experiments with semi-conductors, he followed a trail of discovery which led to his powerful energy converter. By 1939, a unit weighing less than 55 pounds, including its wooden case, converted 50,000 watts of power— enough to run a small factory. He tested it 90 miles from the nearest radio station, at a desolate area now known as the U.S. Army Dugway proving ground, and the device still worked.

Witnesses to his experiments included engineers and curiosity-seekers from other countries as well as local visitors from Utah Power and Light, the Secretary of State's office in Utah and other officials. As far as this author can discover, no one refuted Henry Moray's claim that his Radiant Energy device did run motors, light bulbs and a radio.

The invention had unusual characteristics. Photographers exclaimed over the intensity of the light from the bulbs—remarkably brighter than 100- or 150-watt bulbs normally shone.

While the invention converted energy from the cosmos into light and attracted well-known officials, some people entered Moray's life without leaving a calling card. For example, in 1939 he refused an offer to take his work to Russia. Soon the anonymous threatening phone calls began, telling Henry there was a contract out on his life.

Despite death threats, Henry Moray repeatedly worked on his strange electric generator in front of creditable witnesses. The only threat which stopped him from demonstrations came in the form of advice from his patent attorneys in Washington, D.C.—under patent laws he could have lost his rights to a patent if he showed his invention to just anyone.

The U.S. Patent Office itself was not much help either. That agency rejected seven patent applications for his Radiant Energy Device because the device did not fit the physics known at the time. "Where is the source of energy?" the examiners asked. One rejection notice from the patent office wrongly assumed that the energy was originally electromagnetic. Moray, however, only said it is electrical after it hit his semi-conductors.

BULLETS PIERCE WINDSHIELDS

Henry carried on multiple battles at the same time. Instead of being helped to research the Radiant Energy device, he was hindered. In time-wasting letters he fought the patent office, treachery from business partners, and scientists who witnessed Radiant Energy and later denied it when their employers changed. And he had to be strong to keep his family's morale up in the face of unknown enemies.

John Moray remembers an incident in Salt Lake City when he and the other children were in the family car, with his mother driving. Sitting in the back seat, the boy felt his heart lurch with shock as a bullet crashed through the car and lodged in the windshield in front of his mother. "A classic black sedan with all the shades down almost forced her off the street, then sped away up 21st South."

Within a few weeks, an unknown assailant had also fired shots at Henry's friend S. E. Bringhurst, the first president of his research institute.

Bringhurst did not have bullet-proof glass in his car, and the bullet zipped past his head and out the rear of the car.

Henry bought a 32.20 revolver and a Colt .32 handgun to protect his family, in addition to the bullet-proof glass installed in the windows of his automobiles. The whole Moray family suffered as a result of the mysterious opposition to Henry's work. Mrs. Ella Moray lived in fear that something would happen to one of the children, and the children paid the price of losing a normal childhood. They were forbidden to go anywhere by themselves. Even when the boys were almost teenagers, they could not go out without an escort because of the threat of a kidnapping.

R.E.A. MAN SABOTAGES DEVICE

Violence in Henry's laboratory also shocked the family. A man named Felix Frazer who had been sent by the Rural Electrical Administration to work in Moray's lab went crazy with a sledge hammer (or as some reports say, an axe) one day, and destroyed the Radiant Energy machine. He had not broken into the lab; he had been hired to work there!

What type of person would hammer an important invention into useless pieces—destroy a device which took years to perfect and which contained expensive and almost irreplaceable parts? John Moray describes the saboteur as "a double agent trying to force Dr. Moray to co-operate with the U.S. Department of Agriculture's Rural Electrical Administration and a communist government."

John was 12 years old at the time, and as a grown man he chronicled a related episode in the book *Sea of Energy*:

As a result of the constant threat to his life, my father carried a gun with him at all times. He carried a .32 in his pocket, and whenever he walked from the house to the laboratory at night he would wear a 32.20 revolver. He was an excellent shot in the old Western sense . . .

On three different occasions, he was attacked at his laboratory and shot his way out of the situation.

The incident of March 2, 1940, particularly stands out . . . Late that afternoon a friend of mine and I were playing on the front lawn of our home. My cousin was just starting up his car, which was parked beside my father's car in the garage—the two cars side by side, from the street one could not tell which car was being cranked or who was driving.

Suddenly several men in a sedan turned into the driveway and pulled guns as if they intended to fire on the car that was starting up in the driveway. When my cousin backed out, the men could see that it was not my father, and they quickly drove away.

I told my father about the incident and he laughed, trying to minimize it to prevent my worrying. . . .

Henry Moray later drove John and his two sisters to the Centre Theatre. After the movie, the youngsters phoned home as instructed. They were told to wait there; mother would pick them up.

However, no one came, and we waited for several hours. Finally my cousin Chester picked us up. When we arrived home we discovered that my father had been shot in the leg and the doctor was there . . . the president of the company was also there.

Henry Moray had gone to his laboratory that evening. When he was ready to leave and had the front door open, he remembered to pick up some materials from a locked inner office. As he fumbled with keys in the dark, he had the impression someone was coming up behind him. As he turned to look, a heavy object hit his right shoulder, leaving the arm half-paralysed. With his left arm, he grabbed the assailant's head. While Moray pinned the assailant to his left side, the man's gun became entangled in his overcoat.

As the first man struggled, a second man carrying a gun ran up. Henry Moray kicked the second man, knocking his gun free at the same time as the first man's gun discharged. The bullet travelled downward, grazing the side of Moray's leg, and ricocheted off the concrete floor. Moray's right arm came back to life enough to get his own gun out. He pointed it at the two men and waved them out the front door.

"He was immediately fired upon again by someone at a distance," John Moray writes.

He returned the fire, knocking the third gunman down. A fourth man rushed up to help the wounded gunman. Henry recognized this man as Felix Fraser (Rural Electrification Administration Engineer).

The second man said to the first assailant, "Well, you weren't as quick on the draw as you thought you were," and Henry Moray recognized the voice of a FBI man he had known at one time as a security guard.

At that point, Henry realized he was all alone in a very difficult and dangerous situation.

Bleeding severely, Henry knew he would faint at any moment. If he fainted while the men were there, he would be at their mercy. "So in panic he told them to get out, pretending that he had not recognized any of them, and the men promptly left."

Henry Moray was an excellent shot and could have killed his assailant in his laboratory, but Moray was not a violent man.

WOUNDED AND HARASSED

He believed the harassments were intended to force him to turn over his laboratory notes to Felix Fraser and associates. He tested his theory the next workday. His family helped him to hobble to the laboratory before anyone arrived. Julius Noyes, his assistant at the time, arrived at 8 A.M., greeted Henry, and went to work in the back room, while Henry did not move from behind his own desk. John Moray describes the incident:

> Later, Felix Fraser came in and rushed back to Julius Noyes. Shortly after, Fraser returned to the office and fussed around for a few minutes, looking at the floor. Then he came into my father's office and said, "Henry, why didn't you tell me you were shot?" Immediately Dad asked him how he knew that he had been shot. Fraser said, "Oh, Julius told me." But my father had deliberately prevented Julius from knowing of the shooting.

From then on, trouble multiplied. Henry Moray refused to cooperate further with the REA. John recalls that people attacked his father's credibility. His family later discovered that more than a dozen of Henry's original patent applications had disappeared from the U.S. Patent Office, although the file jackets remained there. "The contents and applications themselves are gone . . . Watergate was not the first great coverup and act of duplicity," John Moray wrote. Who stole the more-than-a-dozen patent applications? John Moray says the question will probably remain unanswered.

Over half a century after Henry Moray's discovery, his sons are still waiting for an investor who will fund the expensive development of the Moray device; engineering problems still have to be solved.

Some researchers believe that T. Henry Moray's secrets died with him and that the family and associates would not be able to replicate his device even if they had funding. After all, a saboteur had destroyed the priceless parts of the Moray device. Moray's sons, however, reply that Henry had built another working model which he later took apart, and that they inherited all his laboratory notes.

John remembers the later model, and he describes a 1942 trip to the mountains of Colorado with his father and the device. Since it was during World War II, Henry had to scrape up enough gas rations for the round trip. He set up the experiment in a park and the unit performed smoothly.

"Well," said their host, "If you leave it here and if my engineers like it, we'll decide if we want to buy it or not."

This is what Moray ran into all the time, John maintained, "My dad said 'thanks but no thanks.'"

The second generation of the Moray family of Utah has lived with the Radiant Energy project close to their hearts for decades. But their experiences make it difficult to trust all too many of the people who seek him out, even today. John tells about a friend of 30 years, with whom Henry left a piece of equipment. It was not even a power component; it was a measuring piece. His friend tore the equipment apart looking for its supposed secret.

WHO WILL HELP INDEPENDENT INVENTOR?

The financial cost of developing the Radiant Energy device was high, considering how difficult it was in the 1920s for even an upper-middle-class family to scrape together $400,000 for materials and equipment, John said. Translating to circa-1995 dollars, the Moray family spent millions on the Radiant Energy Device. Their longtime goal was the development of Radiant Energy, which Henry described in 1958 (letter to Colin Gardner of California) as a source of energy "greater than that coming from the Atom, more unlimited and of no danger to the user whatsoever from radiations, residue, etc."

Gardner was one of countless people who contacted the Moray family after reading Henry's book *Beyond the Light Rays* or the later book. In 1958 Gardner conveyed his enthusiasm to fellow U.S. Navy officers at Point Mugu, California. In a letter to Moray, Gardner offered to connect his superior officers with Moray. Moray's reply illustrates his weariness at that point: "The government has a funny way of going at investigating and/or accepting new ideas . . . That is why we are not first into space . . . Just sending me a form to fill out, treating RE as every other minor discovery, is of no interest to me. That is what all the government branches ever do . . ."

Moray left his laboratory door partly open to the Navy, however; and replied to a second letter from Gardner by saying that he would be delighted to meet Gardner's very top supervisors and discuss his "discoveries which are greater than nuclear fission." However, added Moray, "we have so many hundreds asking for information who take up our time needlessly that we cannot spend the time unless it is with those qualified (to understand Radiant Energy) and with high enough authority to deal."

Moray's office sent his confidential papers to Gardner's boss. After two weeks of silence from the Navy, the message relayed back to Moray was "it is felt that you do not have a commercial product for us to buy and use at our discretion."

This incident is only one example of difficulties facing independent inventors of unorthodox energy devices. Although Moray spent the family's bank account on experiments which produced a laboratory proof-of-concept device, he was expected to somehow without funding take it beyond that stage through the very expensive stage to a commercial product. A final product must be engineered and fine-tuned until it works consistently enough to be mass-produced. Today, the Moray brothers estimate it would take more than $14 million just to build the parts used in a Radiant Energy laboratory model (which is not as refined as a production model); they say that high-priced personnel, expensive test equipment and huge capital outlay would be needed.

Instead of a factory producing Radiant Energy units, Henry Moray had one model which he tore apart—"cannibalized"—to re-use its expensive parts whenever he built an improved model.

Similar in another way to fellow independent inventors throughout the century, Moray's experiences with would-be financiers was discouraging. Moray Products Company, for example, seemed to be going well until Henry found out that the company's treasury was being pilfered from within. Stocks were being sold without benefit to either himself or the company; the thieves kept no records of those stock down-payments and also ignored offers from investors who would have exposed the pilfering. Henry Moray took these associates to court. The costs bankrupted him, and the company broke apart.

To add to his distress, Henry's closest friend, W. H. Lovesy of Utah Oil Refining Company, died under mysterious circumstances in a one-car accident. A hitchhiker who was never identified walked away from the crash.

Hearing the family talk about so many troubling incidents for so many years, John Moray was bound to grow up with a grimly determined set to his jaw. From childhood John lived with the expectation that he would continue the work his father began. As a boy he would be rewarded for good behaviour by being allowed to go downstairs to the basement laboratory in the evening and watch his father experiment. (In 1939 Henry built a 50-foot by 60-foot laboratory with four rooms above it, and the workshop was moved outside of the house permanently).

Around 1950, Henry and his grown sons sat down to brainstorm a plan for financing Radiant Energy development. Richard volunteered to go to Canada and invest in land, and Henry and John stayed in Salt Lake City. Richard and his family found it more difficult than expected—battling bureaucracy in British Columbia in an attempt to develop a subdivision was not always successful. John had planned to go into electrical engineering, but found that the University of Utah physics department was more flexible in allowing him to choose courses.

Nearing the end of his lifetime, Henry Moray became "more and more amazed," wrote John, "for he had never believed he could really be stopped." Dr. T. Henry Moray passed on in 1974.

Interviewed in 1994, John Moray was (in his sixties) a retired army colonel now working full-time as a substitute teacher in Salt Lake City, getting up before 5 A.M. to work on correspondence, and thinking of selling the laboratory. The family had by no means abandoned Radiant Energy, he said; keeping within their budget they contract out work on the project. One time-waster, the family has discovered, is battling at rumours. The latest wild tale which John heard was that there was a Moray device in his basement.

"What a ridiculous statement; that is the last place we would keep one!"

"If I had a machine, what good would it do to show people? If they don't believe the tests that have already been run, they're not going to believe what they see anyhow."

What part did secretiveness play in the fadeout of Moray's Radiant Energy technology? And is secretiveness a result of today's patent office/attorney/competition-oriented setup? Admitting that his father refused to release specifics about his invention without first getting signed and legally-binding contracts, John Moray wrote in *The Sea of Energy* that "If this is carrying an invention as too tight a secret then why do patent laws require it?"

What factors most suppressed the Moray device? John replies, "Finances. And also personal animosity, ego, avariciousness . . ." The violence? "It was always over money."

The ego factor enters when a scientist values a reputation as an expert more than truthfulness. This was underlined when Richard Moray visited Harvey Fletcher Sr. before the eminent scientist died of old age. The man had publicly denied that he had seen a working model of Henry's invention. Now the scientist was well into his nineties and apparently making peace with his life.

"He admitted that, yes, the Radiant Energy device worked just like my father said," Richard said in an interview, with a look of deep frustration. "I asked him 'then why, why did you do what you did?'" Richard measured out his next words flatly. "He said 'because I couldn't admit that I didn't know . . .'"

Ego, greed, excessive pride and distrust. Will enough people rise above these motivations and see themselves connected with all others in a sea of energy? Perhaps then Radiant Energy units could light up this world.

REFERENCES

Burridge, Gaston. "Alchemist 1956?" *Fate magazine*, Sept. 16, 1956.

Davidson, Dan A. *Energy: Breakthroughs to New Free Energy Systems.* Greenville, TX: RIVAS, 1977, 1990.

Davidson, John. *The Secret of the Creative Vacuum*. Essex, England: C.W. Daniel Co. Ltd., 1989.

Kelly, D.A. *The Manual of Free Energy Devices and Systems*. Clayton, GA: Cadake Industries and Copple House, 1987.

King, Moray B. *Tapping The Zero-Point Energy*. Provo, UT: Paraclete Publishing, 1989.

Lindemann, Peter A. *A History of Free Energy Discoveries*. Garberville, CA: Borderland Sciences Research Foundation, 1986.

Moray, John E. and Kevin R. Non-Conventional Energy Symposium, Toronto, 1991.

Moray, John E. "Radiant Energy." 1981.

Moray, T. Henry. *Radiant Energy*. Garberville, CA: Borderland Sciences Research Foundation, 1945.

Moray, T. Henry. Speech given Jan. 23, 1962, Valley State College, Northridge, CA.

Moray, T. Henry and John E. "T. Henry to Colin Gardner, private letters." California: Walter Rosenthal collection, 1960, 1978.

The Sea of Energy. Salt Lake City: Cosray Research Institute, P.O. Box 651045, Salt Lake City, UT 84165.

"The Sea of Energy, A Means for the Preservation of the Environment By Drawing Kinetic Energy From Space." Boston: 26th Intersociety Energy Conversion Engineering Conference. Copyright American Nuclear Society.

Valentine, Tom. "Free Electricity Generated from the Radiant 'Cosmos.'" *NEWSREAL* magazine. Date unknown.

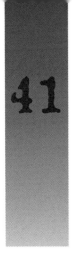

41 Sunbeams From Cucumbers

Richard Milton

*He had been eight years upon a project for
extracting sunbeams out of cucumbers.*

Jonathan Swift, *Gulliver's Travels*

No other scientific endeavor has consumed so much talent, so much cash
and so many years of sustained effort as the race to harness the power that
makes the sun shine. Billions of pounds (and dollars, rubles and yen),
more than four decades of research and the careers of thousands of physicists have been expended on the search for a nuclear reactor that will generate limitless power from the fusion of hydrogen atoms. There are gray-
haired professors with lined faces still poring intently over the equations
they first looked at eagerly with bright young eyes in the 1940s and 1950s.
They will go into retirement with their dreams of cheap, safe power from
fusion still years in the future, for the obstacles in their paths are as formidable now as ever.

Fusion is the process taking place in the sun's core where, at temperatures of millions of degrees, hydrogen atoms are compressed together by
elemental forces to form helium and a massive outpouring of energy in the
thermonuclear reaction of the hydrogen bomb.

It is not difficult, then, to imagine how people who have invested their
talent and their lives in the quest to tame such forces are likely to react
when told that fusion is possible at room temperature, and in a jam jar.

Hydrogen atoms repel each other strongly—so strongly that no known
chemical reaction can persuade them to fuse. There are, though, heavier
isotopes* of hydrogen, such as deuterium, which together with oxygen
makes heavy water and which under the right circumstances can be made
to fuse in nuclear reactions. When they do so, they release energy. However,

---

* Atoms that have the same number of protons—atomic number—but different mass numbers.

the only circumstances so far under which hydrogen atoms have been persuaded to fuse have nothing in common with the measured calm of the laboratory bench but are more like a scene from Dante's *Inferno*. In the center of the sun and other stars, the atoms are squeezed by cataclysmic gravitational forces to form a plasma of the nuclei of hydrogen atoms at a temperature of millions of degrees. These high temperatures kindle a self-sustaining reaction in which hydrogen is "burnt" as the fuel.

The scientific world was thus astonished when, in March 1989, Professor Martin Fleischmann of Southampton University and his former student, Professor Stanley Pons of the University of Utah, held a press conference at which they jointly announced the discovery of "cold fusion"—the production of usable amounts of energy by what seemed to be a nuclear process occurring in a jar of water at room temperature.

Fleischmann and Pons told an incredulous press conference that they had passed an electric current through a pair of electrodes made of precious metals—one platinum, the other palladium—immersed in a glass jar of heavy water in which was dissolved some lithium salts. This very simple set-up (the *Daily Telegraph* later estimated its cost as around £90 [$144 U.S. currency]) was claimed to produce heat energy between four and ten times greater than the electrical energy they were putting in. No purely chemical reaction could produce a result of such magnitude so, said the scientists, it must be nuclear fusion. Further details would be revealed soon in a scientific paper.

Both scientists are distinguished in their field, that of electrochemistry. But in making their press announcement they were breaking with the usual tradition of announcing major scientific discoveries of this sort. The usual process is one of submitting an article to *Nature* magazine which in turn would submit it to qualified referees. If the two chemists' scientific peers found the paper acceptable, *Nature* would publish it, they would be recognized as having priority in the discovery and—all being well—research cash would be forthcoming both to replicate their results and conduct further research.

But the two scientists perceived some difficulties. First, their paper would not be scrutinized by their exact peers because the discovery was unknown territory to electrochemists and indeed everyone else. It would probably be examined mainly by nuclear physicists—the men and women who had grown gray in the service of "hot" fusion. This would be like asking Swift's "Big Endians" to comment objectively on the work of "little Endians." It is not that "hot" fusion physicists could not be trusted to be impartial, or were incapable of accepting experimental facts, but rather that they would be coming from a research background that would naturally give them a quite different perspective.

Fusion Hot and Cold

Fusion is the opposite of fission, although both processes start with atoms. Atoms are the tiny building blocks that make up all matter. An atom consists of a nucleus, which is made up of protons and neutrons, and electrons, which form a cloud around the nucleus. Different atoms contain different amounts of protons, neutrons, and electrons, and form different types of matter.

Fission is the splitting of an atom's nucleus, such as by bombarding it with neutrons. This releases a great amount of energy. An atomic bomb and a nuclear power plant both use fission.

Fusion is the joining together of atomic nuclei. Hot fusion, which is said by some scientists to be what energizes our sun, uses a form of the lightest element, hydrogen.

Textbooks teach that temperatures reaching millions of degrees Fahrenheit are needed before the positively charged hydrogen nuclei can overcome their natural repulsion toward each other, since like charges repel—think of what happens if you attempt to bring the north poles of two magnets together. If the hydrogen nuclei do come close enough together, they form something different—helium nuclei. In the process, tremendous amounts of energy are released.

Instead of using super-heated gas, cold fusion seems to be based on the reaction of a metal such as palladium, which has large spaces between its nuclei, and a liquid form of hydrogen called deuterium. The deuterium seems to move into the spaces within the palladium in the same way that water moves into the open, absorbent surface of a towel. While no one disputes the fact that the metal absorbs the deuterium, cold-fusion proponents cannot prove that the reaction which follows the absorption is a nuclear reaction.

Cold fusion is not without problems. For example, one of the byproducts of cold fusion is the radioactive gas tritium, a rare form of hydrogen. As one new-energy organization has noted, cold fusion introduces concerns about radioactivity, and even a low level of radiation can eventually lead to environmental and health problems.

From *The Coming Energy Revolution* by Jeane Manning (Garden City Park, NY: Avery Publishing Group, 1996).

There was also the problem of money. Whoever develops a working fusion reactor—hot or cold—will be providing the source of energy that mankind needs for the foreseeable future: perhaps for hundreds of years. The patents involved in the technology, and the head start the patent owners will have in setting up a new power industry, will be worth many billions of pounds in revenue. It is potentially the most lucrative invention ever made. With such big sums at stake, the scientists' university wanted no future ambiguity about who was claiming priority, and hence encouraged them to mount a very public announcement.

In the end, the two scientists agreed to a press conference that would stake Utah University's claim to priority in any future patent applications, followed by publication of a joint paper in their own professional journal, *The Journal of Electroanalytical Chemistry.*

There followed a brief honeymoon of a week or two, during which newspaper libraries received more requests from the newsroom for cuttings on fusion than in the previous twenty years, and optimistic pieces about cheap energy from sea-water (where deuterium is common) were penned to keep features editors happy. All over the world, laboratories raced to confirm the existence of cold fusion, although many scientists were unhappy at the lack of scientific detail and at having to learn about such an important event from television news and the popular press. What these researchers were looking for, with their £90-worth of precious metals stuck in test tubes, were one or more of the key tell-tale signs that would confirm cold fusion. When two deuterium nuclei fuse they produce either helium and a neutron particle or tritium. So, if fusion really is taking place, it should be possible to find neutrons being emitted, or helium being formed or tritium being formed. It should also be possible to detect energy being released, probably as heat, that is greatly in excess of any electrical energy being put in. (Of course, if the cell does not do this it is of no use as a power source.)

Despite the experimental difficulties it was not long before confirmations were reported. First were Texas A & M University, who reported excess energy, and Brigham Young University who found both excess heat and measurable neutron flow. Professor Steve Jones of BYU said his team had actually been producing similar results since 1985, but that the power outputs obtained had been microscopically small, too small in fact to be useful as a power source.

One month after the announcement the first support from a major research institute came when professor Robert Huggins of California's Stanford University said that he had duplicated the Fleischmann-Pons cell against a control cell containing ordinary water, and had obtained 50 percent more energy as heat from the fusion cell than was put in as electricity. Huggins gained extra column-inches because he had placed his two reaction vessels in a red plastic picnic cool-box to keep their temperature constant. This kitchen-table flavor to the experiment added even further to the growing discomfort of hot fusion experts, with their billion-dollar research machines.

By the time the American Chemical Association held its annual meeting in Dallas in April 1989, Pons was able to present considerable detail of the experiment to his fellow chemists. The power output from the cell was more than 60 watts per cubic centimeter in the palladium. This is approaching the sort of power output of the fuel rods in a conventional nuclear fission reactor. After the cell had operated from batteries for ten hours producing several watts of power, Pons detected gamma rays with the sort of energy one would expect from gamma radiation produced by fusion. When he turned off the power, the gamma rays stopped too. Pons also told delegates that he had found tritium in the cell, another important sign of fusion taking place.

Pons estimated that the cell gave off 10,000 neutrons per second. This is many times greater than the rate of background level of natural radioactivity, but is still millions or billions of times less than the rate of neutron emission that one would expect from a fusion reaction—a puzzle which Fleischmann and Pons acknowledge as a stumbling block to acceptance of their phenomenon as fusion by any conventional process.

However, despite the reservations, the assembled chemists were ecstatic that two of their number had apparently scooped their traditional rivals from the world of physics, and had, in the words of the American Chemical Society's president, "come to the rescue of fusion physicists."

This was perhaps the high-water mark of cold fusion. Scores of organisations over the world were actively working to replicate cold fusion in their laboratories, and although many reported difficulties a decent number reported success. And by the end of April, Fleischmann and Pons were standing before the U.S. House Science, Space and Technology committee asking for a cool $25 million to fund a centre for cold fusion research at Utah University.

Then things began to go wrong. First, some of the researchers who early on announced confirmation of cold fusion now recanted, citing faulty equipment or measurements. Next, an unnamed spokesman for the Harwell research laboratory—the home of institutional nuclear research in Britain—spoke to the *Daily Telegraph* saying that:

> . . . we have not yet had the slightest repetition of the results claimed by professors Martin Fleischmann and Stanley Pons. Of the other laboratories around the world who have tried to replicate the Pons-Fleischmann result, all but one have recanted, admitting that either their equipment or their measurements were faulty.

> We believe our experiments are much more careful than those con-
> ducted by others. Perhaps for that reason we have been unable to
> observe any more energy coming out of the experiment than was put in.

And by the time the American Physical Society had *its* annual meeting
in Baltimore in May, the opponents of cold fusion were gathering
strength. Steven Koonin, a theoretical physicist from the University of
California at Santa Barbara, received rapturous applause from the physicists when he declared, "We are suffering from the incompetence and perhaps delusion of doctors Pons and Fleischmann."

It was, however, a chemist, Dr. Nathan Lewis of the California Institute
of Technology, who got the loudest applause. Lewis told the delegates that
after exhaustive attempts to duplicate cold fusion, they had found no signs
of unusually high heat. Nor did they detect neutrons, tritium, gamma rays
or helium.

By late May, the headlines in both the popular press and the scientific
press were beginning to carry words like "flawed idea" when the biggest
blow of all hit supporters of cold fusion. Dr. Richard Petrasso of the
Plasma Fusion Center of the ultra-prestigious Massachusetts Institute of
Technology presented the results of a series of intensive investigations
into the Fleischmann-Pons experiment. The fundamental data put forward
by the two men, said Petrasso, was probably a "glitch." The entire gamma
ray signal in the Fleischmann-Pons experiment, he said, might not have
occurred at all.

"We can offer no plausible explanation for the feature other than it is
possibly an instrumental artefact with no relation to gamma ray interaction," he told the same reporters who had clustered around Fleischmann
and Pons only two months earlier.

Dr. Ronald Parker, director of MIT's Plasma Fusion Center, said:
"We're asserting that their neutron emission was below what they thought
it was, including the possibility that it could have been none at all."

Thus within two months of its original announcement, cold fusion had
been dealt a fatal blow by two of the world's most prestigious nuclear
research centres, each receiving millions of pounds a year to fund atomic
research. The measure of MIT's success in killing off cold fusion is that
still today, the U.S. Department of Energy refuses to fund any research
into it while the U.S. Patent Office relies on the MIT report to refuse any
patents based on or relating to cold fusion processes even though hundreds have been submitted.

If Dr. Parker had left his statement there, it is likely that the world
would never have heard of cold fusion again—or not until a new generation of scientists came along. But having been so successful at discredit-

ing MIT's embryonic rival, he decided to go even further and openly accuse Fleischmann and Pons of possible scientific fraud.

According to Dr. Eugene Mallove, who worked as chief science writer in MIT's press office, Parker arranged to plant a story with the *Boston Herald* attacking Pons and Fleischmann. The story contained accusations of possible fraud and "scientific schlock" and caused a considerable fuss in the usually sedate east-coast city. When Parker saw his accusations in cold print and the stir they had caused he backtracked and instructed MIT's press office to issue a press release accusing the journalist who wrote the story, Nick Tate, of misreporting him and denying that he had ever suggested fraud. Unfortunately for Parker, Tate was able to produce his transcripts of the interview which showed that Parker had used the word "fraud" on a number of occasions.

It then began to become apparent to those inside MIT that the research report that Parker and Petrasso had disclosed to the press in such detail was not quite what it seemed; that some of those in charge at MIT's Plasma Fusion Center had embarked on a deliberate policy of ridiculing cold fusion and that they had—almost incredibly—fudged the results of their own research.

The MIT study announced by Parker and Petrasso contained two sets of graphs. The first showed the result of a duplicate of the Fleischmann-Pons cell and did, indeed, show inexplicable amounts of heat greater than the electrical energy input. The second set were of a control experiment that used exactly the same type of electrodes, but placed in ordinary "light" water—essentially no different from tap water. The result, for the control cell should have been zero—if cold fusion is possible at all, it is conceivable in a jar full of deuterium, but not in a jar of tap water. Any activity here, according to current theory, would simply indicate some kind of chemical, not nuclear, process.

But the MIT results for the control showed exactly the same curve as that of the fusion cell. It was the identical nature of the two sets of results that depicted so graphically to the press and scientific community the baseless nature of the Fleischmann-Pons claim and that justified MIT's statement that it had "failed to reproduce" those claims. It was these figures that were subsequently used by the Department of Energy to refuse funding for cold fusion and by the U.S. Patent Office to refuse patent applications. And it is these figures that are used around the world to silence supporters of cold fusion.

But MIT insiders, such as Dr. Eugene Mallove, were deeply suspicious of the published results. It is usual for experimental data to be manipulated, usually by computer, to compensate for known factors.

No one would have been surprised to learn that MIT had carried out legitimate "data reduction." But what they had done was selectively to shift the data obtained from the control experiment, the tap water cell, so that it appeared to be identical to the output from the fusion cell.

When this fudging of the figures became public, MIT came under fire from many directions, including members of its own staff. Eugene Mallove announced his resignation at a public meeting and submitted a letter to MIT accusing them of publishing fudged experimental findings simply to condemn cold fusion. A number of critical papers were published in scientific journals culminating in the paper published by *Fusion Facts* in August 1997 by Dr. Mitchell Swartz in which he concluded,

> What constitutes "data reduction" is sometimes but not always open to scientific debate. The application of a low pass filter to an electrical signal or the cutting in half of a hologram properly constitute "data reduction" but the asymmetric shifting of one curve of a paired set is probably not. The removal of the entire steady state signal is also not classical "data reduction."

In the restrained and diplomatic language of scientific publications this is as close as anyone ever gets to accusing a colleague of outright fiddling of the figures to make them prove the desired conclusion.

Beleaguered and under fire from every quarter (except the other big hot fusion laboratories who simply became invisible and inaudible) MIT backed down. It added a carefully worded technical appendix to the original study discussing the finer points of error analysis in calorimetry.* It also amended its earlier finding of "unable to reproduce Fleischmann-Pons" to "too insensitive to confirm"—a rather different kettle of fish.

Although MIT changed its story, it was its original conclusion that stuck, both in the public memory and as far as public policy was concerned. The *coup de grace* was delivered to cold fusion when the U.S. House committee formed to examine the claims for cold fusion came down on the side of the skeptics. "Evidence for the discovery of a new nuclear process termed cold fusion is not persuasive," said its report. "No special programmes to establish cold fusion research centers or to support new efforts to find cold fusion are justified."

Just where does cold fusion stand four years after the original announcement? The position today is that cold fusion has been experimentally reproduced and measured by ninety-two groups in ten countries around the world. Dr. Michael McKubre and his team at Stanford Research Institute say they have confirmed Fleishmann-Pons and indeed say they can now produce excess heat experimentally at will. Many other

---

* Measurement of the amount of heat absorbed or released in a chemical reaction.

major universities and commercial organisations have also confirmed the reality of cold fusion. U.S. laboratories reporting positive results include the Los Alamos National Laboratory, Oak Ridge National Laboratory (these were the two U.S. research establishments most closely involved in developing the atomic bomb), Naval Research Laboratory, Naval Weapons Center at China Lake, Naval Ocean Systems Center and Texas A & M University. Dr. Robert Bush and his colleagues at California Polytechnic Institute have recorded the highest levels of power density for cold fusion, with almost three kilowatts per cubic centimetre. This is thirty times *greater* than the power density of fuel rods in a typical nuclear fission reactor. Overseas organisations include Japan's Hokkaido National University, Osaka National University, the Tokyo Institute of Technology, and Nippon Telephone and Telegraph Corporation, which has announced that its three-year research programme has "undoubtedly" produced direct evidence of cold fusion. Fleishmann and Pons are working for the Japanese-backed Technova Corporation, a commerical cold fusion company based in France. Eugene Mallove left MIT to become editor of *Cold Fusion* magazine.

The Japanese government, through the Ministry of International Trade and Industry (MITI) has announced a five-year plan to invest $25 million in cold fusion research. The Electric Power Research Institute (EPRI) in California has spent some $6 million on cold fusion already and budgeted $12 million for 1992. In addition, a consortium of five major US utility companies have committed some $25 million for EPRI research.

Some of these research funds are being spent not only on developing a large-scale reactor vessel for use in public utilities but also, because of the inherent simplicity and relative safety of cold fusion, the development of a cheap miniature version for use in the office and even in the home. Even as Harwell and MIT proclaim their impossibility, prototype ten kilowatt cold fusion heating devices are already under test and are likely to find their way to market in the near future.

It is not only the organizations with a vested interest that come out badly from the story of cold fusion. The press, especially the scientific press, has acquitted itself poorly. *Nature* magazine showed how reactionary it can be with coverage that ranged from knee-jerk hostile to near hysterical. Its most intemperate piece was an editorial column in March 1990 headlined "Farewell (not fond) to Cold Fusion," which described cold fusion as "discreditable to the scientific community," "a shabby example for the young," and "a serious perversion of the process of science."

Some sections of the national press were also quick to ridicule Fleischmann and Pons and wrote pieces that have now come back to haunt their consciences. Steve Connor, writing in the *Daily Telegraph*, said that "the now notorious breakthrough in 'cold fusion' only two months ago astonished scientists worldwide, promising a source of limitless energy from a simple reaction in a test tube. Mounting evidence suggests the whole notion is a damp squib." Connor went on to ask "how two respected chemists could apparently make such a blunder?" He provides an answer with the suggestion that Fleischmann and Pons were the victims of "pathological science"—cases where otherwise honest scientists fool themselves with false results.

It is, of course, always fun to read about a good scandal, especially when the detractors who are so free with scorn get their come-uppance so poetically. But the aspect of the cold fusion affair that interests me most is why—exactly why—some scientists felt an overwhelming need to suppress it, even to the extent of behaving in an unscientific way and fudging their results. Money is the most obvious answer, but somehow unsatisfying; they may well have wanted the big research funds to continue to roll in year after year, but that cannot be the whole story. By enthusiastically embracing this possible new field, any of the world's fusion research organizations could have increased their research funds, rather than lost anything.

Injured pride is also plausible—men and women are often driven to extremes of behavior by such feelings, even including murder and suicide. But it is hard to see exactly how and why the feelings of hot fusionists should be so hurt by a simple scientific discovery.

Some interesting clues to this extraordinary behavior come from examining the reasons that several of the institutions themselves gave publicly for wanting to suppress such research during the development of the affair.

The first sounds perhaps the most reasonable. John Maple, a spokesman for the Joint European Torus project at Culham, Oxfordshire, the world's biggest fusion research centre, told the *Daily Telegraph* that a discredited cold fusion might produce a backlash that would damage the funding prospects of hot fusion.

> People in the street often don't know the difference. They confuse cold fusion, which we think will never produce any useful energy, with the experimental work we are doing at Culham, involving temperatures of hundreds of millions of degrees, which is making spectacular progress.

These sound [like] very understandable fears, but look a little closer at the logic underlying them. The people in the street (that's you and me) "can't tell the difference." The difference between what? The difference between hot fusion (which is real) and cold fusion (which John Maple and his colleagues say is not real). But surely, the issue is not whether we, the public, can tell the difference between a nuclear process that is real and

Cold Fusion—Investigations Continue Despite Ridicule From Skeptics

Cold fusion work continues. *Technology Forecasts & Technology Surveys* reports that, in spite of allegations that there is nothing to the observations, a number of labs continue to be intrigued by the unexplained parts of the phenomenon. They report that 50 U.S. labs and 100 labs in other countries are running tests, 60 groups in ten countries have reported results, some of the groups have claimed observation of more than one of the three generally accepted requirements for nuclear fusion, and some tests have produced as much as 600 times more heat than would be accounted for by the input of electrical power.

—Technology Forecasts & Technological Surveys, Vol. 22, No. 9, page 11

one that is not, but whether we, the public, should be asked to entrust millions of pounds of research funds to people who appear resistant to accepting the reality of a process such as cold fusion, for which there is substantial evidence and which may in the long term produce energy far more cheaply than the hot fusion process.

At quite an early stage in the affair, Harwell nuclear research laboratory began to worry about fusion becoming the province of every man. Members of the public were apparently telephoning Harwell and asking for advice on how to perform cold fusion experiments. "I have had many odd calls from people," a spokesman told the *Daily Telegraph* in April, "saying they are going to set it up at home to make it work. One housewife claimed that she already had supplies of heavy water and was asking me for details of how to set up the experiment. I had to tell her it would be extremely unwise." The paper then costed the experiment at £28 [$44.80] for some platinum, £31 [$49.60] for the palladium, £6 [$9.60] for some lithium chloride and £18 [$28.80] for the heavy water. With a few pounds for batteries, test-tubes and the like, the total could come to as little as £90, leading the paper to suggest that concern was mounting for the "retired professors, cranks and housewives" who they thought might be joining the race to produce fusion on their kitchen tables.

It is, of course, touching for Harwell to be so concerned about the safety of the man and woman in the street, but I see another worrying part of the explanation in this amusing reaction. Anyone who interests themselves in cold fusion is immediately labeled as belonging to a group that has either lost its marbles or never had any in the first place—"retired professors, cranks and housewives." Since we, the people in the street, pay many millions each year to fund Harwell, it seems not unreasonable that members of the public should be able to telephone to enquire on scientific matters without being ridiculed, patronized or told, in effect, to mind their own business.

It was not long before Europe's most senior fusion scientist, Dr. Paul Henri Rebut, director of the JET laboratory at Culham (cost, £76 million [$121 million] a year) was offering a word of advice to the man and woman in the street while also, curiously, disclaiming any supernatural powers. "I am not God, and I don't claim to know everything in the universe. But one thing I am absolutely certain of is that you cannot get a fusion reaction from the methods described by Martin Fleischmann and Stanley Pons."

Dr. Rebut clinched his argument with a single decisive stroke. "To accept their claims one would have to unlearn all the physics we have learnt in the last century." Well, we certainly wouldn't want one to have to do that, would we?

Equally illuminating were the remarks of Professor John Huizenga, who was co-chairman of the US Department of Energy's panel on cold fusion and who came down against the reality of the process. In a recent book on the subject, Professor Huizenga observed that "the world's scientific institutions have probably now squandered between $50 and $100 million on an idea that was absurd to begin with."

The question is, what were his principal reasons for rejecting cold fusion? Professor Huizenga tells us: "It is seldom, if ever, true that it is advantageous in science to move into a new discipline without a thorough foundation in the basics of that field."

When you consider that his committee's sole function was to advise whether or not research funds should be spent to investigate an entirely new area of physics/electrochemistry, and that this statement is one of his principal reasons for deciding *not* to invest such research funds, his remarks take on an almost Kafkaesque quality. It is unwise to invest research funds in any new area, unless we already have a thorough foundation in the basics of that new area? How could anyone ever get any money for research out of Professor Huizenga's committee? By proving that they already know everything there is to know?

Cold fusion is the perfect exemplar of the taboo reaction in science. It runs entirely counter to intuitive expectation produced by the received wisdom of physics; it is a discovery by "outsiders" with no experience or credentials in fusion research; its very existence is vehemently denied, even though Fleischmann and Pons have demonstrated a jar of water at boiling point to the world's press and television; and it is inexplicable by present theory: it means tearing up part of the road-map of science and starting again—"unlearning the physics we have learnt."

42 Archie Blue

Years ago, long before the advent of magnetic tape, Archie Blue devised a way of recording on steel wire.

He applied for patents in America, Britain and New Zealand. "Got them okay," he says, turning a screw which tightens a wire on his latest invention.

"I went to the States and tried to get the Victor Talking Machine Company (later RCA Victor and then HMV Victor in England) interested.

"But it wasn't a commercial proposition. You could only record the one wire at a time, whereas with discs they could make as many as they wanted off the one die.

"So I left it. If I'd worked on it further, I would have come up with the tape," he says with a laugh.

There's a fair bit of humour deep down inside Archie Blue, an inventor since he was about nineteen. An electrician by trade, he's applied for dozens of patents, including one which was granted him and his friend Ross Wood in 1939 for "improvements in or relating to TV or like apparatus."

Another, which has been used a fair bit, was for a round corrugated disc to keep speaker cones in shape. He sold the design for an automatic switch to an American company, years ago.

There were other things, too. So long ago it's hard to remember them all without getting all the papers. "I've made hardly anything from my inventions," he says.

But he hopes to make something from his latest, a device originally intended to be a source of cheap fuel . . . He's applied for separate patents for that.

"When I started work on it, four or five years ago, I was investigating the use of hydrogen as fuel for a heater. And then I thought "I might as well try to drive a car with it.'"

From *The Sunday Times*, May 14, 1978

All the evidence says Archie Blue's theories are correct. He says he's proved they are. He came back to Christchurch from Guernsey just before Christmas. There, helped by three retired millionaire friends, he had fitted his device to a small van and driven it around, using only water for fuel. The sceptics had a field day. Scientists admitted it was possible to get the hydrogen from water and use it as a fuel but the cost and equipment needed made it completely impractical. Newspaper reproters, as is their wont, made Archie headlines.

Several, though, took his claims seriously. The motoring man for the Daily Mail, Michael Kemp, made two trips to Guernsey to satisfy his curiosity. He didn't get far the first time. But his second visit of three or four days dispelled initial scepticism.

He reported on the paper's motoring page on August 19 last year that he drove the van himself, in normal traffic, at speeds up to 35 miles an hour.

Until the air blower burned out, the engine was "lively and powerful," he wrote.

The Royal Automobile Club man on the island, one David Hooper has taken a keen interest in all the proceeding and is convinced of its success.

Since his return Archie has worked steadily away in the cramped shed which is his workshop to make a similar device to show New Zealand.

Amid boxes of tangled wires, innards of old radios, bits of television sets, gramophones, Archie has soldered and welded. He's cut copper piping, fitted it to a large jar, set it up with the aircleaner and pump on a base. That little red pump will, say Archie, eventually blow hydrogen through what was once a conventional carburettor [sic]—now cut down, float removed, new controls affixed. Hydrogen is produced in the jar by electrolysis. "It takes very little juice, about $1\frac{1}{2}$ amps," he says, stopping to point out the virtues of his modern multi-function lathe.

It reminds him of the time he lived and worked in New York. In the late fifties, he says, taking off his fur cap to scratch a smooth head, he went to America with one of his inventions. He was working with a German who offered to get him a job so he could stay on and work on the thing. ("It never came to much.")

"It was a machine shop job, turning out one small aeroplane part only.

"The factory was going broke, but the bank kept it going until the contract was over. Then they sold the lot, lathes and machinery went for next to nothing."

With what money he had saved, he rented a general store on 8th street, "just down the road from 14th Avenue." The rent was too high, though, and he barely made a living. After three or four years, he came home.

Home to Christchurch, where Archie Blue was born nearly 74 years ago. Educated at Sydenham Primary, later Christchurch Tech. After working for a while for the Post Office, he took up an apprenticeship with the M.E.D. where he had all the wries he wanted to work with. "You do the lot, switchboards, meters, wiring before you get your ticket."

Later he would move on to State Hydro and the Railways as an electrician in the signals division.

Never had a day's serious illness in his life, he says. Still he was turned down for overseas service with the Army during the war. He served with the Home Guard and later, when he was with the Railways, he did territorial service, going into camp for a fortnight or so each year.

He was attached to a battery as a signaller. More wires as he set up communications between the guns and observation posts and the ike. He says the worst part of all was being called out in the middle of the night to run out some wires.

"Most of the time, I'd get up when called and then head straight back into the tent." He says he wouldn't have liked the job in a real war . . . "right out there, under fire from both sides."

The life story momentarily forgotten as Archie spies an old film projector poking out from under the rubble. He's fixed it with an amplifier so it can take "talking" films. Then an explanation of the secrets of an even older magic lantern.

Strong hands rub a bewhiskered face, lift the glasses over the forehead. Just about time to get the tea on. Archie lives alone in his conventional weather-board home in a typical Spreydon street. Hid wife died about two years ago, just before they were to set off for Guernsey where Archie could work with his "retired millionaires."

In we go through the cluttered porch, resting place for the moulds from which Archie Blue makes plaster figures and ornaments . . . witches, dog's heads, reclining ladies, classic heads, wall hanging-type things, clowns, Snow White, even. It's a hobby he took up while in America. Now the painted models—and a lot as yet untouched by the brush—take up space in every room of the house.

He finds making the pieces and then painting them restful. "Even when I'm taking a break, I've got to do something. I'm not an idle person."

Amongst all the models in the lounge, a huge silver trophy is proof of the young A. H. Blue's athletic prowess. He won it at a long-gone South Island championship meet at which he won the 100 yards, the 220 and the 440 yards. ("They don't have them now, do they?")

Stuffed inside with a lot of other clippings there's a faded piece cut from the *Sun* (or was it the *Star*?). Browned and fragile now, the paper suggests that with a coach, Archie Blue had a great future in the sport.

"But I had too much on my plate to take it up seriously," he says. Inventing.

Any money he might eventually get won't mean all that much to him. He says he could take another trip (he's off to Guernsey again soon anyway) but he'll always come back to Christchurch. That's where his family is. A daughter lives just along the road. He has four grandchildren.

He reckons Guernsey is a dead place most of the year; New York is far too cold in the winter.

After a couple of postponements, Archie Blue hoped to have his device ready for its New Zealand debut some time this week. He's not too fussed by all the publicity, but says that because TV and the papers have asked him to, he cooperates. "But I'll take my car out here without an audience first. Just to make sure all's well," he says.

And when it's all over, he'll be off again.

Could the device be manufactured here?

It could, but "we're not bothering to," he says. "I can't sell the rights here, or market it without my partners. And they don't want to mess about with small concerns. It's too big for that."

What about using his partners' resources to start a factory themselves?

"That would cause far too many headaches. For a start, look at what strikes have done to the big motor firms like Leyland.

"Young men can take these things. It would be pretty hard for us at our age."

43 The Story of Francisco Pacheco and the Suppression of Hydrogen Technology

Karin Westdyk

Francisco Pacheco's patent (#5,089,107) for his invention could revolutionize the field of energy.

The Pacheco Bi-Polar Autoelectric Hydrogen Generator is a unique system which separates hydrogen from seawater (the element's natural storage tank) as it is needed for use. The patent teaches the on demand autoelectrolytic separation of 99.98 percent pure hydrogen from seawater at both electrodes of the generator, and the simultaneous use of the hydrogen's carried energy.

Research and development of hydrogen as an energy source, till now, has been blocked by several factors:

1. It is an extremely volatile element and subject to explosion as happened in the tragic Hindenberg accident and the Challenger;

2. The existing highly pressurized, cryogenic and hydride systems available for storing it are very expensive, cumbersome, and dangerous (there is no need for storage with the Pacheo system);

3. The United States Department of Energy is not interested in promoting or developing new energy sources that compete with the powerful energy monopolies now in place.

Hydrogen is the cleanest burning fuel. When burned, its waste is clean water vapor which can either be recycled back in the system for reuse, or safely released into the environment and returned to the oceans, lakes and rivers—no greenhouse gasses, no atomic pollution, no acid rain, no crippling dependencies on foreign oil, no expensive transportation, or power lines.

The oxides of the two metals used in the system which produce hydrogen at both electrodes are also recyclable (with a minimal 0.25 percent loss). The metals can be produced for use in the system from existing scrap metals actively seeking markets. The infrastructure necessary to develop this clean, safe, and efficient alternative energy source is already in place.

Pacheco has built prototypes which have successfully fueled a car, a motorcycle, a lawn mower, a torch, and a boat (with the ocean serving as its inexhaustible fuel tank). Another prototype system in demonstration energized an entire home in West Milford, New Jersey, providing electric energy and fuel for cooking and heating. In addition, Mr. Pacheco demonstrated his generator to the scientific community at the 1990 Eighth Annual International Hydrogen Energy Conference where he was the only exhibitor actually producing hydrogen. His generator was also exhibited in Canada at the 1990 Green Energy Conference, and at many other notable conferences where he received several awards for his work.

HISTORY OF THE GENERATOR

As a young man in his native country Bolivia, Pacheco was fascinated with the idea of developing a super battery. While experimenting in his makeshift laboratory, he lit a match and the bubbles forming in one of the beakers ignited and blew a hole in the ceiling. He knew that he had made a discovery but was not sure what he had discovered. He abandoned his work with the battery and proceeded to develop his hydrogen generator.

In 1943, while on a Good Will Tour of South America, the Vice President of the United States [Henry Wallace] witnessed the Pacheco generator running an automobile. Wallace invited him to bring the generator to Washington where, later in that year, he demonstrated it to scientists and representatives from the U.S. War Department at the Bureau of Standards. He applied for a patent, but because the United States was at war, all patents were sealed and available only to the military. Later, he was advised to shelve his patents because, at that time, oil was plentiful and cheap and there was no need to develop an alternative source of energy.

Pacheco became a U.S. citizen and brought his family to his newly adopted country, knowing that one day, the time would be right for his invention. He worked in defence plants during the war, and then, until retirement, as a heating engineer in New York City. He discovered the beauty of West Milford, New Jersey, while on a family outing and returned for vacations whenever he could. In 1967, he moved with his family to West Milford and made his home there until his death in 1992.

During the oil shortage in the 70s, Pacheco decided it was time to apply for his patent again and received a U.S. and several foreign patents. But, he soon learned that neither the energy industries, the heavily subsidized utilities, nor the Department of Energy were interested in developing clean, abundant, safe energy from hydrogen.

Determined to bring his invention to the people, he built prototypes and demonstration models to show government and industry officials. Many came and saw, said they were impressed, promised to help, but none ever did. In 1974, with the hopes of acquiring government backing, Pacheco demonstrated his pollution-free hydrogen fuel cell to Congressman Robert Roe, who today speaks often of the wonders of hydrogen fuel. With no outside power source, the self-taught chemical engineer connected the fuel cell to an alternator unit with a 3 horse power, 1000 watt generator with a 4 stroke engine. The demonstration was a success and the excited congressman promised to bring it to the attention of Washington officials. Upon leaving Roe's Paterson office, Pacheco invited him to participate in another demonstration at the Jersey Shore. Roe was invited to take part in a history-making voyage, the first power boat ride "fueled by seawater." Many newspapers were invited as well. But, Roe never showed up, nor did very many newspapers. Pacheco never heard from the Congressman again, but his voyage was a great success. History was made on July 27, 1974 when a 26 foot in-board power boat ran for nine hours using the Pacheco Generator and seawater for fuel, putting back into the ocean, as its waste, clean water.

In an effort to overcome the skepticism he was facing and the Ph.D. he could not add to his name, Pacheco had his invention analyzed by several independent laboratories. It passed all tests but when he tried to introduce it to the automobile industry and the oil companies, the response was either cool or non-existent. After a two hour meeting with one of the oil companies, he was told that developing the generator would be against their interests.

In 1977, Pacheco built a prototype unit which provided hydrogen, electric and thermal energy for a 1000 square foot home in West Milford. The New Jersey Commissioner of Energy, and several of his staff members came to see and were impressed. The Commissioner wrote a letter of recommendation to the Department of Energy, but again nothing happened.

In an effort to bring the generator to the public's attention, Pacheco contacted Geraldo Rivera who expressed great interest after he had read about the power boat demonstration. Rivera wanted to do a TV show about the generator, but the idea was axed by the station.

It was during this time that Pacheco received some recognition for his work at the International Inventor's Exposition. He was the recipient of a plaque and award from the Commissioner of the Patent Department and two consecutive Hall of Fame awards from the Inventor's Club of America in 1978 and 1979.

In 1980, Pacheco was contacted by *60 Minutes*, who promised to help him show his invention to millions of Americans. The *60 Minutes* crew arrived in West Milford and taped the generator producing hydrogen fuel for a bunsen burner, and for a torch which cut through a $\frac{3}{4}$-inch-thick steel plate (indoors). The hydrogen gas inflated a balloon, and produced energy to run an electric motor. The last of his demonstrations involved running a lawnmower with the fuel. Because he was going to be on television, at the last minute, he decided to buy a new one, and did not have time to test it out. The engine choked due to the excessive amount of gas being produced but the *60 Minutes* crew assured him that they had enough material to present an entire show with the successful demonstrations. Later when the show was aired, Pacheco was devastated as the show had a completely different focus. The only demonstration aired was the lawn mower, and it was used to provide an example of an independent inventor's non-working invention.

In 1986, with increasing concerns about the environment, Pacheco wrote to the Department of Energy about the generator but received only a fact sheet in response which provided information on the drawbacks of hydrogen fuel based on the problem of storing it in liquid or gas form. He wrote back explaining that with his system, there is no need to store the hydrogen as it is produced on demand. His detailed response was ignored.

In 1989, after information about the generator was presented at a United Nations Environmental Conference, Pacheco was invited to exhibit a prototype in Canada at the Green Energy Conference. Subsequently, he was asked to participate in the International Hydrogen Energy conference held in Hawaii in 1990. Encouraged by the interest from several scientists, he applied for the new patent which he received in February, 1992.

Though the history of his technology is most intriguing, its future is more important. Its potential clean, renewable, and safe energy source represents long-sought solutions for the environmental degradation caused by existing energy sources. Although Mr. Pacheco had been discouraged and frustrated in his efforts to bring this technology to the people, he remained focused and committed to the end. He strongly believes in the words of an old wise man who once told him, "Son, God put on your shoulders something very big, do not ask yourself 'Why me?' Think, 'Why not me?'"

Edmund Pacheco now owns the rights to his grandfather's patent, which will one day establish Francisco as the true grandfather of the coming hydrogen energy revolution.

44 Amazing Locomotion and Energy Systems Super Technology and Carburetors

John Freeman

The prophecies of our science-fiction writers
have proven more accurate than the
expectations of our scientists and statesmen.
Lord Bertrand Russell

The more radical concepts in this work have good company in the "Buck Rodgers" of yesterday . . . yet they too will be but "tinker toy" technology to the material changes of the future. Some of the more radical concepts here may be incorrect . . . but the goal is the thing of importance. References to some of the exotic technology of the past has been included to help kindle an interest in these areas. In the recurring cycles of life, know that legends will live again and today's dreams will become the reality of our tomorrows. The "when" will be up to you.

SUPER MILEAGE AUTOS AND FUEL SYSTEMS

Carburetors

The most productive inventor in the field of carburetion was probably G. A. Moore. Out of some 1,700 patents that he held, 250 of them were related to the automobile and its carburetion. While industry today relies on his air brakes and fuel injection systems, it has completely ignored his systems for reducing pollution, gaining more mileage, and improving engine performance in general. As far back as the mid 20s, Moore's systems were found to be capable of virtually eliminating carbon monoxide pollution. Persons involved in the automotive field viewed Moore as an authentic genius and could not understand why the industry ignored his advanced automotive designs. (Seventeen of his patents are reprinted in *The Works of George Arlington Moore*).

480

The Bascle Carburetor

[The Bascle carburetor] was developed and patented in the mid 50s. It supposedly raised mileage by 25 percent and reduced pollution by 45 percent. Its inventor, Joseph Bascle, was a well known Baton Rouge researcher who remodified every carburetor in the local Yellow Cab fleet shortly after his arrival there. In the 1970s he was still optimistic and hinted that the time had come for selfishness to be put aside in regard to fuel systems.

Kendig Carburetors

In the early 70s a small concern in the Los Angeles area turned out a number of remarkable Variable Venturi Carburetors. Most of these were hand made for racing cars. Buying one of their less sophisticated prototypes, a young college student mounted it on his old Mercury "gas hog." Entering it in a California air pollution run, the student won easily. Not only did the carburetor reduce pollution; it gave almost twice the mileage. Within the week the student allegedly was told to remove his carburetor—it was not approved by the Air Resources Board. Due for production in 1975, the simpler Kendig model has yet to be produced.

Super Carburetors

In the late 30s there was an inventor in Winnipeg, Canada, who developed a carburetor which got at least 200 miles per gallon by using superheated steam in its system. C. N. Pogue was quite open about this work until very professional thefts indicated his invention was in danger. Local papers of the time stated that his various backers declined many outside offers they received, and, toward the end, used as many as five guards protecting their interests. What eventually happened is still unclear.

In the early 40s there was another inventor who developed a design that cost him many years of heartache and "dead ends." John R. Fish was cut off from every direction, and when he finally resorted to selling his carburetors by mail, the post office stopped him. In tests by Ford, they admitted that his carburetors were a third more efficient than theirs, yet no one helped. As late as 1962 Firehall Roberts used a "Fish" on his winning Indianapolis 500 car.

The Dresserator

In Santa Ana, California, Lester Berriman spent five years designing a pollution reducing carburetor for the Dresser Company. Basically, the Dresserator is able to keep the airflow through its throat, moving at sonic speeds even at small throttle openings. By allowing super-accurate mixture control the device could run a car on up to a 22-to-1 mixture [of air to fuel]. Test cars passed the pollution control standards with ease and got a typical 18 percent mileage gain, besides.

Holley Carburetor and Ford signed agreements to allow them to manufacture the carburetor in 1974.

Water-to-Gas Conversion Powder

One of the most controversial figures of his kind was Guido Franch. In the 70s he created a sensation when he began demonstrating his water-to-gas miracle. Chemists at Havoline Chemical of Michigan and the University [of Michigan] were among the first to test his fuel. According to both, it actually worked better than gasoline. According to Franch, his secret lay in using a small quantity of "conversion powder" which was processed from coal. He stated that he processed coal in a series of barrels containing liquid. Supposedly, as the "processed" coal sank to the bottom, a greenish substance rose to the top. It is this residue that was dried into the mysterious "conversion powder"! Franch said he learned the formula from a coal miner, Alexander Kraft, over 50 years before. While it cost Franch over a dollar a gallon to make his fuel in small quantities, he claimed that it could be produced for a few cents a gallon if mass produced.

A number of private groups tried to deal with Franch for his formula. According to some, the inventor was just too difficult to deal with, and there was just too much gamble involved for the concrete facts they got. Franch continued to put on his demonstrations for years and claimed the auto manufacturers, Government, and private companies just weren't interested in his revolutionary fuel.

Burn Water

Back in the 1930s a number of the early tractors squeezed great economy from a number of simple adaptions. Some simply used a heated manifold to further atomize the gas; others used cheaper fuels. The Rumley Oil Pupp tractor had a carburetor with three chambers and floats in it. One was used for gas, one for kerosene, and the third was for water. After owners started the tractor on gas, they simply switched over to a cheaper mixture of kerosene and water.

With the advent of ultra-sonic devices there were a number of researchers in the early 70s who successfully mixed up to 30 percent water in gasoline—and used to run their automobiles.

Some disgruntled motorists just "spudded" into their carburetors—ran a hose to a container of water and let their engines suck in an extra water ration. Experts claimed this could damage valves if cold water hit them, but few seemed to have trouble.

Gas and Water Mix

In the mid-70s a Dr. Alfred R. Globus of United International Research presented his Hydro-fuel mixture concepts at a meeting of petroleum refiners in Houston, Texas. According to reports, this fuel was a mixture of 45 percent gasoline, 50 percent or more of water, and small percentages of crude alcohol and United's "Hydrelate." This latter chemical was a bonding agent which kept the fuel's ingredients mixed. Even though it was estimated that a hundred million gallons of gas a day could be saved through the use of this product, no one seemed to be interested.

Water and Alcohol Motor

A Paris engineer ran his private cars on a mixture of denatured alcohol and water according to the French magazine *Le Point*. The forty-nine-year-old inventor-mechanical engineer Jean Chambrin maintained that his motor design could be mass produced for only a fraction of the cost of present engines. As publicity surrounded his achievements the inventor took even greater precautions for security.

Super Mileage Additives

L. M. Beam, who had had his super mileage carburetor bought out back in the 20s, worked out a catalytic vegetable compound that produced much the same results. By rearranging the molecules of gas and diesel, he obtained better combustion, mileage, and emission control. At one cent a gallon he guaranteed his W-6 formula would save at least 10 percent in fuel costs. Refused and rejected by State and Federal certification agencies (Air Pollution and Environmental Pollution agencies), Beam was finally forced to survive in the mid-70s by selling his formula abroad.

The Lacco Gas Additive Formula

Eighty percent water, 15 percent gas, 5 percent alcohol, 2 percent lacco.

According to an article in the January 20, 1974 San Bernardino, CA, *Sun Telegram*, a man named Mark J. Meierbachtol of that city patented a carburetor which got significantly greater mileage than was usual. At this time the patent (#3,432,281 March 11, 1969) is being held by attorney T. F. Peterson for the inventor's widow, Ola.

Highway Aircraft Car

One of the more determined crop of radical auto designers was Paul M. Louis of Sidney, Nebraska. For many years he promoted aircraft design, streamlining to provide super economy in his proposed "Highway Aircraft." He called cars of current design "shoeboxes." His first attempt at marketing a car was in the late 30s. He was stopped by the Securities and Exchange Commission, and it was not until his company withered away that he was given a clean "bill of health." In the mid-70s at the age of seventy-eight he again tried to put his unique designs on the road.

Ultrasonic Fuel Systems

With the advent of the fuel crisis of 1973 there were a number of experimenters who found solutions involving the use of ultrasonic fuel systems. Much of this work involved using sonic transducers to "vibrate" existing fuels down to much smaller particles. This procedure simply increased the surface area of the fuel and made it work more efficiently. Using a magneto-strictive or piezo-electric vibrator, conical or cylindrical cones were used at from twenty to forty thousand vibrations per second. An increase in fuel mileage of at least 20 percent was expected of these units.

Eric Cottell was one of the first persons to proclaim the fact that water could be mixed with gas and used as fuel with these units. His customers had been using his commercial units to emulsify foods, paints, and cosmetics for some time. When the word suddenly got out that the super fine S-onized water would mix perfectly with up to 70 percent oil or gas, there was congratulations from many sides (June 17, 1974, *Newsweek*). Later there was nothing but silence again.

Later in 1975, Cottell was interviewed again and explained that Detroit was so myopic that they would probably turn down even the wheel if it were a newly offered invention. Because installations of his reactors was so simple, Cottell ran several of his own cars on a water-gas mixture. He explained that an ultra-sonic unit caused internal stresses so great in gasoline that the molecules can actually absorb water to become a new type of fuel.

Super Mileage from Fuel Vaporization

L. Mills Beam developed a simple heat exchange carburetor back in 1920. In principle it was nothing more than a method of using the hot exhaust gases of an engine to vaporize the liquid gas being burned. Using simple logic Beam reasoned that raw gas going through a normal carburetor simply could not be atomized with high efficiency. As a result there was a waste of fuel when microscopic droplets burned instead of exploded. This, of course, created unnecessary heat and inefficiency.

Since he was easily able to double and triple the gas mileage of the cars he tested, it was not long before Beam was offered a settlement and percentage fee for the rights to his device. Accepting the offer, he never again saw any attempt to market his device or the parties who gained control of his device.

In his "Suppressed Inventions," Mike Brown spoke with Mr. Beam and found that the shadowy trail seemed to lead to a major oil company—but, of course, little could be proved.

Brown tells of a later device which used the same principle. John W. Gulley of Gratz, Kentucky, could supposedly get 115 miles per gallon out of his big 8-cylinder Buick, using his vaporizing arrangement. Typically, this device was assured of obscurity when Detroit interests bought it in 1950.

In the early 70s there was a device made by Shell Research of London that was a bit more sophisticated in design. Vaporizing the gas at around 40°C, a certain amount was allowed to go around the vaporizer to reduce pressure losses. The "Vapipe" unit was supposedly not marketed because it did not meet Federal emission standards.

Another advocate of vaporizing gasoline is Clayton J. Querles of Lucerne Valley, California. According to the *Sun-Telegram* of July 2, 1974, Querles claimed that he could easily develop an engine which could run all day on a gallon of gas. This inventor claims that all he needs to produce such a carburetor is money from an honest backer.

This same inventor said that he took a 10,000 mile trip across the country in his 1949 Buick for ten dollars worth of carbide. Building a simple carbide generator, which worked on the order of a miner's lamp, he said that a half pound of acetylene pressure was sufficient to keep his car running. Because acetylene was dangerous, he put a safety valve on his generator and ran the outlet gas through water to insure there would be no "blow back."

The Alexander Fuelless Car System

Robert Alexander and a partner spent only forty-five days and around five-hundred dollars to put together a car that confounded experts. A small 7/8 twelve-volt motor provided the initial power. Once going, a hydraulic and air system took over and actually recharged the small electric energy drain. The Montebello, California inventors were, at last reports, very determined that the auto industry would not bury their "super power" system. What happened? (U.S. PAT #3913004)

One inventor in the 20s used an electric car which ran off high frequency electricity which he generated at a distance. Using principles similar to the Tesla's ideas, he simply broadcast the re-radiated atmospheric energy from a unit on his house roof.

Henry Ford, acting for himself and the other Detroit oil "powers," quickly bought and quietly shelved this invention. (BSRAJ M-J 1973) John W. Keely reportedly used harmonic magnetic energies from the planet to run his mysterious motor. Later, Harold Adams of Lake Isabella, Ca., worked out a motor thought to be similar to Keely's. In the late 40s it was demonstrated for many persons, including Naval scientists. After a round of "dead ends," it, too, vanished into the pages of the past.

Water to Hydrogen Fuel

The process of converting water to hydrogen has long been known, and the standard electrolysis method was developed back at the turn of the century. The only trouble has been that it takes a great deal of electrical current to convert the water over. With a 40 percent efficiency at best, a lot of people were hoping for the advent of cheap fuel cells which would convert the hydrogen and oxygen to electricity at a much higher efficiency. The standard procedure for the electrolytic extraction included using platinum electrodes in an acidic water solution—with at least 1.7 volts of direct current.

What gives many hope are reports of early experimenters who overcame the conversion problems. W. C. Hefferlin wrote of using a superior conversion method back in 1921. According to the reports, he worked out a method which used a high frequency current passing through steam. Being associated with some unusual projects made him suspect to a degree . . . but there are some who feel he put his discovery to good use despite the continued rejections he faced.

Hydrogen is probably one of the most ideal and easily adaptable forms of fuel that we could ask for at the present. Because it returns to water after it burns in the presence of oxygen, it is also pollution-free, and a joy to work with. Regardless of adverse criticism, it has been proven to be less expensive and dangerous than ordinary gasoline when used in automobiles.

An experiment was mentioned in the *Alternate Sources of Energy Journal* in which a couple of readers ran a car for a short period of time on chemically made hydrogen. Tossing some zinc in a bottle of water and acid (any strong acid), they captured the hydrogen given off in a balloon and manually fed it to their auto later.

Actually, feeding hydrogen to a standard auto engine can be a little involved, depending on one's source. I recall a group of California experimenters who fed their old Model A Ford on straight "tank gases" with not much more than some gas pipe plumbing. Later they developed a more sophisticated (oxyburetor) and allowed their motor suction to feed the correct hydrogen-oxygen mixture. To start the engine on these gases, they allowed the hydrogen to be sucked in first. Later they were in need of a variable Venturi carburetor to aid this procedure.

It is interesting to note that Deuterium, or "Heavy Hydrogen," is what powers the H Bomb. A pound of this fuel at less than a hundred dollars (recent estimates) will produce the power of $75,000 worth of fossil fuels.

The proposed methods of producing cheap deuterium now have already become details of the suppressed past.

A classic case of the "water to auto engine" system was that worked out by Edward Estevel in Spain during the late 60s. This system was highly heralded, then sank among other such "high hope" hydrogen systems. Foul play? Who knows!

Hydrogen Generator

Sam Leach of Los Angeles developed a revolutionary hydrogen extraction process during the mid-70s. This unit was said to easily extract free hydrogen from water and yet be small enough for use in automobiles. In 1976 two independent labs in L.A. tested this generator with perfect results. Mr. M. J. Mirkin who began the Budget car rental system purchased rights for this device and hoped to develop it—against the usual ridicule of a number of scientists. Leach, who was very concerned about his security, was said to be greatly relieved by Mirkin's aid.

Hydrogen Auto Conversions

Rodger Billings of Provo, Utah, headed a group of inventors who worked out efficient methods of converting ordinary automobiles to run on Hydrogen. Rather than rely on heavy cumbersome Hydrogen tanks, his corporation used metal alloys, called Hydrides, to store vast amounts of Hydrogen. When hot exhaust gases passed through these Hydride containers, it released the Hydrogen for use in the standard engines. Billings estimated that the price might run around $500 for the conversions; gas consumption would be greatly reduced.

Because of the nature of this conversion, there even seemed to be favourable interest from various auto and petroleum interests in the mid-70s.

P.S.: In Florence, Italy, an inventor used a special tube to divide water in Hydrogen and Oxygen—without the usual electricity and chemical requirements (unconfirmed 1975 report).

Burning Alcohol

Around 1910 there were a number of automobiles burning alcohol, and for some years it was common to find data on burning it in the popular automobile manuals of the day. A number of carburetors were designed to use alcohol or alcohol and gas. In these earlier days, alcohol was almost as cheap as the various benzenes—or what we now refer to as gasoline. One of the drawbacks to burning alcohol during this early period was the fact that the engines didn't have enough compression to burn the fuel at high efficiency. Today's automobiles, then, are almost perfectly adapted to using not only the alcohol-gas mixtures but pure alcohol.

Over the years, racing car drivers used cheap methanol, or nonbeverage alcohol, in many racing cars, and only the availability of reasonably priced gasoline kept the practice from becoming more popular. In the gas crunch of 1973 only a few (old timers) remembered alcohol as a fuel. Reluctant as the oil companies were to recognise the fact, it remained that alcohol could be made cheaply and used without major problems.

MIT testing at Santa Clara, California, retraced the steps of conversions worked out sixty years earlier. First it was found that the carburetors needed to be heated to properly volatize the methanol. This was done by utilizing the exhaust heat or by running hot water to a jacketed carburetor. Next, because methanol conducts electricity, it can set up an electrolytic action which attracts many modern plastics and metal alloys. Gas tanks, for instance, would often fill with tiny metal particles which required large gasoline line filters to eliminate a plugged up carburetor. Other idiosyncracies included trouble with cars turned to conform to pollution control standards, and difficulty in starting without a heated carburetor.

In the early days a dual carburetor bowl allowed starting on gasoline, but MIT introduced a fog of propane from a small tank and valve, operated manually. In the case of a methanol-gasoline mixture, it was found that only cold weather hampered excellent mixing and performance.

A breakthrough at the Army's Nalick Laboratories in Massachusetts led many persons to believe that a cheap "methanol from waste system" was assured. In the early 70s they discovered and developed certain fungi which could convert a wide variety of cellulose into the sugars necessary for producing alcohol. Researchers felt that a ton of paper scrap, for instance, could produce over 65 gallons of high grade alcohol.

Air Powered Cars

Because air is non-polluting, and does not tend to heat nor contaminate engines it is used in, it is an ideal power source. The one major problem, however, has always been just how to store enough compressed air for lengthy travel.

Air has been used for years to power localized underground mine engines, and even a number of experimental "air autos" have been successful. In 1931, Engineer R. J. Meyers built a 114 pound, 6 cylinder radial air engine that produced over 180 horse power. Newspaper articles reported that the Meyers vehicle could cruise several hundred miles at low speeds. Compressed air stored as a liquid was later used on advanced air auto designs in the 70s. Vittorio Sorgato of Milan, Italy (Via Cavour, 121; 2003 Senago), created a very impressive model that was received with a great deal of interest from Italian sources.

One of the outstanding services for persons wishing to keep up with current scientific discoveries are the *Scientific American Reprints*. They are inexpensive and are listed on current order forms from The S. H. Freeman Co., 660 Market Street, San Francisco, CA, 94104.

While few renegade scientists cared to make themselves conspicuous by divulging "maverick" ideas or "hush-hush" projects, a number of small journals carried very revealing articles. Individuals daring to share data on faster than light radio, exotic space drives, nuclear fission, matter-space-and time theories, New Math, gravity concepts, etc., could often be contacted through current one dollar folios from the publisher.

The Electromatic Auto

Any mention that an electric car could be made which could regenerate its own power as it was driven was a joke to most "experts." Yet, in 1976, this author actually saw such a car function. Using various standard automobile parts and an electric golf cart motor, Wayne Henthron's first model functioned perfectly. Once this remarkable auto reached a speed of 20 miles per hour, it regenerated all of its own electricity. In normal stop and go driving it gave several hundred miles of service between recharges.

The secret to the system lay in the way that the inventor wired the batteries to act as capacitors once the car was moving. Four standard auto alternators acted to keep the batteries recharged. With little official interest shown in this remarkable system, the inventor became involved with other persons of equally far-sighted aims and resolved to make the car available to the public. (World Federation of Science and Engineering, 15532 Computer Lane, Huntington Beach, CA, 92649).

Mixing Water With Gas

Portugese chemist, John Andrews, gave a demonstration to Navy officials that proved his additive could reduce fuel costs down to 2 cents per gallon. It allowed ordinary gasoline to be mixed with water without reducing its combustion potentials. When Navy officials finally went to negotiate for the formula, they found the inventor missing and his lab ransacked. (Saga May, 1974).

INCREDIBLE AND UNUSUAL MOTORS

The Bourke Engine

Russell Bourke was probably one of the true geniuses in the field of internal combustion engines. Upon noting the incredible waste of motion in the standard auto engine, he set about designing his own engine in 1918. In 1932 he connected two pistons to a refined "Scotch yoke" crankshaft and came up with a design using only two moving parts.

For over thirty years this engine was found to be superior in most respects to any competitive engine, yet it was rejected by all of the powers that be. This amazing engine not only burned any cheap carbon-based fuel, but it delivered great mileage and performance. Article after article acclaimed his engine and its test performance results, yet nothing ever came of his many projects except frustration and blockage.

Just before Bourke's passing, he assembled material for a book, and *The Bourke Engine Documentary* is a most revealing work on engine design and on the Bourke engine in particular.

The LaForce Engine

Edward La Force struggled for years in Vermont to get backing to perfect his amazing engine. Ignored for years by the automotive industry, Edward and Robert, his brother, survived on the contributions of several thousand individuals who believed in them. His engine design manages to use even the harder to burn heavy gasoline molecules. Current engines are said to waste these, and, since they make up to 25 percent of the current fuels, the use of the heavy molecules was a great step forward.

According to a *Los Angeles Examiner* article (December 29, 1974), the cams, timing, and so on were altered on stock Detroit engines. These modifications not only eliminated most of the pollution from the motor, but, by completely burning all of the fuel the mileage was usually doubled. One Examiner reporter saw a standard American Motors car get a 57 percent increase in mileage at the Richmond, Vermont, research centre.

With such publicity, the EPA [Environmental Protection Agency] was forced to examine the situation, and of course, they found that the motor designs were not good enough. Few persons believed the EPA, including a number of Senators. A Congressional hearing on the matter in March 1975 still brought nothing to light—except silence.

The LaForces were interviewed by newspapers and auto manufacturers across the world, and even though they only modified the basic Detroit designs; Detroit was not interested. Anyone need 80 percent more mileage?

In his "Suppressed Inventions," author Brown tells of John Gulley of Gratz, Kentucky, who turned down a GM offer of 35 million dollars when they wouldn't guarantee to market his amazing magnetic engine. Gully built his first model from old washing machine parts, and the patent is still available from the patent office file.

Fuelless 15-Cents-Per-Hour Papp Engine

One of the most astonishing engine designs of the 60s was the Papp engine which could run on 15 cents an hour on a secret combination of expandable gases. Instead of burning a fuel, this engine used electricity to expand the gas in hermetically sealed cylinders. Far from being complex, the first prototype used a ninety horsepower Volvo automobile engine with upper end modifications. Attaching the Volvo pistons to pistons fitting the sealed cylinders, the engine worked perfectly and showed an output of three hundred horsepower. In a December 1968 *Private Pilot* article, the inventor, Joseph Papp, claimed that it would cost about twenty five dollars to charge each cylinder every sixty thousand miles. Subscribers couldn't help but wonder why *Private Pilot* soon changed hands, moved across the country, and failed to follow up on this project as promised.

Two Chamber Combustion

Because very lean mixtures of fuel do not ignite easily, there were numerous attempts at solving the problem with a separate and smaller compression chamber. By feeding gas separately to such a chamber, it could easily detonate the very lean mixtures in the larger chamber.

A patent in the early 20s covered this idea and Ford perfected the idea shortly after the war. It actually wasn't until the mid-70s that Honda of Japan used the design to make a joke of the various emission control efforts of the U.S. auto industry. (See numerous *Popular Science* articles, like 768.4.)

Salter's Ducks

While confined to his bed a couple of days, an Edinburgh professor doodled up a method of using ocean wave action to produce an amazing amount of electric energy. Large pods shaped something like a duck simply bobbed up and down in a pumping action that used 90 percent of the waves' energy. Scale models actually functioned perfectly and indicated that larger units should produce hundreds of kilowatts. (*Popular Science,* March, 1977.)

Water-Gas Mix (University of Arizona)

Marvin D. Martin told the press in 1976 that their University funded "fuel reformer" catalytic reactor could probably double auto mileage.

Designed to cut exhaust emissions, the units mixed water with hydrocarbon fuels to produce an efficient Hydrogen, Methane, Carbon Monoxide fuel. Letters to their Aero Building #16 Lab brought replies that indicated little of how the units functioned but gave indications that the hydrogen was responsible for the great efficiency.

From P.O. Box 3146, Inglewood, CA 90304 (1977).

Zubris Electric Car Circuit Design

In 1969 Joseph R. Zubris became disgusted with his ailing automobile and decided to gamble a couple of hundred dollars on putting together an electric car. Using an ancient ten horse electric truck motor, Zubris figured out a unique system to get peak performance from this motor; he actually ran his 1961 Mercury from this power plant. Estimating that his electric car costs him less than $100 a year to operate, the inventor was sure that larger concerns would be very interested, and he could hardly believe the lack of response he received from his efforts. In the early 70s he began selling licenses to interested parties at $500. Thirty-five small concerns were interested enough to respond.

The Zubris invention actually cut energy drain on electric car starting by 75 percent. By weakening excitation after getting started, there is a 100 percent mileage gain over conventional electric motors. The patent probably doubled the efficiency of the series electric motor. (Patent #3,809,978)

Electric Motor

One of the startling electric motors designs of the 1970s was the EMA motor. By recycling energy this astounding motor reportedly was able to get a better than 90 percent efficiency. Using a patented Ev-Gray generator, which intensified battery current, the voltage was introduced to the field coils by a simple programmer. By allowing the motor to charge separate batteries as it ran, phenomenally small amounts of electricity were needed. In tests by the Crosby Research Institute of Beverly Hills, California, a ten horsepower EMA motor ran for over a week on four automobile batteries.

Using conservative estimates, the inventors felt that a fifty horsepower electric car could travel 300 miles at 50 miles per hour without recharging. With such performance the engine could be applied to airplanes, cars, boats, and even electric generators.

According to Dr. Keith E. Kenyon of Van Nuys, California, he discovered a discrepancy in long accepted laws relating to electric motor magnets. When Dr. Kenyon demonstrated his radically different motor to physicists and engineers in 1976, their reaction was typical. They admitted the motor worked remarkably well but since it was beyond the "accepted" laws of physics they chose to ignore it. Because this system could theoretically run an auto on a very small electrical current, entertainer Paul Winchell saw a great potential and began to work with Dr. Kenyon. (Pat. pending.)

Diggs Liquid Electricity Engine

At an inventors workshop (I. W. International) an amazing electrical auto engine was shown by inventor Richard Diggs. Using what he called "liquid electricity," he felt that he could power a large truck for 25,000 miles from a single portable unit of his electrical fuel. Liquid electricity violated a number of the well known physical laws the inventor pointed out. Melvin Fuller, the expositions president, felt that this breakthrough would have a most profound effect upon the world's economy. Some speculated that it only could if . . .

In the June 1973 issue of *Probe* there was an article on an electromagnetic engine that was fuelless.

Magna-Pulsion Engine

A retired electronics engineer named Bob Teal of Madison, Florida, invented a motor which apparently ran by means of six tiny electromagnets and a secret timing device. Requiring no fuel, the engine of course emitted no gases. It was so simple in design that it required very little maintenance and a small motorcycle battery was the only thing needed to get it started. Typically, most persons who had professional background in this field felt that the machine must be a farce and viewed it and the inventor with suspicion. After seeing the machine run a power saw in the inventor's workshop, a number of people were forced to expand their thinking somewhat. Teal dreamed up his engine design after working on a science fiction novel. His first model was made to a large degree of wood and he estimated that it shouldn't cost over a few hundred dollars to put out larger precision models for use in automobiles. Because he lost an estimated $50 million invention while he was working on an earlier government project, he was hoping for a better reward on his "impossible" magnetic motor.

The Hendershot Generator

In the late 1920s there was considerable publicity on a device built by Lester J. Hendershot. Through inspiration and an unusual dream this inventor wove together a number of flat coils of wire and placed stainless steel rings, sticks of carbon and permanent magnets in various positions as an experiment. With later adjustments this device actually produced current. According to the reports the inventor had no idea how the device worked and it was often just a case of working by trial and error to get results. A number of persons speculated that the various magnetic currents of the Earth were used when the resonation of the device was turned to the proper frequency.

Temperature Change Wheel

Wally Minto donated a most remarkable design to the world in 1975. His unique unpatented wheel worked on a change of temperature—as low as $3\frac{1}{2}°$F—and was so simple that anyone with material and a welder could build a full scale model. Using any gas proof tanks around the outer form of the wheel, a simple pipe connection between the upper and lower tanks allows the needed exchange of gas. A warmer lower tank would lighten as the upper tank collected the vaporized propane—or low boiling point gas. While slow, the design gave considerable torque and held great promise for applications in backward areas.

It is interesting to note that some of these perpetual motion machines relied on heavy flywheels. Studies in the 70s concluded that flywheels were about the most efficient energy storing device available. Better than fuel cells, lead acid batteries, or compressed gas, the flywheel could carry the wasted power of high horsepower and save motorists big money.

In 1972 Lockheed reportedly found that an ordinary iron flywheel spinning at around 24,000 revolutions per minute in a reasonable vacuum (anti-friction) worked quite well. In fact, very little research money is required to quickly raise the efficiency of most current motor drive systems, and Cadac Ltd. of Auckland, New Zealand, has one in production in 1993.

Hot and Cold Engine

A sixty-five-year-old Swedish inventor made a major breakthrough in the thermo-electric engine field. Because wires of different metals produce electricity if they are joined and heated, there has long been a potential in this principle. B. Von Platen's secret breakthrough is said to give more than 30 percent efficiency in motors, and, with a radioactive isotope for power, it could free it from fossil fuels. In 1975 Volvo of Sweden obtained rights to his power unit.

Air Fuel

In the 1920s, a Los Angeles (Baldwin Hills) resident worked out a method to run an ordinary automobile on the constituents of ordinary air. Working out a system to keep his motors from melting from the high heat produced by the burning oxygen, he contacted the auto makers. General Motors, acting for the industry, eventually got controlling stock of the small company, and that was the end. A reader of *M-J BSRA Journal* recalled that the motor was warmed up on ordinary fuel and then switched over to air after it became hot.

Air Powered Autos

Air power was used to power rail locomotives and mining equipment for years before the so-called energy crunches. Like the steam engine, the air engine does not need torque converters (transmissions) and lasts for years because of low speeds. Los Angeles Engineer Roy J. Meyers built a 6-cylinder air car in 1931 and it supposedly had a cruising range of several hundred miles at lower speeds. There seems too few reasons why the air system wouldn't work very well in pollution sensitive cities. Air fueling tanks at the strategic spots would be simple.

In 1973 Claud F. Mead of San Diego, California, thought up a simple air car design. Using a scuba bottle full of air, he ran a hose to an air impact wrench. The wrench shaft was, in turn, hooked to the wheel of his small cart. By using a battery to pump up his tank, he was able to go some distance at speeds up to 50 miles per hour.

Air Powered Engines

Back in 1816 a Scottish clergyman, Robert Sterling, designed an external combustion engine that ran on hot air. Since that time, many experiments have been made trying to perfect his idea. In 1975 there was a break-through of some significance in the British Atomic Energy Research Lab at Harwell. There they came up with a working fluid pump which was nothing more than a container with an assortment of pipes and valves. This means that solar energy should be capable of pumping water—or your hot springs or hot air supply can furnish pumping power. A pistonless version of the Sterling motor was designed by the British Atomic Energy Research Lab. It was connected to a linear (nonrotating) alternator and could put out 27 watts of power a day on less than a quart of propane.

There have been a number of Sterling designs for autos. Some European firms have run these designs successfully, and such nonpolluting engines just hum along under a continuous (not instant) combustion. These engines are simple, non-polluting, and will run on anything from charcoal to sunshine.

In the 1930s in Wolvega, Holland, there was a twenty-one-year-old inventor who developed a piston engine which reportedly could run for three months before needing recharging. The engine was remarkable in that it ran on hot compressed air. Before he had a chance to market the engine, he was sent to a mental institution, and his working models disappeared.

In the 60s Louis Michaud designed a simple thermodynamic engine which resembles the internal part of a huge squirrel cage blower. Sitting so that the vanes were horizontal, this machine deflected the air flow path inward and upward to form a miniature hurricane action. Because this system could, theoretically, produce or decrease different types of weather (change temperature and humidity and disperse pollution), it could be a very worthwhile system. Harnessing just a fraction of the energy potential from thermal changes on our planet would supply awesome power.

Hydrogen Car Engines

Many believe that hydrogen is the ideal motive force. Containing no carbon, H_2 can be burned safely in any enclosure and broken up into safe components whatever the conversion.

A number of minor experimental successes proved the worth of these conversions over the years. Some simply hooked up a mixing chamber instead of a carburetor on their car, and they experimented with combinations of oxygen and hydrogen until successful.

In 1972 a UCLA team built an automobile to compete in a "clean air" race. Using a stock gasoline engine, they lowered its compression rate and made a few alterations to allow for a greater heat build-up. Next, they recirculated part of the exhaust gas to decrease the excess oxygen and slow the combustion process slightly. The result was a success. The only real problem was in the bulky, quickly exhausted tanks of fuel.

Billings Energy Research of Provo, Utah, solved the bulky tank problem a couple of years later when they built a hydride storage system. Hydrogen is chemically locked in powdered iron titanium and is released when heat from the engine's cooling fluid warms it. With this, or a less expensive Hefferlin System there is little reason for our continuing dependence on fossil fuels.

Justi and Kalberlah wrote in a 66 French bulletin that they could convert water to hydrogen and oxygen using DC current and simple nickel, double layer, porous electrodes. Their system could store the gases under 100 atm without a pump being used, and they attained a phenomenal 50- to 65-percent energetic efficiency.

In 1975 UCLA experimenters ran liquid hydrogen to a standard propane regulator and mixer atop a standard carburetor. In the carburetor they used water to lower combustion temperatures and to act as a combustion and backfire control. (An "approved" gas mixer or carburetor is necessary in California.)

Electrostatic Cooling

For some reason, when static electricity is played on a red hot object, it will suddenly cool the object. This "electric wind" seems to break up the insulatory boundary layers of air, and it will have numerous applications in our century.

The "tabernacle" [the famous Ark of the Covenant] of Moses in the

Bible was said by Lakovsky to be nothing more than a large electro-static generator. While the friction of air against the silk curtains generated the static electricity, the box condensor stored this energy.

Steam Locomotion

Who could exclude the beloved steam car from a work like this! In 1907 a Stanley Steamer car travelled down a Florida beach at 170 miles per hour before a bump sent it out of control. Losing ground to the cheaper gasoline vehicles, a number of the old steamers were resurrected and run during the World War II fuel shortages. Even in the 50s a Stanley engine carried one researcher and his newer car across the U.S. for six dollars worth of kerosene.

The Doble Steam Auto was probably the first steamer of modern design. Instead of allowing the steam to escape, it recirculated it so that an owner conceivably could drive a thousand miles before refilling the twenty-five gallon water tank. With less than a minute warm-up owners could get performance equal to the best gasoline automobiles.

The amazing Doble engines were guaranteed for 100,000 miles, and some owners reported having got a phenomenal 800,000 miles from them. From his first auto show Doble got $27 million in orders. The War Emergency Board of the period (1917 plus) discouraged production completely, so Doble was forced to survive abroad building steam trucks for an English firm.

Steam power plants have been no problem. Kinetics Inc. of Sarasota, Florida, had a superb engine developed for cars of the late 60s.The Gibbs-Hosick Steam described in *Popular Science* (February, 1966) was to use a tiny piston motor to give it impressive performance. A super efficient steam engine was developed by Oliver Yunick in 1970 (*Popular Science*, December 1971); another, the HBH in *Popular Science*, November 1971. One of the most advanced steam turbine designs came from the DuPont Laboratories in late 1971. They used a recyclable fluid of the freon family. Presumably it contains within its design no need for an external condenser, valves, or tubes. (*Popular Science*, January, 1972.)

Using more basic designs, Sundstrand Aviation put one of their steam power plants in a Dallas city bus. At the same period William M. Brobeck of Berkeley, California, with his assistants, equipped three Oakland Buses with similar "Doble" designs.

Lear Motors Corp. of Reno, Nevada, spent millions on advanced steam designs until it was apparent there would probably be little financial reward in the end. Steam Power Systems of San Diego was another principal experimenter during this period.

About as close as anyone came to putting a production model on the market in the 70s was the attempt by W. Minto. Using Swedish Sullair rotary compressors for motors, he mounted his system on a standard Datsun and got a contract for at least a hundred more. Later modifications included a gyrator engine, which was actually a pump motor working backward.

One of the few new steam engine designs able to be directly tied to the drive wheels of an auto is the KROV design of 1973. Claiming at least a one-third advantage in economy over conventional gas engines, all this engine needed was financing.

I recall that one enterprising gentleman sold a kit to convert gas engines to steam engines during the Second World War. He ran around Los Angeles in a converted Model A Ford until he dropped from sight. In the 60s there was a similar conversion kit put out by a small company in Oregon. Furnishing a smaller cam timing gear sprocket the size of the crankshaft sprocket and a modified camshaft, a normal "gas" engine could easily be converted. This company did not advocate using their units for any but stationary engines, but hinted at a new super fuelless steam power unit coming up.

Another Steam Engine

In the early 70s William Bolon in Rialto, California, developed an unusual steam engine design that was purported to get 50 miles to the gallon. The engine, which used only 17 moving parts, weighed less than 50 pounds and eliminated the usual transmission and drive train in an auto. After contacting Detroit interests, the inventor claimed he was required to sign forms releasing these interests from acknowledging his claims to the design before they would even look at it.

After a *Sun-Telegram* article on the project, his factory was firebombed to the tune of $600,000. After letters to the White House, the inventor finally gave up and let Indonesian interests have the design.

Aside from a token steam project by Ford, the steam auto was ignored right up to the time of various Senate pollution control committees of late 60s. Typically, the representatives of the auto industry alleged that steam systems were not dependable, safe, or necessary—especially since Detroit would soon have good minimal emission designs. So, without funding, the small experimenters of this period tended to fall into obscurity.

A notable exception was Bill Lear, who spent millions perfecting systems in his Reno plant. The complete lack of co-operation and interest from major industries or "powers" eventually discouraged him.

Diesel

Dr. Rudolph Diesel took the crude heavy fuel burning engine designs of those before him and refined them into the major engineering success of the 1900s. His invention immediately threatened the whole steam engine industry, and just as he was plunging into fame and success, he permanently disappeared from the ship on which he was travelling to Europe.

Electrostatic Motors

The modern world's first electric motor was an electrostatic motor invented by Benjamin Franklin in 1748. Through the years, little was done in this field until a Dr. Jehmenko came on the scene. This good physicist felt it was a "waste" not to be using some of the abundant free atmospheric electricity, so he built the most powerful Corona motor so far tested (1974).

He has visions of being able to put his Earth-field antennas on the tops of mountains, where electrostatic energy is particularly concentrated and use an ultraviolet laser beam to ionize the air and send the energy to receiving sites below. To run smaller motors, experimenters find that a few inches of needle pointed music wire will start a Corona. This wire is attached to at least two or three hundred feet of copper lead-in wire held aloft by a balloon, kite, or tower. Tolerances are critical on electrostatic motors, but they are simple to make.

Using more conventional research methods, the Argonne National Labs (Atomic Energy Comm.) spent millions in the early 70s developing numerous "Super Batteries." Somehow, as usual, the public gained little benefit from these breakthroughs.

FLIGHT AND ANTI-GRAVITY CONCEPTS

Anti-Gravity Propulsion

A number of researchers contend that if the isoles of the atomic fields in matter are arranged in a linear polarity, they can produce an anti-gravity effect. This is the principal a magnet works under when its molecules are in alignment. The perfect example of this principle in application is the bumblebee. Flying against all aerodynamic principles, the wings purportedly produce enough electrostatic polarity bands around the bee's body to carry it aloft.

According to some theories anti-gravity can actually come from creating any system which will use the confusion of matter against the orderly flow of energy. In designing a system to use positive and negative (night and day, the Ancients called it) polarities against each other, a Toroid coil with a caduceus winding can be used to separate these fields—and play them against each other. By orienting the poles of the atomic structure of matter instead of the molecular structure (magnet), even nonferrous metals can gain attraction-repulsion qualities.

We should shortly be using propulsion units which are little more than diaphragms of matter sending out discordant vibrations—out of harmony to the resonance of space. (Further data on the working of matter from works by Walter Russell and Geo Van Tassel).

While the electrical resistance of various metals has long been affected by super cold temperatures, it was not until the mid-60s that scientists found a "breakthrough."

Niobium with tin zirconium or titanium were found to produce super-conducting magnets ten times as strong as ordinary magnets.

As with the "live" metals mentioned elsewhere, such super-conductive characteristics could allow a super magnetic shield for space ships. This would, in effect, act as a force field protector against dangerous protons and radiation. Super-conductive wire, of course, could allow frictionless gyros, and ultra small computers and electrical circuits.

When larger super-conductive metals act to repel magnetic fields, we have an actual "levitation."

Vibrations

Besides the well-known oracle caves of antiquity such as Delphi, there were lesser-known objects used for the same purpose. At Dodona there were vases fashioned of metal that supposedly would ring for hours when struck. It would seem logical that certain tonal ranges or octaves would, indeed, assist some to blank out unwanted thought patterns.

Pythagoras was the first person history records as working out a reasonably sound harmonic musical scale. He was also convinced that certain modes or keys had profound effects upon emotions. The "Hard Rock" music of the 70s then was probably far worse on the listener's well being than the less chaotic music he warned his disciples against.

The early Greeks had great knowledge on the use of vibrations, and the priests were able to build highly unique sound chambers to use in their rituals and religious ceremonies.

Many persons have felt that all elements have certain keynotes and, if such a keynote is duplicated, it can disintegrate the compounds into their various parts. The mystical principle that two exact things cannot occupy the same space at the same time is valid whether applied to a mind system or to a wall of Jericho.

According to a number of ancient records, round metal discs of certain shapes and resonance could lift men and objects if sounded. Two such discs were made for the king and queen of Spain by the Aztec ruler Montezuma. About the size of phonograph records, one of these gold discs was said to be thicker than the other. Numerous myths spoke of persons flying when they struck or made songs on plates. Indian Sanskrit records are usually more detailed and indicate a science of acoustics far ahead of ours. The 716 ancient stone discs found in China by the Russians in the 60s were said to vibrate in a peculiar manner when struck.

In a work called *Secrets of the Andes* mention is made of a large disc from ancient Lemuria which was used by the Incas in a sacred temple. If struck in a certain manner it could supposedly cause earthquakes; if tuned to an individual vibrational rate it could transport the person to a distant place. The Spaniards found this disc gone when they finally located the temple.

Well-known occult writer, Annie Basant, explained in some of her works, that the gigantic stones moved by the ancients were rendered weightless by a simple application of natural magnetic law. Legends of almost every continent give accounts of persons striking objects or singing songs to move themselves or other objects around.

In 1971 the conventional spinning gyroscope used in navigation was threatened by a tiny two inch Beryllium copper wire held between the magnetic flux of two electro-magnets. This vibrating wire created a major breakthrough in this field. (Honeywell).

Throughout the ages there have been a certain few who have had the ability to match odours to vibratory levels. Even in the present age there are certain perfumes that are said to use a scale of odours just as a musician uses a harmonic scale of notes.

The mystics of the world have used chants to vibrate areas of the body to fuller efficiency. A typical chant has a mental, love, and power tone, ranging from higher to lower.

Early work by Dr. Oscar Brunler found a direct relationship between the output frequency of the brain and intelligence.

The Energies Science has yet to understand what we could call the "other energies." These energies—or let's say, "this energy" can be operated at great distance without any "grounding" actions by physical bodies. It can even be reflected by mirrors and transported, concentrated, and increased by sound.

The mystics referred to this other energy as "life force," and "Prana." Eeman called it the X force. Reichenbach called it "Odic Force," Paracelsus called it "the mumia," and the ancients referred to it in various ways as the "binding force." Frankly, I suspect it is all a part of "Mind."

UFOs and Propulsion Systems

Back in the 20s a former classmate of Einstein, Townsend Brown, teamed up and discovered a new principal of propulsion. It started with a charged condenser on a string and led into miniature flying saucers.

It was found that the closer the condenser plates, the wider the area they covered, and the more voltage difference between them, the greater the resistance to the effects of gravitation.

Brown continued this propulsion work into the 50s and is thought to have concluded that three large condensers under a saucer (120 degree control) would be sufficient to make practical flight possible. Theoretically, the condensers act in creating a modification of the gravitational field around a craft and, by using a caduceus coil to change field polarity, directional guidance can be attained.

As late as the early 1970s one inventor in the Northwest demonstrated similar anti-gravity discus before Portland TV Channel 8 viewers. His "Sicorsci Aviation" spent seven million dollars on the project before it all faded away.

These and other propulsion systems were all but ignored, officially. They went against the notions of gravity, for one thing, and for another, how would the powers that be make money from them? Jets cost millions . . . these systems were too simple.

A saucer developed by Germany about 1940 consisted of a wing wheel design in which a dozen variable wings acted in principle like a helicopter. The perfect balance required on this design was very difficult to attain, but with jet propulsion it was said to be capable of almost 2,000 kilometers per hour. [1,240 miles per hour].

One of the more advanced German designs was said to be powered by a "Schauberger" flameless, smokeless implosion motor. These power plants ringed the craft and tilted at angles necessary to give direction and speed. By incorporating suction openings at the top of the craft, an added boost in speed came from the vacuum created.

Incredible as it may seem, there were many documentations of all this in various reports at the time. A friend of mine told of tons of Germanium he found in one such plant. It was in some way connected with the drive system of one saucer design. Renato Vesco told an *Argosy* author (issue of August1969) that most of the data on the German saucers was taken by British "T" teams to Bedford, England, to various secret facilities in Australia, and to British Columbia, Canada. After continued work on the better projects, various British sources let it be known in 1946 that Britain would soon have aircraft that would be capable of thousands of miles per hour and need no fuel. By 1960 the Canadians had set aside 125,000 acres of very remote land in BC for "experimental aircraft" and the word sifted through that "Canada had some very advanced aeronautical technology."

Because the U.S. was unwilling to share the nuclear data she came away with after Germany's fall, Britain and Canada were not about to share their aeronautical data with us. It was their ace in the hole. Huge RAF budgets along with continued sightings of slower and more "solid" UFO's has led many persons to suspect that our pilots are ordered not to fire upon such craft with good reason. They are our friends—or maybe even us.

A certain Hefferlin manuscript entitled "Rainbow City" explains that the hero developed a very advanced space ship and offered it to the Hungarians just before Hitler took the country over. Because the Hungarians lacked funds to continue and Germany was closing in, Emery flew two ships to the U.S. and stored them here.

After offering them to the U.S. Government, Emery was rebuffed for a second time; he eventually flew, according to the manuscript, to a small secret protected valley in the Antarctic. Mention is made in this work of a fuelless motor which utilizes water electrolysis.

It is also pointed out that other alien UFOs, having no connection with these projects, commonly exist.* As an example, in unofficial conversations with the various astronauts, certain reports stated that all the early "moon shots" had alien visitors following them for a time. But then, this work is not large enough to go into data on such alien craft.

Alien Triad Propulsion Systems

A number of UFO reports have these crafts' propulsion systems using a triad configuration. A typical case involved a Sgt. Moody who was shown a system which used what appeared to be three large crystals joined by sloping rods. According to Moody the alien had told him that "with a little thought on your own, this could be developed by your people."

Electrostatic Anti-Gravity

With the help of two electrodes charged with 200 kilovolts of direct current, a piece of aluminium foil with a bead of mercury on it can be reportedly suspended between the posts. Mercury engines are described in ancient manuscripts from India.

Anti-Gravity

Henry William Wallace patented an anti-gravity generator in 1971, and many experimenters in this field were given encouragement and help by the new ideas.

Wallace's device uses rotors travelling from 10 to 20 thousand revolutions per minute and the inventor suggests that the intensity of weightlessness can be increased by using mercury—just as is mentioned in the ancient Indian manuscripts.

---

* *When Will Our Govermnent Confide In Us?* As the space projects of earth pushed ahead in the 1960s very few persons were aware of the fact that some of our most sophisticated advances came from duplicating the equipment on "alien" craft. A number of wrecked craft of this period reportedly got rushed to the Wright Patterson installation and thoroughly dissected. As stacks of UFO documents became de-classifed in the 70s, it was still almost impossible to obtain them from the responsible sources.

Dean Space Drive

Norman L. Dean was an amateur experimenter who made modifications to a harmonic drive mechanism known as a "Buehler Drive." Consisting of two counter rotating eccentric masses, the Buehler Drive is used by industry in generating oscillatory motion or vibrations. Because of various complicated circumstances and the death of the persons involved, nothing ever came out of this invention. No government agency ever showed an interest in it, of course.

Early Flight

Surprisingly, a number of legends and records exist concerning pre-historic flight. A number of these stories concern men who learned the art of flying from their more able "Gods." Emperor Shun in China, for instance, allegedly was able to fly after such instruction, and medieval drawings from such early periods are pretty convincing. Hindu writings are filled with "celestial" vehicles which transported the kings and gods. Other accounts of flying machines are recorded in various ancient records and myths including, it would seem, the Bible.

The more mystical works of Phylos and James Churchward tell of how some of these ancient airships worked. One design mentioned by Churchward took power from the atmosphere in what could correspond to a turbine running on atmosphere gases. He claimed that temple records he saw gave specific instructions for building not only a very advanced airship but its power supply as well.

In *A Dweller on Two Planets*, Phylos explains how some early airships ran by a balancing of the day and night sides of nature—anti-gravitational forces were matched against gravity to manoeuvre such ships perfectly. Some feel that such forces are beyond our present grasp because of certain energies unavailable to the Power Sources (Xtals), but that is a story told elsewhere.

The Vedic manuscript, *The Samarangana Sutrachara*, gives no less than forty-nine types of "propulsive fire" used in the wingless flying vehicles of India. This work devoted over 200 pages to describing how to build and fly these advanced ships. Some of the propulsion systems used the power of heated mercury, others that of electrical or magnetic forces.

The "Mahabharata," "Drone Parva," and "Ramayana" also give accounts of these "Vimanas" and their remarkable abilities.

According to Dr. Ruth Reyna, there are sanskrit texts in the University of the Punjab that tell of space flights 3,000 B.C. Commissioned by U.S. Space authorities Reyna found that these flights were considered imperative due to the threat of a deluge on earth.

Gravity Defying Gyroscopes

Edwin Rickman, an English electrical engineer, had recurring dreams about an anti-gravity device in the early 70s. After a patent was obtained on the basic principles, it came to the attention of Prof. Eric Laithwaite of London's Imperial College of Science and Technology. With certain modifications, this scientist declared in 1974 in press releases that this anti-gravity motor should enable us to travel to other solar systems.

Laithwaite Anti-Gravity Machine

Prof. Eric Laithwaite of the Imperial College of Science and Technology in England invented an anti-gravity machine in 1975. Defying the laws of Newton, it depended upon the fact that no energy was required to return its two gyroscopes arms to their starting position.

Flying Suits

The Asian conflicts prodded the development of one man flying suits in the 60s. In his "Gold of the Gods," Von Daniken points out numerous earlier models depicted on monuments, tablets, pots, and even as Polynesian ritual objects.

UFOs

Many strange stories have circulated about the flying saucers being built by various governments on our planet. While there is good reason to believe that alien saucers do exist and do visit our planet, there is a surprising amount of evidence concerning the models of local origin.

Several ancient manuscripts give details on building craft that would fit into the flying saucer category. However, in recent times, the most authentic reports come from records concerning the work of Hitler's scientists.

In the few short years that Hitler gave his scientists free rein to develop technology, there was astonishing progress. Allied teams who rushed into the secret underground bases and projects after the War were dumbfounded by the technological advancement they found. A small plant in central Germany (M-Werke) was on the verge of producing missiles which could destroy entire U.S. cities. Co-operation between G-Works and various other installations produced the "Kugelblitz." This was an advanced lens-shaped craft that destroyed Allied bombers by Electrostatic firing systems. It could travel by remote control, seek a target by infrared detection, and remain undetectable on radar screens. According to a number of later Allied intelligence reports, there were super turbine engines capable of running on liquid oxygen or hydrogen peroxide, a gelatinous, organic-metallic fuel—and on even the atmosphere.

More theoretical was a design based upon the "Lense-Therring effect." Here a torus wrapped in a tube of accelerating dense matter should create a gravity field strong enough to overcome the gravity of Earth. Another device possible under the present accepted laws of physics is built of a thin disc of nuclear matter. Such a device is lightly covered in an August 1975 *Analog-Science Fact Magazine*.

In this issue Dr. Forwards mentioned another system. Because any mass with velocity and acceleration can create force (according to accepted laws), a round torus rotating outward on itself should cancel Earth's gravity. Unfortunately, these machines would require quantities of dense matter.

Because many can't accept the current gravitational theories, there are many theoretical designs which use what we could call negative matter. Because an object of negative would repel an object of positive matter, we would get a principle of great potential. This, of course, would be similar to the "Day and Night" energies supposedly used by the ancients, mentioned elsewhere.

In a similar vein, one could theoretically use the polarity of inertia. By changing inertia from positive to negative—or even redistributing it, one might thus overcome gravity.

Einstein observed that if the UFO occupants had mastered gravity, they would also have overcome inertia. Saucers with anti-gravity screens could ignore both gravity and inertia. They can instantly change direction and speed. Anyone who has observed the darting movements of some UFOs must concede that something is breaking the laws of inertia.

There are those who maintain that we live in a contracting and expanding universe of many dimensions. By using technology which can contract a space craft, for example, the craft cannot only pass into the other dimensions, but pass through less dense materials. Because light rays would be less rapid than the event itself, distortions would result—which seem to be well recorded in documented encounters.

Will our leaders continue to assume that we are too dense to understand? It wouldn't surprise some persons to see the "leaders" looking down from advanced craft, in event there were a major disaster.

Exactly how many Government rooms are filled with data on UFOs could be anyone's speculation. What is well remembered by many is the fact that many samples of strange materials and machinery have been handed over to Government authorities. In all of these cases, the samples have simply disappeared and have been denied to later inquirers. A typical case in 1969 involved a material found by Professor R. Bracewell, the man who solved our spinning satellite problem. Absorbing heat and releasing it slowly by over a period of several days, this material could not be analyzed nor duplicated by our best procedures.

Is there actually an organised force to stamp out rational data on UFOs? An *Argosy* magazine article mentioned dozens of saucer researchers who mysteriously disappeared. Albert Bender, a well-know researcher, told of seven visitations by mysterious "men in black." He felt that with such powers to cloud men's minds as those visitors seemed to possess, they could be of alien origin.

Another well-known UFO researcher with a similar feeling is Laura Mundo. She felt that the "man in black" who contacted her were "front men" for aliens who wished to frighten her out of the work.

Grey Barker wrote his *They Know Too Much About Flying Saucers* during the period when the "men in black" were most active.

What occurs to many persons of open minds in this area is that there are not only UFOs of Earth origin, but there are very sophisticated craft of extraterrestrial "alien" origin as well. With literally hundreds of UFO publications and groups and thousands of sightings, the evidence is pretty overwhelming for either or both craft.

45 The Charles Pogue Story

CARBURETORS AGAINST MILES

Manitoba, Canada, Jan. 24, 1936—If a car weighing a ton and a half will run a mile and a half on a drop and a half of gasoline, people are very likely to forget the famous hen and a half who laid an egg and a half in a day and a half.

Evidence that a Winnipeg inventor's new carburetor gets over 200 miles to the gallon has caused many pencils to be sharpened by amateur physicists. Where and how does he get the miles?

"Gas savers" galore crowd the electric belts and the muscle-builders among the sucker ads in cheap magazines. It is not much of a trick, by getting the motor hot, skinning [leaning] the mixture and holding the car at its most economical speed—about 22 miles an hour—to get 50 or 60 miles out of an old boiler that usually turns in only 18 or 20 miles to the gallon.

But this 200-mile gadget is no gas saver. It is an economizer. It reminds one of the story about the fellow who cascaded two gas savers and had to stop every twenty miles to siphon some gas out of the tank.

In an imperial gallon of gasoline there are 145,000 British thermal units, more or less. This is the equivalent of 113 million foot-pounds, or 57 horsepower-hours. This would lift a 3,000 pound car 37,660 feet straight up in the air; or a little over seven miles—from the bottom of the Dead Sea to the top of Mount Everest, and then some. How far it would pull the same car along a level road depends on how fast you want to go, and how much friction there is in the wheel bearings. A man capable of generating only one-eighth horsepower can keep a car rolling, if he likes that kind of exercise. He will get there sometime: Say you choose to exert a continuous pull of 60 pounds—with a 3,000 pound car that is equivalant to a two-percent grade, a rough approximation of friction-loss plus wind resistance at a moderate speed. At that rate she will roll 356 miles for your gallon of liquid calories.

So, here's luck to a grand new idea. Long may she perk, and far may she fly!

Patents Block Thieves Taking Gas Economizer

Inventor Thinks Theft Is Attempt to Force Invention's Sale

Loss of three models of his 200-mile-per-gallon carburetor sometime Wednesday, was reported today by C. N. Pogue, local inventor. Thieves broke into his workshop, located in the Amphitheatre rink, through a hole in the roof, and escaped undetected.

The thieves will gain nothing by their raid, the inventor told *The Tribune* today. The invention is fully protected by patents in all principal countries of the world, and its theft will result only in delaying Mr. Pogue somewhat in his work of improvement and perfection.

Mr. Pogue believes that the robbers, to whom he gave credit for exceptionally smooth work, did not take the three carburetors they stole for any financial gain. He is of the opinion that their object was to discourage the inventor and his backer, W. J. Holmes, to such an extent that they would be willing to sell their rights.

Offers Turned Down

To date, Mr. Pogue said, he had turned down countless offers to buy the invention, into which they have put thousands of dollars and Mr. Pogue almost twenty years of work. They prefer to bring it to perfection themselves before placing it on the market.

Mr. Pogue described the manner in which the thieves accomplished their purpose, as he sees it.

"There must have been two or three of them, and they probably spent several days in their operations. How they could work here for that time, while the place was guarded day and night, I don't know. I am convinced that they were outsiders, but that they had help from someone who knew the ground here well."

Kept In Workshop

Mr. Pogue kept his carburetors and the car with which tests had been made, in a large workshop inside the Amphitheatre rink. The thieves entered, perhaps through the rink, then climbed to the top of the shop.

Here there were traces indicating a prolonged stay by the raiders. There were footprints in the shavings on the roof, and remains of meals. The raiders gained entrance to the shop through an opening in a switchbox on

Breen Motor Company Limited, Winnepeg

To whom it may concern:

I made a test today of the Pogue Carburettor [sic] installed on a Ford eight-cylinder coupe. The speedometer showed that this car had already run over 9,000 miles. I drove this car 23.2 miles on one pint of gasoline The temperature was averaging round zero with a strong north wind blowing. I drove for 15 miles and back on the same road, and the distance shown by the speedometer mileage was 23.2 miles when the gasoline was exhausted and the car stopped.

The performance of the car was 100 percent in every way. I tested for acceleration, get-away from a standing start and at all speeds, and it performed equal to, if not better, than any car with a standard carburettor.

At very low speeds, under 10 miles per hour, it was smoother in operation than a standard car. In fact, below 5 miles per hour it pulled up a slight grade without labouring of any kind. I stepped on accelerator when the speedometer was below 5 miles per hour and the car got away without a falter.

(signed) T.G. Breen, President [Breen Motor Co. Ltd.]

---

the roof, dropping down and removing the carburetors while Mr. Pogue was away for lunch on Wednesday.

Previous Theft Attempted

A previous attempt to steal the carburetor in April of this year was unsuccessful. At that time thieves stole a car in which the invention had been demonstrated from a garage at the rear of the Amphitheatre. Fortunately, the carburetor had been removed from the car some time before.

The invention was tested officially last December. In below-zero weather, two prominent Winnipeg automobile men, W. S. Kickley and T. G. Breen, reported 209.6 miles to the gallon.

In another test made by Mr. Pogue himself, in February, a car equipped with this carburetor is said to have travelled from Winnipeg to Vancouver on less than 15 gallons of gasoline.

POGUE'S 200-MILES-A-GALLON CARBURETOR IS BEING TRIED OUT THOROUGHLY AT TORONTO

Toronto, Dec. 5—Somewhere within 40 miles of Toronto, generally in a north to northeast direction, engineers are now trying out the new carburetor, invention of John [sic] Pogue, 38-year-old Winnipeg man, which has become the main gossip of engineering and motor car circles throughout the continent.

That was the message imparted to *The Tribune* by John E. Hammell, millionaire mining official and prospector.

Mr. Hammell confirmed the report that a car using the new carburetor traveled 200 miles on a gallon of gasoline.

Just where the old residence and plant at which the carburetor is being tested is located will not be divulged. Gordon Lefebvre, of Toronto, formerly automotive engineer with General Motors, is the personal representative of Mr. Hammell in the final stages of perfecting the carburetor.

"I have not placed any big stake in this invention and won't until it is perfected 100 percent," Mr. Hammell said. "After it is perfected it will take time and it must be proved as an engineering principle."

To date the sum of $150,000 has been expended on the invention, states Mr. Hammell. "I have hardly started to do anything yet—they've got to show me."

W. J. Holmes, Winnipeg sportsman, has backed Pogue.

"But if it clicks," said Mr. Hammell, "there will be all the money required to put it across. I have been approached by some of the biggest oil and motor men on the continent already. They laughed at Pogue when he needed help and now they can talk to me. I have signed up the entire undertaking and have made agreements with both Pogue and Holmes.

"Certainly we have armed guards at the plant where the carburetor is being perfected. Somebody broke into Pogue's shop at Winnipeg months ago, but even if things were stolen now it wouldn't affect matters."

"The carburetor has been tried out on Pogue's own 1934 Ford 8-cylinder car. We have driven the car and got surprising performance—running over 200 miles on a gallon of gasoline. But that doesn't yet prove the thing. It is being installed on one of my own cars of the same make as the inventor's—then it will be tried on larger cars," declared Mr. Hammell.

As yet the invention is crudely made and entirely by hand. It is also costly. It is a slow process in developing. The trying out of the instrument on new cars will proceed under Mr. Hammell's engineers. Then other engineers, a chemist and designers will be called on as part of the undertaking with all the moneys required, states Mr. Hammell.

"I have no illusions in this matter," he remarked. "The principle must

The One That Got Away

Probably the most well-known of all the suppressed carburetors was the one (actually there were several) developed by John Robert Fish. The Fish carburetor was not only an economizer, it was a performer. Fireball Roberts had one on his car when he won the Indianapolis 500 in 1962.

The Fish carburetor was no simple device. The patent for his 1941 model (send 50 cents to the patent office in D.C. for #2,236,595) covers nine pages of explanations and drawings. His carburetor had no choke and wouldn't idle very well, which should have been no problem to solve—had he had a little more money to develop it. However, Fish was so broke before he died in 1958 that he had to have the money for one of his carburetors in advance in order to then turn one out. The U.S. Post Office sent all his mail back with "fraudulent" stamped on his orders when he tried to sell them by mail. "Fraudulent" could hardly have been a legitimate reason when no less a manufacturer than Ford admitted that the Fish exceeded their standard carburetor on two separate road tests by 32.5 percent and 42.8 percent respectively.

Fish went to his grave saying that someone in the auto-oil industry had "bought off" the Postal authorities in order to put him out of business.

What do you think?

be perfected so that it can be a commercial unit and that will take time. It can hardly be said yet that it can be made in commercial quantities—that will be the job of engineers not of the designer."

While the writer was interviewing Mr. Hammell, a long distance call came in from W. J. Holmes, the original backer. "Holmes just confirms that our handshake on the matter goes for good—we're sticking on the deal 100 percent. Holmes has been recently approached by influential interests but our agreement stands. Pogue is now inspecting every detail of the carburetor being installed on machines other than his own. As the affair is made by hand many test runs must be made and many adjustments made.

"If this carburetor is right—and I've got to be shown—it will be a tremendous benefit to mankind not only through automobile and gas engines, which are countless, but more so for aeroplanes, as it cuts gas down to one-tenth. I'm a flying man and everywhere I go these days is practically by plane so I visualize what the thing might mean."

A typical breezy talker, Mr. Hammell replied to some questions.

"How about armed guards?"

"Sure we have a flock of armed guards. They are carrying guns ready to pop at anybody."

"Is the inventor worried?"

"He is nervous and very apprehensive. However, if there was danger before it has now passed. Anything broken or stolen can be replaced without harm to the invention."

"Who owns the thing?"

"All I have is an option for control and full and absolute power to handle it anyway I see fit. Mr. Pogue will receive his interest, as will Mr. Holmes."

ARMED GUARDS, A HOUND AND AN INVENTOR GET ON JACK HAMMELL'S NERVES

"It's Got Me Nuts," Says He, Telling of Pogue Gadget

Toronto, Jan. 28—Inventors are "funny people"—and that goes for C. N. Pogue, of Winnipeg, young mechanical wizard who turned out a gas-saving carburetor. Pogue may be a nice boy, but his invention, armed guards and wolf-hound proved too much as a steady diet for Jack Hammell, millionaire mining magnate.

Hammell, backer of the carburetor, told about his reaction to inventors and inventions in an address here Wednesday to the Kiwanis club. The wealthy mine-owner spoke the day after Professor Alcutt of the Univer-sity of Toronto, had termed "impossible" the claims made for the Pogue invention.

"My engineers like it," said Hammell, "but I don't know. It's got me nuts. Did you ever have anything to do with an inventor? They're funny people. He (Pogue) put this big device on cars for us and we got up to 215 miles a gallon out of them. But it's still got to be proved to me—and then it has to have a little sex appeal put into it for commercial production.

"After our engineers tested it I said, 'I still think it's screwy.' Pogue told me he was going to try to get 450 miles a gallon out of a carburetor and I said 'No you don't—two hundred is as high as I can stand.'"

"There's no reason for the oil companies to worry or for you to sell your oil stocks—it'll be the greatest thing for oil companies that's ever happened, if it works out.

"You could put Amelia Earhart into an aeroplane and let her fly it from Frisco to Berlin and back without refueling—that's what it'd do."

"I had the inventor out at my place and there were two men with revolvers and a big wolf-hound but that wasn't enough for him. He had a revolver himself but he insisted on hiring five more guards with sawed-off shotguns and things and finally I had to send him back to Winnipeg.

"He's spent $35,000 on this carburetor and his backers have spent $150,000—and they haven't got a thing out of it."

"I was glad to get rid of Pogue. He's a nice boy but he's an inventor."

46 News Clips on Suppressed Fuel Savers

WILL MYSTERY ENGINE RUN 300 MILES ON A GALLON OF OIL?

From *Popular Science*, August 1922

Three hundred miles on a gallon of oil—10 times the mileage possible for the usual present-day motor!

Such is the astounding record claimed for a crude-oil engine developed by Harry H. Elmer, of Syracuse, NY, for use in automobiles, airplanes, ships, and lighting systems. In experiments, the engine has generated sufficient power to run a battery of 18 incandescent lamps 18 consecutive hours on $1\frac{1}{4}$ pints of oil, costing less than a cent.

Because this mechanical marvel does not require a cooling system, government officials, it is reported, are studying the possibility of its use in dirigibles.

Among more than 300 radically new features claimed for the engine, the most important are these:

It contains only 64 parts and has only three adjustments.

It has no spark, carburetor, wiring, nor any sort of ignition.

The cylinder has a bore [interior diameter] of $3\frac{1}{4}$ inches and a six-inch stroke, yet the engine, it is said, has developed 200 percent more power than internal combustion engines of the same size, and will pick up almost instantaneously from 100 to 2800 revolutions a minute.

How the Motor Operates

The new engine is described as a four-cycle motor, the cycles being suction, compression, expansion, and exhaust. The crude oil is led through needle valves into [the] mechanism, where it mixes with air and then, through another valve of the same kind, is drawn into the motor head, where it is compressed by the upward stroke of the piston. On compression the oil is "cracked" by chemical process and the expansion of gases takes place. As the piston is forced down the exhaust port is opened, and the incoming charge forces out the expanded gases.

There is no combustion in the cylinder, though hydrocarbon gas, escaping the exhaust, explodes on uniting with atmosphere.

The engine has been operated with equal success on mineral, animal, and vegetable oils.

SEWER GAS SERVES AS FUEL

From *Modern Mechanix*
Municipal trucks and other heavy vehicles operating in Berlin, Germany, are being equipped with motors that enable sewage gas to be used as a fuel. Several sewage gas tanking plants have been erected, where the gas is compressed and stored in tanks for future use by truckmen.

Three of the tanks, each with a capacity of 500 cubic feet of the sewage gas, will operate a five-ton truck at normal speeds for a distance of about 225 miles.

LIFE-LONG BATTERY STARTS TEST CAR 606,969 TIMES IN YEAR-LONG "TORTURE TESTS"

By Dr. J. Morgan Watt
From *Science and Mechanics*
Here is a story typical of how American ingenuity is constantly at work to create new products that make life simpler and easier for all of us. The beginnings of the story go back more than 100 years to the original European invention of the storage battery and its gradual development and improvement. Thomas Edison predicted a lifelong battery as long ago as 1889. But it was not until World War II that a captured Rommel tank revealed that a remarkable battery had been perfected, good for an incredible 40 years or more of service!

Reader's Digest told about this German-type battery in 1948. They also reported on the action of the U.S. Government in breaking up, through the Sherman Anti-Trust Act, the international cartels which had prevented the manufacture of the European battery in this country. About two years ago, an American company finally marketed the Life-Long 10-Year Battery, the logical outgrowth of the most advanced European and American developments. The American-made Life-Long Battery compares favorably with its European predecessor.

In addition to the manufacturer's own rigorous test program to prove this, independent laboratories investigated the new battery thoroughly. For example, a leading commercial laboratory subjected a Life-Long Battery to one of the most unique series of torture tests ever devised. Every 15 seconds, day and night without pause for 12 long months, the battery started a test engine. At the end of the test, the battery had made the fantastic total of 606,969 consecutive starts! This is equivalent to starting your car thirty times every day for fifty years!

The Life-Long Battery was also subjected to extreme heat and cold tests. For example, it was frozen in a block of ice for 48 hours at 70° below zero. The moment it was broken out of its icy prison, it performed with its usual efficiency!

The secret of the Life-Long 10-Year Battery's great power and long life is the sintered Cadalloy plates (derived from Cadmium); indestructible capillary separators which keep plates moist and active; and the special mild-type electrolyte eliminating the strong acids that ordinarily corrode the inside of a battery.The Life-Long Battery differs from ordinary lead-acid batteries in other ways as well.

What does the Life-Long Battery do for you? It makes your car lights up to 50 percent brighter. You get faster, surer starts in all weather which means more efficient use of gasoline. You need add water only once a year, giving you a truly attention-free battery. Most important advantage is the 10-year service period, guaranteed by the factory. The bonded guarantee is backed by a state licensed company and remains in effect no matter how many times you transfer your battery from car to car, for the serial number of each Life-Long Battery is permanently registered for your protection.

Surprisingly enough, Life-Long Batteries cost no more than ordinary premium batteries. 6-volt Life-Long Auto Batteries are just $29.95 and 12-volt, $34.95.

INVENTOR'S MYSTERIOUS MOTOR RUNS ON POWER DRAWN FROM ATMOSPHERE

From *Popular Science*, October 1952
When Cmdr. Ivan Monk, who designs big turbines for the Navy, is pestered by inventors of perpetual-motion machines, he points to a device on his desk that goes them one better—for it works. It is a wheel that spins with no apparent source of power. Commander Monk built it in his spare time, and patented it, since it may find use in clocks, toys and advertising displays.

Actually it is a rotary heat engine, run by temperature difference between its parts. The wheel is at room heat; the long cloth-covered hub, kept dampened with water, is cooled by evaporation. A low-boiling-point liquid, Freon, circulates between wheel and hub, vaporizing in the wheel and condensing in the hub. Valves maintain an unequal weight of liquid on opposite sides of the wheel and gravity does the rest, to turn it.

INQUIRY INTO INVENTION ATTACK LINK

From *New Zealand Press Association*, Wanganui, 1994

Police are investigating whether a man was beaten up to "silence him" over his work on the development of a water-fuelled engine.

Mr. Dylan Whitford, aged 18, was found unconscious in a Wanganui carpark on Saturday night.

He was recently interviewed by Barrie Mitchell-Anyon for an article in a Wanganui community newspaper on his development of a water-fuelled engine.

Mitchell-Anyon said yesterday that after the article was published, he was called by a man who said several Wanganui people knew of the water-fuel technology but were frightened and were lying low.

Mitchell-Anyon said the caller suggested Mr. Whitford should be careful.

The caller said a Nelson man, who knew of the technology, died in mysterious circumstances involving a haybaler.

Mitchell-Anyon said he had passed on all information to the police.

A Maori warden on patrol found Mr. Whitford unconscious in a parking area behind a building at the corner of Taupo Quay and St. Hill Street about 10:40 P.M. on Saturday. He was admitted to Wanganui Hospital initially to the intensive care unit.

Detective Sergeant Dave McEwen said yesterday that a full medical report was not yet available but he understood Mr. Whitford's condition was stable.

"We're looking at all aspects," he said. "It's too early to form any conclusions if he was assaulted or if he has fallen."

Detective Sergeant McEwen said the theory that Mr. Whitford might have been attacked because of his work on a water-fuelled engine was one avenue police would follow up.

WATER VIES WITH GASOLINE AS NEW MOTOR FUEL PASSES TESTS

From *Modern Mechanix and Inventions*

Serious rivalry is being offered the corner filling station by the kitchen water faucet of Baron Alfred Coreth following his invention of a new fuel for his automobile.

The new fuel, called Corethstoff, after its inventor, is composed largely of water and raw alcohol and can be used in any type of internal combustion motor without alteration of engine or carburetor. It is not only cheaper to produce than gasoline, but goes 20 percent farther as well, the inventor claims.

Because of its non-explosive and practically non-inflammable qualities, Corethstoff is being considered as an airplane fuel.

ANOTHER MOTOR FUEL

From *The Model Engineer and Electrician*, May 5, 1921
A new motor car fuel, known as "Penrol" is now making a bid for public recognition says *The Cape Times*, S.A. It is described as being mainly a combination of alcohol, dissolved acetylene, and other hydrocarbons, and it is claimed to have come through practical road tests with complete satisfaction. It is said to exhibit great power on the hills with smooth engine performance and rapid acceleration. The mileage per gallon is given as 9 percent under that of petrol, but "Penrol" will, it is said, be placed on the market at a price which will more than compensate for that. A distillation plant for the production of the necessary alcohol has already been erected in the Transvaal.

AUTOMOBILE ADS

From *Modern Mechanix and Inventions*
200 MILES 35 CENTS. Burn stove distillate in your car. Get parts any junk yard. Detailed information and blueprint 50c. Cecil Carmichael, 5013 2nd Ave, Los Angeles Calif.

SEMI-DIESELIZE TO ECONOMIZE. Convert your car for $2.75 to $5.00 and burn cheap Diesel Oil. Works on any car. No major changes made in motor. Anyone of average mechanical ability can install. Parts available everywhere. 6 big (8½" x 11") pages of Copyrighted drawing and instructions show 2 different, tested, methods of conversion $1.00. Free booklet, stamp apprectiated. Order your copy today. Earl J. Behling, Box 944 Los Angeles, California.

AUTO OWNERS! Adapt your Automobile for cheap Diesel fuel. No radical change. Increased mileage, perfect performance. Parts $4.00 anywhere. Guaranteed satisfaction with our superior plans $1.00 postpaid. Particulars Dime (refundable) Research, 126 Lexington, New York.

WATER-CAR "SOLD FOR $25.6 MILLION"

From the Wellington newspaper, New Zealand
A 30-year-old New Zealand mechanic who says he has developed a car engine fueled by water was reported today to have sold his invention for $25.6 million to an international research group.

The Sunday News of Auckland, said that M. Malcolm Vincent, of

Nelson, had sold the rights to his water-powered rotary engine to the Club of Rome, through its Melbourne representatives.

"They're paying me $NZ600,000 down and $1 million a year for 25 years," Mr. Vincent is reported to have said.

Before the newspaper broke the story today there had been several reports that Mr. Vincent had perfected his water-powered engine, but apart from a television report which showed his car moving down a slight hill the reports had never been substantiated.

After the television film the NZ Government sent Ministry of Transport automotive engineers to appraise the invention.

"No Chance to See It"

"But he never gave us a chance to see it," the chief automotive engineer, Mr. F. D. McWha, said today.

"He hasn't disclosed to us the principle of his engine and he didn't want any help."

Mr. McWha said he understood Mr. Vincent had not patented his invention and had allowed nobody to see it for fear his secret would be stolen.

The Sunday News said Mr. Vincent flew to Melbourne with a crated prototype of his engine a few days ago.

The newspaper quoted Mr. Vincent as saying before he left for Melbourne that, "It didn't take long to negotiate the sale. The first production engine we will be trying to build will be for a car."

Mr. Vincent told the newspaper that in trials round the Nelson district in the South Island, "the best run we had with the engine so far has been in a Holden. We did 150 miles in it around the Nelson area."

According to Mr. Vincent, two manufacturing units will be set up by the Club of Rome: one in Australia and one in New Zealand.

Mr. Vincent died under mysterious circumstances in 1989.

Conclusion

It's conceivable that the next car you buy may be fueled by water. Perhaps, at that time, your income will no longer be eaten up by the electric bill, because free energy will have hastened the extinction of mighty power monopolies. In fact, years from now, your children (or your children's children) may awaken to find their planet Earth unspoiled by toxic wastes, and maybe even fully recovered from the havoc that we have wreaked continuously through the centuries.

However, if we continue along the course that we have mapped out for ourselves, it's not very likely that we'll benefit from these developments any time soon. Too many people have become accustomed to our current state of affairs, accepting the imbalance of our ecosystem as the inevitable byproduct of progress. Too few have questioned the wisdom of this learned helplessness.

What we must ask ourselves, then, is whether or not the bleak outlook that we have constructed really is unavoidable. I believe that our species has the capacity to deviate from the path that we have set out on, and to chart a new course for ourselves, and for future generations. *Suppressed Inventions and Other Discoveries* has highlighted many inventions and ideas that have been developed in the last century of this millennium—the innovations of researchers and inventors who refused to resign themselves to fate. But what has been presented here is only a fraction of what actually exists, because each breakthrough that you have read about actually represents dozens of others that remain obscured.

The suppression of new paradigms and creative thinking in science, technology, and medicine affects us all, and we can no longer afford to ignore the overwhelming evidence that it does exist. What the world could have been and still could be if we encouraged and nurtured non-polluting technologies like hydrogen or free energy is worthy of our deepest consideration. We simply must stop killing the planet, each other, and ourselves and devote our best efforts to repairing the damage that is the legacy of a system devoted to exploitation, greed, and personal aggrandizement.

In my continuing efforts to disclose valuable information, I'm always eager to hear about inventions and discoveries that have been hidden from the public for any number of possible motives. If you, as an informed reader, have proof of a breakthrough that has been ignored or denied, please contact me care of Avery Publishing Group.

Remember to keep your eyes and ears open. An inquisitive mind, coupled with ceaseless determination, is the remedy for the suppression syndrome.

Permissions

19. "Egyptian History and Cosmic Catastrophe." Reprinted by permission of *Nexus*.

20. First appeared in the April–May 1993 issue of *Nexus*. Reprinted by permission of David Hatcher Childress.

23. Reprinted by permission of Stuart H. Troy.

24. First appeared in the February/March 1994 issue of *Nexus*. Reprinted by permission of Patrick Flanagan, 1109 S. Plaza Way, Suite 399, Flagstaff, AZ 86901.

27. From the book *NASA Mooned America!* Reprinted by permission from René.

28. "Extraterrestrial Exposure Law Already Passed by Congress" (August/September 1993). Reprinted by permission of Nexus.

29., 33. From the book *Extra-terrestrial Friends and Foes*, published by Illuminet Press, P.O. Box 2808, Lilburn, GA 30226. Reprinted by permission of George C. Andrews.

30. "UFOs and the U.S. Air Force" (July/August 1991; October/November 1991). Taken verbatim from the United States Air Forces Academy textbook, *Introductory Space Science, Volume II*, Department of Physics, USAF. This volume was being used by the Air Force Academy at Colorado Springs, CO. The Air Force Academy has since pulled this volume from the curriculum in the early 70s because of the controversy it generated. Reprinted by permission of *Nexus*.

31. "UFOs and the CIA: Anatomy of a Cover-Up" (No. 36, May/June 1996, pp. 23–25). Reprinted by permission of *New Dawn Magazine*, GPO Box 3126FF, Melbourne, VIC 3001, Australia.

32. From the book *Above Top Secret* by Timothy Good. Reprinted by permission of Macmillan Publishers, Ltd., London.

34. From the book *UFOs and the Complete Evidence From Space*, published by Pintado Publishing, Walnut Creek, CA 94598. Reprinted by permission of Daniel Ross.

35. Reprinted by permission of Thomas J. Brown, http://www.gold-mountain.co.nz/aereon.

39. Compiled, published, and reprinted by permission of Mark M. Hendershot, P.O. Box 135, Okanogan, WA 98840.

41. From the book *Forbidden Science* by Richard Milton. Reprinted by permission of Fourth Estate Ltd., United Kingdom.

43. First appeared in the June/July 1992 issue of *Nexus*. Reprinted by permission of Karin Westdyk.

44. From *Suppressed and Incredible Inventions* by John Freeman. Reprinted by permission of Health Research, P.O. Box 850, Pomeroy, WA 99347.

46. "Inventor's Mysterious Motor Runs on Power Drawn From Atmosphere" (October 1952). Copyright © 1952 by *Popular Science*, Times Mirror Magazines, Inc. Reprinted by permission.

"A Selection of Alternative Energy Patents" from the book *Exotic Patents*, edited and published by Biagio Conti, P.O. Box 1014, Carmel, NY 10512. Reprinted by permission of the publisher.

The author and publisher's intention was to contact all of the above copyright holders. In certain cases, however, we were not able to do so. Therefore, we invite the copyright holders we could not locate to contact us, so that we can give them proper credit in future editions.

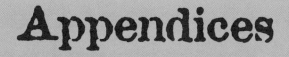

Appendices

Appendix A: A Selection of Alternative Energy Patents

Several inventors have been selected that may be of interest. All inventors listed below carry one or more patents in the United States patent office.

BROWN, TOWNSEND T.

Brown conducted gravity research for over thirty years. He has postulated that there is a definite link between gravity and electricity. There are many other patents listed that involve gravity and electricity and their peculiar relationship (electro-gravitics research). See *Ether-Technology: A Rational Approach to Gravity Control* by Rho Sigma and High Energy Electrostatics Research publications. See patents (1,974,483) (2,949,550) (3,018,394) (3,022,430) (3,187,206) (3,223,887) (3,518,462).

CARR, OTIS T.

In *Return of the Dove* by Margaret Storm, O. T. Carr is mentioned having developed an "anti-gravity flying saucer" and actually started a company which dealt with "free energy." Supposedly, certain secrets of his "free energy" device were incorporated in his "Amusement Device" patent (2,912,244) which resembles a conventional disc-shaped flying saucer. A few years ago, Carr was advertising his model A-X1 energy system that produced 15,000 watts from an input of only 500 watts. *Return of the Dove* can be obtained from Health Research. An article also appeared in *TRUE*, January 1961 titled "Otis T. Carr and the OTC-X1" by Richard Gehrman. Patent and article available from Biago Conti, P.O. Box 1014, Carmel, NY 10512.

FISH, J. R.

This is the famous Fish carburetor which guaranteed at least 20 percent better gas mileage than old-style carburetors. It can also be easily

From the book *Exotic Patents*, available from Biagio Conti, P.O. Box 1014, Carmel, NY 10512.

switched for alcohol. Our sources indicate that the unit was last being sold by Fuel Systems of America, Box 9333, Tacoma, Washington 98401 (206) 922-2228. See patents (2,214,273) (2,236,595) (2,775,818) (2,801,086).

FLANAGAN, PATRICK

Authored an excellent book titled *Pyramid Power*, one of the first on the market researching pyramid energy. His patent (3,393,279) is a device that claims to transmit sound directly to the brain by-passing the audio-neural system. Not only would this be a benefit to the handicapped, but one would be able to communicate with another in a very high noise environment. One could still wear noise mufflers which would not interfere with this type of communication. See patents (3,393,279) (3,647,970). Send inquiries to: 1109 S. Plaza Way, Suite 399, Flagstaff, AZ, 86001.

GRAY, EDWIN V.

Patented a motor that requires no fuel and produces no waste. An article appeared in the June 1973 issue of *Probe the Unknown*, titled "The Engine That Runs Itself" published by Rainbow Publications, 1845 Empire Avenue, Burbank, CA 91504. Unique Technologies, PO Box 56, Richland, MI 49083 also sells "Energy Creation" that describes this invention in layman's terms. See patent (3,890,548).

HIERONYMUS, T. G.

This is the well known patent which uses the force that Hieronymus calls "eloptic energy." This energy has characteristics of both light and electricity. Many radionic units are fashioned after this basic design. Articles written by Hieronymus can be found in the *United States Psychotronics Newsletter*, available at P.O. Box 22697, Louisville, KY, 40252 and also in *Advanced Sciences Advisory* published by Advanced Sciences Research and Development Corporation Inc., PO Box 109, Lakemont, Georgia 30552. See patent (2,482,773).

JOHNSON, HOWARD

Patented a motor #4,151,431 in which the power is generated by magnets alone. Write to: The Permanent Magnet Research Institute, PO Box 199, Blacksburg, Virginia 24063. It took Johnson six years of legal hassles to get this "free" energy motor patented. In our last correspondence with Howard Johnson, he was offering license rights.

LAKHOVSKY, GEORGES

His device produced an electromagnetic field which oscillated in a wide spectrum of wave-lengths. His theory regarded all living organisms as systems of high frequency oscillating circuits, every cell being a simple oscillator vibrating at a specific frequency. Since he believed that harmonious dynamic equilibrium was health, disease was the opposite, and placing the inharmonious cells within the range of the multiple-wave oscillator brought back the cells to equilibrium. See patent (2,351,055). Patent and articles available from Biagio Conti, P.O. Box 1014, Carmel, NY 10512.

MOORE, ARLINGTON M.

Listed are just a few of the 250 automotive patents. Some of his carburetion designs improved engine performance, gained more mileage and virtually eliminated carbon monoxide pollution. If you can get a copy, read *The Works of George Arlington Moore* published by the Madison Company. Seventeen patents of interest can be found between (1,633,791) and (2,123,485).

MORAY, T. H.

Moray wrote a book titled *The Sea of Energy in Which the Earth Floats* which is available from COSRAY, The Research Institute, Inc., 2505 South Fourth East, Salt Lake City, Utah 84115. He was reported to demonstrate as early as 1939 that this "Sea of Energy" can be harnessed. Once operating, his Radiant Energy Device would deliver an output of up to 50 Kilowatts of power with no other input source. See patent (2,460,707).

POGUE, CHARLES N.

Patent #2,026,798 design was used on a 1935 Ford V8 which resulted in 26.2 miles on one pint of gasoline, or in other words about 200 miles per gallon. For more information on this and other high mileage carburetors see *Secrets of the 200 MPG Carburetor* by Roadrunner Publications. See patents (1,938,497) (1,997,497) (2,026,798).

PUHARICH, H. K.

From what I have read about Puharich, he was a doctor who gave up his practice to study closely the works of Tesla and other suppressed inventors. Several of his patents are listed in this book. If you study his patents you will see ideas incorporated from Tesla. Also author of several books on psychic research. See patents (2,995,633) (3,156,787) (3,170,993) (3,267,931) (3,497,637) (3,563,246) (3,586,791) (3,629,521) (3,726,762).

TESLA, NIKOLA

Registering almost a hundred patents, he is the inventor of the Tesla Coil, the poly-phase alternating current system of power generation, basic develops in high voltage, electric motors, oscillators, etc. Almost everything that touches our lives in the twentieth century has been brought out to the world by Tesla. His patents can be found between (334,823) and (1,402,025).

ZUBRIS, V. R.

The circuit described in this patent was designed for an electric automobile. Supposedly, tests have shown that this particular design greatly increases motor efficiency. A few years ago Zubris was selling licenses for his patent at Zubris Electrical Company, 1320 Dorchester Avenue, Boston, MA, 02122. See patent (3,809,978).

United States Patent [19]

Gray

[11]   **3,890,548**

[45] **June 17, 1975**

[54] **PULSED CAPACITOR DISCHARGE ELECTRIC ENGINE**

[75] Inventor: Edwin V. Gray, Northridge, Calif.

[73] Assignee: Evgray Enterprises, Inc., Van Nuys, Calif.

[22] Filed: Nov. 2, 1973

[21] Appl. No.: 412,415

[52] U.S. Cl. 318/139; 318/254; 318/439; 310/46

[51] Int. Cl. H02p 5/00

[58] Field of Search 310/46, 5, 6; 318/194, 318/439, 254, 139; 320/1; 307/110

[56]   **References Cited**
UNITED STATES PATENTS

2,085,708	6/1937	Spencer	318/194
2,800,619	7/1957	Brunt	318/194
3,579,074	5/1971	Roberts	320/1
3,619,638	11/1971	Phinney	307/110

OTHER PUBLICATIONS

Frungel. *High Speed Pulse Technology*, Academic Press Inc., 1965, pp. 140–148.

Primary Examiner—Robert K. Schaefer
Assistant Examiner—John J. Feldhaus
Attorney, Agent, or Firm—Gerald L. Price

[57]   **ABSTRACT**

There is disclosed herein an electric machine or engine in which a rotor cage having an array of electromagnets is rotatable in an array of electromagnets, or fixed electromagnets are juxtaposed against movable ones. The coils of the electromagnets are connected in the discharge path of capacitors charged to relatively high voltage and discharged through the electromagnetic coils when selected rotor and stator elements are in alignment, or when the fixed electromagnets and movable electromagnets are juxtaposed. The discharge occurs across spark gaps disclosed in alignment with respect to the desired juxtaposition of the selected movable and stationary electromagnets. The capacitor discharges occur simultaneously through juxtaposed stationary movable electromagnets wound so that their respective cores are in magnetic repulsion polarity, thus resulting in the forced motion of movable electromagnetic elements away from the juxtaposed stationary electromagnetic elements at the discharge, thereby achieving motion. In an engine, the discharges occur successively across selected ones of the gaps to maintain continuous rotation. Capacitors are recharged between successive alignment positions of particular rotor and stator electromagnets of the engine.

18 Claims, 19 Drawing Figures

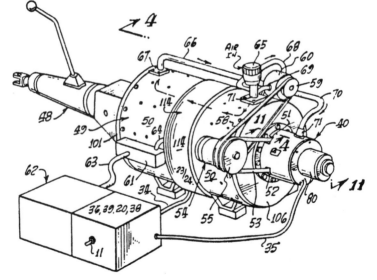

Appendix B: A Selection of Anti-Gravity Patents

April 4, 1967   C. D. LENNON ET AL   3,312,425

AIRCRAFT

Filed Oct. 12, 1965   5 Sheets—Sheet 1

FIG. 1

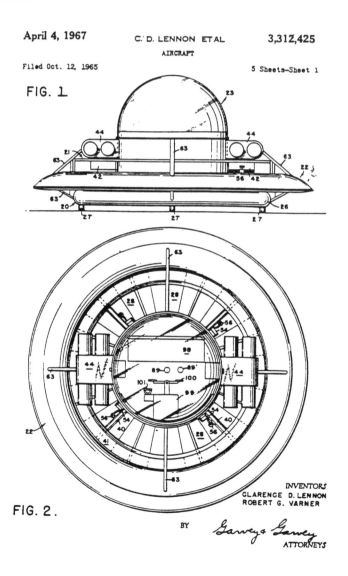

FIG. 2.

INVENTORS
CLARENCE D. LENNON
ROBERT G. VARNER

BY

Garvey & Garvey

ATTORNEYS

April 4, 1967   C. D. LENNON ET AL   3,312,425

Filed Oct. 12, 1965   AIRCRAFT   5 Sheets—Sheet 2

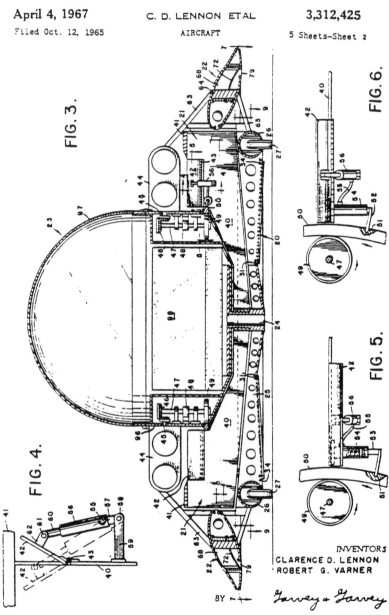

FIG. 3.

FIG. 6.

FIG. 4.

FIG. 5.

INVENTORS
CLARENCE D. LENNON
ROBERT G. VARNER

BY

ATTORNEYS

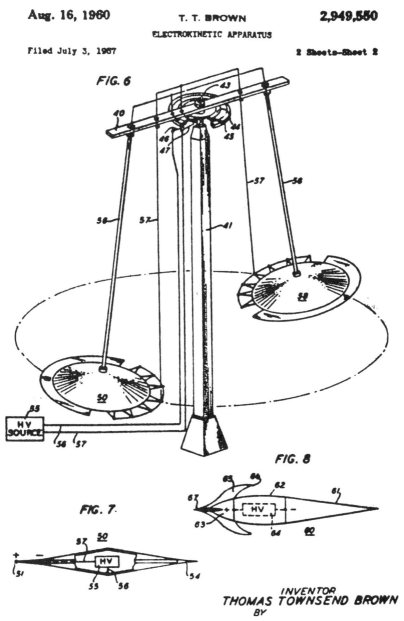

Aug. 16, 1960   T. T. BROWN   2,949,550

ELECTROKINETIC APPARATUS

Filed July 3, 1957   2 Sheets—Sheet 2

FIG. 6

FIG. 8

FIG. 7

INVENTOR
THOMAS TOWNSEND BROWN
BY

ATTORNEYS

Dec. 11, 1962   I. R. BARR   3,067,967
                                FLYING MACHINE

Filed Nov. 19, 1958   3 Sheets—Sheet 1

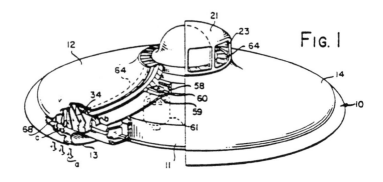

FIG. 1

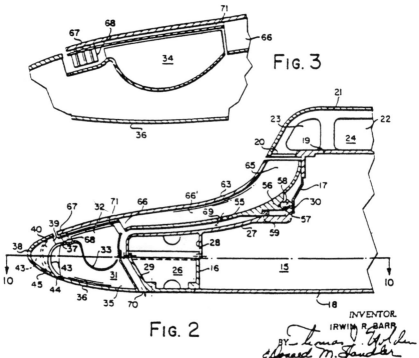

FIG. 3

FIG. 2

INVENTOR.
IRWIN R. BARR

BY
ATTORNEYS

Dec. 11, 1962   I. R. BARR   3,067,967

Filed Nov. 19, 1958   FLYING MACHINE   3 Sheets—Sheet 2

Fig. 4

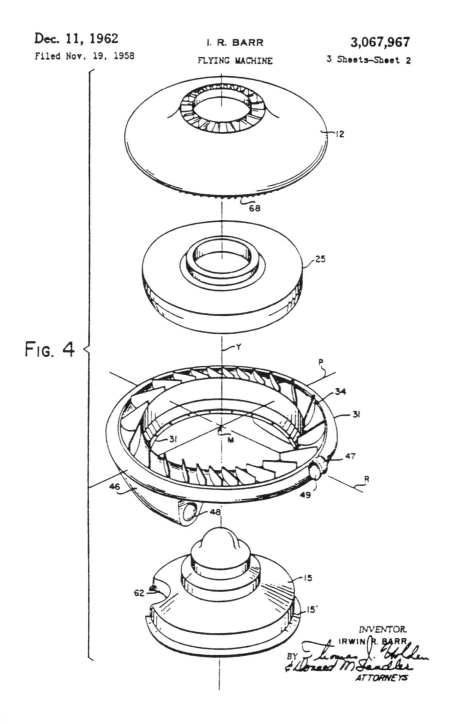

INVENTOR
IRWIN R. BARR,
BY *Thomas J. Holden*
& Donald M. Sandler
ATTORNEYS

March 11, 1969   E. GUERRERO   3,432,120
                             AIRCRAFT

Filed May 20, 1966   Sheet __1__ of 5

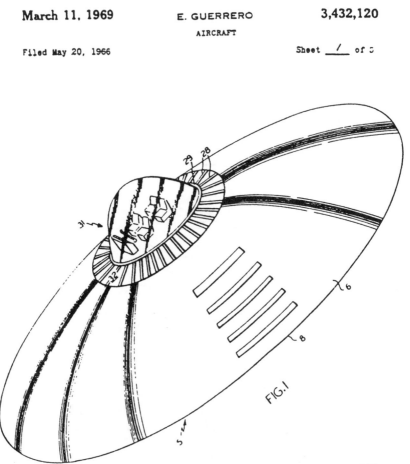

FIG.1

INVENTOR
EFRAIN GUERRERO

BY *Philpitt, Steininger & Paddy*
                    ATTORNEYS

March 11, 1969   E. GUERRERO   3,432,120

Filed May 20, 1966   AIRCRAFT   Sheet _2_ of 3

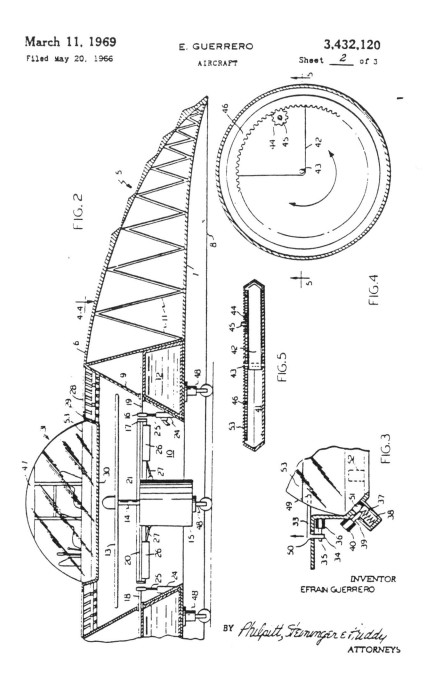

INVENTOR
EFRAIN GUERRERO

BY *Philpitt, Steininger & Ruddy*

ATTORNEYS

Appendix C: Recommended Reading

Access to information is vital in the quest for enlightenment. Now that you are aware of the many critical issues affecting our society today, you may wish to read further about any or all of the innovations presented in *Suppressed Inventions and Other Discoveries*. For this purpose, I have compiled the following list of books and magazines to aid you in your continued educational endeavors.

BOOKS AND COMPILATIONS

Andrews, George C., *Extra-terrestrial Friends and Foes*, Illuminet Press, P.O. Box 2808, Lilburn GA, 30226.

Bird, Christopher, *The Persecution and Trial of Gaston Naessens*, HJ Kramer Inc., P.O. Box 1082, Tiburon CA, 94920.

Carter, James P., *Racketeering in Medicine: The Suppression of Alternatives*, Hampton Roads Publishing Company, 134 Burgess Lane, Charlottesville VA, 22902.
Phone: 1-800-766-8009
E-mail: hrpc@mail.hamptonroadspub.com
Internet: http://www.hamptonroadspub.com

Conti, Biagio, *Exotic Patents*, P.O. Box 1014, Carmel NY, 10512.

Freeman, John, *Suppressed and Incredible Inventions*, reprinted by Health Research, P.O. Box 850, Pomeroy WA, 99347.

Hendershot, Mark M., *From the Archives of Lester J. Hendershot*, compiled and published by Mark M. Hendershot, P.O. Box 135, Okanogan WA, 98840.
This comprehensive collection of articles, correspondence, documents, photos, notes and diagrams is available at $34.95 from Mark M. Hendershot.

Lynes, Barry, *The Cancer Cure That Worked!*, Marcus Books, P.O. Box 327, Queensville, Ontario, Canada.
This book can be obtained for $13 Canadian (includes postage), payable to Marcus Books.

Manning, Jeane, *The Coming Energy Revolution*, Avery Publishing Group, Garden City Park NY.

Ross, Daniel, *Mars—The Telescopic Evidence*, Pintado Publishing, P.O. Box 3033, Walnut Creek CA 94598.
This book can be obtained from Pintado Publishing for $9.95 plus $2 shipping.

Ruesch, Hans, *Naked Empress*, Civis Publications, Via Motta 51, 6900 Massagno, Switzerland.

Walters, Richard, *Options: The Alternative Cancer Therapy Book*, Avery Publishing Group, Garden City Park NY.

Warren, Tom, *Beating Alzheimer's: A Step Towards Unlocking the Mysteries of Brain Diseases*, Avery Publishing Group, Garden City Park NY.

MAGAZINES AND NEWSLETTERS

Adventures Unlimited Press
c/o David Hatcher Childress
Stelle IL 60919
E-mail: aup@azstarnet.com

Ferment!
c/o Roy Lisker
Boughton Place
152 Kisor Road
Highland NY 12528
Ferment! is a newsletter of commentary, poetry, and fiction.

Lost Tech Files
c/o Alan McLaughlin
P.O. Box 96
Piha, Auckland
New Zealand

New Dawn Magazine
GPO Box 3126FF
Melbourne, VIC 3001
Australia
You can subscribe to New Dawn Magazine for $75 for 12 issues, or $40 for 6 issues. Make bank checks or money orders payable to New Dawn.

Nexus
P.O. Box 30
Mapleton, Queensland 4560
Australia
Phone: 61 (0) 7 5442 9280
Fax: 61 (0) 7 5442 9381
E-mail: nexus@peg.apc.org

United States address:
P.O. Box 177
Kempton, IL 60946-0177
Phone: (815) 253-6464
Fax: (815) 253-6300
Internet: http://www.peg.apc.org~nexus/

What Your Doctor Will Never Tell You
Katherine Smith and Jonathan Eisen, editors
Private Bag MBE, P-345
Auckland NZ
This is a bi-monthly international newsletter featuring suppressed medical and health information. A sample newsletter for $2 postage and handling.

Index